ANDREW ELBY

T.A.

A PHYSICS PROBLEM SOLVING GUIDE
VOLUME II

PRENTICE HALL Upper Saddle River, NJ 07458

Executive Editor: *Alison Reeves*
Production Editor: *Mindy De Palma*
Special Projects Manager: *Barbara A. Murray*
Manufacturing Buyer: *Ben Smith*
Supplement Cover Designer: *PM Workshop Inc.*

Printed in the United States of America

10 9 8 7 6 5 4 3

ISBN 0-13-231721-4

Prentice-Hall International (UK) Limited, *London*
Prentice-Hall of Australia Pty. Limited, *Sydney*
Prentice-Hall Canada, Inc., *Toronto*
Prentice-Hall Hispanoamericana, S.A., *Mexico*
Prentice-Hall of India Private Limited, *New Delhi*
Prentice-Hall of Japan, Inc., *Tokyo*
Pearson Education Asia Pte. Ltd., *Singapore*
Editora Prentice-Hall do Brasil, Ltda., *Rio de Janeiro*

Contents

Preface

Last week when they abducted me, space aliens revealed **The Conspiracy** to prevent humans from learning physics.

The Conspiracy focuses on problem-solving. To learn physics, you must practice it, by solving problems and getting immediate feedback. But The Conspiracy tries to make this impossible. For instance, although your textbook probably gives *answers* to some of the problems at the ends of the chapters, it probably doesn't give *solutions*. Therefore, you can't figure out whether you made a trivial algebra mistake or a major conceptual blunder. Sure, the sample problems in the chapters themselves have good solutions. But since those problems "test" material that you've just read, you can't determine whether you're gaining a long-term, integrated understanding. Fiendish, eh?

Also, when you hand in a homework assignment, the solution typically doesn't reach your desk until a day or two later, for logistical reasons. By then, you've moved on to the next topic. And your busy schedule makes it almost impossible for you to go over that solution immediately. The Conspiracy makes sure of it.

Some skeptics deny the existence of The Conspiracy. But look at all the forces trying to prevent you from getting immediate feedback on your conceptual knowledge and problem-solving skills.

Coincidence? I think not.

This book tries to counteract The Conspiracy by providing *lots* of practice problems, with detailed, concept-based solutions. You can gain immediate feedback on your knowledge, and can review abstract concepts in the context of specific problems. If The Conspiracy doesn't stop you.

I'd like to thank my fellow counter-conspirators for their help and encouragement, most notably Bruce Birkett and Jason Zimba.

I dedicate this book to Diana Perry, even though she doesn't believe the stuff about the space aliens.

Andrew Elby
Department of Physics
University of California, Berkeley
elby@physics.berkeley.edu

Introduction

*Hello again, this is Andy, a former graduate student and teaching assistant at the University of California, Berkeley. In case you didn't use the Portable TA volume 1, I need to offer a quick warning about how **not** to use this book. Don't read it the way you'd read a regular book. As many students have discovered, people who simply read the questions and then read the answers don't learn the material well. The concepts and problem-solving skills don't sink in, and don't become integrated with each other or with your prior knowledge, unless you attempt to solve each problem yourself before reading the answer. You probably know this from last semester, but it's worth emphasizing again and again. Below, I'll discuss how students have used the Portable TA effectively. But first, let me address some common questions.*

◆ ***Why should I shell out hard-earned money on a book of practice problems?***

In college-level physics, you cannot learn the concepts solely by hearing and reading about them, even if your lecturer and textbook are excellent. Many students, including myself, learned this the hard way their first semester.

Physics, like driving, can be learned only by *doing* it. After listening to a lecture and reading the corresponding textbook chapter, you may think, "That's not so hard—I understand it." But then, when you try to *apply* those concepts to a hard problem, you can't do it. Did this ever happen to you last semester? It happens even *more* frequently in second-semester physics, because the material gets more abstract.

I often hear students say, "I understand the concepts, but can't do the problems." This really means that the student understands the concepts as well as they can be understood from reading and listening. But you can acquire a *deeper* understanding by solving lots of practice problems. This deeper understanding makes the physics more interesting, and enables you to score better on exams.

◆ ***Won't the homework problems give me enough practice?***

No. Electromagnetism is harder and more abstract than classical mechanics. Most people need to complete several practice problems about any given topic, before it sinks in completely. Besides, most homework problems address *one* discrete topic. Many of the problems in here—especially the review problems, practice midterms, and practice final exams—make you apply several concepts at once.

◆ ***What's special about your practice problems?***

These problems focus on the central concepts, the ones most likely to show up on tests. Equally important, my answers discuss the relevant concepts and problem-solving skills in detail, without assuming that you already understand the material perfectly. When possible, I explain things intuitively. And I've made it easy to "test" yourself. Chapters 48–56 consist entirely of practice exam questions.

I'd appreciate any feedback. Don't hesitate to write me at elby@physics.berkeley.edu.

◆ ***How can I use your practice problems most effectively?***

Find the chapters corresponding to the material covered on this week's (or last week's) homework assignment. For each of my practice problems,

- Read the question carefully, making sure you understand what's being asked and what's given.
- Instead of jumping right to a formula, visualize what's happening, sketching pictures or graphs to organize your thoughts.
- Try to formulate a problem-solving strategy. When possible, think things through before diving into details.
- Even if you can't complete the problem, go as far as you can. Once you've *actively* tried to solve it yourself, the material you learn from reading the answer will "stick" better.

Given all this, here's how I recommend using these chapters.

1) Work through each problem in the way just described.
2) A few weeks later, go back and rework those problems, focusing on the ones that gave you trouble, and skipping the algebra to save time. Most students do this the week before a test.
3) A few days before the test, do the relevant review problems and practice tests. See the table of contents.

◆ ***Should I do your practice problems before or after completing the corresponding homework assignment?***

Both ways work fine, according to my students. These problems range in difficulty from medium-easy to very hard. Some of them will help you with your homework. The hardest ones should probably be postponed until after you've learned the basics and completed the homework assignment. I've labeled some of my problems *very hard, extra credit*, or *challenge problem.* Many students skip these at first, and try them later after completing the rest of the chapter. Unless you're going for an A+, majoring in physics, or looking for a challenge, you can skip the "extra credit" and "challenge problems" entirely.

Do NOT skim through these practice problems the night before a homework deadline, searching for strategies and hints. As mentioned earlier, anything you "learn" in this manner won't stick.

◆ ***How many times does the word "aardvark" appear in this book?***

Three. I think. There's also a duck-billed platypus somewhere.

◆ ***Are there computer simulations accompanying these problems?***

Unfortunately, no. I haven't had time to write any. If you want to write some Interactive Physics simulations, I'll post them on our Web page, http://physics1.berkeley.edu/BPPSG/.

◆ ***What about units and significant digits?***

Units. In most cases, I use standard SI units such as meters, seconds, kilograms, and so on. Sometimes I use other metric units such as centimeters and grams. Although units are important, this book focuses on the central physical concepts. Few of the problems involve unit conversions.

Significant digits. When an integer ends in zero, it's not clear how many digits are significant. For instance, "100 meters" could contain one, two, or three significant digits. In ambiguous cases such as this, the number of significant digits "follows" the other quantities listed in the problem. For instance, if a car travels 100 meters in 5.0 seconds, then the "100 meters" contains two significant digits. But if the car travels 100 meters in 5.00 seconds, then "100 meters" contains three significant digits.

During the intermediate steps of calculations, I usually retain more digits than are significant, but then round to the correct number of significant digits in the final answer. Let me illustrate what I mean. If a car travels 100 meters in 6.0 seconds, how far will it travel in 11 seconds, assuming constant velocity? To answer this, I'll first calculate the car's velocity:

$$v = \Delta x/\Delta t = (100 \text{ m})/(6.0 \text{ s}) = 16.7 \text{ m/s}.$$

Notice that I kept an "extra" significant digit. But when I use this velocity to calculate the distance the car moves in 11 seconds, I'll round to the correct number of significant digits: $D = vt = (16.7 \text{ m/s})(11 \text{ s}) = 180$ meters, not "184 meters."

◆ ***Will this study guide improve my love life?***

In Volume 1, I guaranteed that doing my practice problems would improve your romantic success by *up to* 400%. Of course, sometimes you have to buy two or three copies to make sure the full effect kicks in.

Need proof? After finishing Volume 1 and part of Volume 2, I got married to the most wonderful woman on the planet.

Coincidence? I think not.

33 CHAPTER

Electric Fields And Forces

QUESTION 33-1

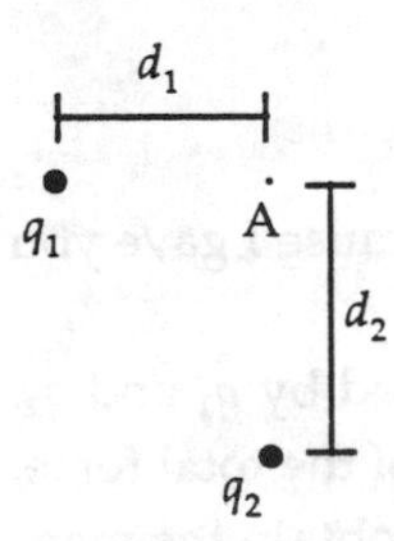

Two point charges, $q_1 = 3.00\ \mu C$ and $q_2 = 2.00\ \mu C$, are attached to the floor as drawn. Here, "μC" stands for microcoulombs.

Point A is located $d_1 = 1.00 \times 10^{-3}$ meters to the right of q_1, and $d_2 = 1.50 \times 10^{-3}$ m above q_2. The electric field at point A has strength $E = 2.8 \times 10^{10}$ newtons per coulomb (N/C). Point A is not a charge; it's just a spot on the floor.

(a) Suppose we place a particle of charge $Q = 0.010\ \mu C$ at point A. What is the magnitude of the electric force felt by Q?

(b) In what direction does that force push Q? Express your answer as an angle above the horizontal.

(c) Where could we place Q so that it would feel no electric force? Be as specific as possible, but use your intuition to get started. How far from q_1 should we put Q? Set up, but do not solve, the relevant equation or equations (unless you want algebra practice).

ANSWER 33-1

(a) You could solve the long way by

(i) calculating $\mathbf{F}_1$, the force exerted on Q by q_1,
(ii) calculating $\mathbf{F}_2$, the force exerted on Q by q_2, and then
(iii) vectorially adding those forces to obtain the total force.

In part (b), we'll use this strategy. But here, a shortcut gets us to the answer more quickly. The shortcut relies on understanding what the electric field *means*, as I'll now explain.

Your first textbook chapter about electrostatics probably says that point charges exert "direct" forces on each other. For instance, q_1 and q_2 directly repel Q. This way of visualizing electrostatic forces works fine. But in many cases, you can gain more insight by visualizing things as follows: q_1 and q_2 generate electric fields that radiate outward from those charges. Q "feels" the total electric field at point A, by which I mean the

A
q_1
q_2

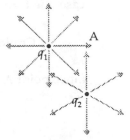

The two point charges spew out electric field lines. Charge Q at point A feels the combined field.

sum of the fields generated by q_1 and q_2. So, instead of picturing the charges as pushing each other directly, we can think of the interaction as mediated by the field. For instance, q_1 and q_2 create an electric field, and this field is what pushes on Q.

The bigger the electric field, the bigger the force felt by Q:

$$\mathbf{F} = Q\mathbf{E},$$

where **E** denotes the electric field generated by the *other* charges (besides Q). So here, **E** is the field created by q_1 and q_2.

In this problem, I *give* you the electric field generated by q_1 and q_2 at point A. Therefore, we can immediately calculate the force on Q:

$$\begin{aligned} F &= QE \\ &= (10. \times 10^{-8}\,\mathrm{C})(2.8 \times 10^{10}\,\mathrm{N/C}) \\ &= 280\ \mathrm{N}. \end{aligned}$$

(b) Now we have no choice but to solve the long way, using Coulomb's law. That's because I gave you the magnitude of the electric field, but not its direction.

To find the total electric force on Q, *separately* calculate the forces generated by q_1 and q_2. Add those forces vectorially, to obtain the total force. Here, we want the direction of the total force.

(Alternatively, you could add the electric fields generated by q_1 and q_2, to obtain the magnitude *and direction* of the field at point A. Since **F** = Q**E**, the force and field point in the same direction.)

Force diagram

$\mathbf{F}_2$, q_1, $\mathbf{F}_1$, y, x, q_2

As usual, begin with a force diagram. Positive charges repel each other. So, q_1 pushes Q rightward with force F_1, while q_2 pushes Q upward with force F_2.

According to Coulomb's law, a point charge q_1 generates an electric field

Coulomb's law (field) $$E = \frac{1}{4\pi\varepsilon_0}\frac{q_1}{r^2},$$

where r is the distance from q, and $1/(4\pi\varepsilon_0) = 9.0 \times 10^9\ \mathrm{N{\cdot}m^2/C^2}$ is a universal constant, analogous to "G" in Newton's law of gravitation. Some books write "k" as an abbreviation for $1/(4\pi\varepsilon_0)$. In any case, the force exerted by q_1 on Q is

Coulomb's law (force) $$F = QE = \frac{1}{4\pi\varepsilon_0}\frac{q_1 Q}{r^2}.$$

Since point A sits a distance d_1 from charge q_1, and a distance d_2 from charge q_2, we get

$$F_1 = \frac{1}{4\pi\varepsilon_0}\frac{q_1 Q}{d_1^2} = \frac{(9.0 \times 10^9\,\mathrm{N{\cdot}m^2/C})(3.00 \times 10^{-6}\,\mathrm{C})(1.0 \times 10^{-8}\,\mathrm{C})}{(1.00 \times 10^{-3}\,\mathrm{m})^2} = 270\ \mathrm{N}$$

$$F_2 = \frac{1}{4\pi\varepsilon_0}\frac{q_2 Q}{d_2^2} = \frac{(9.0 \times 10^9\,\mathrm{N{\cdot}m^2/C^2})(2.00 \times 10^{-6}\,\mathrm{C})(1.0 \times 10^{-8}\,\mathrm{C})}{(1.50 \times 10^{-3}\,\mathrm{m})^2} = 80\ \mathrm{N}.$$

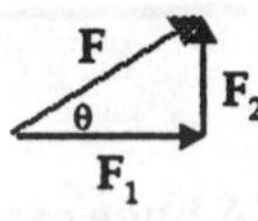

The total force is the vector sum of $\mathbf{F}_1$ and $\mathbf{F}_2$, as indicated on this diagram. From the Pythagorean theorem, $F = \sqrt{F_1^2 + F_2^2}$. You can confirm that this answer for F agrees with the one obtained in part (a). Here, we're looking for the angle θ. From trig, tan $\theta = F_2/F_1$, and hence

$$\theta = \tan^{-1}\frac{F_2}{F_1} = \frac{80\,\text{N}}{270\,\text{N}} = 17°.$$

(c) Intuitively, Q feels no net electric force if q_1 and q_2 push it in opposite directions, equally hard. This cancellation happens somewhere *between* q_1 and q_2. But not midway between. If we place Q right in the middle, the force due to q_1 will "win," because q_1 has more charge and therefore generates a stronger electric field. The forces on Q will balance only if it's closer to q_2 than it is to q_1.

As always, organize your thoughts with a picture. Let L denote the distance between q_1 and q_2. From the above diagram,

$$L = \sqrt{d_1^2 + d_2^2} = \sqrt{(1.00 \times 10^{-3}\text{m})^2 + (1.50 \times 10^{-3}\text{m})^2} = 1.8 \times 10^{-3}\text{m}.$$

Force diagram

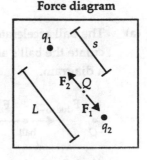

We're solving for s, the distance from q_1 to Q such that the forces on Q cancel. In other words, we're looking for the s such that $F_1 = F_2$ on this diagram.

Well, according to Coulomb's law, the electric force generated by one point charge on another is

$$F = QE = \frac{1}{4\pi\varepsilon_0}\frac{qQ}{r^2}.$$

I've defined $r = s$ as the distance between q_1 and Q. Therefore, the distance from q_2 to Q is $r = L - s$. So, no net force acts on Q when

$$F_1 = F_2$$

$$\frac{q_1 Q}{(4\pi\varepsilon_0)s^2} = \frac{q_2 Q}{(4\pi\varepsilon_0)(L-s)^2}.$$

Since s is the only unknown, we can solve for it. This problem did not require you to finish the algebra. But if you did, here's the answer.

Algebra starts here. Cancel the factors of Q and $1/(4\pi\varepsilon_0)$. Then multiply through by $(L-s)^2$, and divide through by q_1, to get

$$\left(\frac{L-s}{s}\right)^2 = \frac{q_2}{q_1},$$

and hence, $\frac{L-s}{s} = \sqrt{\frac{q_2}{q_1}}$, since both sides must be positive.

Multiply through by s to get $L - s = s\sqrt{\frac{q_2}{q_1}}$. Then add s to both sides: $L = s\left(\sqrt{\frac{q_2}{q_1}} + 1\right)$. Finally, isolate s to obtain

$$s = \frac{L}{1+\sqrt{\frac{q_2}{q_1}}} = \frac{L}{1+\sqrt{\frac{2.0\,\mu\text{C}}{3.0\,\mu\text{C}}}} \approx .55L = (0.55)(1.8 \times 10^{-3}\text{m}) = 9.9 \times 10^{-4}\text{m}.$$

Notice that s is greater than $L/2$, confirming our intuition that Q must be closer to q_2 than it is to q_1.

QUESTION 33-2

My room is filled with a uniform, rightward-pointing electric field of strength E_0. (A huge capacitor creates the field. We'll discuss capacitors in Chapter 36.) Glued to the floor is a particle of charge $-Q$. A distance D to the right of the glued particle, I place a tiny frictionless ball of mass m and change $+Q$.

(a) Immediately after it's released, what is the ball's acceleration?

(b) Assuming the acceleration is rightward, sketch a rough graph of the ball's velocity versus time. I released the ball at $t = 0$.

(c) Could we put the ball somewhere so that it wouldn't move after getting released? Justify your answer, but don't do any detailed calculations.

ANSWER 33-2

(a) The ball accelerates because it feels a net push (force). After finding the net force, we can easily calculate the ball's acceleration using Newton's 2nd law, $a = F_{net}/m$. To obtain the net force, start with a diagram.

$\mathbf{F}_{\text{due to }-Q}$ $\mathbf{F}_{\text{due to }E_0}$ $-Q$ ball

The ball feels two electric forces. The negative particle attracts it leftward. But the electric field created by the capacitor pushes the ball rightward. The leftward force partially cancels the rightward one.

Here's another way to visualize what's happening. The ball "feels" a total electric field $\mathbf{E} = \mathbf{E}_0 + \mathbf{E}_{\text{due to }-Q}$. Since $\mathbf{E}_{\text{due to }-Q}$ points leftward, it partially cancels $\mathbf{E}_0$. In this way, the glued particle "weakens" the total electric field felt by the ball.

Let rightward be the positive direction. The ball feels an electric field

$$E = E_0 - E_{\text{due to}-Q}$$
$$= E_0 - \frac{1}{4\pi\varepsilon_0}\frac{Q}{D^2},$$

where I used Coulomb's law, with $r = D$, to write down the field generated by $-Q$.

Notice that I wrote Q instead of $-Q$, because I *already* introduced a minus sign to indicate the direction of $\mathbf{E}_{\text{due to }-Q}$. A common mistake is to introduce two minus signs that cancel, giving a final answer $F = Q\left(E_0 + \frac{1}{4\pi\varepsilon_0}\frac{Q}{D^2}\right)$. This can't be right, because it says that the two forces "reinforce" each other. In reality, the force due to $-Q$ partially cancels the force due to $\mathbf{E}_0$:

$$F = QE$$
$$= Q\left[E_0 - \frac{1}{4\pi\varepsilon_0}\frac{Q}{D^2}\right]$$
$$= QE_0 - \frac{1}{4\pi\varepsilon_0}\frac{Q^2}{D^2},$$

the same expression you'd get by thinking in terms of forces rather than fields.

In general, when using Coulomb's law, calculate the absolute value of the fields or forces, and then insert plus and minus signs "by hand" to indicate the directions. That way, you're less likely to make a careless error.

Given this net rightward force, we can immediately find the rightward acceleration:

$$a = \frac{F_{net}}{m} = \frac{Q}{m}\left(E_0 - \frac{1}{4\pi\varepsilon_0}\frac{Q}{D^2}\right).$$

(b) Assuming a is positive (rightward), the ball moves rightward, getting faster and faster. Its velocity gradually increases. But does v increase at a steady rate, corresponding to a constant acceleration?

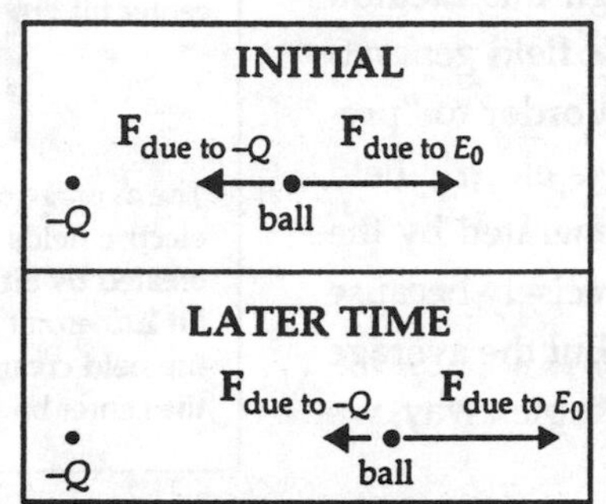

As the ball gets farther from −Q, the leftward force on it decreases. Therefore, the *net* rightward force increases.

To figure this out, visualize the ball's motion. As it slides rightward, the ball gets farther from $-Q$. And according to Coulomb's law, the electric attraction between point charges gets smaller as they get farther apart. Therefore, the leftward force exerted by $-Q$ on the ball *decreases* as time passes. But the rightward force stays constant, because E_0 has the same strength throughout the room.

Let me summarize. As time passes, the rightward force stays constant, while the leftward force (due to $-Q$) decreases. So, the *net* rightward force gets bigger. Therefore, the rightward acceleration increases. In other words, the particle does not speed up steadily. It speeds up at a faster and faster rate, corresponding to an increasing slope on the v vs. t graph. Remember, acceleration is the rate of change of velocity, i.e., the slope of the v vs. t graph.

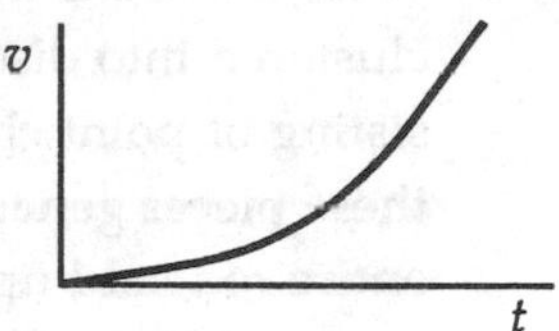

(c) As the above force diagram shows, the ball feels two forces. If those forces cancel, then the ball doesn't move. We can make those forces cancel by putting the ball the "correct" distance R away from $-Q$. The correct R is the one for which

$$F_{\text{due to}-Q} = F_{\text{due to } E_0}$$

$$\frac{1}{4\pi\varepsilon_0}\frac{Q^2}{R^2} = QE_0,$$

You could easily solve for R, though the problem does not require it.

QUESTION 33-3

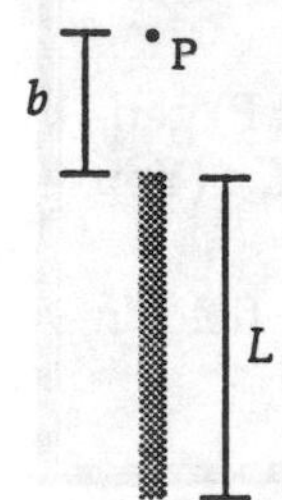

Consider a thin rod of length $L = 0.40$ m, coated with charged paint. Total charge $Q_0 = 1.0 \times 10^{-4}$ C is spread uniformly over the rod. Point P sits a distance $b = 0.20$ m from the end of the rod, as drawn here.

(a) What is the rod's linear charge density, λ?

(b) What is the electric field at point P? Solve symbolically before plugging in numbers.

(c) A point charge $q_1 = -6.0 \times 10^{-7}$ C is placed at P. What force does it exert *on the rod*?

ANSWER 33-3

(a) Linear charge density is charge per length. Intuitively, it's the "concentration" of charge on a thin object. Since the charge is distributed uniformly over the rod, the linear charge density is constant, with value

$$\lambda = \frac{Q_0}{L} = \frac{1.0 \times 10^{-4}\ \text{C}}{0.40\ \text{m}} = 2.5 \times 10^{-4} \text{coulombs per meter.}$$

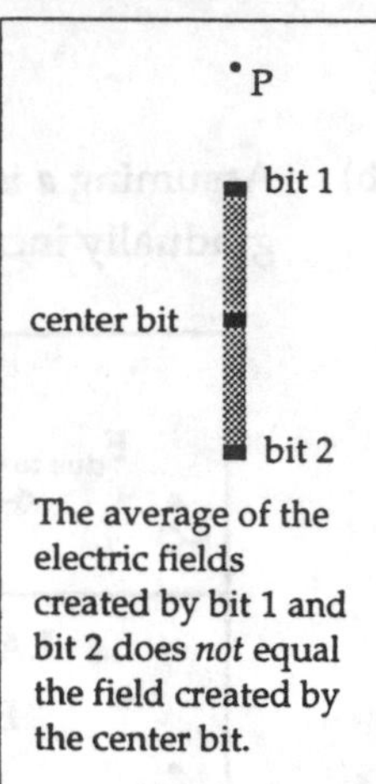

The average of the electric fields created by bit 1 and bit 2 does *not* equal the field created by the center bit.

(b) A common mistake is to use Coulomb's law, $E = \frac{1}{4\pi\varepsilon_0}\frac{Q_0}{r^2}$, with r equal to the distance from point P to the center of the rod. Although this method makes intuitive sense, it fails for a subtle reason. The electric field generated at P by a tiny "bit" of the rod is proportional to $1/r^2$. In order to "pretend" that Q_0 is concentrated at the rod's center, the *average* electric field generated by bits 1 and 2 would have to equal the field generated by the "center bit." But the average does not work out this way, precisely because E depends on $1/r^2$. For instance, the average of 2 and 4 is 3. But the average of $1/2^2$ and $1/4^2$ is *not* $1/3^2$. Because $1/r^2$ "averages" in a complex way, we can't treat the rod as a point mass concentrated at its center.

For this reason, we must find another method of calculating the electric field at point P. Well, the rod is a "continuous charge distribution"; the charge is spread out like a jelly, instead of clustered into discrete point charges. Nonetheless, to solve for E, we must think of the rod as consisting of point charges. Every infinitesimal "piece" (bit) of the rod is a point charge. And each of these pieces generates a tiny electric field at point P. To find the total electric field generated by the entire rod, add up the tiny fields contributed by all the pieces.

Let me flesh out this strategy for finding E at point P. First, find the field produced by an arbitrary infinitesimal piece of the rod. Then, add up (integrate) the contributions from all those pieces, to obtain the total field. Conceptually, this is just like adding up the electric fields created by two or three point charges. But instead, you're adding up the fields created by an infinite number of infinitesimal charges. Here's an outline to keep yourself organized.

Calculating field generated by a continuous charge distribution

1) On a good drawing, color in an arbitrary "piece" of the charge distribution. Don't choose a special point such as the end or the exact middle. The piece should be infinitesimally long and/or infinitesimally wide. So, its length or width will be something like dx or ds.

2) Write down an expression for dq, the infinitesimal charge of that piece, in terms of dx or ds or whatever.

3) Using Coulomb's law, find the infinitesimal electric field at the point of interest (e.g., point P) generated by the piece you colored in. When necessary, break $d\mathbf{E}$ into components, dE_x and dE_y. (You won't need to use components here.)

4) Integrate your dE_x or dE_y over the whole charge distribution, to obtain the total electric field in the x- or y-direction.

This abstract strategy makes more sense once you see it applied.

Step 1: Color in an arbitrary "piece" of the continuous charge distribution.

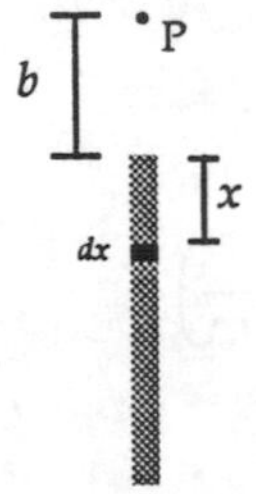

Give a name to the distance between the blackened piece and some reference point, such as the end of the rod. For instance, I'll say that my blackened piece of rod is arbitrary distance x from the end. The length of this piece is dx.

Step 2: Find dq, the charge of the colored-in piece of rod.

The linear charge density λ specifies the charge *per length.* To find the charge of a given piece of the rod, just multiply the charge per length times the length. For instance, if the linear charge density is 3 coulombs per meter, then a 2-meter piece of the rod contains 6 coulombs of charge. Here, the linear density is $\lambda = Q_0/L$ coulombs per meter, and the colored-in piece has length dx. So, the charge of that piece is

$$dq = \frac{\text{charge}}{\text{length}} \times (\text{length of the piece}) = \lambda dx.$$

For now, I'll continue writing "λ" instead of "Q_0/L," to keep my expressions simpler.

Step 3: Write an expression for dE, the electric field at point P generated by this bit of charge.

As the diagram shows, the blackened piece sits a distance $r = b + x$ from point P. Therefore, the electric field generated at point P is

$$dE = \frac{1}{4\pi\varepsilon_0}\frac{dq}{r^2}$$

$$= \frac{1}{4\pi\varepsilon_0}\frac{\lambda dx}{(b+x)^2}.$$

We need not worry about the vector components of this field, because every piece of the rod generates an upward field at point P. By contrast, when the dE's from different pieces of the rod point in *different* directions, we must break those dE's into components. See questions 33-5, 33-6, and 33-9.

Step 4: Add up (integrate) the field contributions from all the "pieces" of charge.

So far, I've found the field produced by a single piece of rod, a distance x from the end. The rod consists of pieces starting at $x = 0$ and ending at $x = L$. So, those are the limits of integration.

$$E = \int dE = \int_0^L \frac{1}{4\pi\varepsilon_0}\frac{\lambda dx}{(b+x)^2}$$

$$= \frac{1}{4\pi\varepsilon_0}\lambda\int_0^L \frac{dx}{(b+x)^2} \qquad \text{[pulling constants out front]}$$

$$= \frac{1}{4\pi\varepsilon_0}\lambda\left(-\frac{1}{b+x}\right)\bigg|_0^L$$

$$= \frac{1}{4\pi\varepsilon_0}\lambda\left(-\frac{1}{b+L}+\frac{1}{b}\right).$$

Let's put in the numbers. The constant $1/(4\pi\varepsilon_0)$, often abbreviated "k," has value $9.0 \times 10^9\ \mathrm{N \cdot m^2/C^2}$. In part (a), we found the linear charge density, $\lambda = Q_0/L = 2.5 \times 10^{-4}\ \mathrm{C/m}$. And the problem tells us that $L = 0.40$ m and $b = 0.20$ m. So, at point P,

$$E = \frac{1}{4\pi\varepsilon_0}\lambda\left(-\frac{1}{b+L}+\frac{1}{b}\right)$$

$$= (9.0 \times 10^9\ \mathrm{N \cdot m^2/C^2})(2.5 \times 10^{-4}\ \mathrm{C/m})\left(-\frac{1}{0.20\ \mathrm{m} + 0.40\ \mathrm{m}} + \frac{1}{0.20\ \mathrm{m}}\right)$$

$$= 7.5 \times 10^6\ \mathrm{N/C}.$$

(c) Since we just found the electric field at point P, we could easily calculate the force exerted *by* the rod *on* the point charge. It's just $F_{elec} = q_1 E$. But the problem asks for the force exerted *by* the point charge *on* the rod.

Well, we could figure out the electric force generated by q_1 at each point on the rod, and then integrate over the rod. But that's unnecessary. According to Newton's 3rd law, the force exerted by q_1 on the rod is equal in magnitude (but opposite in direction) to the force exerted by the rod on q_1. Intuitively speaking, the point charge "pushes back" on the rod just as hard as the rod pushes on the point charge. Those two forces have the same magnitude:

$$F_{q_1\ \text{on rod}} = F_{\text{rod on}\ q_1} = q_1 E_{\text{produced by rod}} = (-6.0 \times 10^{-7}\ \mathrm{C})(7.5 \times 10^6\ \mathrm{N/C}) = -4.5\mathrm{N}.$$

The minus sign indicates that the objects attract instead of repel.

QUESTION 33-4

Q_0 Q_0 P D

Two identical positive charges, Q_0, are nailed down a distance $2D$ apart. In this diagram, point P is midway between the two charges. The dashed line is vertical, i.e., perpendicular to an imaginary line connecting the two charges.

What is the direction and magnitude of the electric force acting on a positive charge Q_1 . . .

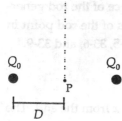

(a) . . . placed at point P?

(b) . . . placed on the dashed line, a distance y above point P?

(c) Where on the dashed line (i.e., how far from point P) should we place Q_1 so that it experiences the biggest force?

ANSWER 33-4

(a) At point P, Q_1 gets pushed in opposite directions by the two Q_0's. Those opposing forces have the same strength, because Q_1 is equidistant from the two Q_0's. Therefore, those two forces cancel. Q_1 feels no net force.

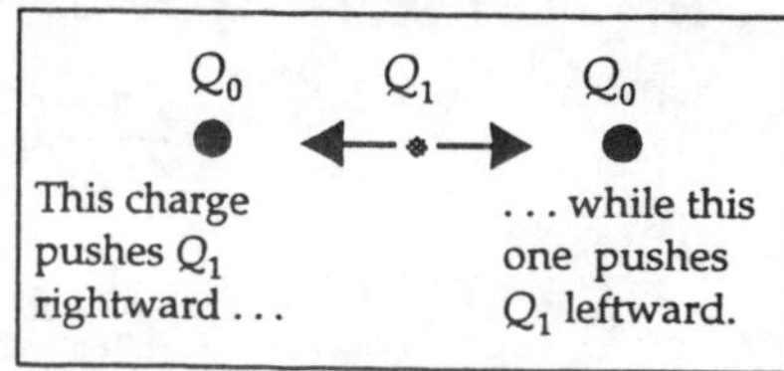

You can also think of this in terms of electric fields. At point P, the fields produced by the two Q_0's cancel each other, making the overall field vanish.

(b) First, I'll figure out **E**, the electric field on the dashed line a distance y above point P. Then, I'll use $\mathbf{F} = Q_1\mathbf{E}$ to find the force on Q_1. You could equally well work the whole problem in terms of forces rather than fields. But you'll have an easier time later in the course if you start thinking in terms of fields.

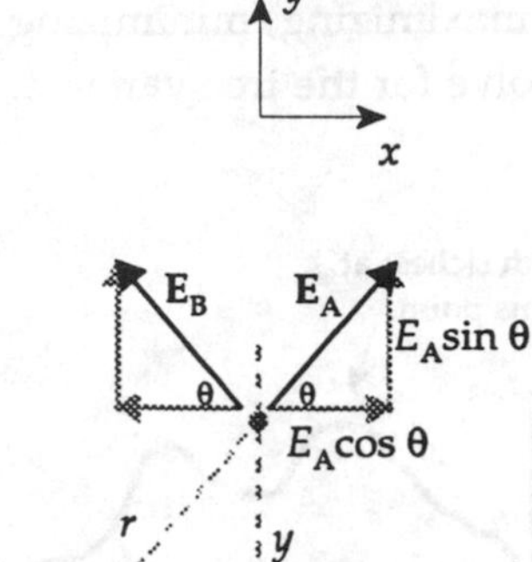

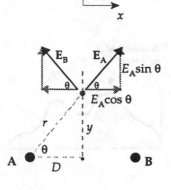

The two Q_0's spew out radial field lines. Let me call those charges "A" and "B," just to give them names. At Q_1, the total field is the sum of the fields generated by A and B: $\mathbf{E}_{tot} = \mathbf{E}_A + \mathbf{E}_B$.

On the dashed line, Q_1 is equidistant from A and B. So, $\mathbf{E}_A$ and $\mathbf{E}_B$ have the same magnitude. For this reason, some students try to solve for E_{tot} by doubling E_A. But $\mathbf{E}_A$ and $\mathbf{E}_B$ point in different directions; we must add them like vectors, not numbers. To sort out vector components, always draw a careful diagram, as illustrated here.

From the diagram, you can see that the x-components of $\mathbf{E}_A$ and $\mathbf{E}_B$ cancel. Therefore, the electric field at Q_1 points entirely in the y-direction. Furthermore, since $\mathbf{E}_A$ and $\mathbf{E}_B$ have equal y-components ($E_{Ay} = E_{By}$), the total E_y is simply twice E_{Ay} or E_{By}.

$$E_{y\ \text{tot}} = 2E_{Ay}$$
$$= 2E_A \sin\theta.$$

From the diagram, $\sin q = y/r$. And by Pythagorean theorem, $r = \sqrt{D^2 + y^2}$. Since a point charge Q_0 generates a field of strength $E = \frac{1}{4\pi\varepsilon_0}\frac{Q_0}{r^2}$, we get

$$\begin{aligned} E_{y\ \text{tot}} &= 2E_A \sin\theta \\ &= 2\left(\frac{1}{4\pi\varepsilon_0}\frac{Q_0}{r^2}\right)\frac{y}{r} \\ &= 2\left(\frac{1}{4\pi\varepsilon_0}\frac{Q_0}{D^2 + y^2}\right)\left(\frac{y}{\sqrt{D^2 + y^2}}\right) \\ &= \frac{Q_0 y}{2\pi\varepsilon_0 (D^2 + y^2)^{3/2}}. \end{aligned}$$

That's the total electric field at Q_1, since the x-components cancel. Therefore, Q_1 feels a y-directed force

$$F = Q_1 E_{\text{tot}} = \frac{Q_1 Q_0 y}{2\pi\varepsilon_0 (D^2 + y^2)^{3/2}},$$

By the way, when $y = 0$, this force equals zero. Since "$y = 0$" corresponds to point P, this confirms our part (a) conclusion; a charge at point P feels no net force.

(c) We're finding the y that maximizes the force F. Here, we can't just guess that y using physical intuition. So, we have a classic maximization problem. Let me review the strategy, and the intuitions underlying it, from Volume 1 of this study guide. (Looks like I'll have to sue myself for plagiarism!)

(1) Figure out what variable you're trying to maximize or minimize. Then figure out which variable you "alter" in order to achieve maximization/minimization. This parameter is called the "free variable."

(2) Write the variable you're maximizing or minimizing as a function of the free variable.

(3) Using your equation from step (2), differentiate the variable you're maximizing/minimizing with respect to the free variable. Set the derivative equal to 0, and solve for the free variable.

To understand where the strategy comes from, look at this graph of \$ vs. t, the money in my bank account. At what time did I have the most? Well, the "peak" of the graph is where the slope reaches zero. In other words, \$ is maximized when the slope hits 0. But the slope of \$ vs. t is $d\$/dt$. Therefore, to find the time at which I had the most money, set $d\$/dt = 0$, and solve for t. This summarizes the strategy outlined above.

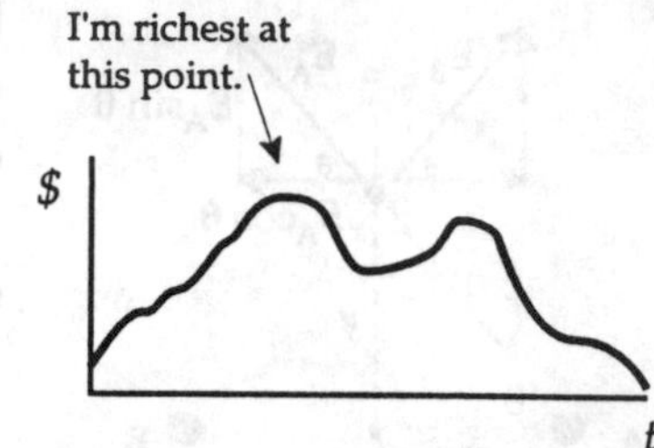
I'm richest at this point.
$
t

I'll now apply this strategy to the current problem.

Step 1: Identify the free variable and the variable we're maximizing.

Here, we're maximizing the electric force, F. We can adjust y, the distance from point P along the vertical dashed line. So, y is the free variable.

Step 2: Write the quantity we're maximizing as a function of the free variable.

We already did this in part (b), where we found

$$F = \frac{Q_1 Q_0 y}{2\pi\varepsilon_0 (D^2 + y^2)^{3/2}}.$$

Step 3: Set the derivative equal to zero.

$$\begin{aligned}
0 &= \frac{dF}{dy} \\
&= \frac{d}{dy}\left[\frac{Q_1 Q_0 y}{2\pi\varepsilon_0 (D^2 + y^2)^{3/2}}\right] \\
&= \frac{Q_1 Q_0}{2\pi\varepsilon_0}\frac{d}{dy}\left[\frac{y}{(D^2 + y^2)^{3/2}}\right] \\
&= \frac{Q_1 Q_0}{2\pi\varepsilon_0}\left[y\frac{d}{dy}\frac{1}{(D^2 + y^2)^{3/2}} + \frac{1}{(D^2 + y^2)^{3/2}}\frac{d}{dy}y\right] \quad \text{(product rule)} \\
&= \frac{Q_1 Q_0}{2\pi\varepsilon_0}\left[y\left(-\frac{3}{2}\frac{1}{(D^2 + y^2)^{5/2}}2y\right) + \frac{1}{(D^2 + y^2)^{3/2}}\right].
\end{aligned}$$

Now divide through by $\frac{Q_1Q_2}{2\pi\varepsilon_0}$, and multiply through by $(D^2 + y^2)^{5/2}$, to get

$$0 = -3y^2 + (D^2 + y^2),$$

and hence

$$y = \frac{D}{\sqrt{2}}.$$

In the above electric field diagram, this corresponds to

$$\theta = \tan^{-1}\frac{y}{D} = \tan^{-1}\frac{d}{\sqrt{2}} = 35.3°.$$

See if you can explain, intuitively, why the force on Q_1 gets smaller if we place it closer to point P *or* farther from point P than this "best value" of y. Hint: Think about the total size of $\mathbf{E}_A$, as well as the percentage of $\mathbf{E}_A$ that points in the y-direction.

QUESTION 33-5

A semicircular plastic wire of radius R has uniform linear charge density λ. A point charge Q is placed at the center of the semicircle. What force does it feel? Give the magnitude and direction.

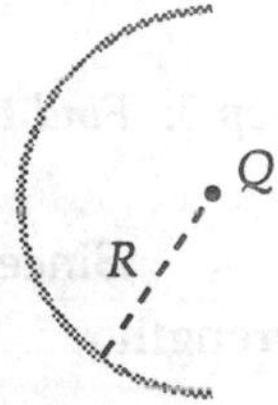

ANSWER 33-5

Welcome to a "continuous charge distribution" problem. Think of the semicircle as consisting of infinitesimal "bits" of charge. Each bit contributes to the electric field at the center. To find the total field, add up the contributions from all those bits. And given the total field, we can easily find the force on Q, using **F** = Q**E**. Let me summarize this strategy, introduced in question 33-3.

Calculating the field generated by a continuous charge distribution

1) Color in an arbitrary infinitesimal "piece" of the charge distribution.
2) Write down an expression for dq, the infinitesimal charge of that piece.
3) Using Coulomb's law, calculate the infinitesimal electric field at the point of interest (e.g., the center of the semicircle) generated by the piece you colored in. When necessary, break $d\mathbf{E}$ into components, dE_x and dE_y.
4) Integrate your dE_x and dE_y over the whole charge distribution, to obtain the total electric field in the x- and y-directions.

Without further ado, I'll carry out this strategy.

Step 1: Color in a tiny piece of the continuous charge distribution.

I've blackened a tiny arc, which "subtends" an infinitesimal angle $d\theta$. On the diagram, $d\theta$ denotes the angle between the two dashed lines. By definition, the angle (in radians) subtended by an arc is related to the arc length s by the formula $\theta = s/r$. Therefore, this colored-in piece has length

$$ds = Rd\theta.$$

Step 2: Write down an expression for dq, the charge of the colored-in piece.

Linear charge density is charge per length. To find the charge contained in the blackened piece, simply multiply the length of that piece by the charge per length:

$$dq = \frac{\text{charge}}{\text{length}} \times (\text{length of the blackened piece}) = \lambda ds = \lambda R d\theta.$$

Step 3: Find the electric field **dE** *generated by that infinitesimal piece of charge.*

Since the charge dq is a distance R from the center of the semicircle, it creates an electric field of strength

$$dE = \frac{1}{4\pi\varepsilon_0}\frac{dq}{r^2} = \frac{1}{4\pi\varepsilon_0}\frac{\lambda R d\theta}{R^2}.$$

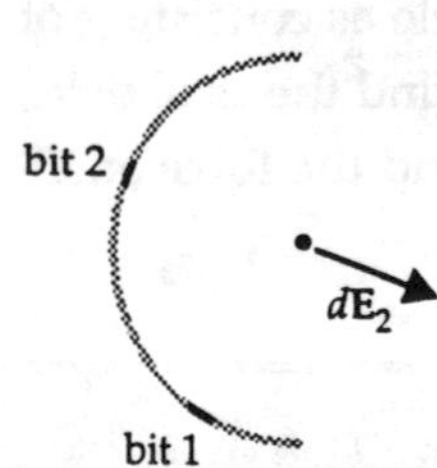

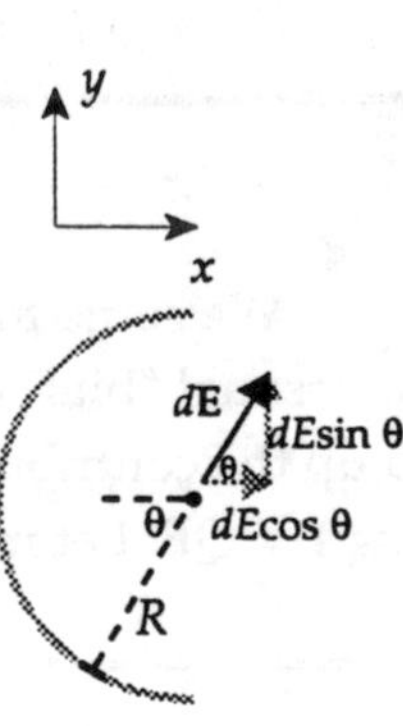

Unfortunately, we can't work directly with this scalar expression. Instead, we must break it into components. To see why, look at this diagram. The fields $d\mathbf{E}_1$ and $d\mathbf{E}_2$ are generated by "bit 1" and "bit 2" of the semicircle. These fields point in *different* directions. Therefore, we must add the $d\mathbf{E}$'s like vectors.

To work out these vector components, let θ denote the angle between the blackened piece of the semicircle and an imaginary horizontal line, as drawn here. So, the bottom and top of the semicircle are at $\theta = +90°$ and $\theta = -90°$, respectively. As the diagram shows,

$$dE_x = dE\cos\theta = \frac{1}{4\pi\varepsilon_0}\frac{\lambda R d\theta}{R^2}\cos\theta,$$

$$dE_y = dE\sin\theta = \frac{1}{4\pi\varepsilon_0}\frac{\lambda R d\theta}{R^2}\sin\theta.$$

Step 4: Add up (integrate over) all the pieces of the semicircle, to find the total electric field

As always, deal with the x- and y-components *separately*. Here, we must evaluate *separate* integrals to figure out E_x and E_y.

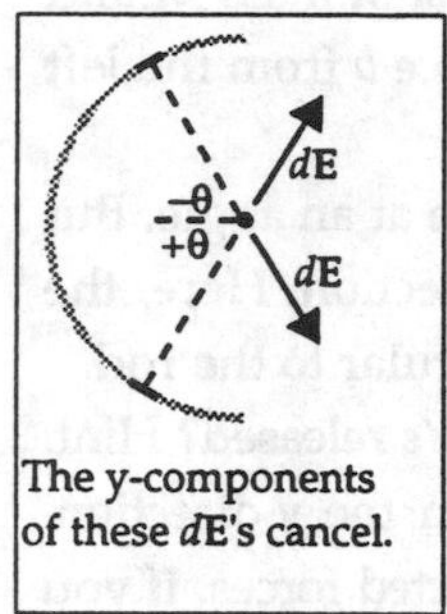

The y-components of these dE's cancel.

Before diving into math, however, you should use physical intuition to guess whether E_x or E_y equals zero. By trusting your intuitions, you can save lots of work! Here, we expect E_y to equal 0. Intuitively, that's because a piece of the semicircle at $+\theta$ and a piece of the semicircle at $-\theta$ generate dE_y's that cancel, as drawn here.

To double-check this conclusion, I'll solve for E_y mathematically. Recall that the top of the semicircle is at $\theta = -90°$, while the bottom is at $\theta = +90°$.

$$E_y = \int dE_y = \int_{-90°}^{+90°} \frac{1}{4\pi\varepsilon_0}\frac{\lambda R\,d\theta}{R^2}\sin\theta$$

$$= \frac{1}{4\pi\varepsilon_0}\frac{\lambda}{R}\int_{-90°}^{+90°}\sin\theta\,d\theta \quad \text{[since } R \text{ doesn't vary]}$$

$$= \frac{1}{4\pi\varepsilon_0}\frac{\lambda}{R}[-\cos\theta]\Big|_{-90°}^{+90°}$$

$$= \frac{1}{4\pi\varepsilon_0}\frac{\lambda}{R}[-\cos 90° - \cos(-90°)]$$

$$= 0,$$

since $\cos 90° = \cos(-90°) = 0$. This result confirms my intuitive guess.

Because $E_y = 0$, the field produced by the semicircle at its center points entirely rightward, with strength

$$E_y = \int dE_y = \int_{-90°}^{+90°} \frac{1}{4\pi\varepsilon_0}\frac{\lambda R\,d\theta}{R^2}\cos\theta$$

$$= \frac{1}{4\pi\varepsilon_0}\frac{\lambda}{R}\int_{-90°}^{+90°}\cos\theta\,d\theta \quad \text{[since } R \text{ doesn't vary]}$$

$$= \frac{1}{4\pi\varepsilon_0}\frac{\lambda}{R}[\sin 90° - \sin(-90°)]$$

$$= \frac{1}{4\pi\varepsilon_0}\frac{\lambda}{R}[1-(-1)]$$

$$= \frac{\lambda}{2\pi\varepsilon_0 R}.$$

Given the electric field at the center, we can easily calculate the force on charge Q. The force points rightward, with magnitude

$$F = QE = Q\frac{\lambda}{2\pi\varepsilon_0 R}.$$

QUESTION 33-6

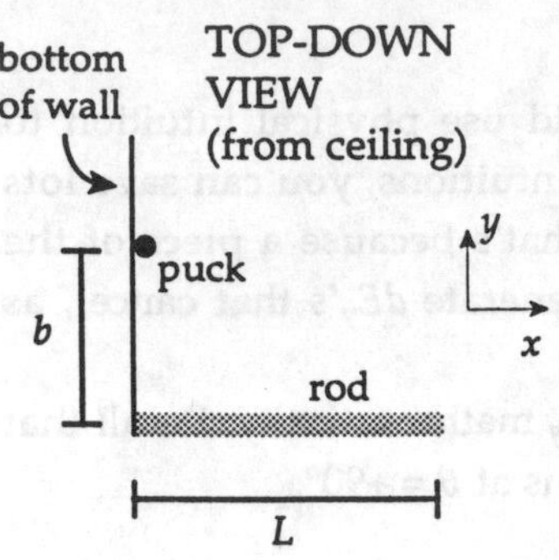

A thin rod of length L, oriented in the x-direction, has uniform linear charge density λ. The rod lies on the floor. The left end of the rod is touching a wall. On the floor, a small frictionless puck of mass M and charge Q is also touching the wall. The puck is a distance b from the left end of the rod.

If the wall were removed, the puck would slide at an angle. But the wall constrains the puck to slide only in the y-direction. Here, the y-direction is *not* vertical. It's along the floor, perpendicular to the rod.

(a) What is the puck's acceleration, immediately after it's released? Hint: The wall constrains the puck to accelerate only in the y-direction. Therefore, you need to worry only about the y-directed forces. If you can't complete the math, that's OK; but set things up completely.

(b) What normal force does the wall exert on the puck, when the puck is first released? Hint: The normal force, which points in the x-direction, must cancel the x-component of the electric force.

(c) If b is *much* larger than L, what is the puck's initial acceleration (to good approximation)? Could you have figured this out without solving part (a)?

ANSWER 33-6

To address parts (a) and (b), we must find the electric force exerted by the rod on the puck. So, let's figure out the electric field. The rod, like the semicircle in the previous problem, is a continuous charge distribution. You can break it into infinitesimal "pieces," and think of each piece as a point charge. The total electric field is simply the sum of the electric fields generated by all those little pieces. I'll summarize the strategy briefly.

> **Finding the field due to a continuous charge distribution**
>
> 1) Color in an arbitrary, infinitesimal piece of the rod.
> 2) Write down dq, the infinitesimal charge of that piece, in terms of dx or ds or whatever.
> 3) Using Coulomb's law, find the infinitesimal electric field at the point of interest generated by the piece you colored in. Break $d\mathbf{E}$ into components when necessary.
> 4) Integrate your dE_x and dE_y to obtain the total electric field in the x- and y-direction.

I'll now use this strategy to solve parts (a) and (b).

(a) Since the puck slides only in the y-direction: $a_x = 0$. Consequently, we need to know only the y-component of the electric force and field. This becomes crucial in steps 3 and 4 below.

Step 1: Draw a careful diagram, with dq colored in.

Always label the distance between your colored-in piece of the rod and a reference point, such as the end of the rod. Here, I've called that distance "x." The rod consists of pieces that run from $x = 0$ to $x = L$.

By Pythagorean theorem, the puck sits a distance $r = \sqrt{b^2 + x^2}$ from the colored-in piece.

Step 2: Write expression for dq, the charge of the colored-in piece.

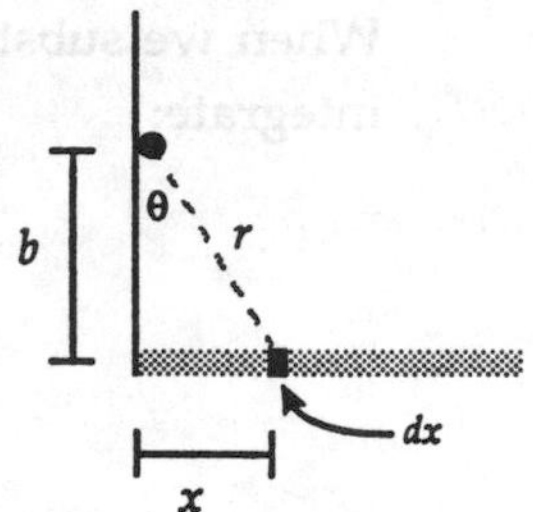

Since we know λ, we can calculate the charge contained within a length dx of the rod:

$$dq = \frac{\text{charge}}{\text{length}} \times (\text{length of the colored-in piece}) = \lambda dx.$$

You can also reach this conclusion by writing the definition of linear charge density in calculus notation. Since λ is charge per length, $\lambda = dq/dx$. Multiply through by dx to get $dq = \lambda dx$.

Step 3: Write an expression for the electric field produced by that infinitesimal charge.

By Coulomb's law, the electric field generated at the puck by the tiny piece of rod has strength

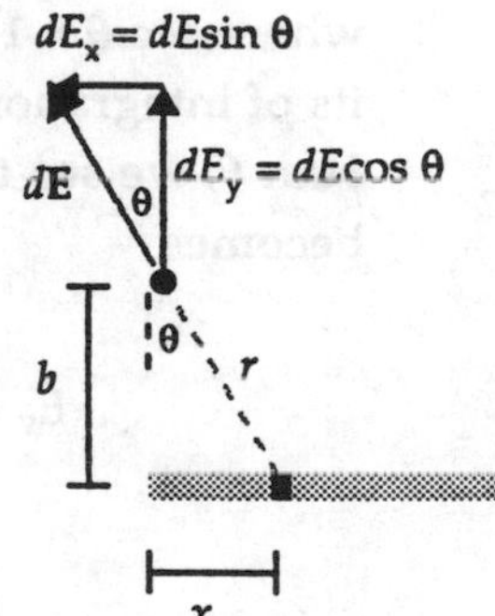

$$dE = \frac{1}{4\pi\varepsilon_0}\frac{dq}{r^2}$$

$$= \frac{1}{4\pi\varepsilon_0}\frac{\lambda dx}{b^2 + x^2},$$

where I used our results from steps 1 and 2 above.

The $d\mathbf{E}$'s contributed by different pieces of the rod all point in different directions. For this reason, we must break $d\mathbf{E}$ into components, as drawn here.

Step 4: Add up (integrate) the infinitesimal electric fields due to all the pieces of rod.

As emphasized above, the wall constrains the puck to accelerate only in the y-direction. Therefore, we care only about E_y. As the diagram shows, $dE_y = dE\cos\theta$. Since the rod extends from $x = 0$ to $x = L$, we get

$$E_y = \int dE_y = \int_0^L \frac{1}{4\pi\varepsilon_0}\frac{\lambda dx}{b^2 + x^2}\cos\theta.$$

But wait. We can't yet integrate, because the integrand contains more than one variable. Although $1/(4\pi\varepsilon_0)$, λ, and b are constants, x and θ are both variables. To evaluate the integral, we must express θ in terms of x, or vice versa.

Fortunately, from the above diagram,

$$\cos\theta = \frac{\text{adjacent}}{\text{hypotenuse}} = \frac{b}{\sqrt{b^2 + x^2}}.$$

When we substitute this into the integral, x becomes the only variable. Therefore, we're ready to integrate:

$$E_y = \int dE_y = \int_0^L \frac{1}{4\pi\varepsilon_0}\frac{\lambda dx}{b^2 + x^2}\cos\theta$$
$$= \int_0^L \frac{1}{4\pi\varepsilon_0}\frac{\lambda dx}{b^2 + x^2}\frac{b}{\sqrt{b^2 + x^2}}$$
$$= \frac{1}{4\pi\varepsilon_0}\lambda b\int_0^L \frac{dx}{(b^2 + x^2)^{3/2}}.$$

We're done with the physics thinking. On a test, some instructors would provide a table of integrals, or would not require you to evaluate the integral. Find out your instructor's policy. For the record, I'll complete the integral.

Clever calculus starts here. Use the trig substitution $x = b\tan\theta$. Then $dx = b\sec^2\theta\, d\theta$, where $\sec\theta = 1/\cos\theta$. Since we're now integrating over θ instead of x, we must change the limits of integration. Because $x = b\tan\theta$, the limit $x = 0$, corresponds to $\theta = 0$. And by setting $x = L = b\tan\theta$, we see that the $x = L$ limit corresponds to $\theta = \tan^{-1} L/b$. So, the trig-substituted integral becomes

$$E_y = \frac{1}{4\pi\varepsilon_0}\lambda b\int_0^L \frac{dx}{(b^2 + x^2)^{3/2}}$$
$$= \frac{1}{4\pi\varepsilon_0}\lambda b^2\int_0^{\tan^{-1} L/b} \frac{\sec^2\theta\, d\theta}{(b^2 + b^2\tan^2\theta)^{3/2}}$$
$$= \frac{\lambda b^2}{4\pi\varepsilon_0 b^3}\int_0^{\tan^{-1} L/b} \frac{\sec^2\theta\, d\theta}{(1 + \tan^2\theta)^{3/2}}$$
$$= \frac{\lambda}{4\pi\varepsilon_0 b}\int_0^{\tan^{-1} L/b} \frac{\sec^2\theta\, d\theta}{(\sec^2\theta)^{3/2}} \quad [\text{Trig identity}: 1 + \tan^2 = \sec^2]$$
$$= \frac{\lambda}{4\pi\varepsilon_0 b}\int_0^{\tan^{-1} L/b} \cos\theta\, d\theta \quad [\text{since } \sec\theta = 1/\cos\theta]$$
$$= \frac{\lambda}{4\pi\varepsilon_0 b}[\sin(\tan^{-1} L/b) - \sin 0]$$
$$= \frac{\lambda}{4\pi\varepsilon_0 b}\frac{L}{\sqrt{L^2 + b^2}},$$

where, to find $\sin(\tan^{-1} L/b)$, I drew a right triangle with legs of length L and b.
End of calculus.

Since $\mathbf{F} = q\mathbf{E}$, the puck experiences a y-directed force

$$F = QE_y = Q\frac{\lambda}{4\pi\varepsilon_0 b}\frac{L}{\sqrt{L^2+b^2}},$$

and therefore speeds up in the y-direction with acceleration $a_y = F_y/M$.

(b) The rod's electric field pushes the puck in the y-direction, and also in the negative x-direction (leftward). But the wall prevents the puck from sliding leftward. In other words, the wall exerts a rightward normal force on the puck that cancels the leftward component of the electric force: $N = F_{x\text{ electric}}$. So, we can obtain the normal force by finding the x-component of the electric force.

I'll recycle my results from the four-step "continuous charge distribution" strategy of part (a). Steps 1 and 2 don't change at all. In step 3, however, I'll pick off the x-component of $d\mathbf{E}$ instead of the y-component.

From the diagram in part (a), $dE_x = dE\sin\theta$, where $\sin\theta = \frac{x}{\sqrt{b^2+x^2}}$. So,

$$dE_x = dE\sin\theta = \frac{1}{4\pi\varepsilon_0}\frac{\lambda dx}{b^2+x^2}\left(\frac{x}{\sqrt{b^2+x^2}}\right).$$

Hence,

$$dE_x = \int dE_x = \int_0^L \frac{1}{4\pi\varepsilon_0}\frac{\lambda dx}{b^2+x^2}\frac{x}{\sqrt{b^2+x^2}}$$

$$= \frac{1}{4\pi\varepsilon_0}\lambda\int_0^L \frac{xdx}{(b^2+x^2)^{3/2}}.$$

This integral is easier than the last one. It evaluates to $-(b^2+x^2)^{-1/2}$. To confirm this, just differentiate $-(b^2+x^2)^{-1/2}$ with respect to x, and you'll get the above integrand. So,

$$E_x = -\frac{1}{4\pi\varepsilon_0}\lambda\frac{1}{\sqrt{b^2+x^2}}\bigg|_0^L$$

$$= \frac{1}{4\pi\varepsilon_0}\lambda\left[\frac{1}{b} - \frac{1}{\sqrt{b^2+L^2}}\right].$$

Therefore, the puck feels a leftward force $F_{x\text{ electric}} = QE_x = Q\frac{1}{4\pi\varepsilon_0}\lambda\left[\frac{1}{b} - \frac{1}{\sqrt{b^2+L^2}}\right]$. As explained above, the wall's normal force cancels this electric force: $N = F_{x\text{ electric}}$. We're done.

(c) You can solve either by using physical intuition, or by taking a mathematical limit of your part (a) answer. I'll demonstrate both techniques.

Physical intuition.

Let's visualize what it means for b to be much larger than L. Maybe the rod is one meter long, and the puck sits a kilometer away. Imagine looking at the rod from the puck. From the puck's perspective, the rod "looks" like a point. When you're that far away, the rod *might as well be a point charge.* (Similarly, when you look at a star in the sky, it might as well be a point source of light.) Therefore, we can use Coulomb's law directly, treating the rod as a single point charge.

Since the rod has linear charge density λ and length L, it carries charge

$$q = \frac{\text{charge}}{\text{length}} \times (\text{length of whole rod}) = \lambda L$$

This "point charge" q sits a distance $r = b$ away from a puck of charge Q. Therefore, from Coulomb's law,

$$F = \frac{1}{4\pi\varepsilon_0}\frac{qQ}{r^2} = \frac{1}{4\pi\varepsilon_0}\frac{\lambda LQ}{b^2}$$

Just divide by M to get the puck's acceleration.

Mathematical limit-taking.

I'll check this result by starting with our part (a) expression for the electric force,

$$F_y = Q\frac{\lambda}{4\pi\varepsilon_0 b}\frac{L}{\sqrt{L^2+b^2}},$$

and taking the limit as b gets much bigger than L. For $b \gg L$, the square root reduces to $\sqrt{L^2+b^2} = \sqrt{b^2} = b$. Hence, in this limit,

$$\begin{aligned} F_y &= Q\frac{\lambda}{4\pi\varepsilon_0 b}\frac{L}{\sqrt{L^2+b^2}}, \\ &= Q\frac{\lambda}{4\pi\varepsilon_0 b}\frac{L}{b} \\ &= \frac{\lambda}{4\pi\varepsilon_0}\frac{\lambda LQ}{b}, \end{aligned}$$

which agrees with the result just obtained using physical intuition and Coulomb's law. Cute, huh?

QUESTION 33-7

(*Hard*) Two particles of charge $+Q$ are nailed to the floor, a distance $2D$ apart. Point A is an imaginary reference point, not a charge. It's midway between the two Q's.

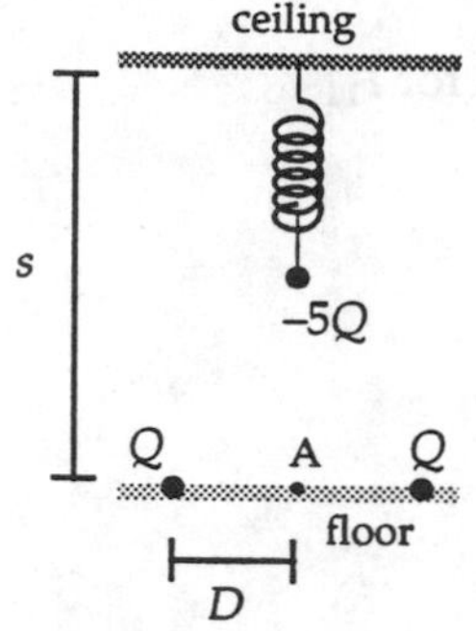

Directly over point A, hanging from a (massless) spring attached to the ceiling, is a third particle of charge $-5Q$ and mass M. The spring has equilibrium length $s/5$, where s is the distance from the floor to the ceiling.

When the hanging particle settles motionless, it's midway between the floor and the ceiling.

What is the spring's spring constant? Be careful not to confuse the spring constant k with the "k" used to abbreviate $1/(4\pi\varepsilon_0)$.

ANSWER 33-7

Ready for some nostalgia? Here's an old-fashioned force problem—more specifically, a statics problem—just like you saw last semester. Since the particle hangs motionless, it has no acceleration. So, the forces on it must cancel, in both the x- and y-directions.

When solving statics problems, always begin with a detailed force diagram, breaking the forces into components. Then write force-balancing and/or torque-balancing equations. Here, forces alone allow us to solve for the spring constant, k.

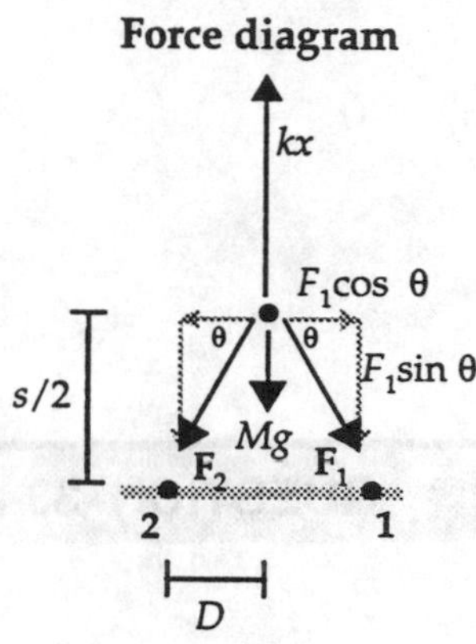

Let me call the two point charges on the floor "1" and "2." They both attract the hanging particle, with forces I'll call $\mathbf{F}_1$ and $\mathbf{F}_2$. Gravity pulls the hanging particle down with force Mg, while the stretched spring pulls it up with force kx. The minus sign in the $F_{\text{spring}} = -kx$ indicates that the spring "wants" to unstretch. So, the minus sign is already built into my force diagram.

Now I'll write expressions for $\mathbf{F}_1$ and $\mathbf{F}_2$. Since charges 1 and 2 have the same charge Q, and since they're the same distance $r = \sqrt{D^2 + (s/2)^2}$ from the hanging charge, those attractive forces have the same strength:

$$F_1 = F_2 \frac{1}{4\pi\varepsilon_0}\frac{Q(5Q)}{r^2} = \frac{1}{4\pi\varepsilon_0}\frac{5Q^2}{D^2 + (s/2)^2}.$$

I left out the minus sign in the "$-5Q$" because I'm calculating the *magnitude* of the forces. The force diagram already specifies the directions.

In the spring force formula, $F_{\text{spring}} = -kx$, a common error is to set $x = s/2$. But the "x" means the distance by which the spring is displaced *from its equilibrium (natural) length*. Since the spring's natural length is $s/5$, while its current length is $s/2$, the displacement is $x = s/2 - s/5 = 3s/10$.

OK, now let's write force-balancing equations. From the diagram, the horizontal components of F_1 and F_2 "automatically" cancel. To solve for k, we must look at the vertical forces. Since the hanging charge doesn't move ($a_y = 0$), the net vertical force is zero:

$$\begin{aligned} 0 = \sum F_y &= kx - Mg - F_1 \sin\theta - F_2 \sin\theta \\ &= kx - Mg - 2F_1 \sin\theta \qquad \text{[since } F_1 = F_2\text{]} \end{aligned}$$

$$= k\frac{3s}{10} - Mg - 2\frac{1}{4\pi\varepsilon_0}\frac{5Q^2}{D^2 + (s/2)^2}\sin\theta \quad \text{[using above expression for } F_1\text{]}$$

$$= k\frac{3s}{10} - Mg - 2\frac{1}{4\pi\varepsilon_0}\frac{5Q^2}{D^2 + (s/2)^2}\frac{s/2}{\sqrt{D^2 + (s/2)^2}},$$

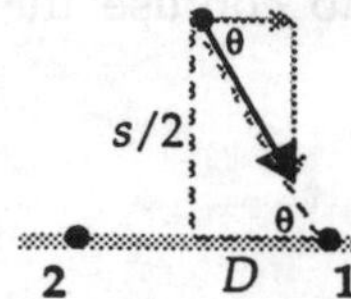

where in the last step I used this diagram to get

$$\sin\theta = \frac{\text{opposite}}{\text{hypotenuse}} = \frac{s/2}{\sqrt{D^2 + (s/2)^2}}.$$

We're done with the physics. All that remains is to isolate k. Doing so yields

$$k = \frac{10}{3s}\left\{Mg + \frac{1}{4\pi\varepsilon_0}\frac{5Q^2 s}{\left[D^2 + (s/2)^2\right]^{3/2}}\right\}$$

$$= \frac{10Mg}{3s} + \frac{1}{12\pi\varepsilon_0}\frac{50Q^2}{\left[D^2 + (s/2)^2\right]^{3/2}}$$

QUESTION 33-8

Consider two small conducting balls of mass m inside a large hemispherical bowl of radius R, where R is much bigger than the radius of the balls. By "conducting," I mean that charges are free to move around on the surface of the balls, and to flow from one conducting ball to the other when they touch. Initially, ball 1 carries charge $+Q$ while ball 2 carries no charge. Otherwise, the balls are identical.

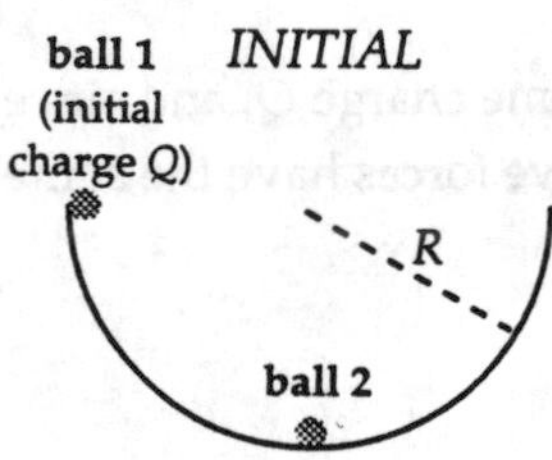

With ball 2 initially at rest at the bottom, ball 1 gets released from rest from the lip, as drawn here. It rolls down and crashes into ball 2. After bouncing off each other and rolling around a bit, the balls eventually settle motionless as shown in the "final" picture, equidistant from the bottom of the bowl.

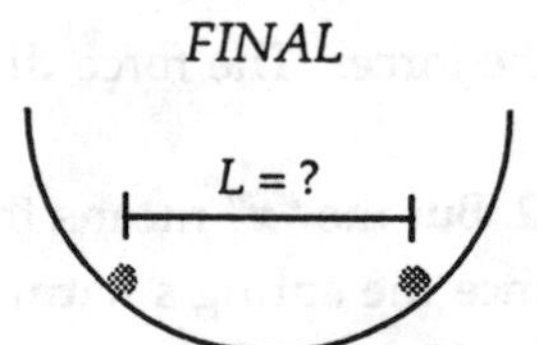

(a) [*Ask your instructor if you need to understand part (a). Parts (b) and (c) you definitely need to understand.*] As ball 1 rolls towards ball 2 (before they touch), do the two balls electrostatically attract each other, electrostatically repel each other, or neither? Justify your answer.

(b) Explain qualitatively why the balls end up with some distance between them, instead of settling right next to each other at the bottom of the bowl. Remember that ball 2 starts off uncharged.

(c) Set up, but don't solve, the equation or equations needed to solve for L, the distance between the two balls when they settle. Make sure the equations would allow you to find L in terms of Q, m, R, and any constants you need.

ANSWER 33-8

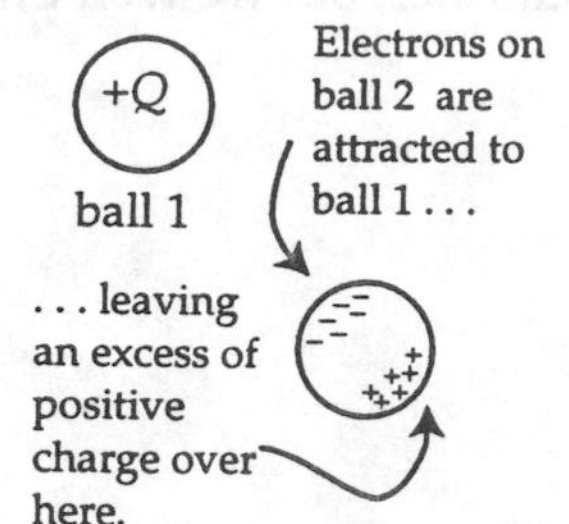
+Q
ball 1
Electrons on ball 2 are attracted to ball 1 . . .
. . . leaving an excess of positive charge over here.

(a) At first glance, you might think there's no electric force between the balls, because ball 2 carries no net charge, and therefore Coulomb's law spits out $F = 0$. But for a subtle reason called *electrostatic induction*, the balls weakly attract each other. This effect kicks in *only* when the objects conduct. Here's how it works:

A conductor contains "unbound" electrons, which aren't attached to individual atoms. These electrons move around freely on the surface of the conductor. Since ball 1 carries positive charge, it attracts the unbound electrons on ball 2. Therefore, many of those unbound electrons clump together on the edge of ball 2 closest to ball 1, giving that edge a small negative charge. This migration of electrons to one side of the ball leaves a *shortage* of electrons on the other side. On the "back" side of ball 2, there aren't enough electrons to cancel out all the positively-charged atomic nuclei. Therefore, the back side acquires a small positive charge. If this seems weird, remember that ball 2 as a whole carries no net charge. Therefore, if one side of the ball becomes slightly negative, the other side must become slightly positive.

Now I can explain why electrostatic induction causes balls 1 and 2 to attract. Because ball 1 is positive, it attracts the negative charge on the "near" side of ball 2, and repels the positive charge on the far side of ball 2. Do these attractions and repulsions cancel? No, because ball 1 is *closer* to the negative charge on ball 2 than it is to the positive charge on ball 2. See the above diagram. The attractive force between ball 1 and the "near" (negative) side of ball 2 overcomes the repulsive force between ball 1 and "far" (positive) side of ball 2. The two balls feel a small net attractive force.

This same kind of interaction, happening at the molecular level, helps to explain the "van der Waal's forces" you may have learned about in chemistry.

(b) Gravity "wants" the balls to settle right next to each at the bottom of bowl. Since the balls don't end up at the bottom, they must be repelling each other electrostatically. But why? After all, ball 2 is uncharged, isn't it?

Not quite. When the *conducting* balls touch, they can exchange charge, because unbound electrons can move from one ball to the other. (Nonconductors cannot exchange charge in this way.) Since the balls are identical, they end up sharing the total charge equally. Ball 1 gives half its charge to ball 2. So, both balls end up with positive charge $Q/2$. As a result, they repel.

(c) As explained in part (b), both balls end up with charge $Q/2$ after they crash. Since each ball ends up motionless, the net force on each ball must vanish. With luck, we can solve for L by setting the net force on ball 1 (or ball 2) equal to 0.

As always, draw a force diagram, breaking the forces into components and labeling the relevant angles. We must choose coordinate axes. Here, we can sensibly choose either "regular" axes (horizontal and vertical) or "tilted" axes (parallel and perpendicular to the surface of the bowl on which the ball rests). I'll pick tilted coordinate axes, because the algebra works out easier. You get the same final answer either way, of course.

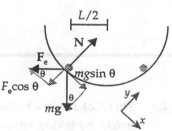

Let N denote the normal force exerted by the bowl on ball 1, and $\mathbf{F}_e$ denote the electric force exerted by ball 2 on ball 1. Notice how I broke $\mathbf{F}_e$ and gravity into components. Since the ball doesn't accelerate, the net x-directed force must vanish. (The net y-force also vanishes; but we won't need to exploit that fact.) Using Newton's 2nd law in the x-direction, I get

$$\sum F_x = ma_x$$

$$mg\sin\theta - F_e\cos\theta = 0$$

$$mg\sin\theta - \frac{(Q/2)^2}{4\pi\varepsilon_0 L^2}\cos\theta = 0, \quad (1)$$

where in the last step, I used Coulomb's law, with both charges equal to $Q/2$, and with $r = L$, the distance between the charges.

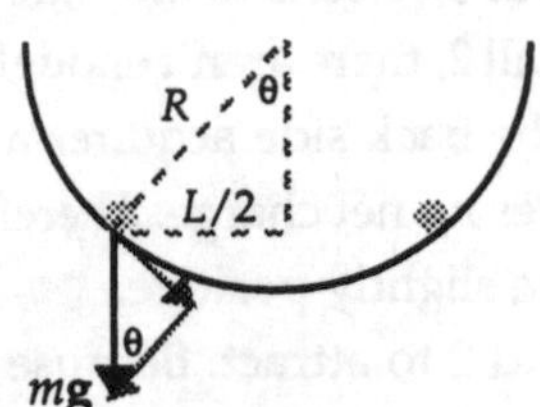

We can't yet solve for L , because Eq. (1) contains a second unknown, namely θ. From this new diagram, however, we get

$$L/2 = R\sin\theta. \quad (2)$$

Now we now have two equations in two unknowns, L and θ. So, it's just a matter of algebra to solve.

QUESTION 33-9

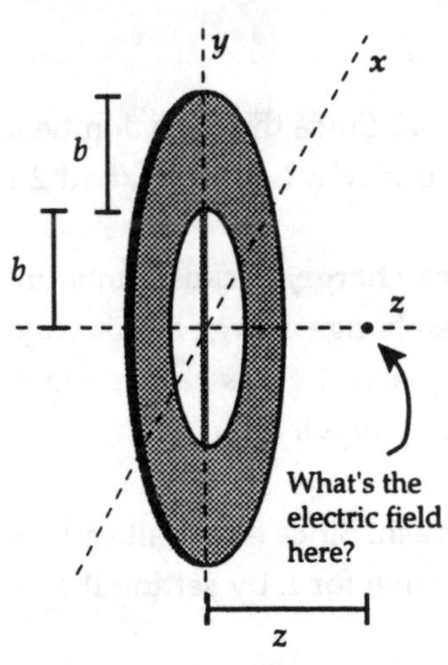

(*Challenge problem*) Consider this cardboard decoration. It consists of (i) A ring of inner radius b and outer radius $2b$; and (ii) a very thin strip of length $2b$ and width w, centered at the center of the ring. The strip and the ring "live" in the x-y plane. The z-axis passes through the center.

The ring and strip have surface charge density σ. Surface charge density is charge per area.

(a) What is the electric field at a point on the z-axis that's an arbitrary distance z from the center of the ring? Derive your answer "from scratch" using basic laws of physics. Don't use any pre-derived formulas from the textbook. Math hint:

$$\int_{-b}^{b}\frac{dy}{(a^2+y^2)^{3/2}} = \frac{2b}{a^2\sqrt{a^2+b^2}}$$

(b) At what point on the z-axis is the electric field strongest? Don't complete the math; but describe with words and formulas exactly how you could figure this out. Also take a guess using physical intuition.

ANSWER 33-9

(a) This looks overwhelming at first. But we can break the overall problem into manageable "subproblems," by taking advantage of the superposition principle. According to this principle, the total electric field produced by a charge distribution is simply the sum of the fields created by the different "pieces" of the charge distribution. Here, we can think of the overall decoration as consisting of a ring and strip. So,

$$E_{total} = E_{ring} + E_{strip}.$$

Therefore, I can first solve for the electric field on the *z*-axis generated by the strip. Then, in a separate subproblem, I can find the field created by the ring. Each of these two subproblems consists of many challenging steps. But at least we have an overall strategy.

Subproblem 1: Field due to the strip

From previous problems, recall the strategy for this kind of problem:

Finding the electric field due to continuous charge distribution

1) Color in an arbitrary infinitesimal piece of the charge.
2) Write down *dq*, the charge of that piece, in terms of *dx* or *dy* or whatever.
3) Using Coulomb's law, calculate the infinitesimal electric field at the point of interest due to your *dq*. Break that *d*E into components if needed.
4) Integrate your dE_x or dE_y or dE_z to obtain the total electric field.

For now, I'll deal only with the strip, *not* with the ring.

Step 1: Draw a careful diagram, with a piece of the charge distribution colored in.

The colored-in piece of the strip has infinitesimal length *dy*. As always, I've labeled in the distance of that piece from a reference point, in this case the center of the strip.

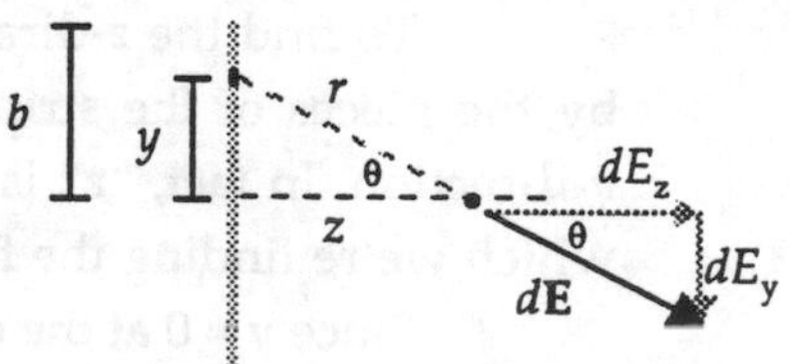

Step 2: Write an expression for dq, the charge of that colored-in piece.

The piece of strip under consideration has length *dy* and width *w*. So, it has infinitesimal area $dA = wdy$. Therefore, its charge is

$$dq = \frac{\text{charge}}{\text{area}} \times (\text{area of that piece}) = \sigma dA = \sigma w dy.$$

Step 3: Write an expression for the electric field created by dq.

Since the electric field is a vector, we must deal separately with its different components. Fortunately, symmetry often allows us to guess which components don't matter. For instance, in this case, the electric field created by the strip points solely in the *z*-direction, with no *y*-component.

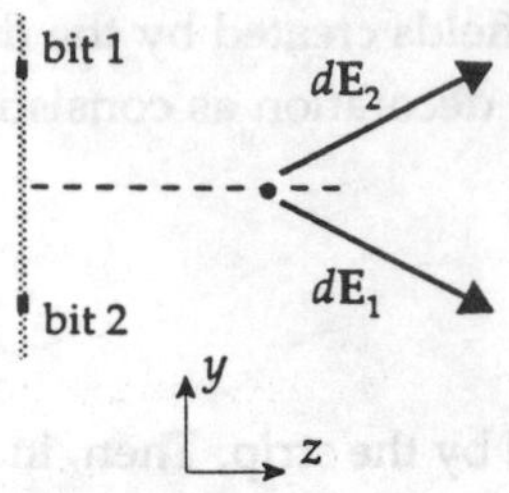

To see why, look at this diagram. The y-components of the fields generated by "bit 1" and "bit 2" cancel. And we can divide the whole strip into paired pieces like bit 1 and bit 2. Therefore, the overall field has no y-component. For this reason, I'll worry only about z-components.

From Coulomb's law, the colored-in piece of the strip generates a field

$$dE = \frac{1}{4\pi\varepsilon_0}\frac{dq}{r^2} = \frac{1}{4\pi\varepsilon_0}\frac{\sigma w dy}{y^2+z^2},$$

where I used Pythagorean theorem to get r, and I used my dq from step 2.

From the diagram, $dE_z = dE\cos\theta$, where

$$\cos\theta = \frac{\text{adjacent}}{\text{hypotenuse}} = \frac{z}{\sqrt{z^2+y^2}}.$$

Putting all this together yields

$$\begin{aligned} dE_z &= dE\cos\theta \\ &= \frac{1}{4\pi\varepsilon_0}\frac{\sigma w dy}{y^2+z^2}\frac{z}{\sqrt{z^2+y^2}} \\ &= \frac{\sigma w z}{4\pi\varepsilon_0}\frac{dy}{(y^2+z^2)^{3/2}}. \end{aligned}$$

Step 4: Sum up the electric field contributions from all the pieces of charge.

To find the z-directed field due to the whole strip, add up (integrate) the dE_z's generated by the pieces of the strip. Here, the integral is over dy, not dz, because the strip extends in the y-direction. In fact, "z" is a *constant*, the fixed distance from the center of the strip to the point at which we're finding the field.

Since $y = 0$ at the center of the strip, the strip extends from $y = -b$ to $y = b$. Those are the limits of integration. So,

$$\begin{aligned} E_z \int dE_z &= \int_{-b}^{b}\frac{\sigma w z}{4\pi\varepsilon_0}\frac{dy}{(y^2+z^2)^{3/2}} \\ &= \frac{\sigma w z}{4\pi\varepsilon_0}\int_{-b}^{b}\frac{dy}{(y^2+z^2)^{3/2}} \\ &= \frac{\sigma w z}{4\pi\varepsilon_0}\frac{2b}{z^2\sqrt{z^2+b^2}} \qquad \text{[using math hint, with } a = z\text{]} \\ &= \frac{\sigma w b}{2\pi\varepsilon_0 z\sqrt{z^2+b^2}}. \end{aligned}$$

So far, we've found the field on the z-axis generated by the strip. Now we must find the field created by the ring at that same point.

Subproblem 2: Field due to the ring

Your textbook probably demonstrates how to calculate the electric field generated by a very thin ring of charge. But this ring isn't infinitesimally thin. It's thick. Fortunately, you can use the standard calculus technique of breaking the thick ring into a bunch of concentric, infinitesimally thin rings.

This suggests a strategy. First, we can derive the electric field produced by a *single* infinitesimally thin ring of charge. Then, we can add up the electric field contributions from all the thin rings that comprise the thick ring. Notice that I've divided this subproblem into further subproblems! But hey, it's the best way to stay organized.

Find the field generated by a single infinitesimally thin ring of charge.

We can reuse the continuous charge distribution strategy from subproblem 1 above.

Step 1: Draw a careful diagram with a "piece" of the charge colored in.

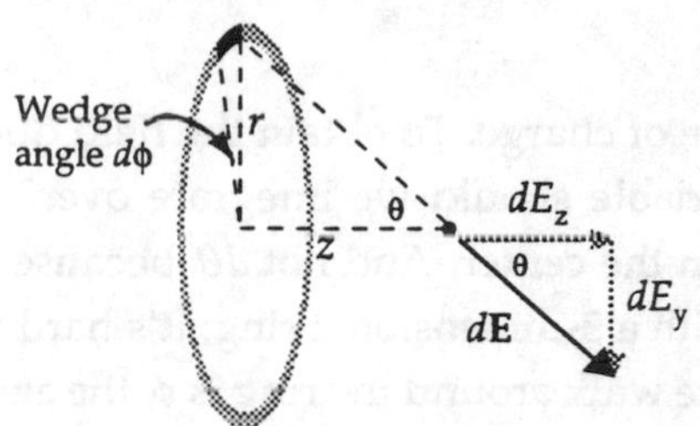

This ring has inner radius r and outer radius $r + dr$. So, it's infinitesimally thin, with width dr. I've blackened a tiny wedge of the ring. The wedge subtends infinitesimal angle $d\phi$.

Step 2: Write an expression for dq, the charge of the colored-in piece.

How big is the blackened piece of the ring? Well, by definition, angle = arc length/radius. So, the blackened piece has arc length $ds = rd\phi$. And its width, as just noted, is dr. Therefore, the blackened piece has area dA = length × width = $rd\phi dr$. Multiply that area by the surface charge density to obtain the charge of the colored-in piece:

$$dq = \frac{\text{charge}}{\text{area}} \times (\text{area of the infinitesimal piece}) = \sigma dA = \sigma r d\phi dr.$$

Step 3: Write an expression for the infinitesimal electric field generated by that piece of charge.

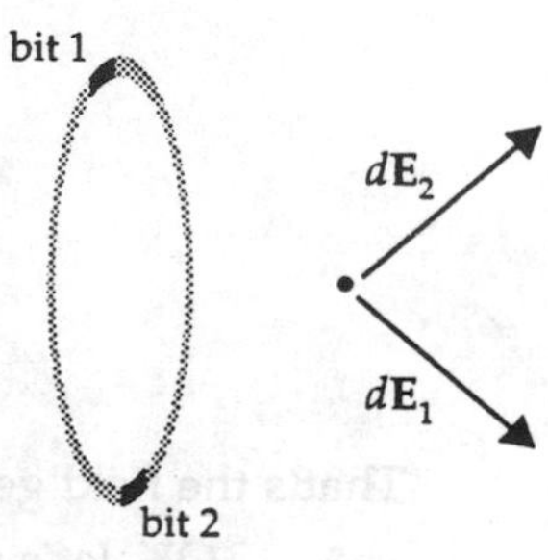

Only the rightward (z) components don't cancel.

Once again, symmetry shows us that the x- and y-components of the electric field created by the ring vanish. Here's why. Consider any two bits of the ring 180° apart. The x- and y-components of their electric fields cancel, as this drawing shows. Because we can break the whole ring into pairs of bits that are 180° apart, this cancellation argument applies to the whole ring. Because the x- and y-components of the field cancel away, I'll focus only on dE_z.

From the above diagram, the infinitesimal z-directed field generated by the colored-in piece has strength $dE_z = dE\cos\theta$, where

$$\cos\theta = \frac{z}{\sqrt{z^2 + r^2}}.$$

Don't confuse r, the ring's radius, with the "r" in Coulomb's law. The r in Coulomb's law denotes the distance from the charge creating the field to the point at which we're "measuring" the field. By Pythagorean theorem, that distance is $\sqrt{z^2 + r^2}$. So,

$$\begin{aligned} dE_z &= dE\cos\theta \\ &= \frac{1}{4\pi\varepsilon_0}\frac{dq}{r^2 + z^2}\cos\theta \\ &= \frac{1}{4\pi\varepsilon_0}\frac{\sigma r dr d\phi}{r^2 + z^2}\frac{z}{\sqrt{z^2 + r^2}}, \end{aligned}$$

where in the last step, I used our dq from step 2, and our expression for $\cos\theta$ from above.

Step 4: Add up the field contributions from all the pieces.

That's the z-field generated by the blackened piece of charge. To obtain the field due to the whole thin ring, integrate around the ring. But what variable should we integrate over? Not dr, because all points on the ring are the same distance r from the center. And not $d\theta$, because θ is the same for all "bits" of the ring. (To see this, play around with a 3-dimensional ring; it's hard to visualize in two dimensions.) The only variable that *varies* as we walk around the ring is ϕ, the azimuthal angle in the x-y plane. Don't confuse ϕ with θ. When integrating around the ring, ϕ varies from 0 to 2π radians, while θ, r, and z stay constant. We'll integrate over dr later, to build the thick ring out of infinitesimally thin rings. But here, we're considering *one* thin ring:

$$\begin{aligned} E_z \text{ due to thin ring} &= \int_{\phi=0}^{\phi=2\pi} dE_z = \int_0^{2\pi}\frac{1}{4\pi\varepsilon_0}\frac{\sigma r dr d\phi}{r^2 + z^2}\frac{z}{\sqrt{z^2 + r^2}} \\ &= \frac{1}{4\pi\varepsilon_0}\frac{\sigma r dr}{r^2 + z^2}\frac{z}{\sqrt{z^2 + r^2}}\int_0^{2\pi} d\phi \\ &= \frac{1}{4\pi\varepsilon_0}\frac{\sigma r dr}{r^2 + z^2}\frac{z}{\sqrt{z^2 + r^2}}2\pi \\ &= \frac{1}{2\varepsilon_0}\frac{\sigma r dr}{r^2 + z^2}\frac{z}{\sqrt{z^2 + r^2}} \end{aligned}$$

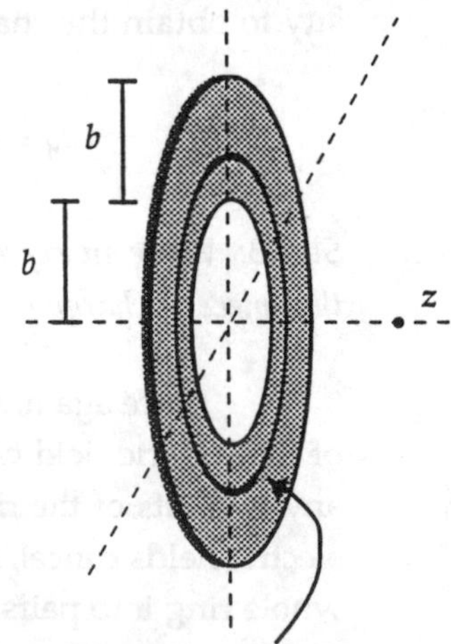

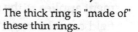

The thick ring is "made of" these thin rings.

That's the field generated by a *single* infinitesimally thin ring.

OK, let's step back to get our bearings. In this subproblem, we're finding the electric field generated by the thick ring. So far, we've found the field generated an infinitesimally thin ring. But the thick ring is "made of" concentric infinitesimally thin rings. By adding up the field contributions from all the thin rings contained in the thick ring, we can obtain the total electric field generated by the thick ring.

Find the field generated by the thick ring.

As the diagram shows, the smallest thin ring has radius $r = b$, while the largest one has radius $r = 2b$. So,

$$\begin{aligned}
E_z \text{ thick ring} &= \int_{r=b}^{r=2b} (E_z \text{ due to thin ring} \\
&= \int_b^{2b} \frac{1}{2\varepsilon_0} \frac{\sigma r dr}{r^2 + z^2} \frac{z}{\sqrt{z^2 + r^2}} \\
&= \frac{\sigma z}{2\varepsilon_0} \int_b^{2b} \frac{r dr}{(r^2 + z^2)^{3/2}} \\
&= \frac{\sigma z}{2\varepsilon_0} \left(-\frac{1}{\sqrt{z^2 + r^2}} \Bigg|_b^{2b} \right) \\
&= \frac{\sigma z}{2\varepsilon_0} \left[\frac{1}{\sqrt{z^2 + b^2}} - \frac{1}{\sqrt{z^2 + (2b)^2}} \right].
\end{aligned}$$

In this subproblem, we evaluated two integrals. The first gave us the z-directed electric field generated by an infinitesimally thin ring. In that integral, we held r constant, and integrated around the ring over the azimuthal angle ϕ. The second integral added up the contributions from all the thin rings comprising the thick ring. In that integral, we varied r, the radius of the thin rings, by integrating over dr.

Multivariable calculus fiends often combine those two integrals into a double integral. The answer comes out the same, of course. For me, breaking the double integral into two single integrals highlights the physical reasoning underlying the mathematics.

End of subproblem 2.

At this point, we've figured out the electric fields produced by the strip (in subproblem 1) and by the thick ring (in subproblem 2). According to the superposition principle,

$$\begin{aligned}
E_{\text{total}} &= E_{\text{strip}} + E_{\text{ring}} \\
&= \left\{ \frac{\sigma w b}{2\pi\varepsilon_0 z \sqrt{z^2 + b^2}} + \frac{\sigma z}{2\varepsilon_0} \left[\frac{1}{\sqrt{z^2 + b^2}} - \frac{1}{\sqrt{z^2 + (2b)^2}} \right] \right\} \hat{\mathbf{z}},
\end{aligned}$$

where the unit vector $\hat{\mathbf{z}}$ specifies the field direction.

(b) In this standard "maximization" problem, we're trying to maximize the electric field. The free variable, by which I mean the quantity we can adjust in order to raise or lower E, is the distance z from the center of the ring (along the z-axis). As explained in question 33-4c, we can solve for the z that maximizes E by setting $dE/dz = 0$.

The derivative comes out messy. On a test, many instructors would award nearly full credit for setting up the equation, even if you don't complete the math. Find out your instructor's policy.

Here, we can take an intelligent guess at the answer. I think it's $z \approx 0$, i.e., *extremely* close to the strip. When you're extremely close to the strip, the strip "looks" like an infinite plane of charge, from your perspective. I'll discuss this subtle point more fully in the next chapter.

Electric Flux And Gauss' Law

34 CHAPTER

QUESTION 34-1

Consider a (nonconducting) wire bent into a rectangle of length l and width w. It's placed inside a uniform electric field $\mathbf{E} = E_0\,\hat{\mathbf{z}}$, where $\hat{\mathbf{z}}$ is a unit vector in the z-direction. What is the electric flux through the rectangle if

(a) the plane of the rectangle "faces" the z-direction, i.e., the rectangle "lives in" the x-y plane? (When I talk about the direction in which a surface "faces," I mean the direction perpendicular to the surface. For instance, your floor faces up, and your wall faces sideways.)

(b) the plane of the rectangle faces the x-direction?

(c) the rectangle "faces" at a 30° angle to the z-direction?

(d) Now suppose a hemicylindrical mesh is attached to the rectangular wire, as drawn here. The plane of the wire rectangle again faces the z-direction, as in part (a). What electric flux flows through the mesh?

Part (d)

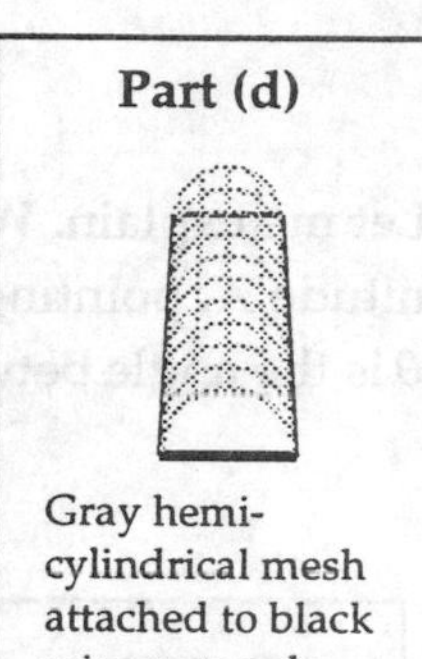

Gray hemi-cylindrical mesh attached to black wire rectangle from part (a)

ANSWER 34-1

This problem gives me an excuse to review the intuitions underlying electric flux. Of course, you could answer parts (a) through (c) by plugging 'n' chugging. But a deeper understanding helps you to avoid common mistakes, and enables you to complete part (d) in 15 seconds, something you couldn't do formulaically.

I'll now draw an extended analogy between electric field lines and water stream lines. Electric field lines correspond to tiny streams of water. Here, the electric field points in the z-direction, with uniform strength. Therefore, we can think of the room as filled with a steady stream of water flowing in the z-direction. A stronger field corresponds to a faster stream of water.

Now we hold the wire rectangle inside that stream. How much water per time flows through the rectangle? Intuitively, the answer depends on two factors: the strength (speed) of the stream, and the area of the rectangle. For instance, if we double the rectangle's area, then twice as much water flows through—not because the stream gets stronger, but simply because an enlarged rectangle "captures" more water. On the other hand, if the stream is very strong (fast), then lots of water would flow through even a small rectangle.

In summary, the water per time passing through the rectangle (i.e., the water *flux*) is proportional to the strength of the stream, and also proportional to the area of the rectangle. Let W denote the strength of water, and A denote the area of the rectangle. Then the water flux, Φ, is given by $\Phi = WA$.

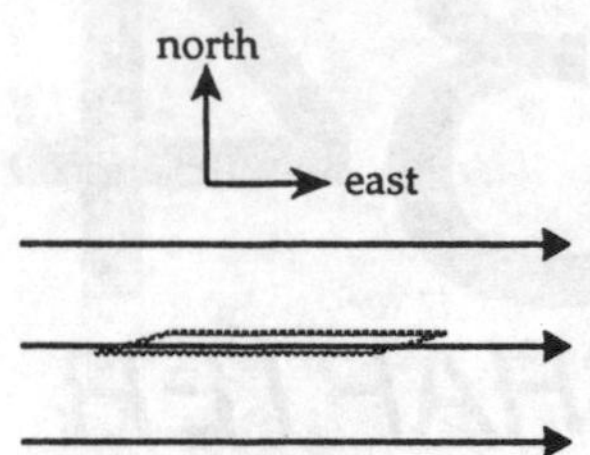

On second thought, that formula can't be quite right. To see why, imagine holding the rectangle so that it faces perpendicular to the flow-direction of the water. For instance, maybe the stream flows east, and the plane of the rectangle faces north, as drawn here. Although water "skims along" the rectangle, water does not flow *through* the rectangle. So, the water flux is $\Phi = 0$, even though the rectangle has nonzero area. This means we must modify our flux formula, $\Phi = WA$.

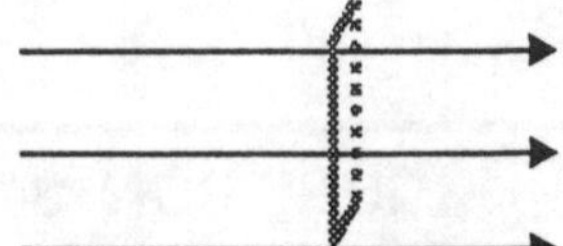

The most water flows through the rectangle when it faces the direction of flow

As the east-north example shows, that formula fails because it doesn't take into account the rectangle's orientation. The most water flows through the rectangle when it faces the same direction in which the water flows. In that case, the water flux is indeed $\Phi = WA$. But if we face the rectangle in some other direction, then less water flows through it. And if we face it perpendicular to the water flow, the flux is zero.

To capture this intuition, we must use the dot product,

$$\Phi = \mathbf{W} \cdot \mathbf{A} = WA\cos\theta.$$

Let me explain. **W** is a vector of magnitude W, pointing in the water-flow direction. **A** is a vector of magnitude A, pointing in the direction that the surface *faces*. For instance, your floor's **A** points upward. And θ is the angle between those vectors.

This formula captures our intuitions perfectly. When the rectangle faces the same direction that the water flows ($\theta = 0$), then $\Phi = WA\cos 0 = WA$, exactly as expected. By contrast, when the surface faces perpendicular to the water flow ($\theta = 90°$), $\Phi = WA\cos 90° = 0$, as we saw above.

So far, we've found the formula for water flux, in the case of a uniform stream of water. (I'll talk about non-uniform streams later.) Now we just need to "analogize" that formula to electric fields. Well, the water strength **W** corresponds to the electric field strength **E**, because water stream lines correspond to electric field lines. Therefore,

$$\Phi_E = \mathbf{E} \cdot \mathbf{A} = EA\cos\theta \qquad \textbf{for constant electric field.}$$

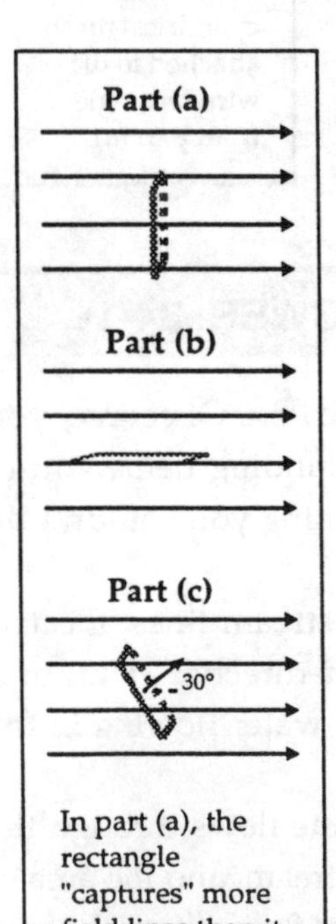

In part (a), the rectangle "captures" more field lines than it does in part (c)

Given this result, we can solve parts (a) through (c) immediately. In these drawings, the z-direction is rightward.

(a) Here, the rectangle faces the direction in which the electric field "flows." So, $\theta = 0$. Therefore,

$$\Phi_E = \mathbf{E} \cdot \mathbf{A} = EA\cos 0 = E_0 lw,$$

since the rectangle has area $A = lw$.

(b) Now the rectangle faces perpendicular to the electric field lines, analogous to the north vs. east water example from above. Since $\theta = 90°$,

$$\Phi_E = \mathbf{E}\cdot\mathbf{A} = EA\cos 90° = 0.$$

Intuitively speaking, no field lines "pierce" through the rectangle.

(c) Because of the 30° angle, the rectangle "captures" fewer field lines than it did in part (a):

$$\Phi_E = \mathbf{E}\cdot\mathbf{A} = E_0 lw\cos 30° = (0.87)E_0 lw.$$

END-ON VIEW
We're looking at the end of the hemicylinder. It extends into the page.

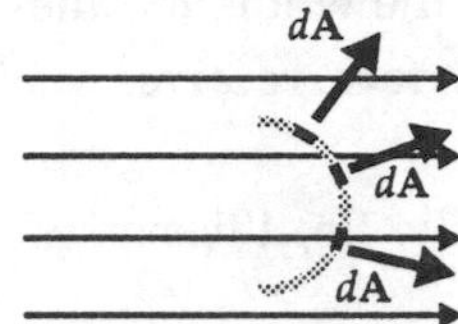

(d) At first glance, things appear to get more complicated, because the relevant surface makes a non-constant angle with the electric field. Think of the hemicylinder as consisting of tiny "patches." I've blackened three infinitesimal patches, each of area dA. Unfortunately, each of those area elements makes a different angle with the electric field. There's no one angle we can plug into $\Phi E = EA\cos\theta$.

Instead, we could consider each patch separately. Specifically, we could write down the infinitesimal flux through one of those patches: $d\Phi E = \mathbf{E}.d\mathbf{A}$. Then we could add up (integrate) the flux contributions from every patch, to flux contributions from every patch, to obtain the total flux through the hemicylinder. This strategy works fine. But it takes awhile.

Fortunately, we can find a short cut, by thinking *intuitively* about flux. In the water analogy, flux tells us how much water flows through a surface. Consider the water stream lines flowing through this rectangle with the hemicylindrical mesh attached. Any stream line that pierces the rectangle also passes through the hemicylinder. In other words, any water droplet that goes through the rectangle also goes through the hemicylindrical mesh. Therefore, the water flux through the rectangle equals the water flux through the hemicylinder.

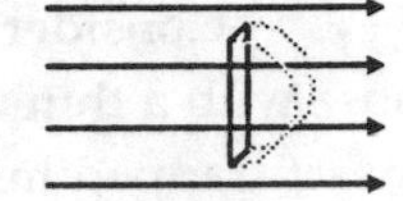
Any field line through the rectangle also goes through the hemicylinder.

The same reasoning applies to electric flux. Any electric field line passing through the rectangle also pierces the hemicylinder. Since electric flux corresponds (roughly speaking) to the number of field lines flowing through the surface, the flux through the rectangle equals the flux through the hemicylinder. We already found ΦE through the rectangle in part (a). For this reason, we need not mess around with integrals. Intuitive physical reasoning got us to the answer!

$$\Phi_{\text{through hemicylinder}} = \Phi_{\text{through rectangule in part (a)}} = E_0 lw.$$

By the way, these same physical intuitions underlie *Gauss' law*, which I'll discuss soon.

QUESTION 34-2

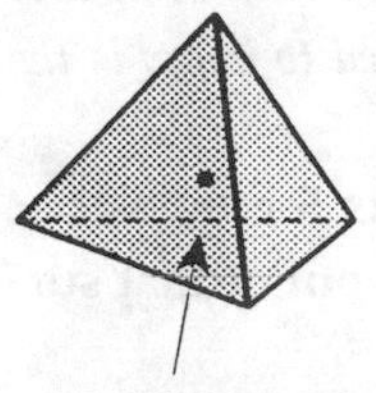
Charge Q_0 at center

A tetrahedron is a pyramid made of four identical equilateral triangles. Suppose a charge Q_0 is suspended inside a tetrahedron, at the center. So, Q_0 is the same distance s from all four vertices of the tetrahedron.

What is the electric flux through the bottom face of the tetrahedron? Hint: If you find yourself grinding through a lot of math, you're missing a shortcut.

ANSWER 34-2

After outlining how you could solve this the "long way" by integrating, I'll explain how you can jump to the answer using Gauss' law.

In principle, you can use Coulomb's law to find the electric field at each point on the bottom face of the tetrahedron. Unfortunately, those different points are different distances from the charge Q_0. Therefore, you can't use FE = E.A directly, because the electric field isn't constant over the whole bottom face. Since the electric field varies in strength (and direction) over the bottom face, you'd have to divide that face into tiny "patches" of area dA, and then integrate E.dA. What a mess!

Fortunately, there's a shortcut. First, use Gauss' law to find the flux through the whole tetrahedron (all four faces). Then, use symmetry to argue that each face, including the bottom face, receives one fourth of that total flux.

This "trick" seems mysterious, unless you understand Gauss' law intuitively. So, I'll explain Gauss' law in terms of the water analogy developed above.

Electric field lines correspond to water stream lines. Water stream lines come from a source, such as a faucet or a sprinkler. The "strength" of a water source is simply the amount of water per second that it squirts out.

Consider a water source, say, the end of a hose. Suppose you completely surround the end of the hose with a thin wire mesh that water can flow through. The surface must be closed, in the sense that an insect trapped inside could not escape. For instance, a bowl is an open surface. But if you cover the top with plastic wrap, the bowl-and-plastic become a closed surface. A disk or rectangle can't be closed, because it can't "enclose" anything.

A closed surface, in the insect-trapping sense, is called a *Gaussian surface.* Actually, Gaussian surfaces are usually imaginary, not made of mesh. But you get the idea. Here, we're enclosing the end of the hose inside a Gaussian surface. The arrows represent water stream lines.

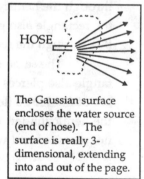

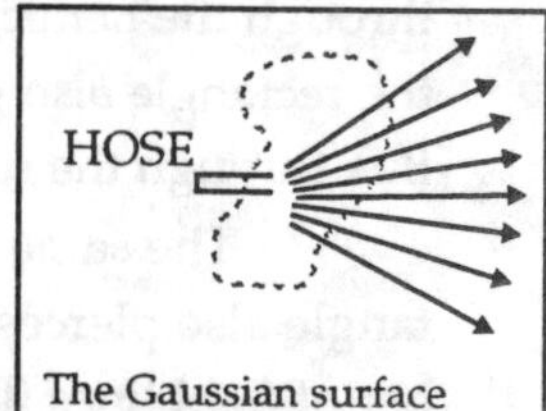

The Gaussian surface encloses the water source (end of hose). The surface is really 3-dimensional, extending into and out of the page.

Let's say that the water source strength is 5 gallons per minute. That's how much water per time the hose squirts out. Here's the crucial question: What's the water flux out of the Gaussian surface? In other words, how many gallons of water per minute pass through the Gaussian surface? Well, since water doesn't "build up" inside the mesh, and since it can't get anywhere without passing through the surface, every drop of water squirted by the hose eventually passes through the mesh. In other words, *all* the water streamlines eventually pass through the Gaussian surface. They've got nowhere else to go. Since the hose spurts out 5 gallons per minute, it follows that 5 gallons per minute pierce through the Gaussian surface. So, *the water source strength,* in this case 5 gallons per minute, *must equal the water flux through a Gaussian surface enclosing that source.*

This result, though seemingly obvious, has deep implications. Notice that the size and shape of the Gaussian surface don't matter. To predict the water flux out of that surface, *all we need to know is the strength of the enclosed water source (or sources).*

For instance, if a Gaussian surface encloses two hose ends, one of which squirts 5 gallons per minute and the other of which squirts 3 gallons per minute, then we know the water flux out of that surface. It's 8 gallons per minute, end of story. Let me express this result mathematically:

Gauss' law for water: $\Phi_{water} = (\text{water source strength})_{encl},$

where Φ_{water} denotes the water flux out of a Gaussian surface, and (water source strength)$_{encl}$ denotes the total strength of all the water sources *inside* that surface.

What if the water source sits *outside* the Gaussian surface? At first glance, Gauss' law for water appears to be violated. After all, the enclosed water source strength is zero, since the surface encloses no water sources. But Φ_{water} seems not to be 0, since water passes through the surface.

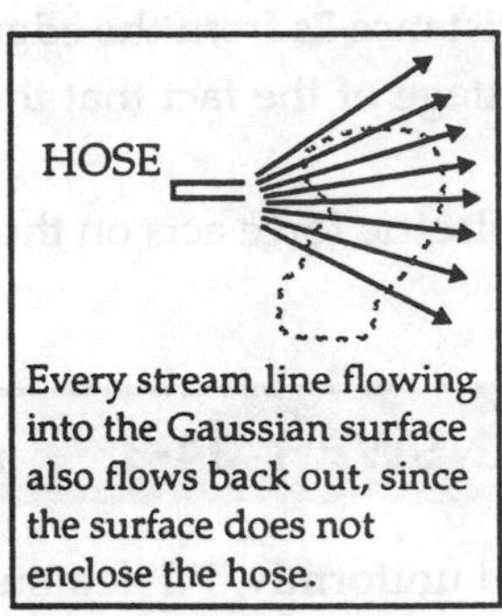

Every stream line flowing into the Gaussian surface also flows back out, since the surface does not enclose the hose.

But here's the thing: When water flows *into* the surface, the water flux counts as *negative*. And when water passes *out of* the surface, the flux counts as positive. From this diagram, notice that every water stream line flowing into the Gaussian surface eventually comes back out (the other side). So, if you're a water droplet, you "contribute" negative flux when you enter the Gaussian surface, but then you contribute positive flux when you leave. Those two contributions cancel. Your net flux contribution is 0. This is true of all the droplets contained in all the stream lines. So, the net flux through the surface, taking into account the positive and negative contributions, is indeed $\Phi_{water} = 0$, exactly as predicted by Gauss' law for water.

Now that we understand Gauss' law for water, we can "analogize" to obtain Gauss' law for electrostatics. The "source" of electric field lines is *charge*. A bigger charge produces more field lines. So, water source strength corresponds to q, the charge of the "field line source."

Since charge has different units from electric flux, we must introduce the "unit-conversion" constant ε_0 into Gauss' law. But don't let that constant distract you. Conceptually, Gauss' law for electrostatics is equivalent to Gauss' law for water:

Gauss' law: $$\Phi_E = q_{encl} / \varepsilon_0.$$

$$\int \mathbf{E} \cdot d\mathbf{A} = q_{encl} / \varepsilon_0.$$

In the second line, I wrote down the expression for electric flux in terms of electric field. By q_{encl}, I mean the sum of the charges enclosed (trapped) by the Gaussian surface.

Make sure you understand the intuitions behind Gauss' law, in terms of the water analogy. These intuitions can turn seemingly horrendous problems into easier ones. For instance, according to Gauss' law, if we surround the charge Q0 with a tetrahedron, or with any other closed surface, the total electric flux out of the surface is

$$\Phi_{\text{through tetrahedron}} = \frac{Q_0}{\varepsilon_0}..$$

That's the total flux through all four faces of the tetrahedron. But what's the flux through the bottom face? Well, a point charge spews out field lines equally in all directions. Since each face is the same size and shape, and since Q_0 sits at the center (equidistant from each face), *the same flux flows through each of the four faces*. In other words, each face receives the same "share" of the total flux. Therefore, the bottom face receives exactly one fourth of the total flux:

$$\Phi_{\text{through bottom face}} = \frac{1}{4}\Phi_{\text{through tetrahedron}} = \frac{Q_0}{4\varepsilon_0}.$$

QUESTION 34-3

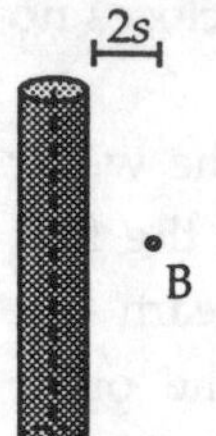

Consider a long thin cylindrical rod of uniform charge density (charge per volume) ρ. The rod has radius s and length $50s$.

(a) Find the rod's linear charge density.

(b) What is the (approximate) electric field at point B? Point B is a distance $2s$ from the edge of the rod, near the middle. Hint: Use Gauss' law, taking advantage of the fact that the rod is *very* long as compared to B's distance from the rod.

(c) Suppose a small particle of charge q_1 is placed at point B. What electric force acts on that particle?

ANSWER 34-3

(a) You can solve either with or without calculus, since the charge is distributed uniformly. I'll demonstrate both ways.

Method 1: No calculus

Given the charge per volume, we can figure out the total charge of the rod, Q. Since linear charge density is charge per length, just divide Q by the rod's length, $l = 50s$.

To figure out the rod's total charge, we need its volume. Well, since the base of the rod is a disk of radius s, it has area πs^2. The volume is therefore $V = \text{(base)} \times \text{(height)} = \pi s^2(50s)$. Therefore,

$$Q = \frac{\text{charge}}{\text{volume}} \times \text{(total volume)} = \rho V = \rho\pi s^2(50s).$$

So, the rod has linear charge density

$$\lambda = \frac{\text{charge}}{\text{length}} \times \frac{Q}{50s} = \frac{\rho\pi s^2(50s)}{50s} = \rho\pi s^2.$$

Method 2: Calculus

The following method would work even if the rod carried non-uniform charge density. By definition, charge density is charge per volume: $\rho = dq/dV$. And linear charge density is charge per length, $\lambda = dq/dl$. We can manipulate these expressions to obtain λ in terms of ρ. For instance, divide λ by ρ to get

$$\frac{\lambda}{\rho} = \frac{dq/dl}{dq/dV} = \frac{dV}{dl},$$

and hence, $\lambda = \rho \dfrac{dV}{dl}$.

If we let l denote the rod's length, then it has volume $V = \text{(base)} \times \text{(length)} = \pi s^2 l$. So,

$$\lambda = \rho\frac{dV}{dl} = \rho\frac{d}{dl}(\pi s^2 l) = \rho\pi s^2,$$

in agreement with method 1 above.

Enlarged picture of part of the rod, with an infinitesimally thin disk colored in.

To gain physical insight into this calculus, and also to learn an alternate problem-solving technique, ask yourself the following question: How much charge dq is contained in an infinitesimal piece of the rod, of length dl? Well, a slice of the rod of length dl is an infinitesimally thin disk, with area πs^2. So, that disk has infinitesimal volume $dV = \pi s^2 dl$. Since $\rho = dq/dV$, the disk contains charge

$$dq = \rho dV = \rho \pi s^2 dl.$$

Now just divide both sides by dl to get the linear charge density.

(b) You could divide the rod into infinitesimal pieces, calculate the electric field generated by each piece, and then add up those field contributions. But it's quicker to use Gauss' law.

Gauss' law, though *always* true, does not always help us solve problems. For reasons we'll see below, Gauss' law can help you find an electric field only when the charge distribution is highly symmetric. Here's the idea. By surrounding part of the rod with a Gaussian surface, we can invoke Gauss' law's to calculate the electric flux through that surface. Using the definition of electric flux, $\Phi_E = \int \mathbf{E} \cdot d\mathbf{A}$, we can then find the electric field—but only if we can pull **E** outside the flux integral. This happens only when the field is constant over (part of) the Gaussian surface, which happens only when the field is highly symmetric. As we'll see below, a very long rod produces a sufficiently symmetric field for this approach to work.

Let me translate these theoretical considerations into a strategy. Although the strategy seems hopelessly abstract at first, it makes sense once you practice using it a few times.

Gauss' law to calculate electric fields

1) Using symmetry and intuition, draw the electric field lines.
2) Enclose all or part of the charge distribution with a Gaussian surface. The electric field should have the same strength at all points on (at least part of) the surface.
3) Apply Gauss' law: $\Phi_E = \int \mathbf{E} \cdot d\mathbf{A} = q_{\text{encl}} / \varepsilon_0$. If E is constant over (part of) the Gaussian surface, you can pull it outside the integral. This simplification is what allows you to solve for the field.

If you can't pull E outside the flux integral, $\int \mathbf{E} \cdot d\mathbf{A}$, then Gauss' law won't get you to the answer. In those cases, use the "continuous charge distribution" strategy illustrated in Chapter 33.

Luckily, the Gauss' law strategy works here, as I'll now show.

Step 1: Draw field lines.

Let's say the rod extends along the z-axis. Because point B sits very close to the rod (as compared to its length), the rod looks *very* long from B's perspective. So, we can approximate the rod as

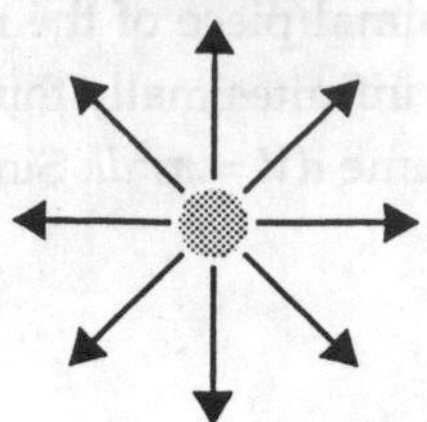

infinitely long. So, the electric field lines emanate radially outward from the rod. Only a negligible electric field component points in the z-direction. Here's one way of seeing why. If the rod extends infinitely far "up" and infinitely far "down," then half the rod is "above" point B and half is "below" point B. In that case, symmetry guarantees that the field points radially outward, with no z-component, as shown in Chapter 33.

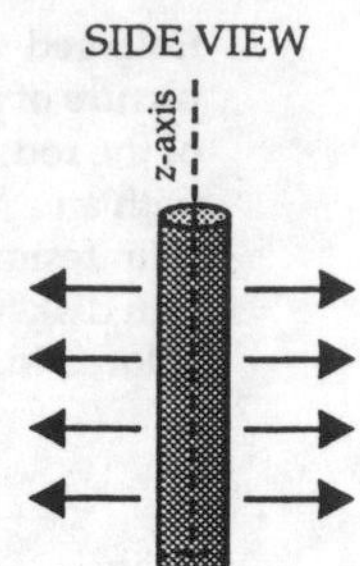

Step 2: Draw a "good" Gaussian surface

We want the electric field to have the same strength at all points on (at least part of) the Gaussian surface. That way, we can pull E outside the flux integral, $\int \mathbf{E} \cdot d\mathbf{A}$.

Well, as drawn above, the rod spews out field lines equally in all (radial) directions. Therefore, by symmetry, the field has the same strength at any two points the same distance form the rod. So, if our Gaussian surface consists of points equidistant from the rod, then the electric field strength E will be constant over that whole surface. For this reason, we should surround the rod with a circle, as drawn here.

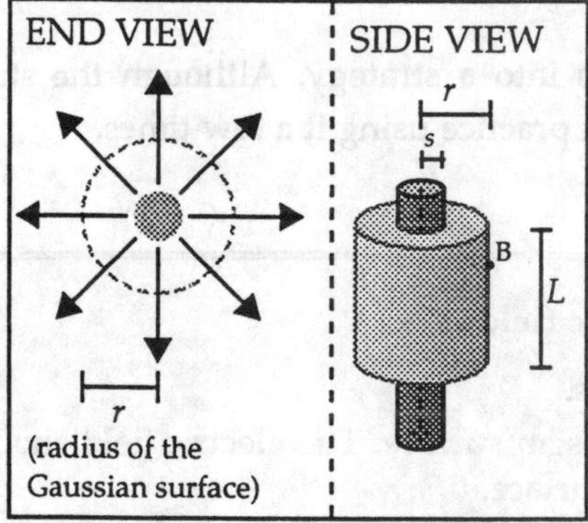

But wait. A circle can't be a Gaussian surface. Recall from the water hose example that a Gaussian surface must be closed. So, the Gaussian surface extends an arbitrary distance L into the page (in the end view), forming a cylinder. You can set L equal to the rod's length, if you choose. Notice that the top and bottom of this "Gaussian cylinder" cut through the rod.

I've given the Gaussian surface arbitrary radius r. In this problem, we want the electric field at point B, a distance $r = 3s$ from the center of the rod. So eventually, we'll set $r = 3s$. But to illustrate the reasoning more clearly, let me work in terms of a general "r."

Step 3: Use Gauss' law to solve for the electric field.

By Gauss' law, the flux through the Gaussian surface is proportional to the charge enclosed by that surface:

$$\oint \mathbf{E} \cdot d\mathbf{A} = \frac{q_{end}}{\varepsilon_0}.$$

I'll now simplify both sides of this equation, starting with the enclosed charge.

In part (a), we found the rod's linear charge density, i.e., the charge per length. It's $\lambda = \pi s^2 \rho$. Since my Gaussian surface encloses length L of the rod, the charge trapped inside the Gaussian surface is simply

$$q_{\text{end}} = \frac{\text{charge}}{\text{length}} \times (\text{length of rod enclosed by Gaussian surface}) = \lambda L = \pi s^2 \rho L.$$

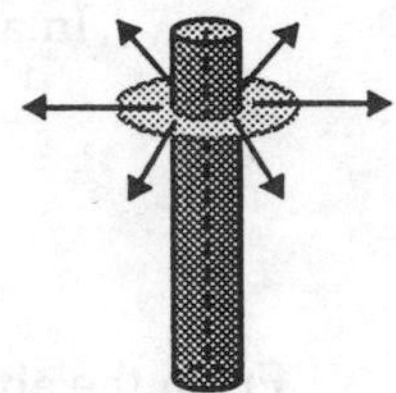
No flux through top of Gaussian surface

Now let's work on the flux integral, $\int \mathbf{E} \cdot d\mathbf{A}$. It consists of three separate parts; the flux through the top of the Gaussian surface, through the bottom of the Gaussian surface, and through the "tube" (side) of the Gaussian surface. Since the field lines "flow" radially outward, with no z-component, the field lines skim along the top and bottom of the Gaussian surface, without piercing *through* those surfaces. In other words, *no* flux passes through the top or the bottom. To see this mathematically, note that the top and bottom of the Gaussian surface "face" up and down along the z-direction, while the electric field points perpendicular to the z-direction. Consequently, $\mathbf{E} \cdot d\mathbf{A} = EdA\cos 90° = 0$. *Therefore, the only "contribution" to the flux integral comes from the tube (side) of the Gaussian surface.*

As the above diagrams show, the tube of the Gaussian surface faces radially outward, the same direction in which the field lines point. Therefore, the dot product $\mathbf{E} . d\mathbf{A}$ reduces to a regular product, because $\theta = 0$. Furthermore, as emphasized above, the electric field has the same strength over the whole tube, because the tube consists of points equidistant from the rod. Therefore, we can pull E outside the integral. So,

$$\begin{aligned}\oint_{total} \mathbf{E} \cdot d\mathbf{A} &= \int_{top} \mathbf{E} \cdot d\mathbf{A} + \int_{bottom} \mathbf{E} \cdot d\mathbf{A} + \int_{tube} \mathbf{E} \cdot d\mathbf{A} \\ &= 0 \quad + \quad 0 \quad + \quad E \int_{tube} dA \\ &= EA_{\text{tube}} \\ &= E(2\pi rL),\end{aligned}$$

where I used the formula for the surface area of the side (tube) of a cylinder. Crucially, "A" refers to the area of the Gaussian surface, *not* the area of the charged rod.

OK, now that we've simplified both sides of Gauss' law, let's tie it all together.

$$\oint \mathbf{E} \cdot d\mathbf{A} = \frac{q_{\text{end}}}{\varepsilon_0}$$

$$E(2\pi rL) = \frac{\rho\pi s^2 L}{\varepsilon_0}.$$

The L's cancel, proving that the length of the Gaussian surface makes no difference. Solve for E to get

$$E = \frac{s^2\rho}{2\varepsilon_0 r}.$$

To confirm that this result agrees with your textbook's formula for the electric field due to a line charge (or thin rod), multiply the numerator and denominator through by π, and remember that $\lambda = \pi s^2\rho$, to get $E = \dfrac{\pi s^2\rho}{2\varepsilon_0 r} = \dfrac{\lambda}{2\varepsilon_0 r}$.

In any case, to find the field at point B, just substitute in $r = 3s$ to get

$$E_B = \frac{s^2\rho}{2(3s)\varepsilon_0} = \frac{s\rho}{6\varepsilon_0}.$$

From the above Gaussian diagrams, note that r is measured from the center of the rod, not from the edge of the rod. That's why r equals $3s$, not $2s$.

(c) In general, the electric force acting on a charge is $F = qE$, where q refers to the charge being acted upon, and E denotes the field created by the *other* charges. So, to find the force on q_1, use $F = q_1E$, where E is the field created by the rod:

$$F = q_1E_B = q_1\frac{s\rho}{6\varepsilon_0}.$$

QUESTION 34-4

A kitchen floor, which is 3.0 meters long by 4.0 meters wide, has charge $Q_0 = -2.0\ \mu C$ spread uniformly over its whole area.

A point charge $q_1 = -5.0\ \mu C$ is held a distance 10 centimeters above the floor, near the middle of the room.

(a) What electric force acts on the point charge? Derive any formulas you use; don't plug in pre-derived textbook equations. Hint: Use Gauss' law, taking advantage of the fact that the floor is very long and very wide, as compared to the distance from the floor to the point charge.

(b) When released, the point charge just floats in place. What is its mass?

ANSWER 34-4

(a) The floor is very long and very wide in comparison to q_1's distance from the floor. From q_1's perspective, the floor extends very far in all directions. For this reason, when calculating the electric field, we can treat the floor as an infinite plane of charge. (Similarly, in the previous problem, we treated the rod as infinitely long.) These "infinity" approximations typically yield results accurate within 1%.

Let me summarize and then implement the strategy introduced in the previous problem.

Gauss' law to calculate electric fields

1) Using symmetry and intuition, draw the field lines.

2) Enclose all or part of the charge distribution within a Gaussian surface. The electric field should have the same strength over (at least part of) the Gaussian surface.

3) Use Gauss' law: $\int \mathbf{E} \cdot d\mathbf{A} = q_{encl} / \varepsilon_0$. If $\mathbf{E}$ is uniform over (part of) the surface, then you can pull it outside the integral, allowing you to solve for it.

Step 1: Draw the field lines.

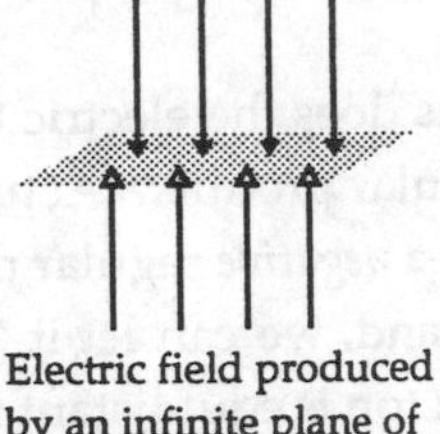
Electric field produced by an infinite plane of negative charge

Symmetry implies that the field near the floor points vertically. Here's why. If the floor extends infinitely to the right and infinitely to the left, then the electric field generated by the right half of the floor equals the electric field generated by the left half. The horizontal (leftward and rightward) components of these fields cancel, leaving only a vertical field. This vertical field is *uniform*, unless you get too near the edge of the floor. See your textbook for more discussion about this.

Since the floor carries negative charge, the field lines point inward instead of outward.

Step 2: Draw a "good" Gaussian surface enclosing some (or all) of the charge.

To solve for E in step 3 below, we must be able to pull it outside the flux integral, $\int \mathbf{E}\cdot d\mathbf{A}$. In other words, E must be constant over (at least part of) the Gaussian surface. For this reason, I'll use a Gaussian "box," as drawn here. The box top and box bottom, both of which have arbitrary area A, face parallel to the electric field. Also, as I've drawn them, the top and bottom are the same distance z from the plane of charge. Therefore, the electric field strength at the box top equals the field strength at the box bottom. In this problem, we want the electric field at $z = 10$ cm above the plane.

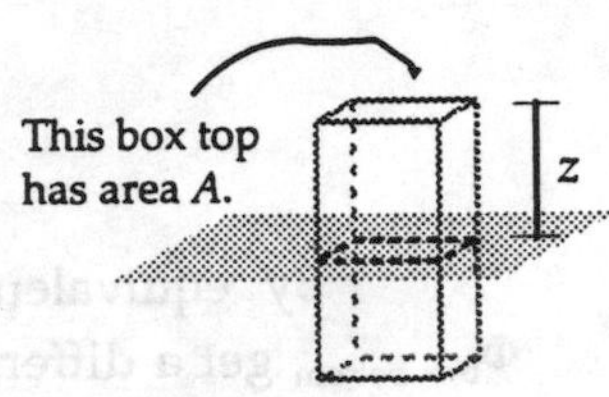

Gauss' law Step 3: Use Gauss' law to figure out the electric field.

To solve for the electric field at the box top, we must simplify both sides of Gauss' law,

$$\oint \mathbf{E}\cdot d\mathbf{A} = \frac{q_{end}}{\varepsilon_0}.$$

Let's start with the right-hand side. The Gaussian box encloses area A of the charged plane, where A is also the area of the box top and bottom. Given the enclosed area, we need to know the plane's *charge per area* to calculate the enclosed charge. Well, $Q_0 = -2.0 \times 10^{-6}$ C of charge is uniformly spread out over area $A_{floor} = (3.0 \text{ m}) \times (4.0 \text{ m}) = 12 \text{ m}^2$. So, the floor has surface charge density (charge per area)

$$\sigma = \frac{\text{charge}}{\text{area}} = \frac{Q_0}{A_{floor}} = \frac{2.0 \times 10^{-6}\,\text{C}}{12 \text{ m}^2} = -1.7 \times 10^{-7}\,\text{C/m}^2.$$

Therefore, the Gaussian surface encloses charge

$$q_{end} = \frac{\text{charge}}{\text{area}} \times (\text{area enclosed by Gaussian surface}) = \sigma A.$$

Now I'll work on the left-hand side of Gauss' law, $\int \mathbf{E}\cdot d\mathbf{A}$. The Gaussian box has six faces. But no field lines pierce through any of the four "sides" of the box. As shown here, the field lines skim along the sides, without actually passing *through* the sides. If the electric field lines were water stream lines, no water would flow *through* the sides of the Gaussian box. So, all the flux through this Gaussian surface flows through the top and bottom.

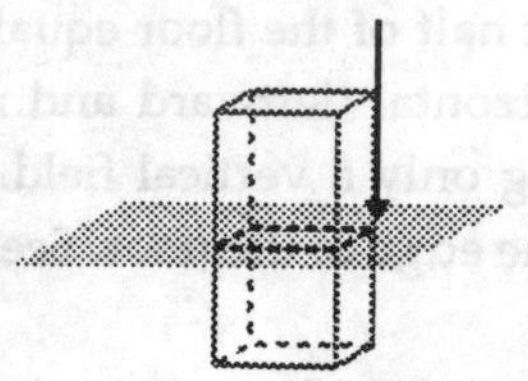

Consider the box top. It faces vertically, as does the electric field. Therefore, the dot product $\mathbf{E}\cdot d\mathbf{A}$ reduces to a regular product. (Actually, since $\mathbf{E}$ and $d\mathbf{A}$ point in opposite directions, we get a *negative* regular product. But instead of inserting the minus sign by hand, we can let it "pop out" of Gauss' law.) Crucially, since the whole box top is equidistant from the charged plane, the electric field is uniform over the entire box top. For this reason, we can pull E outside the flux integral:

$$\begin{aligned}\Phi_{\text{boxtop}} &= \oint_{\text{box top}} \mathbf{E}\cdot d\mathbf{A} \\ &= \int_{\text{box top}} E\,dA \quad \text{[since E and } dA \text{ are (anti)parallel]} \\ &= E\int_{\text{box top}} dA \quad \text{[since } E \text{ is uniform over whole box top]} \\ &= EA\end{aligned}$$

By equivalent reasoning, the same flux passes through the box bottom. But does $\Phi_{\text{box bottom}}$ get a different sign? You might think so, because the electric field points in the opposite direction. But remember, the flux through a Gaussian surface is positive or negative based entirely on whether it flows *into* or *out of* the Gaussian surface. At the box top and at the box bottom, field lines flow *into* the Gaussian surface. Therefore, both of those flux contributions are negative. To see how the minus sign emerges from the math, substitute all the above results into Gauss' law:

$$\oint \mathbf{E}\cdot d\mathbf{A} = \frac{q_{\text{end}}}{\varepsilon_0}$$

$$\oint_{\text{box top}} \mathbf{E}\cdot d\mathbf{A} + \oint_{\text{box top}} \mathbf{E}\cdot d\mathbf{A} = \frac{\sigma A}{\varepsilon_0}$$

$$EA \quad + \quad EA = \frac{\sigma A}{\varepsilon_0}$$

$$2EA = \frac{\sigma A}{\varepsilon_0}.$$

The arbitrary area A of the box top cancels out. Solve for E, and substitute in the numbers, to get

$$E = \frac{\sigma}{2\varepsilon_0} = \frac{-1.7\times10^{-7}\ \text{C/m}^2}{2(8.85\times10^{-12}\ \text{C}^2/\text{N}\cdot\text{m}^2)} = -9400\ \text{N/C}.$$

As noted above, the minus sign indicates that field lines flow *into* the Gaussian surface.

Notice that the height z of the Gaussian box does not appear in our answer. The electric field near the floor has the *same* strength no matter whether you're 5 cm, 10 cm, or 20 cm away—assuming you're not *so* far away that the infinite-floor approximation breaks down.

Given this electric field, we can calculate the force acting on point charge q_1. Since the field at q_1 points downward, but q_1 is negative, the force on it points upward. In other words, the negatively-charged floor *repels* the negative point charge.

$$F = q_1 E = (-5.0 \times 10^{-6}\,\text{C})(-9400\,\text{N/C}) = 0.047\ \text{N}.$$

This positive answer indicates an "outward" force (pointing away from the floor).

(b) In order for the point charge to float ($F_{net} = 0$), the upward electric force must cancel the downward gravitational force. In other words, since $F_{net} = q_1 E - mg$ must equal 0,

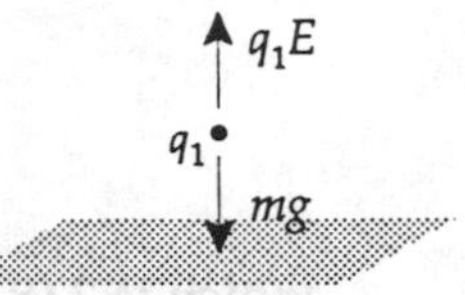

FORCE DIAGRAM

$$q_1 E = mg.$$

Solve for m to get

$$m = \frac{q_1 E}{8} = \frac{0.047\ \text{N}}{9.8\ \text{m/s}^2} = 0.0048\ \text{kg} = 4.8\text{g rams}.$$

QUESTION 34-5

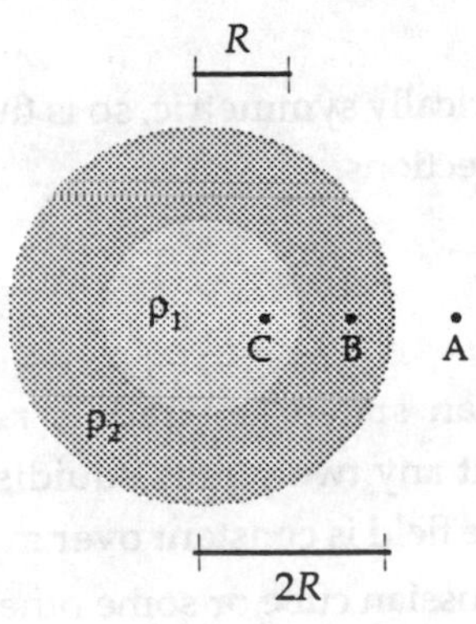

Consider a two-layered spherical ball, of total radius $2R$. The inner **core** of the ball has radius R and charge density ρ_1. The "**mantle**," by which I mean the outer shell extending from R to $2R$, has charge density ρ_2. I've labeled points A, B, and C for later reference. Charge density is charge per volume.

(a) What is the total charge of this ball?

(b) What is the electric field at point A, which is a distance r_A from the center of the ball? As the diagram indicates, $r_A > 2R$.

(c) What is the electric field at point B, a distance r_B from the center? Since B sits in the mantle, $R < r_B < 2R$.

(d) What is the electric field at point C, a distance r_C from the center? Because point C lives in the core, $r_C < R$.

ANSWER 34-5

(a) Let V_1 denote the volume of the core, and V_2 denote the volume of the mantle. Since charge density is charge per volume ($\rho = q/V$), the core carries charge $q_1 = \rho_1 V_1$, while the mantle carries charge $q_2 = \rho_2 V_2$. Therefore, the whole ball carries charge

$$Q_{total} = q_1 + q_2 = \rho_1 V_1 + \rho_2 V_2.$$

So, to find Q_{total}, we need to figure out the volume of the core and the volume of the mantle.

Since the core is a sphere of radius R, we know that $V_1 = \frac{4}{3}\pi R^3$. But the mantle is more complicated. Think of it as a big sphere with a small spherical hole cut out. As this diagram indicates, the volume of the mantle is simply the volume of the entire big sphere *minus* the volume of the hole. Since the big sphere has radius $2R$, while the hole has radius R,

$$V_2 = \tfrac{4}{3}\pi(2R)^3 - \tfrac{4}{3}\pi R^3$$

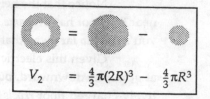

$$\begin{aligned} V_2 &= \frac{4}{3}\pi(2R)^3 - \frac{4}{3}\pi R^3 \\ &= \frac{4}{3}\pi 7R^3 \\ &= \frac{28}{3}\pi R^3, \end{aligned}$$

which is 7/8 of the volume of the whole ball.

Putting all this together,

$$Q_{\text{total}} = \rho_1 V_1 + \rho_2 V_2 = \frac{4}{3}\pi R^3 \rho_1 + \frac{28}{3}\pi R^3 \rho_2.$$

(b) With spherically symmetric charge distributions, you can always use Gauss' law productively. Let me demonstrate why, relying upon the usual Gauss' law strategy.

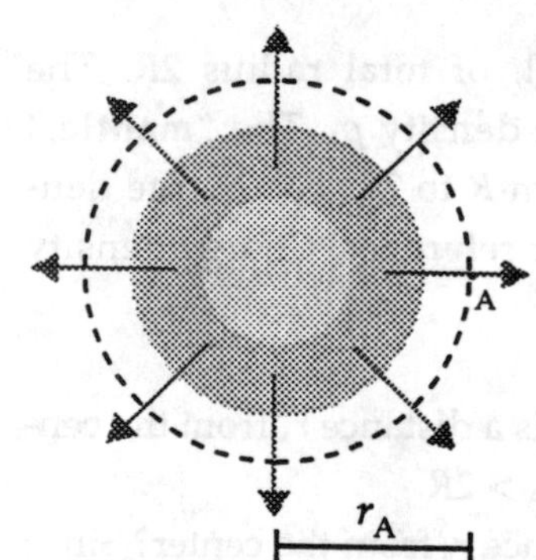

Step 1: Draw the electric field lines.

Because the charge distribution is spherically symmetric, so is the electric field. It "radiates" out equally in all directions.

Step 2: Draw a "good" Gaussian surface.

I'll surround the ball with a "Gaussian sphere" of radius r_A. By symmetry, the field has the same strength at any two points equidistant from the center of the sphere. Therefore, the field is constant over my entire Gaussian surface. *That's why I chose a Gaussian sphere* instead of a Gaussian cube or some other shape. Remember, when E is constant over your Gaussian surface, you can pull it outside the flux integral $\oint \mathbf{E}\cdot d\mathbf{A}$ in Gauss' law.

Step 3: Use Gauss' law to solve for E.

According to Gauss' law, the electric flux out of a Gaussian surface is proportional to the charge enclosed by that surface. This makes sense, in light of the water hose analogy. Mathematically,

Gauss' law $$\oint \mathbf{E}\cdot d\mathbf{A} = \frac{q_{end}}{\varepsilon_0}.$$

Here, the Gaussian surface encloses the entire charged ball. Therefore, q_{encl} is simply Q_{total} our part (a) answer.

Now I'll simplify the other side of Gauss' law, the flux integral. As the diagram indicates, **E** points radially outward and therefore pierces "head-on" through the Gaussian surface. As a result, **E**. d**A** reduces to a regular product. Also, as I emphasized above, the field has the same strength over the entire Gaussian surface. Consequently, we can pull E outside the integral:

$$\begin{aligned}\oint \mathbf{E}\cdot d\mathbf{A} &= \oint_{\text{Gaussian sphere}} E\,dA \\ &= E \oint_{\text{Gaussian sphere}} dA \\ &= EA_{\text{Gaussian sphere}} \\ &= E(4\pi\varepsilon r_A^2),\end{aligned}$$

where in the last step I used the surface area of a sphere. Crucially, in this flux integral, we're talking about the surface area of the *Gaussian surface*, not the surface area of the charged ball. That's why I used r_A instead of $2R$.

OK, now that we've simplified both sides of Gauss' law, let's tie it all together.

$$\begin{aligned}\oint \mathbf{E}\cdot d\mathbf{A} &= \frac{q_{\text{encl}}}{\varepsilon_0} \\ E(4\pi\varepsilon r_A^2) &= \frac{Q_{\text{total}}}{\varepsilon_0}.\end{aligned}$$

Isolate E to get

$$E = \frac{Q_{\text{total}}}{4\pi\varepsilon_0 r_A^2},$$

which is Coulomb's law! *We now see why the electric field due to a point charge—or any other spherically symmetric charge—is given by Coulomb's law. We just derived Coulomb's law from Gauss' law.*

For this particular ball, $Q_{\text{total}} = \frac{4}{3}\pi R^3\rho_1 + \frac{28}{3}\pi R^3\rho_2$, and hence $E = \frac{R^3}{3\varepsilon_0 r_A^2}\rho_1 + \frac{7R^3}{3\varepsilon_0 r_A^2}\pi R^3\rho_2$.

(c) To find the field at point B, again use Gauss' law,

$$\oint \mathbf{E}\cdot d\mathbf{A} = \frac{q_{\text{encl}}}{\varepsilon_0}.$$

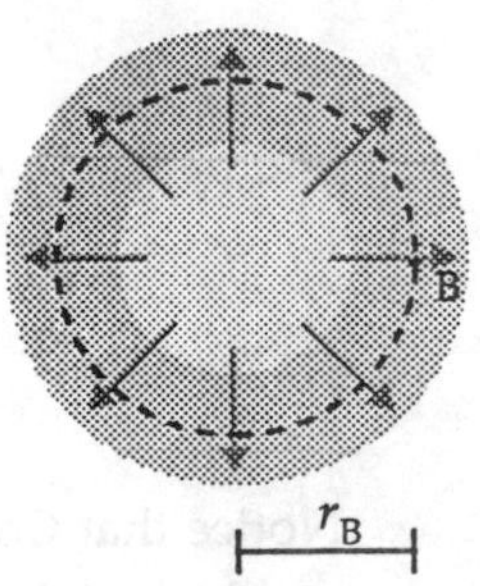

But now the Gaussian sphere sits *inside* the charge sphere, as drawn here. Let's see how this modifies our reasoning from part (b).

I'll start with the left-hand side of Gauss' law. Since the electric field still pierces head-on through the Gaussian surface, and since (by spherical symmetry) it still has the same strength over the entire Gaussian surface, the flux integral $\oint \mathbf{E}\cdot d\mathbf{A}$ simplifies in the same way it did above. But now it has radius r_B instead of r_A:

$$\oint \mathbf{E}\cdot d\mathbf{A} = E \oint_{\text{Gaussian sphere}} dA = E(4\pi r_B^2).$$

The right-hand side of Gauss' law gets more complicated, because q_{encl} is *not* the whole charge of the ball. This Gaussian surface encloses the entire core, but only part of the mantle. To finish using Gauss' law, we must figure out the new q_{encl}.

Subproblem: Find the charge enclosed by the Gaussian surface of radius r_B.

The core, all of which is enclosed, carries charge

$$q_1 = \rho_1 V_1 = \frac{4}{3}\pi R^3.$$

Let $V_{2\,\text{encl}}$ denote the volume of the mantle that's inside the Gaussian sphere (of radius r_B). The enclosed portion of the mantle carries charge $\rho_2 V_{2\,\text{encl}}$. If we visualize this part of the mantle as a solid sphere of radius r_B with a hole of radius R cut out, then

$$V_{2\,\text{encl}} = \frac{4}{3}\pi r_B^3 - \frac{4}{3}\pi R^3 = \frac{4}{3}\pi(r_B^3 - R^3).$$

So, the Gaussian surface encloses charge

$$\begin{aligned} q_{\text{encl}} &= \rho_1 V_1 + \rho_2 V_{2\,\text{encl}} \\ &= \rho_1 \frac{4}{3}\pi R^3 + \rho_2 \frac{4}{3}\pi(r_B^3 - R^3). \end{aligned}$$

End of subproblem

Substitute all the information we've gathered into Gauss' law, to get

$$\oint \mathbf{E}\cdot d\mathbf{A} = \frac{q_{\text{encl}}}{\varepsilon_0}$$

$$E(4\pi r_B^2) = \frac{q_{\text{encl}}}{\varepsilon_0},$$

and hence

$$E = \frac{q_{\text{encl}}}{4\pi\varepsilon_0 r_B^2} = \frac{R^3}{3\varepsilon_0 r_B^2}\rho_1 + \frac{r_B^3 - R^3}{3\varepsilon_0 r_B^2}\rho_2.$$

Notice that Coulomb's law again popped out; but the "q" now refers to the charge *inside* the point at which we're finding the electric field.

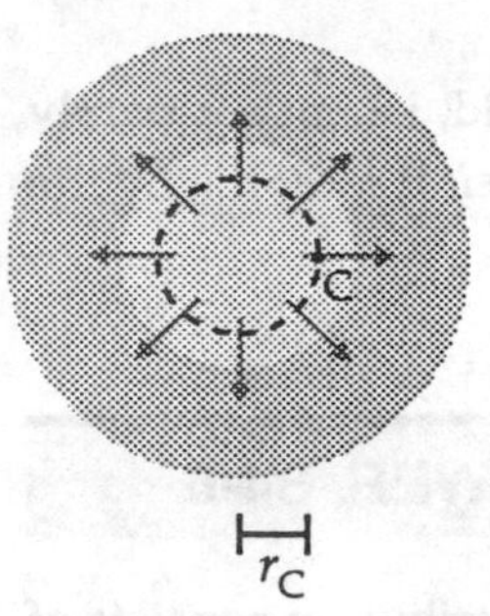

(d) Here, the Gaussian surface encloses part of the core but *none* of the mantle. The electric flux simplifies in the same way it did above, with a smaller radius:

$$\oint \mathbf{E} \cdot d\mathbf{A} = E(4\pi r_C^2).$$

The part of core enclosed by this Gaussian has volume $V_{1\,\text{encl}} = \frac{4}{3}\pi r_C^3$. Therefore,

$$q_{\text{encl}} = \rho_1 V_{1\text{encl}} = \rho_1 \frac{4}{3}\pi r_C^3.$$

Gauss' law therefore gives us

$$\oint \mathbf{E} \cdot d\mathbf{A} = \frac{q_{\text{encl}}}{\varepsilon_0}.$$

$$E(4\pi r_C^2) = \frac{\frac{4}{3}\pi r_C^3 \rho_1}{\varepsilon_0}.$$

Only in part (d) do we get the same "*r*" on both sides of Gauss' law. Usually those *r*'s differ, because one refers to the radius of the *Gaussian surface*, while the other refers to the radius of (part or all of) the *charge distribution*. To keep the *r*'s straight, draw a good picture and think conceptually, not formulaically.

We can solve the above equation for *E* to get

$$E = \frac{\rho_1 r_C}{3\varepsilon_0},$$

the standard textbook result for the field inside a uniformly charged sphere. But don't memorize these kinds of results. Instead, practice using Gauss' law so that you can *derive* E, even in cases you've never seen before.

QUESTION 34-6

In this problem, you may use any results that you derived previously in this study guide. *You need not rederive those earlier results.*

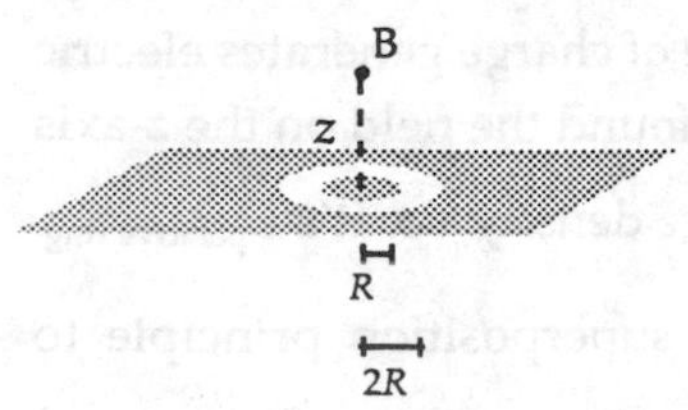

A soccer field, which lies in the *x-y* plane, is coated with surface charge density σ, EXCEPT for a small ring of inner radius *R* and outer radius 2*R*. That ring has no surface charge density. In this drawing, gray represents surface charge density. The soccer field is actually *much* bigger than the ring. The ring is centered on the *z*-axis, which is perpendicular to the soccer field.

Consider point B, located on the *z*-axis a small distance *z* from the soccer field.

(a) What is the direction and magnitude of the electric field at point B?

(b) (*Very hard*) Suppose point B were on the z-axis but right next to the soccer field, i.e., at a z barely greater than zero. What would be the electric field at that point? Explain the *physical* reason why the answer comes out simple.

ANSWER 34-6

(a) This is a classic superposition problem. Here's the crucial insight: If a charge distribution consists of different "parts," the total electric field is simply the sum of the fields generated by each separate part. Here, you can think of the overall charge distribution as consisting of a plane of charge of surface charge density σ, plus a ring of surface charge density $-\sigma$. The ring's negative charge density cancels the plane's positive charge density, resulting in a ring of zero charge density. Let me graphically represent what I just said:

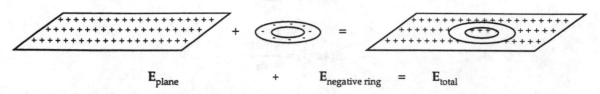

$$\mathbf{E}_{\text{plane}} + \mathbf{E}_{\text{negative ring}} = \mathbf{E}_{\text{total}}$$

You can arrive at the same equation by thinking of the charge distribution as a plane of charge with a *positive* ring "cut out" (i.e., subtracted) from the plane:

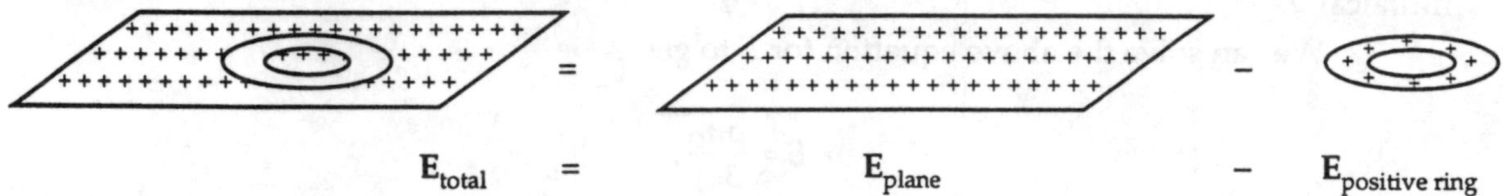

$$\mathbf{E}_{\text{total}} = \mathbf{E}_{\text{plane}} - \mathbf{E}_{\text{positive ring}}$$

Make sure you understand why these two graphical representations, and the corresponding equations, say the same thing. Starting with a plane of charge, *adding* a *negative* charged ring is equivalent to *subtracting* a *positive* charged ring.

So, to figure out the total electric field at point B, we must

(i) calculate the field generated by a plane of charge,
(ii) calculate the field produced by a positive (or negative) charged ring; and then
(iii) subtract (or add) those two fields to obtain the total electric field.

Fortunately, we completed steps (i) and (ii) earlier in this study guide! Since the soccer field is huge compared to the distance between that field and point B, we can treat the soccer field as an infinite plane of charge. In question 34-4, we found than an infinite plane of charge generates electric field $E_{\text{plane}} = \dfrac{\sigma}{2\varepsilon_0}$, perpendicular to the plane. And in question 33-9, we found the field on the z-axis generated by a ring of inner radius R, outer radius $2R$, and surface charge density $+\sigma$. It's $E_{\text{positive ring}} = \dfrac{\sigma z}{2\varepsilon_0}\left[\dfrac{1}{\sqrt{z^2+R^2}} - \dfrac{1}{\sqrt{z^2+(2R)^2}}\right]$, pointing in the z-direction. Apply the superposition principle to these earlier results, in order to solve the current problem:

$$E_{total} = E_{plane} - E_{positive\ ring}$$

$$= \frac{\sigma}{2\varepsilon_0} - \frac{\sigma z}{2\varepsilon_0}\left[\frac{1}{\sqrt{z^2+R^2}} - \frac{1}{\sqrt{z^2+(2R)^2}}\right].$$

That's it! Since those two electric fields both point in the z-direction (perpendicular to the soccer field), I subtracted them without worrying about vectors components.

Before proceeding to part (b), see if you can derive the electric field due to an infinite plane of charge. After you've tried it, check your derivation against the following brief review.

Brief review: Using Gauss' law to find the electric field generated by an infinite plane of charge

Gaussian "box." The box top has area A.

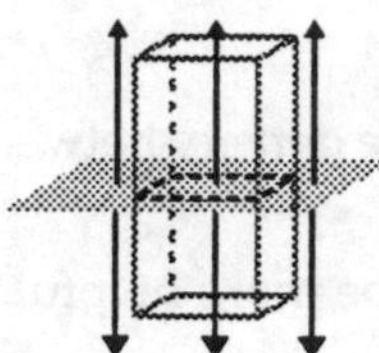

The field is uniform (unless you get too far away or too close to an edge), and points perpendicular to the plane of charge. By using a Gaussian "pill box," I ensure that the field has the same strength over the entire top and bottom of that box, both of which have arbitrary area A. Therefore, we can pull the field outside the flux integral. Furthermore, no field lines pierce through the sides of the box; the field lines "skim along" those sides instead. Therefore,

$$\oint \mathbf{E}\cdot d\mathbf{A} = \oint_{\text{box top}} \mathbf{E}\cdot d\mathbf{A} + \oint_{\text{box bottom}} \mathbf{E}\cdot d\mathbf{A} = EA + EA = 2EA.$$

This Gaussian box encloses area A of the charged plane. The enclosed charge is therefore

$$q_{\text{encl}} = \frac{\text{charge}}{\text{area}} \times (\text{area enclosed by Gaussian surface}) = \sigma A.$$

Substitute all this information into Gauss' law to get

$$\oint \mathbf{E}\cdot d\mathbf{A} = \frac{q_{\text{encl}}}{\varepsilon_0}$$

$$2EA = \frac{\sigma A}{\varepsilon_0}$$

Solve for E to get $E = \dfrac{\sigma}{2\varepsilon_0}$.

End of review of how to find the electric field generated by an infinite plane

(b) **To find the $z \to 0$ limit of the field, we don't need calculus. Just substitute $z \approx 0$ into our answer for E_{total} from part (a). The E_{ring} term vanishes, leaving us with**

$$E = \frac{\sigma}{2\varepsilon_0},$$

the field due to an infinite plane of charge with no ring cut out.

This result makes sense because, when you get *very* close to the plane, the part of the plane inside the ring seems "infinitely" big. Let me explain.

Under what circumstances is a plane of charge "infinite," to good approximation? When the plane is much longer and much wider than *your distance from the plane*. For instance, your head is probably about 1 or 2 meters above your floor right now. And your floor is maybe 3 or 4 meters long. So, from your perspective, the floor is not an infinite plane. But to an ant, whose head is less than a millimeter above the floor, the floor "looks" like it extends (nearly) infinitely far in all directions.

If point B is only a microscopic distance above the blackened disk (of radius R), then from point B's perspective, that disk alone is an infinite plane.

Now suppose that point B is *extremely* close to the charged plane, only a microscopic distance away. From B's perspective, the region of the floor within just a few centimeters of point B is, all by itself, a nearly infinite plane. So, the disk of positive charge density, colored black on this diagram, is an effectively infinite plane, as far as B is concerned. The electric field produced by that disk alone is $E = \frac{\sigma}{2\varepsilon_0}$, when B is only a microscopic distance above the soccer field. For this reason, the "missing" surface charge density between R and 2R makes no difference.

This is hard to explain in writing. Discussing it with your teacher might be more helpful.

QUESTION 34-7

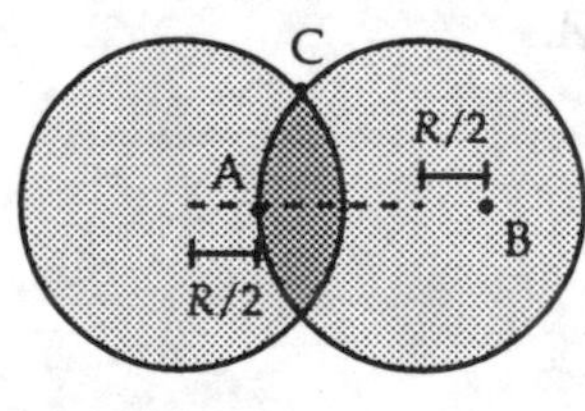

(*Very hard*) Consider a charge distribution that consists, roughly speaking, of two overlapping spheres of radius R. In the region where the spheres don't overlap, the charge density is $-\rho$. In the region where the spheres *do* overlap, the charge density is -2ρ. ρ is positive. On this diagram, the imaginary dashed line connects the centers of the two spheres. Notice that the edge of the right-hand sphere is a distance $R/2$ from the center of the left-hand sphere; and vice versa.

(a) If a point charge Q were placed at point A, what electric force would it feel? Hint: Superposition.

(b) If that point charge were placed at point B, a distance $R/2$ to the right of the center of the right-hand sphere, what electric force would it feel?

(c) (*Challenge problem*) Same question, at point C?

ANSWER 34-7

As the hint suggests, you should use the superposition principle: Divide the overall charge distribution into "parts," calculate the electric field generated by each separate part, and then add up those field contributions to obtain the total electric field.

At first glance, you might want to break things into three pieces: Two chipped spheres and a wedge.

Not-so-helpful way to break the system into pieces

But it's nearly impossible to find the electric field generated by these pieces, due to their odd shapes. You can solve quicker by breaking the system into two spheres. When you "add" two spheres of charge density $-\rho$, the charge density in the overlap region is indeed -2ρ. That's why you can think of this system as consisting of two overlapping spheres.

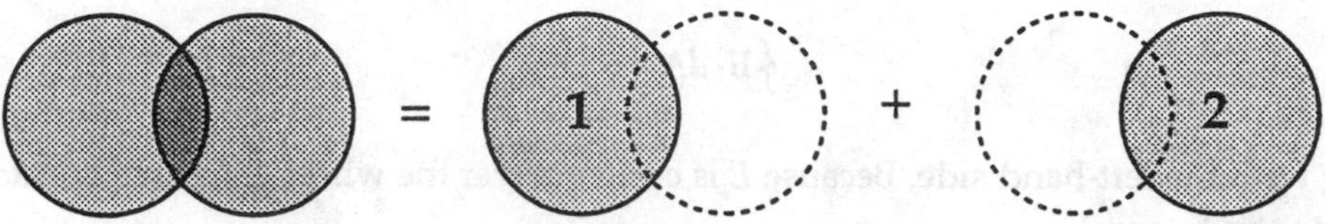

At any point, $\mathbf{E}_{total} = \mathbf{E}_1 + \mathbf{E}_2$, where $\mathbf{E}_1$ and $\mathbf{E}_2$ denote the fields produced by spheres 1 and 2, respectively.

(a) Point A is outside sphere 2 and inside sphere 1. Both spheres carry negative charge. So, intuitively, we expect sphere 2 to pull the positive point charge rightward, while sphere 1 pulls it leftward. We don't yet know which of those opposing forces "wins." But at least we have a strategy:

(i) Calculate the electric field *outside* sphere 2,
(ii) calculate the electric field *inside* sphere 1, and then
(iii) add those two fields to obtain the total field.

The addition is really a subtraction, since those two fields point in opposite directions. In any case, given the total electric field, we can use $F = QE$ to find the electric force.

Subproblem 1: Calculate electric field outside sphere 2.

In this subproblem, pretend sphere 1 doesn't exist. Focus all attention on sphere 2. Here, we care only about the field at the edge of sphere 2. But the same Gaussian reasoning allows us to find the electric field due to sphere 2 at *any* distance $r \Delta R$ from the center of that sphere. This more general result will help us solve parts (b) and (c) below.

I'll follow the usual Gauss' law strategy. Because I dealt with charged spheres in question 34-5 above, I'll abbreviate my presentation here. See that earlier problem for more details.

Gauss' law Step 1: Draw in electric field lines.

Since sphere 2 carries negative charge, the field lines flow inward instead of outward. My two-dimensional drawing doesn't capture the fact that field lines come out of and go into the page as well.

Gauss' law Step 2: Draw a "good" Gaussian surface.

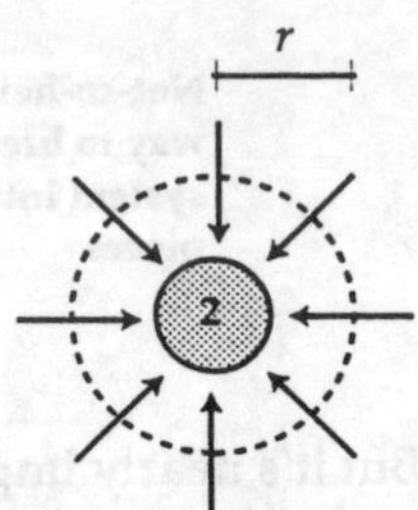

Gaussian sphere surrounds the charged sphere.

By symmetry, the field has the same strength at any two points equidistant from the charged sphere. So, by using a "Gaussian sphere," I ensure that the field is constant over my entire Gaussian surface. Let me give this Gaussian sphere arbitrary radius r, where $r \Delta R$. On this diagram, the dashed line represents a three-dimensional sphere, not a two-dimensional circle.

Gauss' law Step 3: Use Gauss' law to solve for the electric field on the Gaussian surface.

Now I'll simplify both sides of Gauss' law,

$$\oint \mathbf{E} \cdot d\mathbf{A} = \frac{q_{\text{end}}}{\varepsilon_0},$$

starting with the left-hand side. Because E is constant over the whole Gaussian surface, we can pull it outside the integral:

$$\oint \mathbf{E} \cdot d\mathbf{A} = E \int_{\text{Gaussian sphere}} dA = EA_{\text{Gaussian sphere}} = E(4\pi r^2),$$

where I've used the radius of the Gaussian sphere, not of the charged sphere. In part (a), those two radii happen to coincide. But I'm deriving a more general result.

Since the Gaussian surface encloses the whole charged sphere (of volume $V = \frac{4}{3}\pi R^3$),

$$q_{\text{end}} = \frac{\text{charge}}{\text{volume}} \times (\text{volume of charge enclosed}) = -\rho\frac{4\pi R^3}{3},$$

Notice that I used the volume of the charged ball, not the volume of the Gaussian sphere.

OK, now that we've simplified both sides of Gauss' law, let's finish it up:

$$\oint \mathbf{E} \cdot d\mathbf{A} = \frac{q_{\text{end}}}{\varepsilon_0}$$

$$E(4\pi r^2) = \rho\frac{4\pi R^3}{3\varepsilon_0},$$

and hence

$$E = \frac{q_{\text{encl}}}{4\pi\varepsilon_0 r^2} = -\frac{\frac{4}{3}\pi R^3 \rho}{4\pi\varepsilon_0 r^2}. \tag{1}$$

This formula for the electric field applies at any point outside the charged sphere. Point A sits a distance $r = R$ from the center of sphere 2. So, to find the field generated by sphere 2 at point A, just substitute $r = R$ into Eq. (1):

$$E_{\text{A due to sphere 2}} = -\frac{\frac{4}{3}\pi R^3 \rho}{4\pi\varepsilon_0 R^2} = -\frac{R\rho}{3\varepsilon_0}.$$

The minus sign indicates *towards sphere 2*, i.e., rightward.

So far, we've figured out the electric field at point A due to **sphere 2**. Now we must find the field due to sphere 1.

Subproblem 2: Find the electric field inside sphere 1.

In this subproblem, pretend sphere 2 doesn't exist.

Since we're finding the field inside a charged sphere, we **must draw** the Gaussian surface inside the charged sphere. Since point A sits a distance $r = R/2$ from the center of sphere 1, we could give our Gaussian surface radius $r = R/2$. But hey, we might as well find the electric field at arbitrary distance r from the center. After finding $E(r)$, we can then set $r = R/2$ to address this particular problem.

Gauss' law steps 1 & 2:

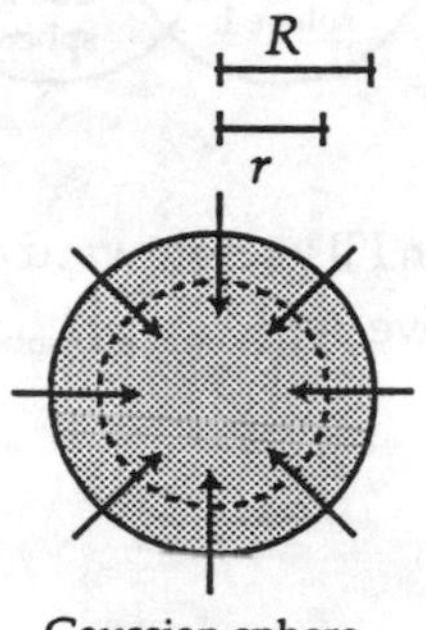

Gaussian sphere *inside* the charged sphere

By symmetry, the electric field has the same strength at all points equidistant from the center, i.e., at all points on a "Gaussian sphere." The field lines point radially inward, since the sphere carries negative charge. Crucially, the Gaussian surface has radius r, which is less than the charged sphere's radius, R.

Step 3: Use Gauss' law to solve for electric field on the Gaussian surface.

The flux integral simplifies exactly as it did in subproblem 1 above:

$$\oint \mathbf{E}\cdot d\mathbf{A} = E \int_{\text{Gaussian sphere}} dA = EA_{\text{Gaussian sphere}} = E(4\pi r^2)\,.$$

But this time, the Gaussian surface does not enclose the entire charged sphere. It captures only a "subsphere" of radius r. On the diagram, this subsphere is the region inside the dashed Gaussian surface. It has volume $V_{\text{encl}} = (4/3)\pi r^3$. Therefore, the Gaussian surface encloses charge

$$q_{\text{encl}} = \frac{\text{charge}}{\text{volume}} \times V_{\text{end}} = -\rho\frac{4\pi r^3}{3}\,.$$

Substituting all this information into Gauss' law yields

$$\oint \mathbf{E}\cdot d\mathbf{A} = \frac{q_{\text{end}}}{\varepsilon_0}$$

$$E(4\pi r^2) = -\rho\frac{4\pi R^3}{3\varepsilon_0}\,,$$

and hence

$$E = -\frac{r\rho}{3\varepsilon_0}\,. \qquad (2)$$

That's the general expression for the field strength a distance r from the sphere's center, when you're inside the charged sphere. Since point A sits a distance $r = R/2$ from the center of sphere 1,

$$E_{\text{A due to sphere 1}} = -\frac{R\rho}{2(3\varepsilon_0)},$$

where the minus sign indicates *toward* the center of sphere 1.

End of subproblem 2.

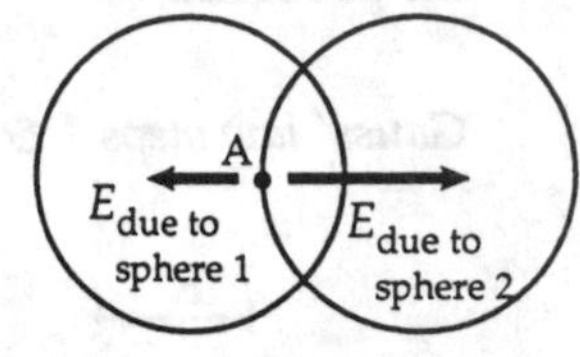

At this point, we've calculated the electric field at point A generated by spheres 1 and 2. Those fields point in opposite directions, towards the centers of the respective spheres.

In our above expressions for $E_{\text{A due to sphere 1}}$ and $E_{\text{A due to sphere 2}}$, the minus signs indicate "towards the center" of the sphere. So, the minus sign in $E_{\text{A due to sphere 1}}$ indicates "leftward," while the minus sign in $E_{\text{A due to sphere 2}}$ indicates "rightward." To avoid this inconsistency, from now on I'll let rightward count as positive. Therefore, according to the above diagram, $E_{\text{A due to sphere 2}}$ is positive, and $E_{\text{A due to sphere 1}}$ is negative. The net rightward field at point A is

$$\begin{aligned} E_{\text{A}} &= E_{\text{A due to sphere 2}} - E_{\text{A due to sphere 1}} \\ &= \frac{R\rho}{(3\varepsilon_0)} - \frac{R\rho}{2(3\varepsilon_0)} \\ &= \frac{R\rho}{6\varepsilon_0}. \end{aligned}$$

Therefore, the electric force on a particle of charge Q located at point A is

$$\begin{aligned} F &= QE_A \\ &= Q\frac{R\rho}{6\varepsilon_0}. \end{aligned}$$

(b) Once again, we can use the superposition principle: $E_{\text{B}} = E_{\text{B due to sphere 1}} + E_{\text{B due to sphere 2}}$. Indeed, we've already done most of the dirty work. In part (a), we derived formulas for the electric field both outside and inside a charged sphere. See Eqs. (1) and (2) above. We'll recycle those formulas.

Subproblem 1: Field at point B due to sphere 1

Since point B sits *outside* sphere 1, we must use Eq. (1), derived in subproblem 1 above. According to that formula, the field a distance r from the center of the sphere is

$$E = \frac{q_{\text{encl}}}{4\pi\varepsilon_0 r^2} = -\frac{\frac{4}{3}\pi R^3\rho}{4\pi\varepsilon_0 r^2}.$$

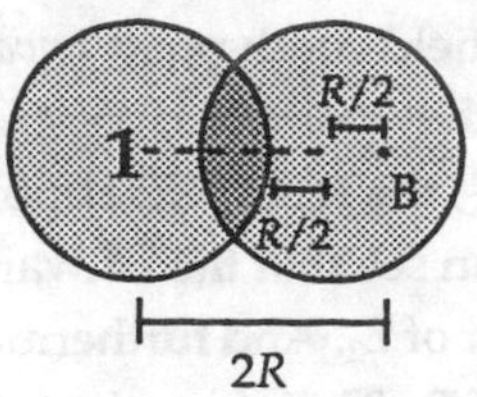

From this diagram, notice that B sits a distance $R/2$ from the center of sphere 2. And the center of sphere 2 is distance $R/2$ from the edge of sphere 1. So, the distance from the edge of sphere 1 to point B is $R/2 + R/2 = R$. Therefore, B sits a distance $2R$ from the *center* of sphere 1. Setting $r = 2R$ gives

$$E_{\text{B due to sphere 1}} = -\frac{\frac{4}{3}\pi R^3 \rho}{4\pi\varepsilon_0 r^2} = -\frac{\frac{4}{3}\pi R^3 \rho}{4\pi\varepsilon_0 (2R)^2} = -\frac{R\rho}{12\varepsilon_0}.$$

The minus sign means "towards sphere 1," i.e., leftward.

Subproblem 2: Field at point B due to sphere 2

Since point B lives *inside* sphere 2, we must use Eq. (2), derived in part (a) subproblem 2. According to that formula, the electric field a distance r from the center of a charged sphere, is

$$E = -\frac{r\rho}{3\varepsilon_0},$$

for $r < R$. Since point B is distance $r = R/2$ from the center of sphere 2,

$$E_{\text{B due to sphere 2}} = -\frac{R\rho}{6\varepsilon_0}.$$

The minus sign means "toward the center of sphere 2," i.e., leftward. *End of subproblem 2.*

Since $\mathbf{E}_{\text{B due to sphere 1}}$ and $\mathbf{E}_{\text{B due to sphere 2}}$ both point leftward, the total field at point B also points leftward, with magnitude

$$\begin{aligned} E_{\text{B}} &= |\, E_{\text{B due to sphere 1}} + E_{\text{B due to sphere 1}} \,| \\ &= \frac{R\rho}{12\varepsilon_0} + \frac{R\rho}{6\varepsilon_0} \\ &= \frac{R\rho}{4\varepsilon_0}. \end{aligned}$$

Therefore, a charge Q at point B would experience a leftward force $F = QE = Q\dfrac{R\rho}{4\varepsilon_0}$.

(c) Point C sits on the edge of both sphere 1 and sphere 2. Therefore, both spheres generate equally strong electric fields at that point.

To find that field strength, you can use the above formula for the field outside a charged sphere. *Or* you can use the formula for the field inside a charged sphere. I didn't mention this before, but those two formulas must agree with each other at the boundary between "outside" and "inside"—i.e., at the edge of a sphere. In other words, plugging $r = R$ into Eq. (1) and Eq. (2) must give the *same* result. And indeed it does, as you can confirm. If we let $\mathbf{E}_1$ and $\mathbf{E}_2$ denote the fields generated at point C by sphere 1 and sphere 2, then

$$E_1 = E_2 = E_{\text{at edge of charged shpere of radius } R} = -\frac{R\rho}{3\varepsilon_0}.$$

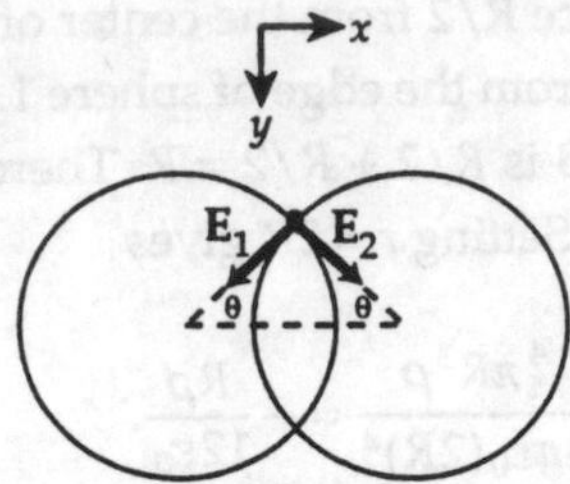

The minus sign indicates that those fields both point *towards* the centers of the spheres, as drawn here. Since they point in different directions, we must add them *vectorially* to find $\mathbf{E}_C$, the total field at point C.

Just by looking at the diagram, we can see that the leftward component of $\mathbf{E}_1$ cancels the rightward component of $\mathbf{E}_2$. And furthermore, the y-component of $\mathbf{E}_1$ equals the y-component of $\mathbf{E}_2$. Therefore, the total electric field points in the y-direction, with twice the magnitude of E_{1y} or E_{2y}.

Let me summarize this reasoning mathematically.

$$\begin{aligned} E_c &= E_1 + E_2 \\ &= 0\,\hat{x} + E_{1y}\,\hat{y} + E_{2y}\,\hat{y} \\ &= 2E_{1y}\,\hat{y}\,, \end{aligned}$$

since E_{1x} cancels E_{2x}, and since $E_{1y} = E_{2y}$. (Recall that $\hat{x}$ and $\hat{y}$ denote unit vectors, which tell you everything about direction, but nothing about length.)

So, we just need to find E_{1y} or E_{2y} individually, and then double it to obtain the total field. From this diagram,

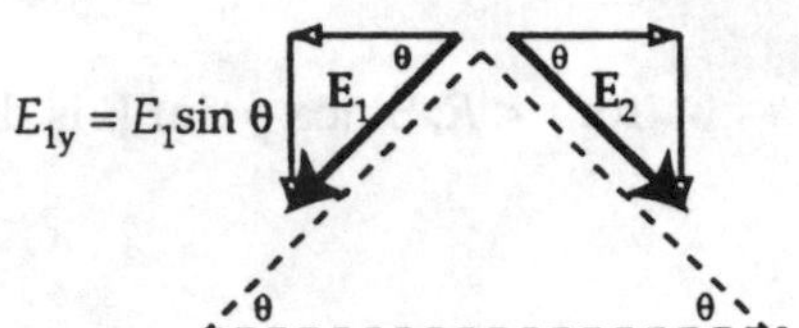

$$E_{1y} = E_1 \sin\theta = \frac{R\rho}{3\varepsilon_0}\sin\theta\,.$$

To complete this problem, we must find θ. Well, as this new diagram shows, the distance from the center of either sphere to point C is R, the sphere's radius. And the centers of the spheres are separated by a "radius and a half," $3R/2$. That's because the edge of one sphere is half a radius away from the center of the other sphere. So, if we divide the big dashed triangle into two right triangles, each right triangle has a base of length $3R/4$, which is half the distance between the centers of the spheres. Looking at one of those right triangles, we immediately get

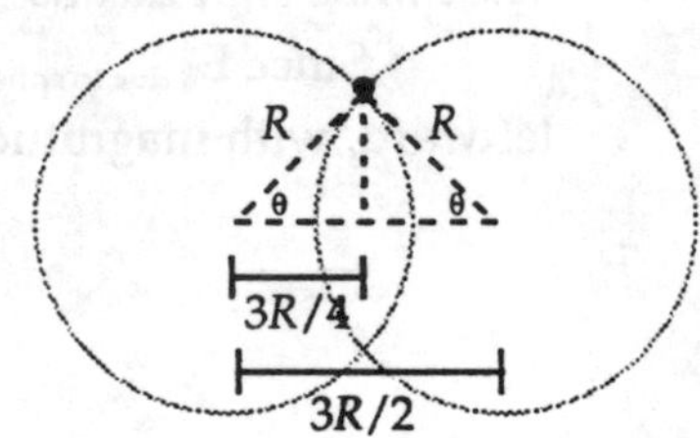

$$\cos\theta = \frac{\text{adjacent}}{\text{hypotenuse}} = \frac{3R/4}{R} = \frac{3}{4}\,,$$

and hence $\theta = \cos^{-1}(3/4) = 41.4°$. So, the y-component of either field contribution is

$$\begin{aligned} E_{1y} = E_{2y} &= E_1 \sin\theta \\ &= \frac{R\rho}{3\varepsilon_0}\sin 41.4° \\ &= (.66)\frac{R\rho}{3\varepsilon_0}\,. \end{aligned}$$

Just double this to get the total field at point C,

$$E_C = 2E_{1y}\; 2\frac{R\rho}{3\varepsilon_0}\sin 41.4° = (1.32)\frac{R\rho}{3\varepsilon_0}\,.$$

35 CHAPTER

Electric Potential

QUESTION 35-1

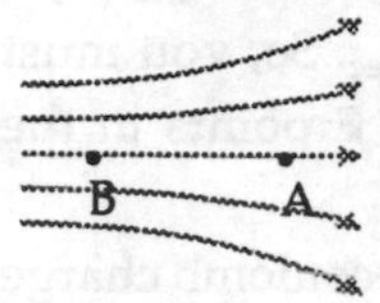

Points A and B are inside an electric field, as drawn to the left. An experiment reveals that it takes 100 joules of work to push a 2-coulomb charge along a straight line from point A to point B. But I don't know the distance from A to B, or the electric field strength.

(a) Let's say the potential energy of the charge at point A was 0. What is its potential energy at point B?

(b) If I repeated this experiment with a 4-coulomb charge, what would its potential energy be at point B?

(c) What's the electrostatic *potential*, *V*, of point B?

(d) To move a 2-coulomb charge from A to B along the dashed triangular path drawn here, would it take more than 100 joules, less than 100 joules, or exactly 100 joules? Justify your answer.

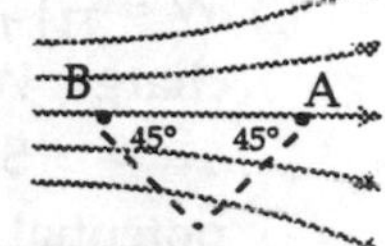

ANSWER 35-1

(a) Despite all the missing information, you can solve using the relationship between work and potential energy. To review that relationship, I'll discuss a gravitational example. But the same intuitions apply to electrostatics.

Suppose I lift a block from the floor to my desk. In the process, I expend 50 joules of work fighting gravity. How much potential energy does the block now have, compared to its potential energy on the floor? Well, since I did 50 joules of work lifting the block, it now has 50 joules of potential energy. Potential energy is the energy "stored" in an object because work has been done on it. For instance, if you spend 40 joules of work stretching a spring, it now has 40 joules of potential energy stored up.

This connection between work and potential energy is not just a rough intuition. It's the *definition* of potential energy:

Definition of potential energy $$W_{\text{by you}} = \Delta U,$$

where ΔU denotes the change in potential energy. If you see this equation with a minus sign, the "W" refers to the work done by the field, not the work done by you *fighting* the field.

Since it takes $W = 100$ joules of work to push the charge from A to B, it ends up with $\Delta U = 100$ more joules of potential energy than it started with. Since it started with $U_A = 0$, it ends up with $U_B = 100$ J.

(b) Intuitively, doubling the charge doubles the work needed to push it from A to B, thereby doubling the potential energy at point B. That's correct! To see why, in more rigorous terms, we must think physically about the meaning of "work."

Moving a positive charge leftward from A to B takes work *because* you're "fighting" the electric field, which pushes the charge rightward. The field exerts a rightward electric force on the charge, a force you must counteract. The bigger the electric force opposing you, the more work it takes to push the particle from A to B.

The definition of work confirms this intuition:

Definition of work $W = \int \mathbf{F}_{\text{by you}} \cdot d\mathbf{s} (= -\int q$

Here, your leftward force, $\mathbf{F}_{\text{by you}}$, must cancel the rightward electric force on the particle, $\mathbf{F}_{\text{elec}} = q\mathbf{E}$. In other words, your force is equal and opposite to the electric force: $\mathbf{F}_{\text{by you}} = -\mathbf{F}_{\text{elec}}$. So, you must expend work $W = \int \mathbf{F}_{\text{by you}} \cdot d\mathbf{s} (= -\int q\mathbf{E} \cdot d\mathbf{s}$. This work comes out positive, because **E** points in the opposite direction from $d\mathbf{s}$, the direction of motion.

OK, let's tie this together. When you push a 4-coulomb charge instead of a 2-coulomb charge from A to B, you're fighting against the same electric field in either case. But by doubling q, you double the electric force $q\mathbf{E}$ against which you're fighting. For this reason, it takes twice as much work to push the 4-coulomb charge as it took to push the 2-coulomb. The formula we just derived, $W = -\int q\mathbf{E} \cdot d\mathbf{s}$, confirms that work is proportional to the charge you're pushing. For the 4-coulomb charge, $W = 200$ J instead of 100 J.

Since it takes 200 joules (instead of 100 joules) to push the 4-coulomb charge to point B, its potential energy at point B is simply $U_B = 200$ joules instead of 100 joules.

(c) I could just write down the definition of "potential." But first I'll show you *why* it's useful to introduce this new physical quantity. Understanding the motivation will help you understand the concept itself.

Suppose someone asks for the value of the "potential energy at point B." Maybe she wants to paint that number next to point B, or to record that value in her lab book. Unfortunately, I can't give her a straight answer. As we saw in part (b), U_B depends on whether we're using a 2-coulomb charge or a 4-coulomb charge. More specifically, **the work required to move the charge from A to B—and therefore, the potential energy at point B—is proportional to the charge you're pushing.** If you're pushing a 2-coulomb charge, the potential energy at B is $U_B = 100$ J. If you're pushing 4 coulombs, then $U_B = 200$ J, as shown in part (b). If $q = 6$ coulombs, then $U_B = 300$ J. And so on. This table displays the proportionality between q and U.

q (coulombs)	U_B (joules)
2	100
4	200
6	300
8	400
10	500

Now let's return to my colleague, who wants to write down a *single* number related to "the potential energy at B." Looking at the table, what number would you tell her?

Well, "50" pops out at me. Although U_B depends on q, the ratio U_B/q always equals 50. This ratio tells us something general about the potential energy at point B, something that applies to *any* charge.

And indeed, the potential of point B is $V_B = U_B/q = 50$ joules per coulomb. By definition,

Definition of potential $$V = \frac{U}{q}.$$

From this definition, and also from the above table, notice that V is *the potential energy of a 1-coulomb charge*. Since it takes 50 joules to push a 1-coulomb charge from A to B, the potential of point B is $V_B = 50$ volts. ("Volts" are joules per coulomb.)

"Potential" is one of the hardest concepts in this whole course. Here, I've discussed one intuitive way of understanding it. In later problems, I'll develop the concept further.

(d) The question asks whether pushing a charge from A to B along a triangular path, instead of a straight path, takes more work, less work, or the same work.

To gain physical insight, let's ask the same question in a gravitational context. Suppose you can lift a block one meter off the floor either by hoisting it straight up along path 1, or by moving it along triangular path 2, using frictionless ramps. In which case would you do more work fighting gravity?

I have conflicting intuitions, because work equals $\int \mathbf{F} \cdot d\mathbf{s}$; it's proportional to the pushing force, and also to the distance. Along path 1, you directly oppose the downward gravitational force. In other words, you fight gravity head on.

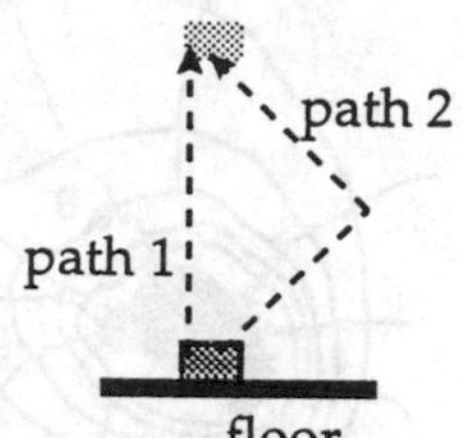

Along path 2, by contrast, you don't fight gravity head-on. Pushing the block up some ramps requires less force than hoisting it *straight* up. Since you exert a smaller force along path 2 than you did along path 1, using path 2 might take less work. But path 2 is longer. Since work is proportional to distance, path 2 might involve *more* work.

In summary: Along path 1, you exert a greater force, but cover less distance. Along path 2, you exert a smaller force, but cover more distance. It turns out—indeed, it's a deep result—that $\int \mathbf{F} \cdot d\mathbf{s}$ comes out the same either way. Along path 2, the bigger distance exactly compensates for the smaller average force. The work needed to go from arbitrary point A to arbitrary point B is the *same* no matter which path you take, in *any* gravitational field. Pithily put, we've got "path independence in a conservative force field." This may sound familiar from last semester.

Electric fields, like gravitational fields, are conservative in this sense. (In other senses, electric fields are liberal, favoring socialized medicine and voting Democratic.) It takes 100 joules of work to push the 2-coulomb charge from A to B, no matter whether you move along a straight line, a triangle, or any other path.

You can use this insight to simplify complicated-looking problems, as we'll see below.

QUESTION 35-2

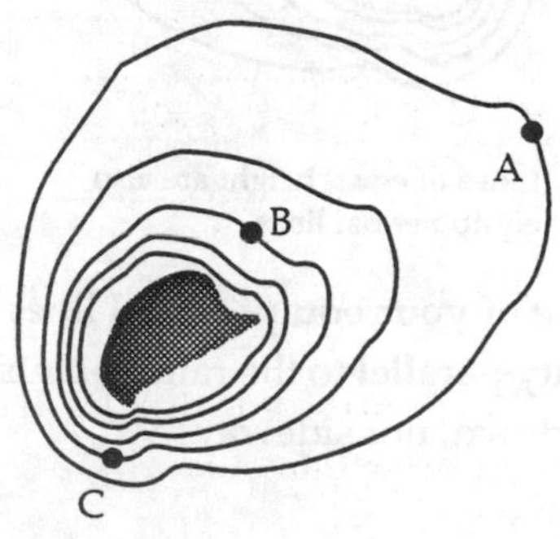

A funky gray charge distribution generates an electric field correspond ing to the following equipotential surfaces. (All points on a given equipotential surface have the same potential.) The potentials at points A and B are $V_A = 3.0$ V and $V_B = 1.0$ V.

(a) On this diagram, sketch some of the electric field lines emanating from the charge distribution. Is the charge distribution positive or negative? (Yes, you have enough information to tell.)

(b) Where is the electric field strongest? Is it stronger to the right of thecharge distribution or to the left of it? Above it or below it? Justify your answer.

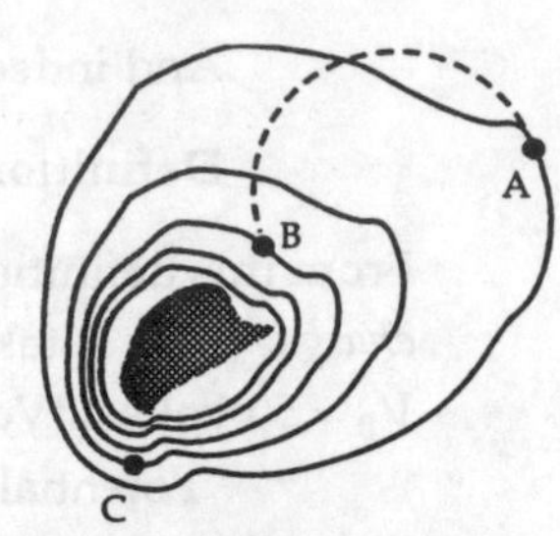

(c) How much work would it take to move a $Q = 0.50$ C point charge along a straight line from B to A?

(d) Now consider a semicircular path from B to A. To move the $Q = 0.50$ C charge along this path, would it take more work, less work, or the same work, as compared to part (c)? Explain your answer.

(e) Which takes more work: Moving charge Q from point C to point A, or moving it from point B to point A? Justify your answer.

ANSWER 35-2

This question requires you to understand the concept of electric potential, not just formulaically, but at a deeper intuitive level.

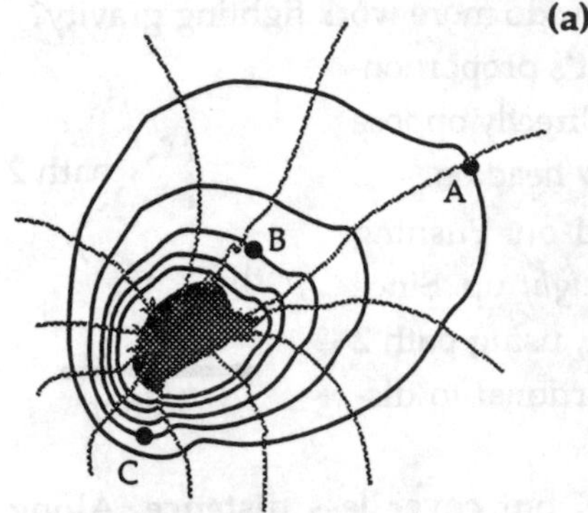

Electric field lines (gray) point perpendicular to equipotential surfaces. Since the charge distribution must be negative (for reasons discussed below), the field lines point inward.

(a) Electric field lines always point *perpendicular* to equipotential lines. I'll give a brief explanation, and then a longer one.

Moving a charge around an electric field takes work, because the field can "fight" or "help" the motion. But if two points have the same potential, then it takes no work to move a charge from one point to the other. So, if you move the charge *along* an equipotential surface, you do no work on the particle, and it does no work on you. Therefore, you are neither fighting the electric field nor getting helped by it. In other words, you are pushing the particle neither *with* the field nor *against* the field. So, you must be pushing it *perpendicular* to the field. Therefore, electric field lines point *perpendicular* to equipotential sur faces, as drawn here.

In order to make this reasoning more intuitive, let me discuss equipotential surfaces in a more familiar context, namely, gravity. Even if you got this problem right, you might want to keep reading.

Consider this contour map of a crater. The curves represent points of equal height. Let's say point A has altitude 3 meters and point B has altitude 1 meter. Since $U_{grav} = mgh$, lines of equal height are *also* lines of equal potential energy—equipotential lines! I'll now explore some properties of equipotential lines.

CONTOUR MAP OF CRATER

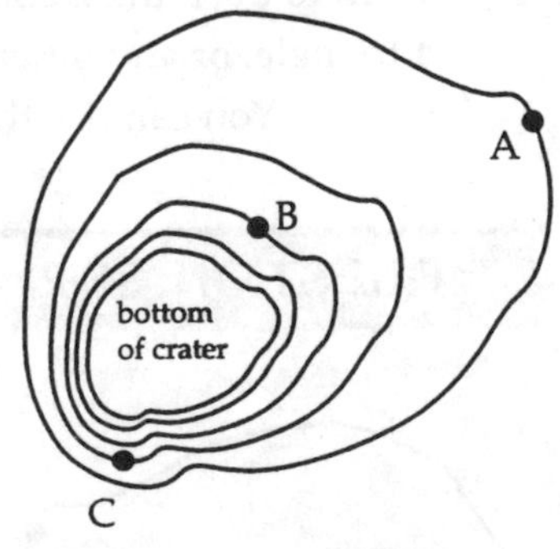

Lines of equal height are also equipotential lines.

If you place a ball on the ground at point A, which way does it roll? Downhill, of course. But which direction is "downhill?" Well, if the ball rolls along an equipotential line, its height stays the same. To lower its height at the largest possible rate—i.e., to go "downhill"—the ball must roll perpendicular to the equipotential line. In other words, gravity pushes the ball perpendicular to the equipotential lines. So, if we painted gravitational field lines on the hill, they would point perpendicular to the equipotential lines.

If this doesn't seem right, then take a ramp, draw in some equipotential lines (i.e., lines of equal height), and place a ball on one of your equipotential lines. It rolls perpen dicular to the equipotential line. The component of gravity parallel to the ramp points perpendicular to the equipotential lines. In other words, the ball rolls down, not sideways.

RAMP
Thick black lines are equipotential lines.

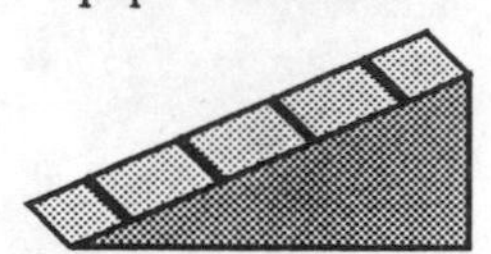

The gravitational field along the ramp points perpendicular to the equipotential lines.

The same reasoning applies to electric field lines and equipotential surfaces. Field lines point perpendicular to equipotential surfaces. In other words, if you "drop" a charge at point A, the electric field pushes it perpendicular to the equipotential line through point A. Similarly for B, C, and all other points.

Here's another way to reach the same conclusion. Suppose that a component of the electric field points along an equipotential line. Then you'd have to fight against that field when pushing a charge along the equipotential line. In other words, you'd have to do work. When work gets done on the particle, its potential energy changes. But the particle's potential energy *can't* change when you push it along an *equi*potential line. This contradiction shows that *no* component of the electric field points along an equipotential line. The field must point entirely perpendicular to the equipotential lines.

In this particular case, do the field lines point inward or outward? Well, just as a ball wants to fall from higher potential energy (greater height) to lower potential energy (lower height), a positive test charge wants to "fall" from higher to lower potential. Since $V_B < V_A$, the potential gets lower as you get closer to the charge distribution. Therefore, the electric field lines must point inward. The funky charge distribution must be negative.

(b) The field is strongest where the equipotential lines are most closely spaced, in this case the lower left part of the diagram. To understand why, we can exploit the gravitational analogy.

The *strength* of the electric field around the charge distribution corresponds to the *steepness* of the ground around the crater. A hiker must work extra hard to "fight" a steep hill. Similarly, you must work extra hard to push a charge against a strong electric field. So, if we can use a contour map to figure out where the ground is steepest, then we automatically know how to use an equipotential map to figure out where the electric field is strongest. Let's think about the contour map . . .

A landscape is "steep" when its height changes rapidly, i.e., when you change your altitude substantially by walking only a short distance. On the contour map, suppose that two adjacent equipotential lines correspond to a height difference of 1 meter. If those two equipotential lines are very close together (say, 3 meters apart), then the ground is steep. You'd have to walk only three or four steps to climb a meter. By contrast, if those two equipotential lines are far apart (say, 50 meters), then the ground is very flat. You'd have to walk sixty or seventy steps to climb a mere meter. From this example, we see that closely-spaced lines on a contour map correspond to steep ground. When the contour lines are close together, the altitude changes rapidly, i.e., the landscape is steep.

By analogy, the electric field is "steepest," by which I mean strongest, where the equipotential lines are closely spaced.

(c) So far, I've talked about electrostatic potential as if it were the same thing as electrostatic potential energy. Actually, they're proportional to each other:

$$\Delta U = q\Delta V.$$

In addition, recall that the change in an object's potential energy equals the work you performed on it: $W = \Delta U$. Here, you're moving the particle from a lower to a higher potential. Therefore, you must perform positive work:

$$
\begin{aligned}
W = \Delta U &= q\Delta V \\
&= Q(V_A - V_B) \\
&= (0.50\,\text{C})(3.0\,\text{V} - 1.0\,\text{V}) \\
&= 1.0\,\text{joule}.
\end{aligned}
$$

Notice that "ΔV" denotes the *final* minus the *initial* potential. Also, I used $W = +\Delta U$ instead of $W = -\Delta U$ because we're calculating the work done by you "fighting" the electric field, not the work done by the electric field itself. Here, the field point inwards. That's why it takes positive work to push the particle outward from B to A.

Notice that, if we use a $q = 1$-coulomb charge, then the above formula reduces to $W = \Delta U = \Delta V$. In words, the potential difference between two points is the work needed to move a 1-coulomb charge from one point to the other.

Also, notice that we solved this problem without integrating force over distance. By telling you the potential at various points, I've told you what that integral works out to be, for a 1-coulomb charge.

(d) To gain physical insight, ask the same question in a gravitational context. You can lift a block one meter off the floor either by hoisting it straight up along path 1, or by moving it along curvy path 2 using a some curved ramps. In which case would you do more work fighting gravity?

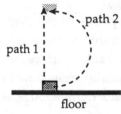

path 2
path 1
floor

Well, along path 2, you don't fight gravity "head on." This suggests that path 2 involves less work than path 1. But path 2 is longer. Since work depends on distance $\left(W = -\int \mathbf{F}\cdot d\mathbf{s}\right)$, this suggests that path 2 involves more work than path 1. Actually, these two effects "cancel." Along path 2, the bigger distance exactly compensates for the smaller average force. Therefore, the work does not depend on which path you take.

This same result holds for electric fields. The work needed to move charge Q from point B to point A is the value obtained in part (c), no matter what path you take.

(e) Don't get distracted by the fact that C-to-A is a longer distance than B-to-A. While going from C to A, you need not fight the electric field head on, as you can confirm from the electric field lines drawn in part (a). So, pushing the particle from C to A may take less work than pushing it from B to A.

In fact, $W_{\text{C to A}}$ *has* to be less than $W_{\text{B to A}}$. The equipotential line containing C is *between* the equipotential lines containing points A and B. Therefore, the potential of point C must lie between $V_A = 3.0$ V and $V_B = 1.0$ V. So, the potential *difference* between points C and A, $V_A - V_C$, is less than the potential *difference* between points B and A. For instance, if $V_C = 1.3$ V, then $V_A - V_C = 1.7$ V. That's less than $V_A - V_B = 2.0$ V. Therefore, since $W = \Delta U = q\Delta V$, we know that $W_{\text{C to A}}$ is less than $W_{\text{B to A}}$.

We can understand this intuitively by thinking about the contour map. To get from C to A, you can walk along C's equipotential ("equiheight") line until you're directly "under" point A. While walking along that ridge, your potential energy doesn't change. It changes only when you climb from that equipotential line to point A. So, although walking from C to A involves covering more distance, walking from B to A involves more *climbing*.

QUESTION 35-3

A charge distribution creates an electric potential $V = (1\ \text{V/m}^2)x^2$. Ignore gravity.

(a) If a positively charged particle is released from rest from $x = 1$ meter, does it move in the $+x$ direction or the $-x$ direction? Justify your answer.

(b) A positive particle has charge $Q = 0.10$ C and mass $M = 0.020$ kg. It's released from rest from $x = 1.00$ meter. How fast is the particle moving after it has covered 0.60 meters?

(c) What is the strength and direction of the electric field at $x = 2.0$ meters?

(d) *(Challenge problem)* In part (c), you found that $E_x = -2x$, in units of volts per meter. Describe the part (b) particle's motion in the long run. Does it keep traveling forever in one direction? Does it speed up or slow down? Be specific, and if possible, quantitative.

ANSWER 35-3

(a) To answer this, I'll rely upon physical intuitions about potential energy. In part (c), we'll see how these intuitions correspond to a mathematical formula relating potentials and fields.

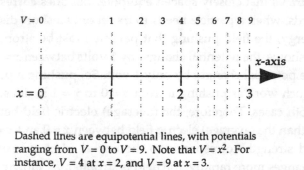

To take advantage of the "contour map" analogy from question 35-2, draw some equipotential surfaces. Since $V = x^2$ (omitting the units), equipotential lines are lines of constant x. The equipotential lines become more closely spaced as x increases, corresponding to a stronger field. But here we don't care about field strength. We just want to know the field direction, i.e., the direction in which a positive test charge would get pushed.

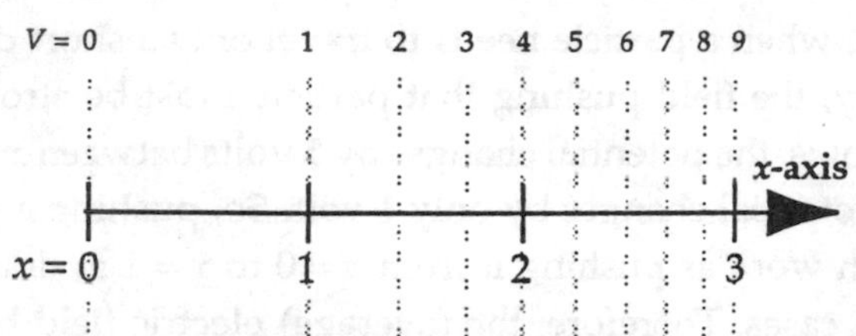

Dashed lines are equipotential lines, with potentials ranging from $V = 0$ to $V = 9$. Note that $V = x^2$. For instance, $V = 4$ at $x = 2$, and $V = 9$ at $x = 3$.

In general, an object "wants" to lower its potential energy. For instance, a stretched spring "wants" to unstretch, thereby lowering its potential energy. In other words, the forces inside the spring tend to make the spring lose potential energy. Similarly, gravity pulls a marble down a ramp, from points of higher potential energy to points of lower potential energy. And an electric field pushes a charged particle so as to lower the particle's potential energy.

Since $U = qV$, a positive particle lowers its potential energy by lowering its potential. From the above diagram, the particle must move leftward, in the $-x$ direction, to decrease its potential. So, the particle naturally "falls" leftward. In other words, the electric field pushes the particle leftward. Because the particle is positive, this goes to show that the electric field points leftward.

In part (c), I'll hook up these intuitions to a rigorous formula. Stay tuned.

(b) As just noted, the particle "falls" leftward, just like a block falls in a gravitational field. Let's keep thinking in terms of this gravitational analogy. If I ask you for a block's speed after falling through 0.60 meters, how would you solve? You could use old-fashioned kinematics. Or, you could use energy conservation. Indeed, if the gravitational field isn't uniform (e.g., planetary gravity), energy conservation works best. Intuitively, as the block falls, its gravitational potential energy converts into kinetic energy.

The same intuitions and problem-solving strategies apply here. As the charged particle falls leftward, some of its electric potential energy converts into kinetic energy We can calculate its velocity at $x = 0.40$ m by setting the total initial energy (at $x_0 = 1.00$ m) equal to the total "final" energy (at $x_f = 0.40$ m):

$$K_0 + U_0 = K_f + U_f$$

$$0 + QV_0 = \frac{1}{2}Mv^2 + QV_f$$

$$Q(1\,\text{V/m}^2)x_0^2 = \frac{1}{2}Mv^2 + Q(1\,\text{V/m}^2)x_f^2,$$

where I used the given expression for potential, $V = (1\ \text{V/m}^2)x^2$. Isolate v, and substitute in the numbers, to get

$$v = \sqrt{\frac{2Q(1\,\text{V/m}^2)(x_0^2 - x_f^2)}{M}} = \sqrt{\frac{2(0.10\,\text{C})(1\,\text{V/m}^2)[(1.00\ \text{m})^2 - (0.40\ \text{m})^2]}{0.020\ \text{kg}}} = 2.9\,\text{m/s}.$$

(c) Instead of just plugging in the relevant formula, I'll first show where it comes from.

The contour map analogy demonstrates that closely spaced equipotential lines correspond to a stronger ("steeper") field. In other words, when a particle needs to travel only a short distance to undergo a large change in potential energy, the field pushing that particle must be strong. For instance, as the above equipotential "map" shows, the potential changes by 5 volts between $x = 2$ and $x = 3$ m. But between $x = 0$ and $x = 1$ m, the potential changes by only 1 volt. So, pushing a particle from $x = 2$ to $x = 3$ m takes five times as much work as pushing it from $x = 0$ to $x = 1$ m does, even though the particle travels one meter in both cases. Therefore, the (average) electric field between $x = 2$ and $x = 3$ m must be *five times stronger* than the (average) electric field between $x = 0$ and $x = 1$ m.

Here's the point. The electric field strength corresponds to the *rate at which the potential changes over distance*. Where the potential changes more rapidly, the field is stronger, as shown in the previous paragraph. Specifically, between $x = 2$ and $x = 3$ m, the average field strength is 5 volts per meter. Between $x = 0$ and $x = 1$ m, the average field strength is only 1 volt per meter.

At this stage, we can guess that $E = dV/dx$, or something like that. As your textbook shows, doing the vector calculus properly yields

$$E_x = -\frac{\partial V}{\partial x}, \quad E_y = -\frac{\partial V}{\partial y}, \quad E_z = -\frac{\partial V}{\partial z}.$$

The minus sign indicates that the field points toward the region of lower potential. For instance, in this problem, $\partial V/\partial x$ is positive, because V increases as x increases. But as we saw in part (a), the electric field must point in the $-x$ direction, towards the region of lower potential. Hence, the minus sign in front of $\partial V/\partial x$.

Don't get psyched out by the partial derivatives. They're just like a regular derivatives, except you "turn off" part of the chain rule. Specifically, when evaluating $\partial V/\partial x$, you "pretend" that y and z are constants. Similarly, when calculating $\partial V/\partial y$, you "pretend" that x and z are constants.

We now have a formula relating the potential to the electric field. But don't just plug 'n' chug. Make sure you understand the underlying intuitions, as discussed above.

In this problem, since V depends only on x, E_y and E_z are zero. The field points in the x direction, with magnitude

$$E_x = -\frac{\partial V}{\partial x} = -(1\,\text{V/m}^2)\frac{\partial}{\partial x}x^2 = -(2\,\text{V/m}^2)\,x.$$

To obtain the field at $x = 2$ m, just substitute $x = 2$ m into this formula:

$$E_x(x = 2) = -(2\,\text{V/m}^2)(2\text{m}). = -4\,\text{V/m}.$$

(d) As just found, the electric field has strength $E_x = -2x$, where I've left out units for brevity. Therefore, the electric force on the particle, in newtons, is

$$F = QE = -2Qx.$$

So, the particle always gets pushed toward $x = 0$. And the force is proportional to x. As the particle nears $x = 0$, the force gets smaller and smaller. It passes through $x = 0$, at which point the force reverses direction. For instance, at $x = -1.0$ m, the particle feels a force $F = -2Qx = +0.20$ volts/meter, a rightward (positive) force that tries to push the particle back toward $x = 0$. So, when the particle reaches the negative x's, it slows down, eventually turning around and heading back toward the positive x's. When it again reaches the positive x's, it slows down, eventually turning around and heading back to the negative x's. In this way, the particle oscillates back and forth.

Indeed, the particle oscillates in simple harmonic motion, exactly like a pendulum or a block on a spring. That's because the net force takes the form $F = -(k_{\text{eff}})x$, where in this case the effective spring constant is $k_{\text{eff}} = 2Q$. Consequently, the particle oscillates *as if* it were attached to a spring of spring constant $2Q$. In Volume 1 of this study guide, I used the term "stuff" instead of k_{eff} to denote the effective spring constant.

If you didn't study simple harmonic motion last semester, then you can't get any more precise. If you *did* study simple harmonic motion, then you know that the particle's position oscillates according to

$$x = A\cos\omega t,$$

where A is the amplitude and ω is the angular frequency, given by $\omega = \sqrt{k_{\text{eff}}/m}$. Here, since the particle starts motionless one meter from the origin, the amplitude is $A = 1$ meter. In words, the particle oscillates back and forth between $x = +1$ and $x = -1$ meter. And

$$\omega = \sqrt{\frac{k_{\text{eff}}}{m}} = \sqrt{\frac{2Q}{M}} = \sqrt{\frac{2(0.10)}{0.020}} = 3.2\,\text{s}^{-1},$$

where I've continued to leave out the units. So,

$$x = A\cos\omega t = (1.0\,\text{m})\cos 3.2t,$$

with t in seconds.

QUESTION 35-4

Along the z-axis, indicated by the dashed line on this diagram, the electric field is

$$\mathbf{E} = -Cz^2\hat{\mathbf{z}},$$

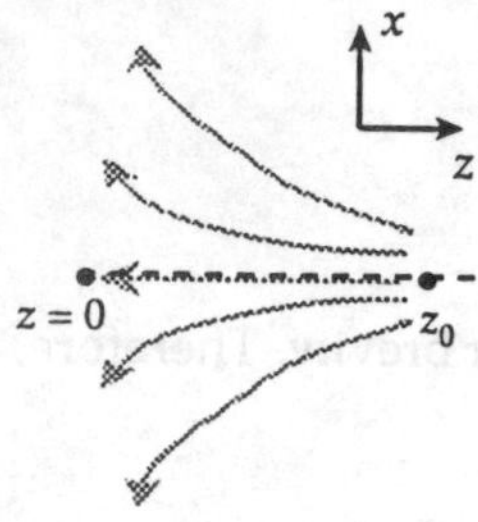

where C is a positive constant, and $\hat{\mathbf{z}}$ is a unit vector. The minus sign indicates that the field points "backwards" towards z = 0. Ignore gravity.

A particle of charge $+q_0$ and mass m is released from rest from $z = +z_0$, on the z-axis.

(a) What is the particle's speed when it reaches z = 0? Answer in terms of q_0, m, C, and z_0.

(b) At z = 0, what is the particle's acceleration?

ANSWER 35-4

(a) We could try to solve using forces and acceleration. Since $F = qE$, we can calculate the particle's acceleration using Newton's 2nd law, $a = F_{net}/m$. We could then plug that acceleration into one of those old kinematics equations. Right?

Not quite. Those old kinematic equations, such as $v^2 = v_0^2 + 2a\Delta x$, apply only when the acceleration is constant. Here, the electric force gets weaker as the particle approaches z = 0. Therefore, the acceleration gets smaller and smaller. Because a changes, we need another problem-solving technique.

As we saw last semester, you can sidestep messy forces by invoking conservation laws. Here, the particle generates no heat or other dissipative energy. So, we can solve using energy conservation.

Initially, the particle has no kinetic energy. It starts with potential ("stored up") energy only. When "dropped," it loses potential energy while gaining kinetic energy. Energy conservation demands that

Energy conservation $$K_0 + U_0 = K_f + U_f$$

$$0 + q_0V_0 = \frac{1}{2}mv^2 + q_0V_f,$$

where V_0 denotes the initial potential (at $z = z_0$), V_f denotes the final potential (at $z = 0$), and v denotes the final velocity. Solve for v to get

$$v = \sqrt{\frac{2q_0(V_0 - V_f)}{m}} = \sqrt{\frac{2q_0(-\Delta V)}{m}},$$

where in the last step, I noticed that the answer depends only on the potential *difference*, $\Delta V = V_f - V_0$, not on either individual potential. (In general, an individual potential or potential energy has no meaning, because you can set the "zero point" anywhere you want. For instance, when using $U_{grav} = mgh$, you can measure your heights from the floor or from the roof. Physically meaningful answers depend only potential energy *differences*, which come out the same no matter what zero point you choose.)

Since we know q_0 and m, we can solve for v as soon as we figure out the potential difference.

Subproblem: Find the potential difference between $z = z_0$ and $z = 0$.

From the given electric field, we must derive an expression for potential. Let's start with the relationship between work and potential energy, $W = \Delta U$, where the work is $W = -\int \mathbf{F}_{\text{elec}} \cdot d\mathbf{s}$. The minus sign indicates that I'm talking about the force exerted by the field, not the force exerted by you "fighting" the field. So,

$$\Delta U = W$$

$$q_0 \Delta V = -\int \mathbf{F}_{\text{elec}} \cdot d\mathbf{s}$$

$$= -\int q_0 \mathbf{E} \cdot d\mathbf{s}.$$

Now cancel the q_0's to get the formula we're seeking,

$$\Delta V = \int \mathbf{E} \cdot d\mathbf{s}.$$

When using this equation to calculate ΔV, you may integrate **E** along *any path* that connects the starting point to the ending point. The integral is path-independent. See question 35-1d for more discussion of this. Here, we might as well use a straight line connecting $z = z_0$ to $z = 0$. In that case, $d\mathbf{s}$ turns into dz. And since the electric field points in the $\pm z$-direction, the dot product reduces to a regular product:

$$\begin{aligned} \Delta V = -\int \mathbf{E} \cdot d\mathbf{s} \quad &= -\int_{z_0}^{0} E\,dz \\ &= -\int_{z_0}^{0} (-Cz^2)\,dz \\ &= C\int_{z_0}^{0} z^2\,dz \\ &= C\frac{z^3}{3}\Big|_{z_0}^{0} \\ &= -\frac{Cz_0^3}{3}. \end{aligned}$$

ΔV is negative because the particle *loses* potential energy while falling from z_0 to 0.

End of subproblem

Substitute this expression for ΔV into the above equation for the particle's final velocity, the equation we found using energy conservation:

$$v = \sqrt{\frac{2q_0(-\Delta V)}{m}} = \sqrt{\frac{2q_0 \frac{Cz_0^3}{3}}{m}}.$$

(b) To find the acceleration at $z = 0$, use $F = ma$. Since $E = Cz^2$, the electric field—and hence, the electric force—vanishes at $z = 0$. Consequently, the acceleration is zero at that point.

If this seems weird, recall that acceleration is the *rate of change* of velocity. As the particle "falls" from z_0 to 0, it speeds up (accelerates). And it keeps speeding up until reaching $z = 0$. But when the particle hits $z = 0$, the acceleration vanishes. Although the particle doesn't stop, it stops speeding up, at least momentarily.

QUESTION 35-5

A solid spherical ball of uniform charge density has total charge $Q_0 = 3.0 \times 10^{-12}$ C and radius $R =$ 0.010 m.

As always, we're free to decide the "zero point" of potential energy. Please follow the usual textbook convention of saying a particle has zero potential energy when it is infinitely far away. Mathematically speaking, $V = 0$ at $r = \infty$.

Solve everything symbolically *before* plugging in numbers.

(a) What is the potential a distance $2R$ from the ball's center? Derive any formulas you use.

(b) A tiny particle of charge $q_1 = 4.0\ \mu$C and mass $m_1 = 2.0 \times 10^{-6}$ kg is released from rest from the surface of the ball. What velocity does the particle have when it reaches a distance $2R$ from the ball's center?

(c) (*Very hard*) What is the potential inside the ball, at radius $r_2 = 0.0075$ m from the ball's center? Remember to solve symbolically before cranking the numbers.

(d) (*Very hard*) What is the potential at the ball's center?

ANSWER 35-5

(a) Let me figure out the general formula for the potential outside the ball, a distance r from its center. Then, to answer part (a), I'll set $r = 2R$.

As shown in question 34-4, the potential difference between points A and B is $\Delta V = \int_A^B \mathbf{E} \cdot d\mathbf{s}$. Therefore, to calculate the potential outside the sphere, we need to know the electric field outside the sphere.

Subproblem: Use Gauss' law to find the electric field outside the charged sphere

According to Gauss' law, the electric flux out of a closed surface is proportional to the charge "trapped" inside that surface:

$$\oint_{\text{Gaussian surface}} \mathbf{E} \cdot d\mathbf{A} = \frac{q_{\text{encl}}}{\varepsilon_0}.$$

Since we've already dealt with charged spheres, I'll race through the usual 3-step Gauss' law strategy for finding electric fields:

Step 1: Draw the electric field lines.

By symmetry, the field generated by charged sphere points radially.

Step 2: Draw a "good" Gaussian surface.

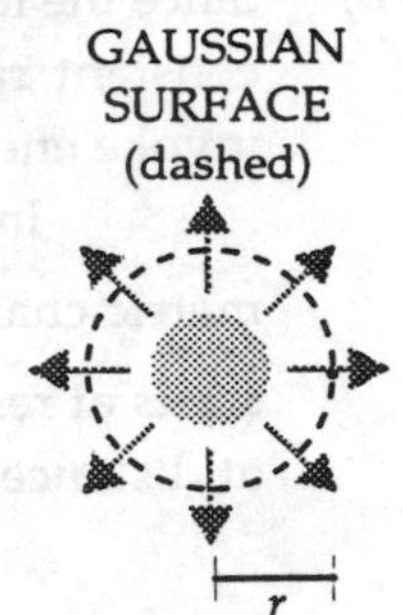

If we use a Gaussian sphere, then the field always points in the same direction that the Gaussian surface faces, namely radially outward. And by spherical symmetry, the field has the same strength at all points on the Gaussian surface. . .

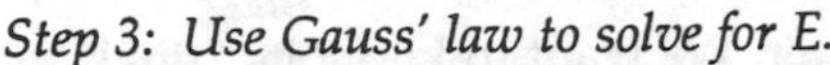

Step 3: Use Gauss' law to solve for E.

. . . Therefore, the dot product $\mathbf{E}\cdot d\mathbf{A}$ reduces to a regular product, and we can pull E outside the flux integral:

$$\oint_{\text{Gaussian sphere}} \mathbf{E}\cdot d\mathbf{A} = E \oint_{\text{Gaussian sphere}} dA = E(4\pi r^2),$$

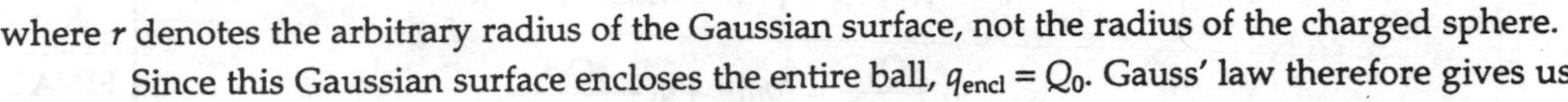

where r denotes the arbitrary radius of the Gaussian surface, not the radius of the charged sphere.

Since this Gaussian surface encloses the entire ball, $q_{\text{encl}} = Q_0$. Gauss' law therefore gives us

$$\oint_{\text{Gaussian surface}} \mathbf{E}\cdot d\mathbf{A} = \frac{q_{\text{encl}}}{\varepsilon_0}$$

$$E(4\pi r^2) = \frac{Q_0}{\varepsilon_0}.$$

Solve for E to get the familiar result for point charges and other spherically symmetric charge distributions,

$$E = \frac{Q_0}{4\pi\varepsilon_0 r^2}.$$

End of Gauss' law subproblem

Given this electric field, we can find a formula for the potential using the ***indefinite*** integral,

$$\begin{aligned} V &= -\int \mathbf{E}\cdot d\mathbf{s} \\ &= -\int \frac{Q_0}{4\pi\varepsilon_0 r^2}\,dr \\ &= \frac{Q_0}{4\pi\varepsilon_0 r} + C, \end{aligned}$$

where C is the integration constant. To figure out C, recall that the problem wants us to set $V = 0$ at $r = \infty$. So, substitute $r = \infty$ and $V = 0$ into this equation, and solve for C. You get $C = 0$. By setting the "zero point" of potential an infinite distance away, you get rid of the messy integration constant, leaving us with the textbook result,

Potential due to point charge or charged sphere $\quad V = \dfrac{Q_0}{4\pi\varepsilon_0 r}.$

To find the potential a distance $2R$ from the ball's center, just substitute $r = 2R$ into this equation. Recalling that $1/(4\pi\varepsilon_0) = 9.0 \times 10^9\ \text{V}\cdot\text{m/C}$, and keeping an "extra" significant digit, we get

$$V(2R) = \frac{Q_0}{4\pi\varepsilon_0(2R)} = \frac{(9.0\times 10^9\ \text{V}\cdot\text{m/C})(3.0\times 10^{-12}\ \text{C})}{2(0.010\ \text{m})} = 1.35\ \text{V}.$$

(b) Since the force on the particle varies as it travels from $r = R$ to $r = 2R$, the particle accelerates at a non-constant rate. Therefore, you can't use those old constant-acceleration kinematic formulas. Let's invoke energy conservation instead.

In part (a), we found the general expression for the potential created by a spherically symmetric charge distribution, at points *outside* the charge distribution. It's $V(r) = \frac{Q_0}{4\pi\varepsilon_0 r}$. The particle starts at rest on the ball's surface, a distance R from its center. We want to know its "final" velocity at distance $2R$ from the center. According to energy conservation,

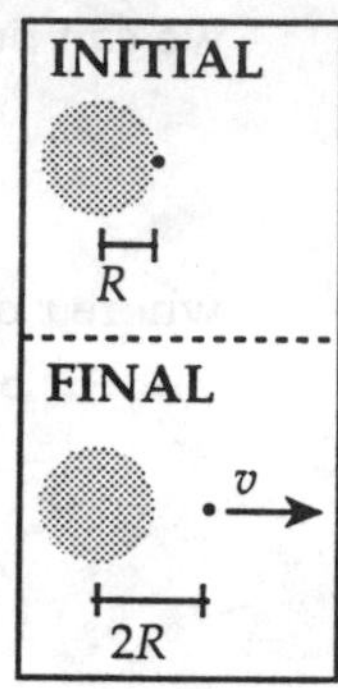

$$K_0 + U_0 = K_f + U_f$$

$$0 + q_1 V(R) = \frac{1}{2}mv^2 + q_1 V(2R)$$

$$q_1 \frac{Q_0}{4\pi\varepsilon_0 R} = \frac{1}{2}mv^2 + q_1 \frac{Q_0}{4\pi\varepsilon_0 (2R)}.$$

Solve for v to get

$$v = \sqrt{\frac{2q_1}{m}\left[\frac{Q_0}{4\pi\varepsilon_0 R} - \frac{Q_0}{4\pi\varepsilon_0 (2R)}\right]}.$$

Notice that the answer depends only on the negative *change* in potential, not on the initial or final potential individually. We could have found this *change* in potential by evaluating the definite integral, $\Delta V = -\int_R^{2R} \mathbf{E} \cdot d\mathbf{r} = -\int_R^{2R} \frac{Q_0}{4\pi\varepsilon_0 r^2} dr = -\left[\frac{Q_0}{4\pi\varepsilon_0 R} - \frac{Q_0}{4\pi\varepsilon_0 (2R)}\right]$, which agrees with our answer. By choosing a different integration constant, we would have obtained different values for $V(R)$ and $V(2R)$. But the potential *difference* would have come out the same as it did above.

Now let's substitute in the numbers. You can save time by exploiting the fact that we already calculated $V(2R)$ in part (a). Since $V(r)$ is proportional to $1/r$, $V(R)$ must be *twice* $V(2R)$, namely, 2.7 V. So,

$$v = \sqrt{\frac{2q_1}{m}\left[\frac{Q_0}{4\pi\varepsilon_0 R} - \frac{Q_0}{4\pi\varepsilon_0 (2R)}\right]} = \sqrt{\frac{2q_1}{m}\left[\frac{Q_0}{4\pi\varepsilon_0 R} - \frac{Q_0}{4\pi\varepsilon_0 (2R)}\right]} = 2.3\,\text{m/s}.$$

(c) The formula we've been using for potential, $V(r) = \frac{Q_0}{4\pi\varepsilon_0 r}$, applies only *outside* the charged ball. That's because we derived $V(r)$ by integrating over $E(r) = \frac{Q_0}{4\pi\varepsilon_0 r^2}$ the field *outside* the ball. But the field *inside* the ball is different. For this reason, we must go back to basics to derive the formula for $V(r)$ inside the charged ball.

Let's think about the meaning of potential. Since we've set the zero point at $r = \infty$, the "potential" at point A is really the potential *difference* between $r = \infty$ and point A:

$$V_A = -\int_\infty^A \mathbf{E} \cdot d\mathbf{s}$$

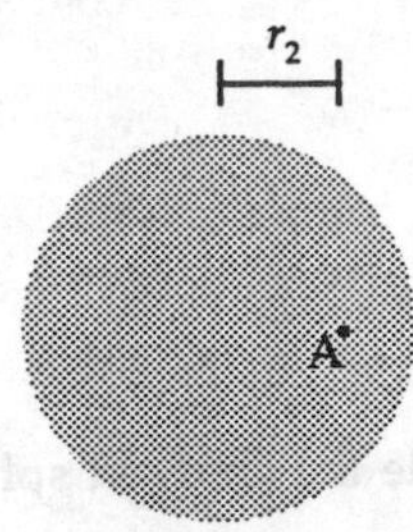

So, physically, the potential at point A is the work needed to push a 1-coulomb test charge from *very* far away to point A. But you can think of this work as consisting of two parts. First, you push the test charge from very far away to the surface of the charged ball. Then you push the test charge from the surface of the ball to point A (inside the ball). In mathematical terms,

$$V_A = -\int_{\infty}^{\text{surface of bal}} \mathbf{E}\cdot d\mathbf{s} = \quad + \quad - \int_{\text{surface of ball}}^{A} \mathbf{E}\cdot d\mathbf{s}.$$

In the first integral, **E** denotes the field *outside* the charged ball. In the second integral, **E** denotes the field *inside* the ball. We must evaluate the two integrals separately, because those two fields differ in form.

To make further progress, we must find the field inside the ball.

Subproblem: Use Gauss' law to find the electric field inside the charged ball.

Step 1: Draw field lines.

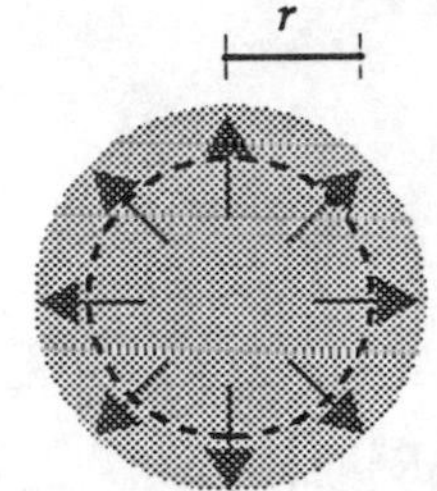

Step 2: raw "good" Gaussian surface.

As in part (a), a spherical Gaussian surface ensures that E has the same strength over the whole surface. Therefore . . .

Step 3: Use Gauss' law, $\oint_{\text{Gaussian surface}} \mathbf{E}\cdot d\mathbf{A} = \frac{q_{\text{encl}}}{\varepsilon_0}$, *to find the field.*

. . . we can pull E outside the flux integral to get

$$\oint_{\text{Gaussian sphere}} \mathbf{E}\cdot d\mathbf{A} = E \oint_{\text{Gaussian sphere}} dA = E(4\pi r^2).$$

To find the charge enclosed by my Gaussian surface, I'll first find ρ, the charge per volume. Then, I'll use q_{encl} = (charge per volume) × (volume enclosed by Gaussian surface).

Since the ball contains charge Q_0 uniformly spread out over volume $V_{\text{ball}} = \frac{4}{3}\pi R^3$, the charge density is ρ = charge/volume = $Q_0/\left(\frac{4}{3}\pi R^3\right)$. Therefore, the Gaussian sphere encloses charge

$$q_{\text{encl}} = \frac{\text{charge}}{\text{volume}} \times (\text{volume enclosed by Gaussian surface}) = \frac{Q_0}{\frac{4}{3}\pi R^3} \times \frac{4}{3}\pi r^3 = Q_0\left(\tfrac{r}{R}\right)^3.$$

Now that we've simplified both sides of Gauss' law, tie it all together to get

$$\oint_{\text{Gaussian surface}} \mathbf{E}\cdot d\mathbf{A} = \frac{q_{\text{encl}}}{\varepsilon_0}$$

$$E(4\pi r^2) = \frac{Q_0}{\varepsilon_0}\left(\tfrac{r}{R}\right)^3,$$

and hence, the field inside the charged sphere is

$$E_{\text{in}} = \frac{Q_0 r}{4\pi\varepsilon_0 R^3}.$$

End of Gauss' law subproblem

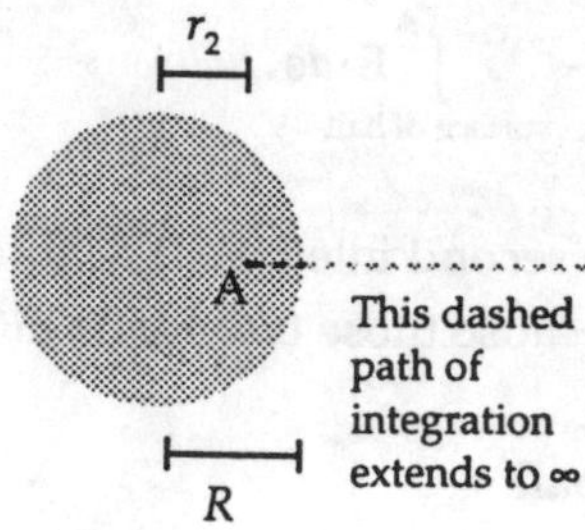

From part (a), we know the field outside the charged sphere. It's $E_{\text{out}} = \frac{Q_0}{4\pi\varepsilon_0 r^2}$. And remember, we're finding the potential at point A, a distance r_2 from the ball's center, by integrating from infinitely far away to the surface of the ball (at $r = R$), and then from the surface of the ball to point A. Those two integrals correspond to the gray and black segments of this dashed line, respectively:

$$\begin{aligned}
V_A &= -\int_{\infty}^{\text{surface of ball}} \mathbf{E}\cdot d\mathbf{s} \quad + \quad -\int_{\text{surface of ball}}^{A} \mathbf{E}\cdot d\mathbf{s} \\
&= -\int_{\infty}^{R} E_{\text{out}}\,dr \quad + \quad -\int_{R}^{r_2} E_{\text{in}}\,dr \\
&= -\int_{\infty}^{R} \frac{Q_0}{4\pi\varepsilon_0 r^2}\,dr \quad + \quad -\int_{R}^{r_2} \frac{Q_0 r}{4\pi\varepsilon_0 R^3}\,dr \\
&= +\frac{Q_0}{4\pi\varepsilon_0 r}\bigg|_{\infty}^{R} \quad - \quad \frac{Q_0 r^2}{2(4\pi\varepsilon_0)R^3}\bigg|_{R}^{r_2} \\
&= \left[\frac{Q_0}{4\pi\varepsilon_0 R} - \frac{Q_0}{4\pi\varepsilon_0 \infty}\right] \quad - \quad \left[\frac{Q_0 r_2^2}{2(4\pi\varepsilon_0)R^3} - \frac{Q_0 R^2}{2(4\pi\varepsilon_0)R^3}\right] \\
&= \frac{Q_0}{4\pi\varepsilon_0 R} \quad + \quad \frac{Q_0(R^2 - r_2^2)}{2(4\pi\varepsilon_0)R^3}.
\end{aligned}$$

That's $V(r_2)$, the potential a distance r_2 from the ball's center. Intuitively, the first term specifies how much work it takes to push a 1-coulomb charge from infinitely far away to the surface of the charged ball. The second term tells us the additional work needed to push the 1-coulomb charge inside the ball until it reaches r_2. Notice that the first term is just $V(R)$, the potential at the ball's surface, as found in part (b).

Indeed, in part (b), we already found that $V(R) = \frac{Q_0}{4\pi\varepsilon_0 R} = 2.7$ V. We also know that $1/(4\pi\varepsilon_0) = 9.0 \times 10^9$ V·m/C. So, at $r_2 = 0.0075$ m from the center,

$$\begin{aligned}
V &= \frac{Q_0}{4\pi\varepsilon_0 R} + \frac{Q_0(R^2 - r_2^2)}{2(4\pi\varepsilon_0)R^3} \\
&= 2.7\,\text{V} + \frac{(9.0\times 10^9\,\text{V}\cdot\text{m/C})(3.0\times 10^{-12}\,\text{C})[(0.010\,\text{m})^2 - (0.0075\,\text{m})^2]}{2(0.010\,\text{m})^3} \\
&= 2.7\,\text{V} + 0.59\,\text{V} \\
&= 3.3\,\text{V}.
\end{aligned}$$

When evaluating the above integrals, a common mistake is to plug in a specific value for r, such as R or r_2. But since we're integrating over dr, we must leave r as a variable in the integrand. We must allow E to vary over the path of integration. Physically, these integrals specify the work needed to push a test charge along the path of integration, "fighting" the electric field. If you substitute r_2 into the second integral (i.e., $\int \frac{Q_0 r_2}{4\pi\varepsilon_0 R^3} dr$), then you're calculating the work needed to fight a *constant* electric field $\frac{Q_0 r_2}{4\pi\varepsilon_0 R^3}$, not the work needed to fight a *variable* electric field $\frac{Q_0 r}{4\pi\varepsilon_0 R^3}$.

(d) To find the potential at the center, we need not derive any new formulas. In part (c), we already found the formula for the potential inside the ball, a distance r_2 from its center. Just recycle that formula, setting $r_2 = 0$:

$$\begin{aligned} V(0) &= \frac{Q_0}{4\pi\varepsilon_0 R} + \frac{Q_0(R^2 - 0^2)}{2(4\pi\varepsilon_0)R^3} \\ &= \frac{Q_0}{4\pi\varepsilon_0 R} + \frac{1}{2}\frac{Q_0}{4\pi\varepsilon_0 R} \\ &= 2.7\text{ V} + \frac{1}{2}(2.7\text{ V}) \\ &= 4.0\text{ V}. \end{aligned}$$

From this calculation, notice that pushing the particle from infinity to the surface of the ball takes twice as much work as pushing it from the surface to the center.

QUESTION 35-6

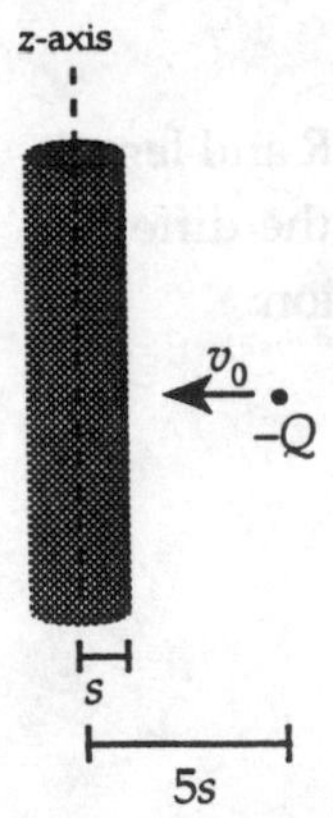

Consider a very long nonconducting charged cylinder of radius s, centered on the z-axis. The cylinder is *much* longer than it is wide. The charge is not uniformly distributed in the cylinder. Instead, the charge density is $\rho = +br$, where r is the radial distance from the z-axis. This charge density applies only between $r = 0$ and $r = s$. Outside $r = s$, the charge density drops to 0.

A small particle of charge $-Q$ and mass m is "shot" directly toward the cylinder, at initial speed v_0, from a distance $5s$ away from the center of the cylinder.

(a) What is the linear charge density of the cylinder? Check your answer against mine before proceeding to part (b).

(b) (*Hard, but an excellent exam-type question*) How fast will the particle be movin when it reaches the edge of the cylinder? Don't use any pre-derived text-book fomulas about charged cylinders. If you didn't complete part (a), use my answer.

(c) (*Challenge problem*) The charged cylinder isn't very viscous. Therefore, the particle can burrow into the cylinder, encountering only a negligible drag force. How fast is it moving when it reaches the center of the cylinder?

(d) Suppose the particle, instead of getting shot directly toward the cylinder, were shot "into the page" (in the above drawing), again from distance $5s$ away from the cylinder's center. With what speed must we shoot the particle so that it traces a circular path around the cylinder?

ANSWER 35-6

(a) First, I'll find the charge q of an arbitrary length L of the cylinder. Then I'll divide that charge by the length, to obtain the linear charge density: $\lambda = q/L$. The "L" had better cancel out.

Since the charge density is not uniform, we can't just multiply it by the volume. (Similarly, when velocity isn't constant, you can't use $\Delta x = vt$.) Instead, we must break the cylinder into infinitesimal "pieces," and calculate the charge of each piece. Then we can add up (integrate) those charges to obtain the total charge.

Because the charge density is radially symmetric, break the charged cylinder into infinitesimally thin cylindrical shells. A given shell has radius R and "width" dR. Notice that R is less than s, the cylinder's radius. In this diagram, I've darkened one of these shells (rings).

END-ON VIEW (looking down the z-axis at the top of the cylinder)

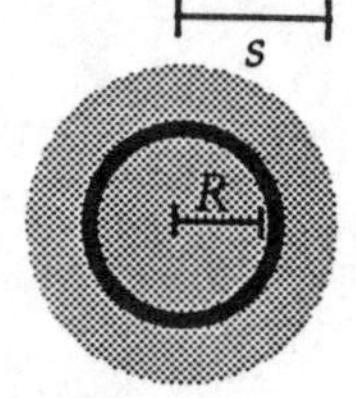

Since the ring has length (circumference) $2\pi R$ and width dR, its "area" is dA = length × width = $2\pi R dR$. But this "ring" actually extends distance L into the page, forming a cylindrical shell. This shell has volume $dV = L dA = 2\pi L R dR$.

To find the infinitesimal charge dq contained in that shell, multiply the charge density (charge per volume) by the volume of the shell:

$$dq = \frac{\text{charge}}{\text{volume}} \times (\text{volume of the shell})$$

$$= \rho dV$$

$$= (bR)(2\pi L R dR)$$

$$= 2\pi b L R^2 dR,$$

where in the third line, I used the given expression for charge density, $\rho = br$.

So far, we've found the charge contained in a *single* infinitesimal shell of radius R and length L. To find q, the total charge of length L of the cylinder, we must add up the dq's of the different shells. The shells range in radius from $R = 0$ to $R = s$. So, those are the limits of integration:

$$q = \int dq = \int_0^s 2\pi b L R^2 dR$$

$$= 2\pi b L \int_0^s R^2 dR$$

$$= 2\pi b L \frac{s^3}{3}.$$

That's the charge of a segment of the cylinder of length L. Divide this charge by L to obtain the linear charge density:

$$\lambda = \frac{\text{charge}}{\text{length}} = \frac{q}{L} = \frac{2}{3}\pi b s^3.$$

The L canceled out.

(b) We're looking for the particle's velocity after it has "fallen" from $r = 5s$ to the surface of the cylinder at $r = s$. By "r," I mean the distance from the center of the cylinder, the z-axis. We can solve using energy conservation. (Forces and old-fashioned kinematics work less well, because the electric force gets stronger as the particle approaches the cylinder, and hence, the acceleration is not constant.)

The positive cylinder attracts the negative particle. So, as the particle approaches the cylinder, it loses potential energy, and therefore gains kinetic energy. To calculate how much speed the particle gains, we'll have to figure out how much potential energy it loses.

By definition, this change in potential energy relates to the change in potential by

$$\Delta U = q\Delta V = (-Q)\Delta V,$$

since the particle has charge $-Q$. And as we saw in the previous problem, $\Delta V = -\int q\mathbf{E}\cdot d\mathbf{s}$. Therefore, to calculate ΔV, the potential difference between $r = 5s$ and $r = s$, we must find the electric field outside the cylinder. Which we can accomplish using Gauss' law.

So, this problem involves many steps of reasoning. Each individual subproblem, though challenging, shouldn't psych you out. The hardest part is formulating the overall game plan and keeping organized. Let me summarize the strategy discussed above.

1) Using Gauss' law, calculate the electric field outside the cylinder.
2) Using $\Delta V = -\int q\mathbf{E}\cdot d\mathbf{s}$, calculate the potential difference between the particle's starting point (at $r = 5s$) and its ending point (at $r = s$).
3) Given that potential difference, use $U = qV$ and energy conservation to solve for the particle's "final" velocity at the surface of the cylinder.

Subproblem 1: Find the electric field outside the cylinder, using Gauss' law.

Using Gauss' law to find an electric field involves many steps. These potential problems are difficult precisely because they can contain multi-step problems "nested" inside other multi-step problems.

To excellent approximation, we can treat the cylinder as infinitely long. In question 34-3, I demonstrated in detail how to derive the electric field generated by an infinitely long charged cylinder. If the following brief presentation skips too many steps, please check out that earlier problem.

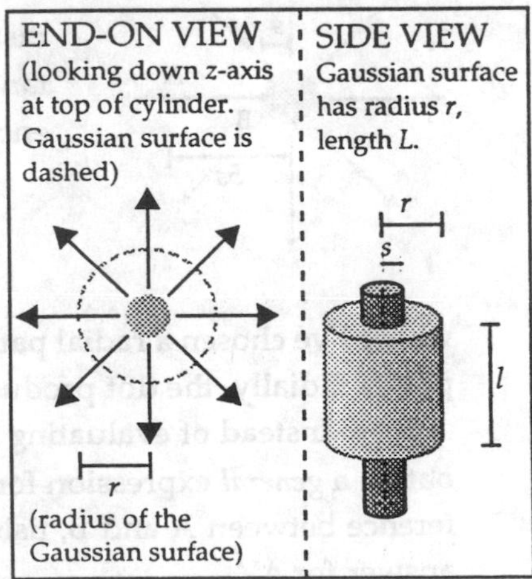

Gauss' law step 1: Draw in field lines.

By symmetry, the field lines point radially outward.

Gauss' law step 2: Draw "good" Gaussian surface.

We want the electric field to have the same strengt over (at least part of) the Gaussian surface. That way, we can pull $\mathbf{E}$ outside the flux integral, $\int q\mathbf{E}\cdot d\mathbf{s}$. For this reason, I've drawn a "Gaussian cylinder," of arbitrary radius r and length L. By symmetry, the field has the same strength at any two points equidistant from the charged cylinder. Therefore, E is constant over the entire side (tube) of my Gaussian surface.

Gauss' law step 3: Use Gauss' law to solve for E.

For this reason, we can pull it outside the flux integral, which simplifies to

$$\oint \mathbf{E}\cdot d\mathbf{A} = \int_{\text{side of the Gaussian cylinder}} E\,dA = E(2\pi rL),$$

where I've used the surface area of the side (tube) of a cylinder. No flux flows through the top or bottom of the Gaussian cylinder, because field lines "skim along" those faces.

Now I'll calculate the enclosed charge, using our part (a) expression for the linear charge density:

$$q_{\text{encl}} = \frac{\text{charge}}{\text{length}} \times (\text{length of the charged cylinder enclosed by Gaussian surface}) = \left(\frac{2}{3}\pi bs^3\right)L.$$

We can now apply Gauss' law:

$$\oint \mathbf{E}\cdot d\mathbf{A} = \frac{q_{\text{encl}}}{\varepsilon_0}$$

$$E(2\pi rL) = \frac{\frac{2}{3}\pi bs^3 L}{\varepsilon_0}.$$

The L's cancel. Isolate E to get

$$E = \frac{bs^3}{3\varepsilon_0 r}.$$

That's the field outside the charged cylinder, a distance r from its center.

Let's step back to reorient ourselves. Finding E is merely subproblem 1 of our game plan, which is to (1) find the electric field outside the charged cylinder, (2) use that field to calculate the potential difference ΔV between the "initial" and "final" points, and then (3) invoke energy conservation to solve the problem. We're ready for subproblem 2.

Subproblem 2: From the electric field, calculate the relevant potential difference.

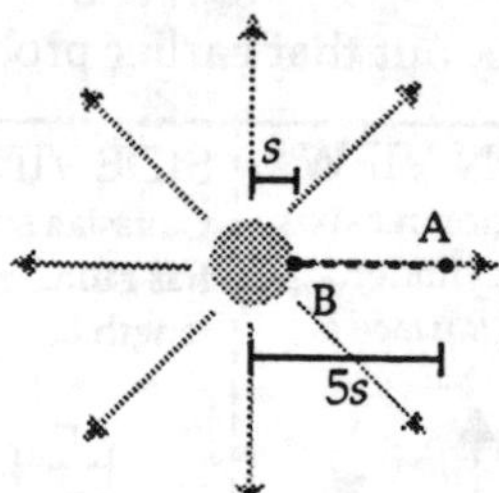

Let me introduce points A and B. The particle starts at point A, a distance $5s$ from the center of the cylinder, and ends up at point B, a distance s from the center of the cylinder. We can find the potential difference between those points by evaluating the definite integral

$$\Delta V = -\int_A^B \mathbf{E}\cdot d\mathbf{s} = -\int_{5s}^{s} E\,dr,$$

where I've chosen a radial path of integration, the dashed line in the diagram. Since the electric field points radially, the dot product $\mathbf{E}\cdot d\mathbf{r}$ reduces to a regular product.

Instead of evaluating this definite integral, you could take the indefinite integral $-\int \mathbf{E}\cdot d\mathbf{s}$ to obtain a *general* expression for the potential, $V(r)$. Given $V(r)$, you could then find the potential difference between A and B, using $\Delta V = V_A - V_B = V(5s) - V(s)$. These two techniques yield the same answer for ΔV.

I'll take the indefinite integral. Be sure to use the general expression for $E(r)$, instead of plugging in a *particular* r (such as s or $5s$), because the integral must "sample" the electric field at different r's.

$$V(r) = -\int \mathbf{E}\cdot d\mathbf{s} = -\int E dr$$

$$= -\int \frac{bs^3}{3\varepsilon_0 r} dr$$

$$= -\frac{bs^3}{3\varepsilon_0}\ln(r).$$

I set the integration constant equal to zero. Notice that the potential decreases, i.e., becomes a bigger and bigger negative number, as you get farther away from the charged cylinder. So, a positive particle would fall *away* from the charged cylinder. But a negative particle falls toward it.

Subproblem 3: Using the potential from step 2, invoke energy conservation.

I'll equate the total energy at point A to the total energy at point B. Recall that $U = qV$, where q denotes the charge acted upon by the electric field, not the charge creating that electric field. So here, "q" is actually $-Q$.

$$U_A + K_A = U_B + K_B$$

$$-QV_A + \frac{1}{2}mv_0^2 = -QV_B + \frac{1}{2}mv_B^2,$$

and hence, $v_B = \sqrt{v_0^2 + \frac{2Q(V_B - V_A)}{m}}$, proving that the potential *difference* between B and A is all that matters.

In step 2, we found that $V(r) = -\frac{bs^3}{3\varepsilon_0}\ln(r)$. So, $V_A = -\frac{bs^3}{3\varepsilon_0}\ln(5s)$, and $V_B - \frac{bs^3}{3\varepsilon_0}\ln(s)$. Therefore,

$$\Delta V = V_B - V_A = -\frac{bs^3}{3\varepsilon_0}\ln(s) + \frac{bs^3}{3\varepsilon_0}\ln(5s)$$

$$= \frac{bs^3}{3\varepsilon_0}[\ln(5s) - \ln(s)]$$

$$= \frac{bs^3}{3\varepsilon_0}\ln\left(\frac{5s}{s}\right)$$

$$= \frac{bs^3}{3\varepsilon_0}\ln(5),$$

the same expression you'd obtain by evaluating the definite integral $-\int_{5s}^{s} E dr$. Substitute this potential difference into the above expression for v_B, to get

$$v_B = \sqrt{v_0^2 + \frac{2Q(V_B - V_A)}{m}} = \sqrt{v_0^2 + \frac{2Q\left(\frac{bs^3}{3\varepsilon_0}\ln 5\right)}{m}}$$

(c) We need to find how much the particle speeds up as it burrows from the edge of the cylinder (point B) to the center of the cylinder, which I'll call point C. Once again, invoke energy conservation. While "falling" from the edge to the middle of the cylinder, the particle loses potential energy, and therefore gains kinetic energy. To figure out the potential energy loss, we can find the potential difference between points B and C, ΔV_{BC}. To calculate this potential difference, we must integrate over the electric field *inside* the cylinder: $\Delta V_{BC} = -\int_B^C E dr$. And to find the field inside the cylinder, we can use Gauss' law.

Let me summarize the strategy I just laid out. It replicates the game plan from part (b), except now we're dealing with the field *inside* the charged cylinder.

> 1) Using Gauss' law, calculate the electric field *inside* the charged cylinder.
> 2) Using $\Delta V = -\int q\mathbf{E}\cdot d\mathbf{s}$, calculate the relevant potential difference between the "starting" and "ending" points, by which I mean points B and C.
> 3) Then use energy conservation to solve for the particle's "final" velocity at point C.

A common mistake is to recycle the potential formula from part (b), $V = -\frac{bs^3}{3\varepsilon_0}\ln(r)$. But we found that formula by integrating over the field *outside* the cylinder. Since the field *inside* the cylinder takes a different form, the potential comes out different, too.

Here goes . . .

Subproblem 1: Find electric field inside cylinder.

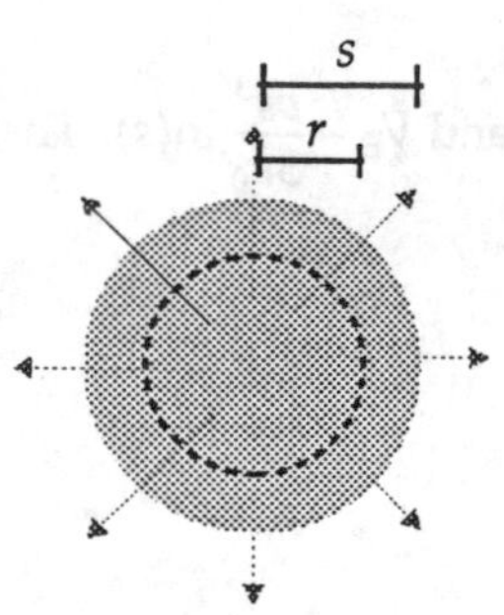

In this "end-on" view, we're looking down at the top of the cylinder. The dashed Gaussian surface is actually a cylinder extending distance *l* into the page.

Gauss' law step 1: Draw in field lines.

The field lines point radially outward from the center.

Gauss' law step 2: Draw "good" Gaussian surface.

Again by symmetry, the electric field has the same strength at any two points equidistant from the center. So the field strength stays constant over the entire side (tube) of this "Gaussian cylinder" inside the charged cylinder.

Gauss' law step 3: Solve for the electric field using Gauss' law.

Therefore, the flux integral in Gauss' law simplifies to

$$\oint \mathbf{E}\cdot d\mathbf{A} = \int_{\text{side of the Gaussian cylinder}} E\,dA = E(2\pi rl),$$

where I've used the surface area of the side (tube) of a cylinder. Notice that *r*, the radius of the Gaussian cylinder, is smaller than *s*, the radius of the charged cylinder.

To finish using Gauss' law, we must calculate the charge enclosed by my Gaussian surface, a cylinder of length *l* and radius *r*. To figure this out, divide the charged cylinder into infinitesimal cylindrical shells, as shown in part (a). When adding up the dq's of these shells, include *only* the shells with radius less than or equal to r. That way, we'll find the charge *inside* the Gaussian surface.

In part (a), we found the charge dq contained in one these infinitesimal shells. Let me briefly review the reasoning. A given shell, drawn as a black circle in this diagram, has radius R and "width" dR. Since the ring has length (circumference) $2\pi R$ and width dR, its area is $dA = \text{length} \times \text{width} = 2\pi R dR$. But this "ring" actually extends distance l into the page, forming a cylindrical shell of volume $dV = l dA = 2\pi l R dR$. Therefore, the shell contains charge

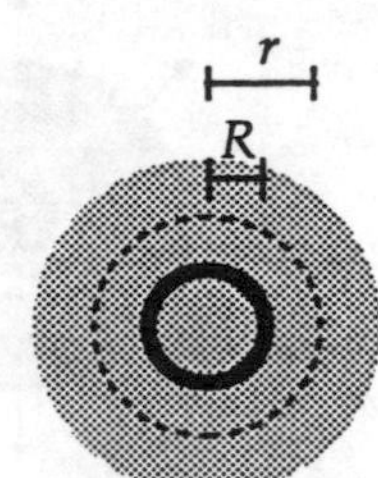

The blackened cylinrical shell has radius R and width dR.

$$\begin{aligned} dq &= \frac{\text{charge}}{\text{volume}} \times (\text{volume of the shell}) \\ &= \rho dV \\ &= (bR)(2\pi l R dR) \\ &= 2\pi b l R^2 dR, \end{aligned}$$

where in the third line I used the given expression for charge density, $\rho = bR$.

To find the charge enclosed by my Gaussian surface, we must add up the dq's for the shells *inside* the dashed line on the diagram. In other words, we want the shells whose radii range from $R = 0$ to $R = r$. *Don't* integrate all the way out to the edge of the charged cylinder at $R = s$.

$$\begin{aligned} q_{\text{encl}} &= \int_{R=0}^{R=r} dq = \int_0^r 2\pi b l R^2 dR \\ &= 2\pi b l \frac{r^3}{3}. \end{aligned}$$

Now we we've simplified the flux integral and calculated the enclosed charge, we can apply Gauss' law:

$$\begin{aligned} \oint \mathbf{E} \cdot d\mathbf{A} &= \frac{q_{\text{encl}}}{\varepsilon_0} \\ E(2\pi r l) &= \frac{\frac{2}{3}\pi b r^3 l}{\varepsilon_0}. \end{aligned}$$

Solve for E, the field inside the charged cylinder, to get

$$E = \frac{br^2}{3\varepsilon_0}.$$

This differs from the field outside the charged cylinder, which drops off as $1/r$. Inside the cylinder, the electric field increases as you get farther from the center, because there's more charge "inside" you when you get farther from the center.

So far, we've calculated the electric field inside the charged cylinder. That was subproblem 1. Now we must (2) use that field to calculate the potential difference between the edge and middle of the cylinder, and then (3) invoke energy conservation to find how fast the particle is moving when it reaches the center. Time for subproblem 2.

Step 2: From the electric field, calculate the potential.

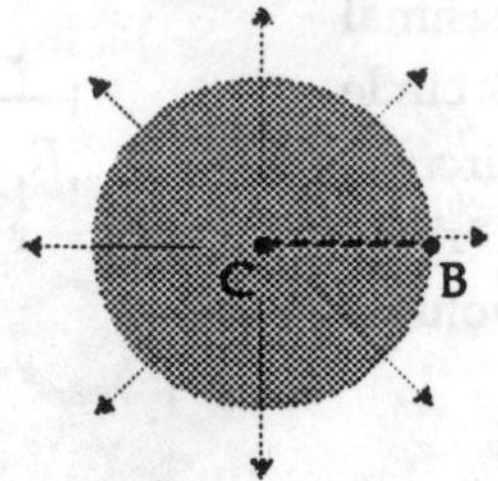

In this problem, we can consider particle's trip from point B to point C, since we already know the velocity at B. I'll find the potential difference between those two points by taking the definite integral along the radial path I've drawn:

$$V_C - V_B = \Delta V = -\int_B^C \mathbf{E}\cdot d\mathbf{s}$$

$$= -\int_s^0 E dr$$

$$= -\int_s^0 \frac{br^2}{3\varepsilon_0} dr$$

$$= -\frac{br^3}{9\varepsilon_0}\bigg|_s^0$$

$$= -\frac{bs^3}{9\varepsilon_0}.$$

You obtain this same expression by taking the indefinite integral $\int \mathbf{E}\cdot d\mathbf{s}$ to get $V(r) = -\int E dr = -\frac{br^3}{9\varepsilon_0}$, and then subtracting $V(s)$ from $V(0)$. Notice that $V(r)$ inside the cylinder differs from $V(r)$ outside the cylinder.

Advanced theoretical digression. Our expressions for $V(r)$ inside and outside the cylinder should agree at $r = s$, the boundary between "inside" and "outside." In fact, those formulas for $V(s)$ disagree, but only because I chose my integration constants sloppily. By tweaking the integration constants, we could make sure our expressions for $V(r)$ inside and outside the cylinder agree at the boundary. But we need not worry about this, because we're dealing *separately* with the potential difference outside the cylinder (point A to point B) and inside the cylinder (point B to point C).

Subproblem 3: Invoke energy conservation.

Set the total energy at point B equal to the total energy at point C:

$$U_B + K_B = U_C + K_C$$

$$-QV_B + \frac{1}{2}mv_B^2 = -QV_C + \frac{1}{2}mv_C^2,$$

and hence,

$$v_C = \sqrt{v_B^2 + \frac{2Q(V_C - V_B)}{m}}$$

$$= \sqrt{v_B^2 + \frac{2Q\left[0 - \left(-\frac{bs^3}{9\varepsilon_0}\right)\right]}{m}} \quad \text{[using subproblem 2]}$$

$$= \sqrt{v_0^2 + \frac{2Q\left(\frac{bs^3}{3\varepsilon_0}\ln 5\right)}{m} + \frac{2Q\frac{bs^3}{9\varepsilon_0}}{m}},$$

where in the last line, I used our part (a) answer for v_B.

(d) Welcome to an old-fashioned circular motion problem, something you probably didn't expect to encounter here. When moving in a circle of radius r at speed v, a particle experiences a radial acceleration $a_{\text{radial}} = v^2/r$. By "radial," I mean towards the center of the circle. Therefore, by Newton's 2nd law, the particle must feel a net radial force $\sum F_{\text{radial}} = ma_{\text{radial}} = mv^2/r$.

In this problem, we know the radius of the circle, $r = 5s$. We can also find the net radial force. It's the attractive electric force exerted on the particle by the charged cylinder. This force has magnitude $F = |-QE| = QE$, where E is the field at distance $r = 5s$ from the cylinder's center. In part (b), we found that $E_{\text{outside cylinder}} = \frac{bs^3}{3\varepsilon_0 r}$. Tie all this together to get

$$\sum F_{\text{radial}} = m\frac{v^2}{r}$$

$$QE = m\frac{v^2}{r}$$

$$Q\frac{bs^3}{3\varepsilon_0 r} = m\frac{v^2}{r}.$$

We're solving for v. Notice that the r's cancel! No matter how far the particle is from the charged cylinder, we must give it the same speed to achieve circular orbit. (This peculiar result applies only to forces that fall off as $1/r$.) Solving for that speed yields

$$v = \sqrt{\frac{Qbs^3}{3\varepsilon_0 m}}.$$

QUESTION 35-7

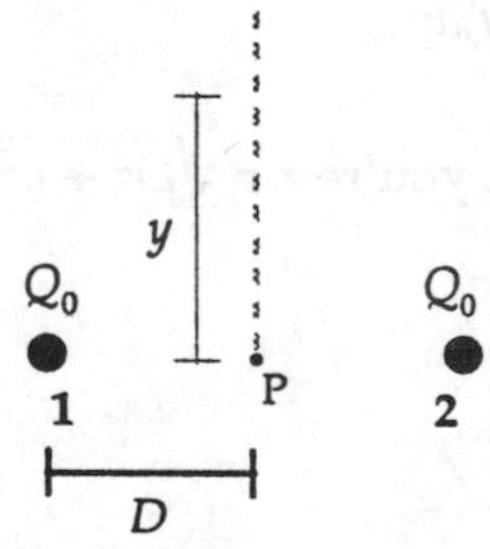

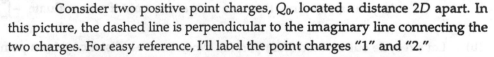

Consider two positive point charges, Q_0, located a distance $2D$ apart. In this picture, the dashed line is perpendicular to the imaginary line connecting the two charges. For easy reference, I'll label the point charges "1" and "2."

As usual, take the "zero point" of potential to be at infinity. In other words, set $V = 0$ at $r = \infty$.

In this problem, you may reuse any formulas previously encountered in this study guide, without rederiving them.

(a) What is the potential at point P, midway between the two charges?

(b) What is the potential at the point on the dashed line a distance y from point P?

(c) How could you use your part (b) answer to calculate the electric field on the dashed line a distance y from point P? Try it!

ANSWER 35-7

(a) I see two ways of approaching this, both of which rely on the superposition principle. On the one hand, we could (vectorially) add the electric *fields* generated by the two point charges, and then integrate that field over distance to get potential at point P:

$$V_P = -\int_{\infty}^{\text{point P}} \mathbf{E}\cdot d\mathbf{s}.$$

In question 33-4, we found $E(y)$, the electric field on the dashed line, an arbitrary distance y from point P. That field points in the y-direction. So, $V_P = -\int_{\infty}^{\text{point P}} \mathbf{E}\cdot d\mathbf{s} = -\int_{\infty}^{0} E(y)dy$.

But it's easier to apply the superposition principle *directly* to potential:

$$V_P = V_{\text{P due to charge 1}} + V_{\text{P due to charge 2}}.$$

Intuitively, this equation says that the total work needed to push a 1-coulomb charge from infinity to point P is simply the work you exert fighting the field generated by charge 1, plus the work you exert fighting the field generated by charge 2.

Adding potentials is *easier* than adding fields, because potentials are *scalars*, while fields are vectors. Furthermore, as we saw in question 35-5a above, a point charge Q_0 generates a potential

$$V_{\text{point charge}} = -\int E_{\text{point charge}}\, dr = -\int \frac{Q_0}{4\pi\varepsilon_0 r^2}dr = \frac{Q_0}{4\pi\varepsilon_0 r}.$$

Since point P sits the same distance $r = D$ from both point charges,

$$\begin{aligned} V_P &= V_{\text{P due to charge 1}} + V_{\text{P due to charge 2}} \\ &= \frac{Q_0}{4\pi\varepsilon_0 D} + \frac{Q_0}{4\pi\varepsilon_0 D} \\ &= \frac{Q_0}{4\pi\varepsilon_0 D}. \end{aligned}$$

As you can confirm, the same answer pops out if you evaluate $-\int_{\infty}^{0} E(y)dy$.

(b) Let's recycle our method from part (a). When you're a distance y from point P, you're $r = \sqrt{D^2+y^2}$ from either point charge. Therefore,

$$\begin{aligned} V &= V_{\text{due to charge 1}} + V_{\text{due to charge 2}} \\ &= \frac{Q_0}{4\pi\varepsilon_0\sqrt{D^2+y^2}} + \frac{Q_0}{4\pi\varepsilon_0\sqrt{D^2+y^2}} \\ &= \frac{Q_0}{2\pi\varepsilon_0\sqrt{D^2+y^2}}. \end{aligned}$$

(c) You can find an unknown potential from a known electric field by integrating: $V = -\int \mathbf{E}\cdot d\mathbf{s}$. Inversely, you can find an unknown electric field from a known potential by differentiating:

$$E_y = -\frac{\partial V}{\partial y}$$

$$= -\frac{\partial}{\partial y}\left(\frac{Q_0}{2\pi\varepsilon_0\sqrt{D^2+y^2}}\right)$$

$$= -\frac{Q_0}{2\pi\varepsilon_0}\left[-\frac{1}{2}\frac{1}{(D^2+y^2)^{3/2}}\right](2y)$$

$$Q_0y$$

which agrees with our answer to question 33-4b.

By switching back and forth between potentials and fields, you can simplify long problems. Mathematically, since $V = -\int q\mathbf{E}\cdot d\mathbf{s}$, and since differentiation is the inverse (opposite) of integration, it follows that you can find the field by differentiating the potential: $E_x = -\partial V/\partial x$, etc. But don't settle for this purely mathematical understanding. The formula $E_x = -\partial V/\partial x$ makes *physical* sense, too. Here's why. When the potential changes a lot over a small distance—i.e., when $\partial V/\partial x$ is large—then it takes a lot of work to push a charged particle over a short distance. Therefore, the electric field you're "fighting" must be strong. In summary, when $\partial V/\partial x$ is big, the field E_x must be large as well. So, the formula $E_x = -\partial V/\partial x$ makes sense. See question 35-3c for a deeper discussion.

QUESTION 35-8

Two large square conducting sheets are hooked to a 1.5-volt battery, with the lower plate attached to the negative terminal. So, the potential difference between the upper and lower plates is 1.5 volts. And the lower plate has the lower potential.

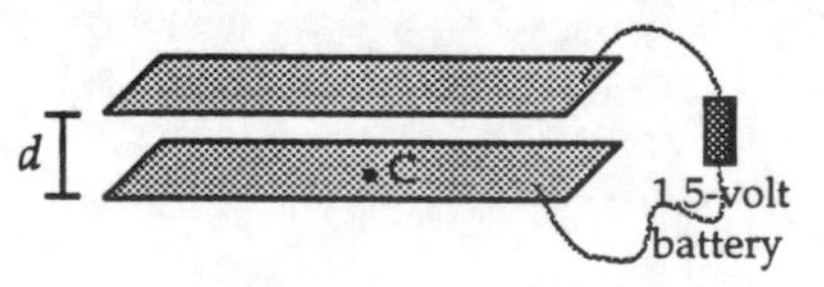

The plates, both of $A = 0.50$ square meters, are fixed a distance $d = 0.10$ meters apart.

A particle of charge $q = 4.0\ \mu C$ and mass $m = 6.0 \times 10^{-6}$ kg is fired straight upwards from a "particle gun" at point C on the lower plate.

What is the smallest initial speed with which the particle can be fired, so that it (barely) reaches the upper plate? In this problem, do *not* neglect gravity. Hint: You might not use all the numbers provided.

ANSWER 35-8

Let's begin by figuring out whether the electric field points up or down. That way, we'll know whether the field "helps" or "hinders" the particle as it travels upward.

The lower plate has lower potential. Therefore, since an object "wants" to lower its potential energy, a positive particle "wants" to move towards the lower plate. In other words, the electric field points

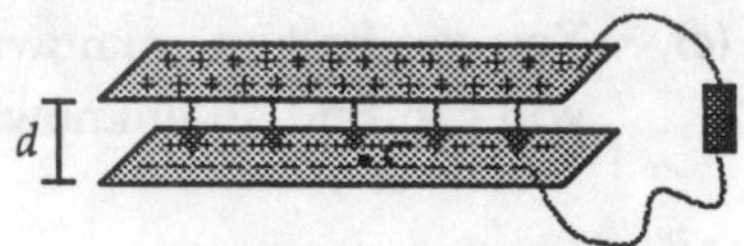

The electric field points downward.

downward. The battery must have given the upper plate a positive charge, and the lower plate a negative charge, so that the field lines point downward.

As you know, the gravitational field *also* points downward. Therefore, as the particle travels upward, both gravity *and* the electric force slow it down. Since these forces happen to be constant, we can solve using forces and acceleration and constant-acceleration kinematics. But let's practice using energy conservation instead, because it works even when the forces *aren't* constant.

First, I'll think about the gravitational and electric potential energy. As the particle rises, its gravitational potential energy increases. From last semester, recall that $U_{grav} = mgh$. If we set $h = 0$ at the lower plate, then the particle begins with no gravitational potential energy, but ends up with $U_{grav} = mgd$ at the upper plate.

While rising, the particle also gains electrostatic potential energy. Recall that $U_{elec} = qV$. Here, the upper plate's potential is 1.5 volts higher than the lower plate's potential. So, if we set the "zero point" at the lower plate ($V_{lower} = 0$), then the upper plate has $V_{upper} = 1.5$ volts. Of course, all that matters is $\Delta V = V_{upper} - V_{lower}$, which equals 1.5 volts no matter where you set the zero point.

What about kinetic energy? At the lower plate, it's $K_0 = \frac{1}{2}mv_0^2$, where v_0 is the unknown initial velocity. As the particle travels toward the upper plate, the electric and gravitational forces slow it down. We want to know the *smallest* v_0 needed to ensure that the particle reaches the upper plate. Well, if we make v_0 as small as possible, the particle *barely* reaches the upper plate. In other words, the particle is hardly moving when it reaches the upper plate. So, at the upper plate, $v_f \approx 0$. Therefore, the final kinetic energy is $K_f = 0$.

Tie all this together to get

Initial energy at lower plate = "Final" energy at upper plate

$$K_0 + U_{0\,grav} + U_{0\,elec} = K_f + U_{f\,grav} + U_{f\,elec}$$

$$\frac{1}{2}mv_0^2 + 0 + qV_{lower} = 0 + mgd + qV_{upper}.$$

Solve for v_0 to get

$$v_0 = \sqrt{2gd + \frac{2q(V_{upper} - V_{lower})}{m}}$$

$$= \sqrt{2(9.8\,m/s^2)(0.10\,m) + \frac{2(4.0 \times 10^{-6}\ C)(1.5\ V - 0)}{6.0 \times 10^{-6}\ kg}}$$

$$= 2.0\,m/s.$$

QUESTION 35-9

A square consists of four positively-charged rods, each of length l and charge Q. On each rod, the charge is uniformly distributed. In this problem, neglect gravity.

(a) (*Very hard*) How much work would it take to move a small particle of charge $+q$ to the center of the square, if the particle starts *very* far away from the square? If you can't complete all the math, that's OK; but set up an equation that could be solved for the work.

(b) About how much work is required to move a small particle of charge $+q$ to a point a distance $100l$ away from the center of the square (on the axis perpendicular to the plane of the square), if the particle starts *very* far away. Hint: You may use approximations, if you justify them adequately. You can solve in under 3 minutes.

ANSWER 35-9

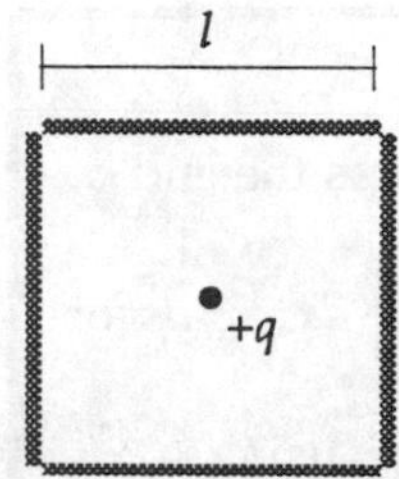

As you know from last semester, begin with pictures, not formulas. The particle ends up at the center of the square. While pushing the particle to this point, you must fight against the electric fields generated by the four rods. Here, each rod is identical, and the particle ends up the same distance from each rod. Therefore, the work needed to push the particle to the center of the square is *four* times the work needed to push it a perpendicular distance $l/2$ from the midpoint of *one* rod. Intuitively, each rod generates an electric field that repels the particle. Because you're fighting *four* of these fields, you must work four times as hard. So, we can solve the "one-rod" problem, and then multiply by four.

To confirm that this "trick" works, think about the relationship between work and potential. If you expend 20 joules of work pushing a charge against an electric field, then you've "stored" 20 joules of electrostatic potential energy in the charge: $\Delta U = W_{\text{by you}}$. So, by pushing the positive particle to the center of the square, you raise its potential energy, and hence, its potential:

= 4

Work to push particle from ∞ to center of square = 4 times the work needed to push it from ∞ to this position (with one rod)

$$
\begin{aligned}
W &= \Delta U \\
&= q\Delta V_{\text{due to all four rods}} \\
&= 4q\Delta V_{\text{due to one rod}}.
\end{aligned}
\tag{1}
$$

In the last step, I used the superposition principle:

$$V_{\text{due to four rods}} = V_{\text{due to rod 1}} + V_{\text{due to rod 2}} + V_{\text{due to rod 3}} + V_{\text{due to rod 4}} = 4V_{\text{any one of the rods}}.$$

In general, $V_{\text{due to rod 1}}$ would *not* equal $V_{\text{due to rod 2}}$. But here, the rods are identical, the particle starts out the same distance from each rod, and the particle ends up the same distance from each rod. Because of this symmetry, each rod contributes the same potential. Eq. (1) confirms our earlier conclusion that the work needed to fight the field generated by the square is *four* times the work needed to fight the field generated by one rod.

The particle starts *very* far away, where the potential is essentially zero. So, particle has initial potential $V_0 = 0$. Therefore, the potential change is simply the final potential: $\Delta V = V_f - V_0 = V_f$. For instance, if the particle gains $\Delta V = 10$ volts of potential, then its final potential is $V_f = 10$ volts, because it starts with *no* potential ($V_0 = 0$). Given this insight, we can rewrite Eq. (1) as

$$W = 4qV_{\text{due to one rod}}. \tag{1'}$$

So, to complete this problem, we just need to find the potential generated by one rod of length l and charge Q at a perpendicular distance $l/2$ from its midpoint.

I see two ways to proceed. We could calculate the electric field generated by one rod, and then integrate over it: $\Delta V = -\int_{r=\infty}^{r=l/2} \mathbf{E} \cdot d\mathbf{s}$. But it's easier to calculate the rod's potential "directly" using the superposition principle. Here's how. In your mind, divide the rod into a bunch of tiny "pieces" that you can treat as point charges. We know how to write down the potential generated by a single point charge. Therefore, to find the potential created by the whole rod, just sum up (i.e., integrate over) the potentials created by each of those pieces. You employed a similar strategy back in Chapter 33, to calculate the electric field generated by a continuous charge distribution. The same strategy applies here, except we'll sum up the potential instead of the electric field. Let me summarize.

Finding the potential due to a continuous charge distribution

1) Blacken an arbitrary infinitesimal "piece" of the rod. Don't pick a special point such as the end or the middle.

2) Write an expression for dq, the infinitesimal charge of that "piece," in terms of dx or ds or whatever.

3) Write down the infinitesimal potential generated by that infinitesimal piece of charge, using $dV = \frac{dq}{4\pi\varepsilon_0 r}$. (That's the formula for the potential produced by a point charge.)

4) Integrate over all the "pieces" of the rod, to obtain the total potential $V_{\text{due to one rod}}$.

(a) Now I'll implement this strategy, to find the potential at the center of the square generated by *one* rod. The center of the square is located a distance $l/2$ from the midpoint of a rod. Given this $V_{\text{due to one rod}}$, we can substitute it into Eq. (1'), to solve for the work required to push the particle to the center of the square.

Step 1: Blacken an infinitesimal piece of the charge distribution.

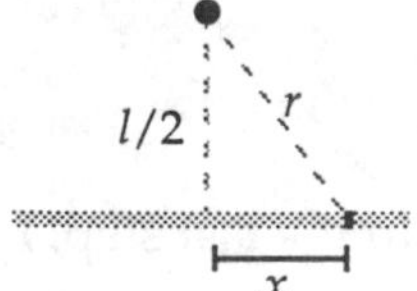

I've blackened a tiny piece of the rod. As always, I've labeled the distance from that piece to a reference point, in this case the midpoint of the rod. This piece has horizontal length dx.

Step 2: Write an expression for dq.

Each rod carries charge Q uniformly distributed over length l. So, the linear charge density is l = charge/length = Q/l. Therefore, the blackened piece of rod contains charge

$$dq = \frac{\text{charge}}{\text{length}} \times (\text{length of blackened piece}) = \lambda dx = \frac{Q}{l}dx.$$

Step 3: Write down the infinitesimal potential generated by that piece of charge.

From Pythagorean theorem, this blackened piece of rod sits a distance $r = \sqrt{x^2 + (l/2)^2}$ from the center of the square. So,

$$dV = \frac{dq}{4\pi\varepsilon_0 r} = \frac{\frac{Q}{l}dx}{4\pi\varepsilon_0\sqrt{x^2+(l/2)^2}}.$$

Step 4: Integrate over the charge distribution to obtain V.

Since I measured "x" from the midpoint of the rod, the rod extends from $x = -l/2$ to $x = l/2$. When integrating over dV, do *not* break the potential into x- and y-components. V is a scalar; it has no direction. That's why calculating potentials is easier than calculating electric fields.

$$\begin{aligned} V_{\text{due to one rod}} &= \int dV \\ &= \int_{-l/2}^{l/2} \frac{\frac{Q}{l}dx}{4\pi\varepsilon_0\sqrt{x^2+(l/2)^2}} \\ &= \frac{Q}{4\pi\varepsilon_0 l}\int_{-l/2}^{l/2}\frac{dx}{\sqrt{x^2+(l/2)^2}}. \end{aligned}$$

The integral turns out to be tricky. But once you solve it, just substitute the result into Eq. (1′), to find the work needed to push the particle from very far away to the center of the square: $W = 4qV_{\text{due to one rod}}$. Remember, the factor of 4 reflects the fact that fighting the field generated by the whole square demands four times as much work as fighting the field produced by a single rod.

On a test, many instructors would supply a relevant table of integrals, or would allow you to leave the integral unevaluated. Find out your instructor's policy. For the record, I'll evaluate the integral.

Funky calculus starts here. Staring at the denominator of this integral, experienced physicists and mathematicians know to look for a trig substitution. Here, we must use hyperbolic trig functions. If this doesn't sound familiar from math class, skip to the next problem.

Use the substitution

$$x = (l/2)\sinh\theta.$$

Then $dx = (l/2)\cosh\theta\, d\theta$.

To find the upper limit of the trig-substituted integral, set $x = l/2 = (l/2)\sinh\theta$. So, the upper limit is the θ such that $l/2 = (l/2)\sinh\theta$, i.e., such that $\sinh\theta = 1$. So, $\theta = \sinh^{-1}(1) = 0.88$. By equivalent reasoning, the lower limit of integration is $\theta = \sinh^{-1}(-1) = -0.88$.

Putting all this together, I get

$$\begin{aligned} \int_{-l/2}^{l2}\frac{dx}{\sqrt{x^2+(l/2)^2}} &= \int_{\sinh^{-1}(-1)}^{\sinh^{-1}(+1)}\frac{(l/2)\cosh\theta\, d\theta}{\sqrt{(l/2)^2\sinh^2\theta+(l/2)^2}} \\ &= \int_{\sinh^{-1}(-1)}^{\sinh^{-1}(+1)}\frac{(l/2)\cosh\theta\, d\theta}{(l/2)\sqrt{\sinh^2\theta+1}} \\ &= \int_{\sinh^{-1}(-1)}^{\sinh^{-1}(+1)}\frac{\cosh\theta\, d\theta}{\sqrt{\sinh^2\theta+1}} \end{aligned}$$

$$= \int_{\sinh^{-1}(-1)}^{\sinh^{-1}(+1)} \frac{\cosh\theta\, d\theta}{\cosh\theta} \qquad [\text{since } \cosh^2\theta - \sinh^2\theta = 1]$$

$$= \int_{\sinh^{-1}(-1)}^{\sinh^{-1}(+1)} d\theta$$

$$= \sinh^{-1}(+1) - \sinh^{-1}(-1)$$

$$= 2\sinh^{-1}(+1)$$

$$= 1.76.$$

End of funky calculus.

Substitute this result into the above formula for W to get

$$W = 4qV_{\text{due to one rod}} = 4q\frac{Q}{4\pi\varepsilon_0 l}\int_{-l/2}^{l/2}\frac{dx}{\sqrt{x^2+(l/2)^2}} = \frac{qQ}{\pi\varepsilon_0 l}2\sinh^{-1}(+1).$$

(b) You might think that we need to repeat the strategy of part (a). To get an *exact* answer, you would. But you can obtain a 99.99% accurate answer by reasoning as follows:

When you stand a distance $100l$ away from a square of length l, the square looks very small. For instance, if you look at a meter stick from 100 meters away (the length of a soccer field), the meter stick looks tiny. This goes to show that we can treat the square as a point charge.

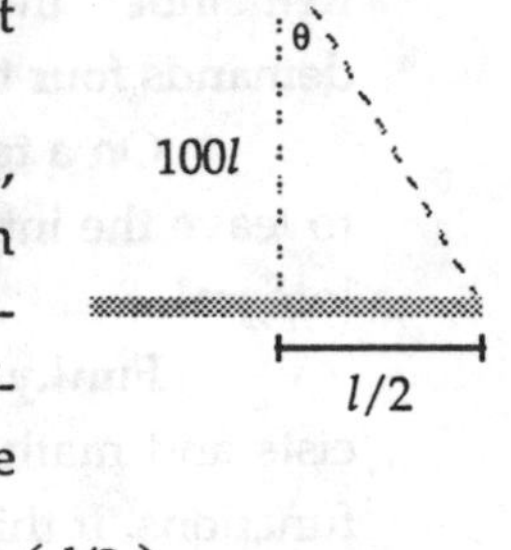

Not convinced? Then consider the following diagram of a single rod, *not* drawn to scale. The black point sits a perpendicular distance $r = 100l$ from the rod's mid point. By Pythagorean theorem, the rod's end is located a distance $r = \sqrt{(100l)^2 + (l/2)^2} = (100.001)l$ from the black point, imperceptibly different from $100l$. So, every point on the rod is almost exactly $100l$ from the black point.

Furthermore, the angle subtended by (half) the rod is tiny: $\theta = \tan^{-1}\left(\frac{l/2}{100l}\right) = 0.3°$. For this reason, to excellent approximation, we can treat the rods as point charges.

In summary, when you're a distance $r = 100l$ from the square, it "behaves" like a point charge of charge $4Q$. So, at that distance, the square generates potential

$$V_{\text{at } 100l} = \frac{4Q}{4\pi\varepsilon_0(100l)}.$$

By contrast, infinitely far from the square, the potential is $V_0 = 0$, as explained above. So, pushing a particle of charge q from infinitely far away to $100l$ away from the square requires work

$$W = \Delta U$$

$$= q\Delta V$$

$$= q(V_{\text{at } 100l} - V_0)$$

$$= q\left(\frac{4Q}{4\pi\varepsilon_0(100l)} - 0\right)$$

$$= \frac{qQ}{100\pi\varepsilon_0 l}.$$

CHAPTER 36

Capacitors and Dielectrics

QUESTION 36-1

(Before attempting this problem, please read your textbook's introductory sections on capacitance and dielectrics.) Usually, a capacitor consists of two conducting objects with equal and opposite charge. But a single conductor can also be considered a capacitor. Think of its "partner" object as being located infinitely far away.

Consider the following conducting objects. Black represents a solid conductor. Objects 1 and 5 are hollow conductors. Gray represents a nonconducting dielectric material.

Which objects have higher capacitance, and which have lower capacitance? Rank them from highest to lowest capacitance, and briefly justify your rankings. Don't get bogged down in formulas. Answer by thinking *physically* about what "capacitance" means, and what a dielectric material does.

(1) Hollow sphere

(2) Solid sphere

(3) Hollow sphere filled with dielectric

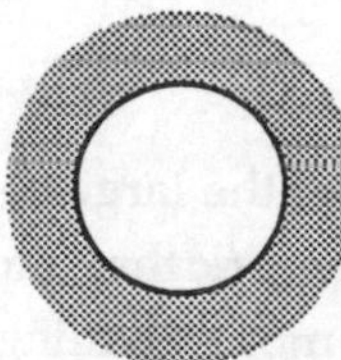
(4) Sphere enveloped by dielectric

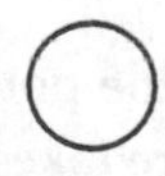
(5) Smaller hollow sphere

ANSWER 36-1

To answer, we must develop a *physical* understanding of "capacitance." Roughly speaking, capacitance is the capacity (ability) of a system to hold charge. If you can pile lots of charge onto an object with minimal effort, then it has high capacitance. If it's hard to pile on lots of charge, then the object has low capacitance. In the next few paragraphs, I will show intuitively that "high capacitance" means we can pile lots of charge onto the object without increasing its potential very much. By contrast, if you place the same charge on a low-capacitance object, its potential increases drastically, making it difficult to pile on any additional charge.

Let's think about why it's "hard" to pile charge onto an object. Imagine starting with an uncharged conducting sphere, and pushing charges onto it, one coulomb at a time. The charges come from an "ion reservoir" very far away. Suppose you've already piled 9 coulombs onto the sphere. How hard is it to push the 10th coulomb onto the sphere? Well, those first 9 coulombs give the sphere a positive charge. As a result, the sphere generates an electric field that repels the 10th electron. For this reason, you must expend work "fighting the field" to push the 10th coulomb onto the sphere.

How much work? As discussed in Chapter 35, the work needed to move an extra coulomb of charge onto the sphere from very far away is the *potential* of the sphere, V. So, if the sphere with 9 coulombs generates a low potential, then it's comparatively easy to push the 10th coulomb onto the sphere. By contrast, if the 9-coulomb sphere produces a high potential, then it's relatively difficult to push the 10th coulomb of charge onto the sphere. In summary: If we place equal charge on two objects, then the object with lower potential has higher capacitance, because it's more "willing" to accept additional charge.

This theoretical discussion gives us a way to address the problem. Pretend that each of the five capacitors carries the same charge. Then ask which objects have higher potential, and which objects have lower potential. The ones with higher potential have lower capacitance, because they're less willing to "take on" additional charges.

Trying deal with all five objects at once would psych me out. Instead, let's compare them two at a time.

I'll start by comparing the bigger hollow sphere (1) to the smaller hollow sphere (5). When both spheres hold equal charge, the charge is more concentrated (tightly packed) on the smaller sphere. In other words, the smaller sphere carries a higher surface charge density, because the charge has less area over which to spread out. Since more concentrated charges generate a stronger electric field, the field near the small sphere is bigger than the field near the large sphere. Therefore, the potential of the small sphere,

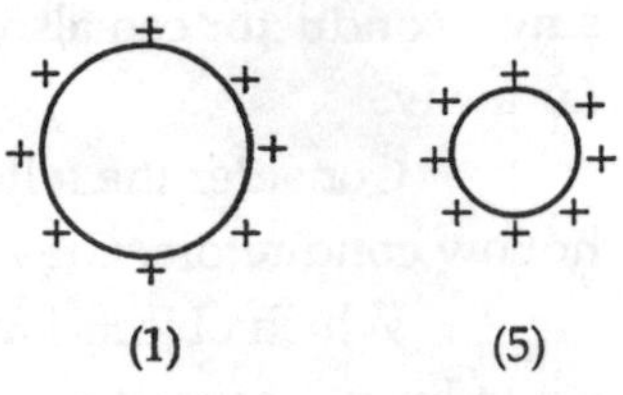

$$V_{sphere} = -\int_{\infty}^{\text{edge of sphere}} \mathbf{E} \cdot d\mathbf{s},$$

is bigger than the potential of the large sphere. Therefore, moving an additional coulomb of charge onto the small sphere takes more work than moving an additional coulomb onto the large sphere.

Since sphere (1) is more "willing" to accept additional charge, it has higher capacitance than sphere (5). From this argument, you can see that bigger objects have higher capacitance than smaller objects, other things being equal.

Now let's compare a hollow to a solid sphere, (1) vs. (2). You might think that the charges on (2) spread themselves out more, because they have access to the whole interior of the sphere. But actually, **charges always settle on the surface of a conductor.** The charges settle on (2) exactly the same way they settle on (1). So, both spheres generate the same electric field, and hence, the same potential. Consequently, they have the same capacitance; both spheres are equally "willing" to accept additional charge.

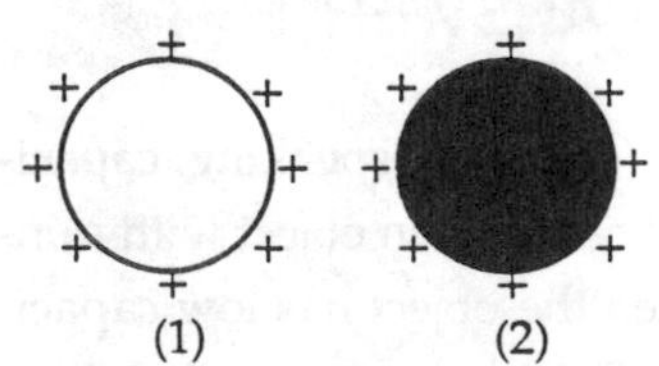

Digression: Why do charges always settle on the surface of a conductor?

Before continuing with this problem, let me explain why charges always migrate to the surface of a conductor. People make sense of this fact in multiple ways. For instance, using a computer simulation, you can show that the charges minimize their potential energy by spreading themselves out on the conductor's *surface*. Alternatively, you can consider the electric field inside the conductor. It must be zero, for the following reason. A conductor contains a huge supply of unbound electrons, which aren't attached to individual atoms. If the electric field inside a conductor were nonzero, then these unbound electrons would move around, rearranging themselves. But ironically, those electrons getting pushed around by the electric field inside the conductor also *create* the electric field inside the conductor. Consequently, as long as excess charges slosh around *inside* the conductor (as opposed to on the *surface*), they generate electric fields inside the conductor, which cause charges to slosh around inside the conductor, which generate electric fields inside the conductor, which . . . . This vicious cycle ends only when the excess charges settle on the conductor's *surface* and spread themselves out in a certain way. When that happens, the electric field inside the conductor drops to zero, and therefore electrons stop moving around. The process of electron rearrangement stops; everybody's happy!

This explains why a hollow conducting sphere and a solid conducting sphere "behave" equivalently, when charged up. The excess charges settle entirely on the surface of the solid sphere, as if it were hollow.

End of digression

OK, let's return to the capacitance problem. So far, we've figured out that a bigger sphere has higher capacitance than a smaller sphere, and that hollow and solid spheres have the same capacitance, other things being equal. Now we must think about dielectric material.

As your textbook explains, a dielectric becomes "polarized" in an electric field, creating its own electric field that points oppositely to the original field. So, in the presence of an externalelectric field, a dielectric generates its own field which partially cancels the external field. In this way, **a dielectric "dampens" the electric field**, making it smaller than would otherwise be the case.

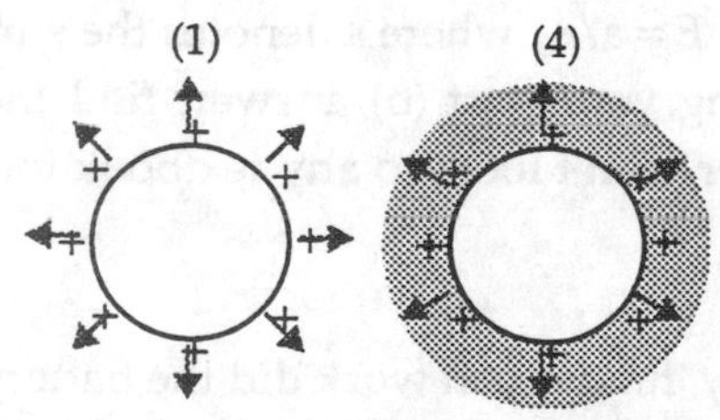

The field inside the dielectric is weaker than the field in the air near sphere (1). Therefore, (4) has lower potential than (1).

Let's see how this affects object (4), the sphere enveloped by dielectric. Suppose we give spheres (1) and (4) equal charge. Both spheres generate equivalent electric fields, except for one thing: the dielectric decreases the electric field iinside of it. The field inside the dielectric near sphere (4) is smaller than the field in the air near sphere (1). Therefore, less work is required to push an additional coulomb of charge onto (4), because you're fighting a smaller electric field (during part of the push). In other words, (4) has lower potential, and hence, higher capacitance.

Does a similar argument apply to sphere (3), which is filled with dielectric? No. When piling charge onto the sphere, you fight the electric field *outside* the sphere. In other words, the potential on the surface of a sphere depends only on the electric field *outside* the sphere. But the dielectric material inside sphere (3) can dampen only the field inside that sphere. The dielectric has no effect on the field *outside* sphere (3). For this reason, filling sphere (3) with dielectric can't change its potential. In other words, pushing an extra coulomb of charge onto sphere (1) or sphere (3) takes equal work. So, both spheres have the same capacitance.

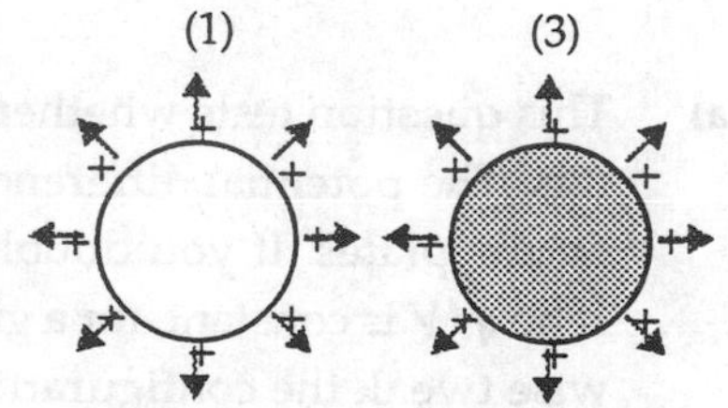

Since the dielectric *inside* sphere (3) does not affect the field (and potential) *outside* that sphere, both spheres have the same capacitance.

(As it turns out, the electric field inside the sphere vanishes, with or without the dielectric in place.)

Let me summarize the capacitance rankings.

Highest: (4). The dielectric dampens the electric field outside the sphere, thereby lowering the sphere's potential, making it easier to bring in extra charge.

Tied for 2nd highest: (1), (2), and (3). Since charges settle on the surface of a conductor, it doesn't matter that sphere (2) is solid. And since the dielectric *inside* sphere (3) doesn't affect the field *outside* the sphere, that dielectric has no effect on the capacitance.

Lowest: (5). Small objects generate large potentials even when charged up only a little, because the charge gets "concentrated," thereby producing a large electric field. For this reason, it's harder to pile additional charge onto a smaller object, other things being equal.

QUESTION 36-2

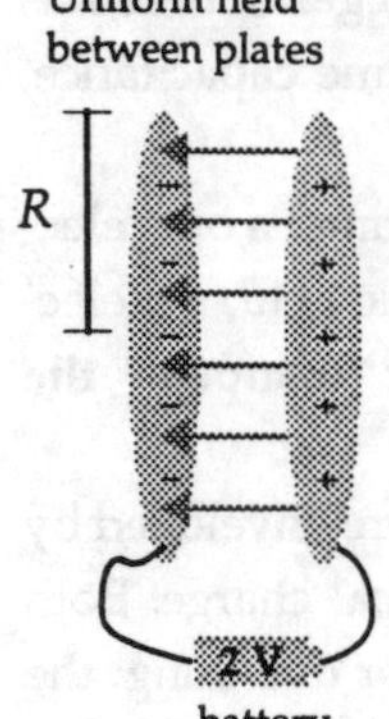

Consider a parallel-plate capacitor consisting of two circular disks of radius $R = 0.10$ m. When a 2-volt battery is connected to the plates, 2.510^{-11} coulombs builds up on the right plate. An equal negative charge builds up on the left plate. When solving this problem, do not look up any formulas from the "parallel plate capacitor" section of your textbook, other than the definition of capacitance.

(a) If we hooked this same capacitor to a 4-volt battery, what charge would build up on the right plate?

(b) What is the capacitance of this capacitor?

(c) The electric field between two parallel plates is $E = s/e_0$, where s denotes the surface charge density of the positive plate. Using your part (b) answer, find the distance s separating the two plates. Remember, don't look up any textbook formulas about parallel plates.

(d) When the uncharged capacitor was connected to the 2-volt battery, how much work did the battery do while charging the capacitor? In other words, how much chemical energy did the battery "use up" while charging the capacitor?

ANSWER 36-2

(a) This question tests whether you understand the conceptual insight underlying "capacitance." Here it is: The potential difference (voltage) between two conducting plates is proportional to the charge on the plates. If you double the charge on the plates, you double the voltage. In other words, the ratio q/V is constant, for a given capacitor, unless you change the distance between the plates or otherwise tweak the configuration.

Therefore, by replacing the 2-volt battery with a 4-volt battery, we double the charge on the plates. It's now 5.010^{-11} coulombs, instead of 2.510^{-11} coulombs.

Let me explain why the ratio q/V stays constant. When charged, the plates generate an electric field. As you know from Gauss' law, the electric field strength E is proportional to the charge generating the field. Double the charge, and you automatically double the field. And since the potential difference (voltage) between the plates is $V = -\int_{\text{negative plate}}^{\text{positive plate}} \mathbf{E}\cdot d\mathbf{s}$ doubling the electric field doubles the voltage. Physically, that's because doubling the field strength makes it twice as difficult to push a 1-coulomb test charge from one plate to the other. In summary, doubling the charge doubles the field, which doubles the voltage. So V is proportional to q, and hence, the ratio q/V stays constant.

(b) As just hinted, "capacitance" is the proportionality constant relating charge to voltage: $q = CV$. We can calculate it using $V = 2.0$ volts and $q = 2.510^{-11}$ coulombs. Or we can use $V = 4.0$ volts and $q = 5.010^{-11}$ coulombs. The whole point of part (a) is that the ratio $C = q/V$ stays the same:

$$C = \frac{q}{V} = \frac{2.5\times 10^{-11}\ \text{C}}{2.0\ \text{V}} = 1.2\times 10^{-11}\ \text{farads} = 12\ \text{picofarads (pF)},$$

since a picofarad equals 10^{-12} farads.

(c) To solve this, we need to find the relationship between the capacitance, C, and the distance between the plates, s. Given that equation, we can solve for s, since we already found C. Before deriving the mathematical relationship, let me make it plausible that the capacitance C depends on the plate separation s.

By definition, $C = q/V$, where q denotes the charge on the positive plate, and V denotes the potential difference between the plates, i.e., the work needed to push a 1-coulomb test charge from one plate to the other. Suppose we fix a certain charge on the plates, but then pull them apart, increasing s. By changing the distance between the plates, we've altered how much work is needed to move a test charge from one plate to the other. In other words, we've altered V, for a given q. This goes to show that changing the plate separation changes the capacitance.

This discussion suggests a strategy for finding the precise mathematical relationship between C and s. For an arbitrary charge q on the plates, I'll calculate the corresponding potential difference V, as a function of the plate separation, s. Of course, as we saw in parts (a) and (b), this voltage will also depend on the charge q. But the capacitance—the ratio q/V—remains the same no matter what q we use. In other words, when we divide q by V, the charge should cancel out, leaving us with an equation for capacitance in terms of s. Given that equation, we can solve for s, since we know C.

Subproblem: For an arbitrary charge q on the plates, find the potential difference, V.

To make sure V comes out positive, we must integrate the electric field over a path from the negative to the positive plate. (A positive test charge traveling in this direction experiences a potential rise, not a potential drop.) As my path of integration, I'll choose the dashed line in this drawing. Since the path "points" rightward while the electric field points leftward, the dot product E.*ds* reduces to a *negative* regular product, –*Eds*. And this path has length s. Therefore,

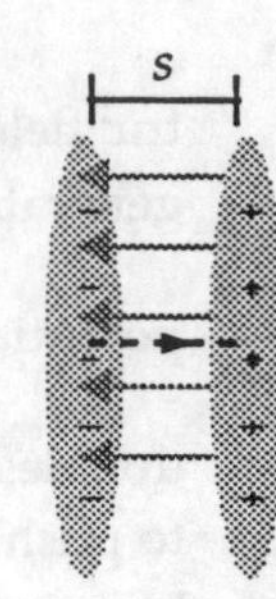

$$V = -\int_{\text{negative plate}}^{\text{positive plate}} \mathbf{E} \cdot d\mathbf{s},$$

$$= \int_0^S E ds \qquad [\text{since } \mathbf{E} \cdot d\mathbf{s} = -E ds]$$

$$= \int_0^S \frac{\sigma}{\varepsilon_0} ds \qquad [\text{since } E = \sigma / \varepsilon_0, \text{ as the problem tells us}$$

$$= \frac{\sigma}{\varepsilon_0} s.$$

Since charge q is spread uniformly over plate area $A = \pi R^2$, the plate carries surface charge density $\sigma = \frac{q}{A} = \frac{q}{\pi R^2}$. So,

$$V = \frac{\sigma}{\varepsilon_0} s = \frac{qs}{\pi R^2 \varepsilon_0}.$$

End of voltage subproblem

Now we can substitute this voltage into the definition of capacitance, to get

$$C = \frac{q}{V} = \frac{q}{\left(\frac{qs}{\pi R^2 \varepsilon_0}\right)} = \varepsilon_0 \frac{\pi R^2}{s}.$$

Notice that the q's cancel, proving that the ratio q/V stays the same no matter what charge the plates carry. We just derived the textbook equation for a parallel-plate capacitor, $C = \varepsilon_0 A/s$, where A is the plate area and s is the plate separation.

Normally, we'd solve for C in terms of a known s. But here, we're solving for s in terms of a known C, the C we found in part (b).

$$s = \frac{\varepsilon_0 \pi R^2}{C} = \frac{(8.85 \times 10^{-12} \text{ F/m})\pi(0.10 \text{ m})^2}{1.2 \times 10^{-11} \text{ F}} = 0.022 \text{ m},$$

about 2 centimeters.

(d) Think in terms of energy conservation. The work expended by the battery (i.e., the chemical energy "lost" by the battery) must equal the potential energy stored in the charged-up capacitor. So, we just need to find the potential energy stored in the capacitor.

The **common mistake** is to reason as follows: since $U = qV$, the potential energy stored in the capacitor is simply the charge times the voltage,

Incorrect equation $$U_{\text{stored in capacitor}} = qV = (2.5 \times 10^{-11} \text{ C})(2.0 \text{ V}) = 5.0 \times 10^{-11} \text{ J}.$$

As I'll prove below, however, the potential energy stored in the capacitor—and hence, the work needed to charge up the capacitor—is

Correct equation $$U = \frac{1}{2}qV = \frac{1}{2}(2.5 \times 10^{-11} \text{ C})(2.0 \text{ V}) = 2.5 \times 10^{-11} \text{ J}.$$

OK, let's see where the unexpected factor of a half comes from. My explanation takes a few pages. But it's worth the time, because some interesting physical insights pop out.

When we connect the battery to the two plates, what happens? Most people think the battery "supplies" the positive and negative charges that end up on the plates. But actually, the battery supplies no charges. It simply moves charges from one plate to the other. Since the plates begin uncharged, you might wonder where those "charges" come from.

Well, each plate consists of positively-charged atomic nuclei paired with negatively-charged electrons. Since the plates are conductors, many of those electrons are not "bound" to individual nuclei. Rather, they're free to move around the plate, and to flow off the plate through wires. The battery pushes some of those unbound electrons along a wire from one plate to the other. In this way, one plate acquires excess negative charge, while the other plate ends up with an equal "deficit" of negative charge—and hence, a net positive charge.

Now, if the battery were to push charge q from one plate to the other, and the potential difference between the plates were V during that entire push, then the battery would indeed perform work $W = qV$. But that's not what happens. The battery does not move charge q all at once. Rather, the battery transfers the charge one electron at a time. Of course, the battery accomplishes this so fast, it *seems* like charge q gets transferred "all at once." But if we could watch the process in super slow motion, we'd see electrons flowing off the positive plate and onto the negative plate, one at a time.

To find the work done during this process, consider the uncharged plates right when the battery first gets connected. The battery starts moving electrons from one plate to the other. The first electron is easy to move, because the plates aren't yet charged. Therefore, they create no electric field that must be "fought" against; there's no potential difference between the plates.

By contrast, after 1000 electrons have been transferred, the plates are partly charged up. They create an electric field, and hence, a potential difference between the plates. For this reason, it takes more work to move additional electrons from one plate to the other, in order to "fight" this field.

In summary, as the plates become more charged up, moving additional electrons from one plate to the other becomes harder and harder, due to the electric field created by the previously-moved electrons. For instance, if the battery transfers a billion electrons total, the first couple million take very little work to move, while the last couple million take lots of work. The following graphic should clarify this.

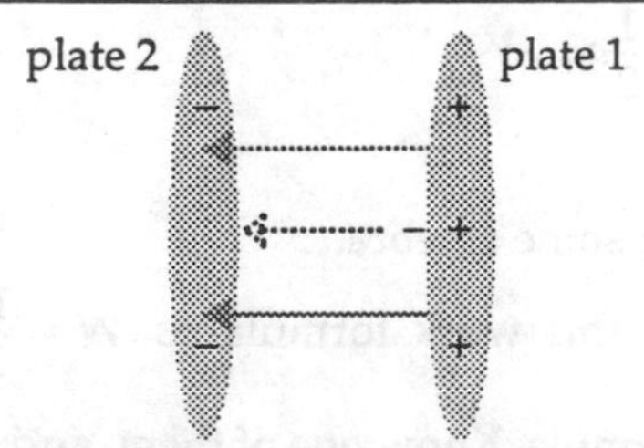

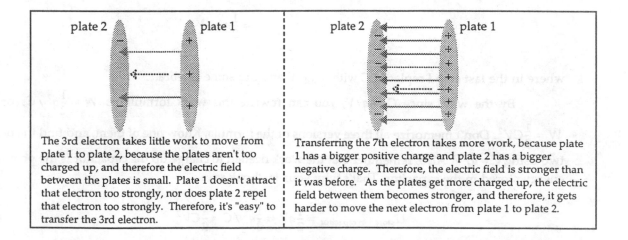

The 3rd electron takes little work to move from plate 1 to plate 2, because the plates aren't too charged up, and therefore the electric field between the plates is small. Plate 1 doesn't attract that electron too strongly, nor does plate 2 repel that electron too strongly. Therefore, it's "easy" to transfer the 3rd electron.

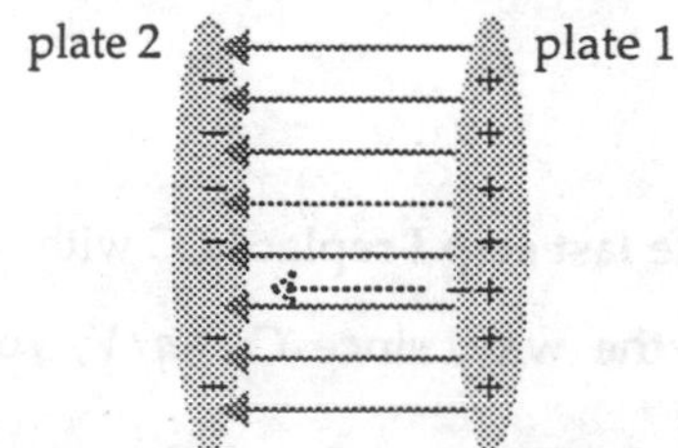

Transferring the 7th electron takes more work, because plate 1 has a bigger positive charge and plate 2 has a bigger negative charge. Therefore, the electric field is stronger than it was before. As the plates get more charged up, the electric field between them becomes stronger, and therefore, it gets harder to move the next electron from plate 1 to plate 2.

At this stage, we're ready to understand why the work needed to charge up the capacitor is *less* than qV. Remember, V denotes the final potential difference between the plates, *after* they're fully charged (i.e., *after* the whole q has been transferred). As the plates charge up, the potential difference between them is usually less than V, because they're not yet fully charged. Indeed, since the plates begin with no potential difference and end with potential difference V, the *average* potential difference during the "charge-up" process turns out to be $V/2$. So, charge q gets pushed across a potential difference that, on average, equals $V/2$. Therefore, the battery does work

$$W_{\text{to charge plates}} = qV_{\text{average during the charge-up process}} = \frac{1}{2}qV.$$

So much for rough intuitive reasoning. Let me obtain this result rigorously using calculus. How much work is required to move a tiny "bit" of charge dq from one plate to the other? It's

$$dW = Vdq \tag{1}$$

where V denotes the potential difference between the plates *while dq gets moved*. So, the V in Eq. (1) is determined by how much charge has already been transferred, *not* by the voltage of the battery. See the above graphic. Quantitatively, if we let q denote the charge that's already been transferred, then V is proportional to q. Indeed, from the definition of capacitance, $V = q/C$. So, Eq. (1) becomes

$$dW = \frac{qdq}{C}. \tag{1'}$$

To find the total work needed to charge up the capacitor from 0 to q_{final}, we must add up (integrate over) the work it takes to transfer each bit of charge:

$$\begin{aligned} W = \int dW &= \int_0^{q_{\text{final}}} \frac{qdq}{C} \\ &= \frac{q_{\text{final}}^2}{2C} \\ &= \frac{1}{2}q_{\text{final}}V, \end{aligned}$$

where in the last step I replaced C with q_{final}/V and did some algebra.

By the way, since $C = q/V$, you can rewrite this work formula as $W = \frac{1}{2}q^2/C$, or as $W = \frac{1}{2}CV^2$. Don't memorize all three versions of the formula. Know one of them, and find the other two using $C = q/V$. Also, remember that the work needed to charge up the capacitor is the potential energy stored in the capacitor. That's why your textbook says

$$U_{\text{stored in capacitor}} = \frac{1}{2}qV = \frac{1}{2}q^2/C = \frac{1}{2}CV^2.$$

QUESTION 36-3

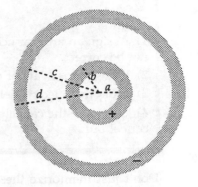

Consider a capacitor consisting of two concentric conducting spherical shells. Both shells are noticeably thick. The inside shell has inner radius a and outer radius b. The outside shell has inner radius c and outer radius d. Initially, neither sphere is charged.

The spheres are hooked up to a battery of voltage V_0. Consequently, the inside sphere acquires positive charge, while the outside sphere acquires an equal negative charge.

(a) (*Quite hard, but a typical capacitance question*) What is the capacitance of this system? Don't look up a pre-derived textbook formula. Figure out the capacitance from scratch.

(b) What is the electric field a distance $2d$ from the center of the spheres?

(c) What is the surface charge density on the inner edge of the small shell (at distance $r = a$ from the center)? What's the surface charge density on the outer edge of the small shell (at $r = b$)? On the inner edge of the large shell (at $r = c$)? On the outer edge of the large shell ($r = d$)?

(d) How much work is needed to move a point charge $+q_1$ from very far away to the center of the spheres? Small holes have been drilled into the conducting shells, so that the point charge can be pushed inside. Neglect gravity.

(e) Suppose we replace the original battery with a double-strength battery, which applies a potential difference $2V_0$ between the two shells. Is the new capacitance bigger than, smaller than, or the same as the value you obtained in part (a)?

ANSWER 36-3

(a) Capacitance is a system's "willingness" to get charged up. Specifically, it's the charge a system is willing to hold *per volt* of applied potential difference:

$$C = \frac{q}{V},$$

where $+q$ and $-q$ are the charges acquired by the two "plates," in this case, the two spherical shells. So, we must divide the charge by the potential difference between the plates. (For some reason, textbooks write "V" instead of "ΔV" to denote the potential *difference* between two capacitor plates.)

Most students approach this problem by trying to find the unknown charge q corresponding to the known potential difference V_0. Unfortunately, this strategy turns out to be extremely difficult. Even though you know V but not q, it's *easier* to pretend you know q but not V. In other words, you can calculate the capacitance more easily by "pretending" that q is given, and solving for the corresponding "unknown" V.

Why is this legal? Because the *ratio* q/V comes out the same no matter what charge or voltage you start with. In other words, when you divide q by V to obtain the capacitance, the q cancels out. So, it doesn't matter if you use a "wrong" q.

Let me summarize and flesh out this strategy:

Capacitance-finding strategy

1) Assign an arbitrary charge $+q$ and $-q$ to your two plates. As just explained, this charge need not be correct.

2) Using Gauss' law or other techniques, calculate the electric field between your two plates.

3) From that electric field, calculate the potential difference between the plates, $V = -\int \mathbf{E} \cdot d\mathbf{s}$.

4) Calculate the capacitance using $C = q/V$. The arbitrary charge q you chose in step 1 should cancel out.

Don't just memorize these steps. Make sure you understand the underlying idea. This strategy quantifies the qualitative reasoning used in question 36-1 above. There, you pretended that each object carries the same charge, and you figured out which objects generated higher vs. lower potentials. In that way, you decided which objects would be most "willing" to accept additional charge, i.e., which objects have higher capacitance. Similarly, in the capacitance-finding strategy, you also pretend that the system carries a certain charge, and you calculate the corresponding potential.

I'll now apply the strategy to this problem.

Steps 1 and 2: Assign charge $\pm q$ to the plates, and figure out the electric field between the plates.

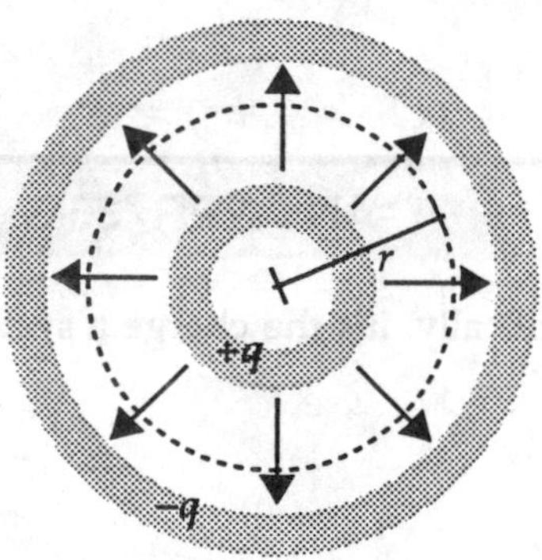

Dashed line is "Gaussian sphere" of radius r.

From now on, assume the inner shell carries charge $+q$, and the outer shell carries charge $-q$.

We care only about the electric field *between* the conducting shells, because that field completely determines the potential difference between the shells. Intuitively, the work needed to push a test charge from one sphere to the other depends only on the field between the shells, not on the field outside both spheres or inside both spheres.

To find the field, use Gauss' law. As always, start by drawing in some field lines, and picking a "good" Gaussian surface. Choosing a "Gaussian sphere" ensures (by symmetry) that the electric field has the same strength over the entire surface. I've given this Gaussian sphere arbitrary radius r, where $b < r < c$.

Now we can apply Gauss' law,

$$\oint \mathbf{E} \cdot d\mathbf{A} = \frac{q_{\text{encl}}}{\varepsilon_0}.$$

Since the Gaussian surface encloses only the inside shell, $q_{\text{encl}} = +q$. The flux integral also simplifies, for two reasons. First, the electric field points in the same direction that the Gaussian surface "faces," radially outward. So, the dot product reduces to a regular product. Furthermore, we deliberately chose a Gaussian surface over which the field stays constant, thereby allowing us to pull E outside the integral:

$$\oint_{\text{Gaussian sphere}} \mathbf{E} \cdot dA = \oint_{\text{Gaussian sphere}} E\,dA = E \oint_{\text{Gaussian sphere}} dA = EA_{\text{Gaussian sphere}} = E(4\pi r^2).$$

Put all this information into Gauss' law to get

$$\oint \mathbf{E} \cdot d\mathbf{A} = \frac{q_{\text{encl}}}{\varepsilon_0}$$

$$E(4\pi r^2) = \frac{q}{\varepsilon_0},$$

and hence $E = \frac{q}{4\pi\varepsilon_0 r^2}$, which is Coulomb's law. So, the electric field between the conducting *spherical* shells "behaves" as if it were generated by a point charge.

Step 3: Find the potential difference between the plates, using the electric field just calculated.

Before calculating the potential difference between the shells using

$$V = -\int_{\text{negative shell}}^{\text{positive shell}} \mathbf{E} \cdot \mathrm{d}\mathbf{s},$$

let me briefly review where this formula comes from. By definition, the potential difference between two points is the potential energy change, ΔU, that a 1-coulomb charge would undergo while moving from one point to the other. A potential energy difference is the work needed to push the charge from one point to the other: $\Delta U = W$. Therefore, *a potential difference is the work needed to push a 1-coulomb charge from one point to the other.* Since $W = -\int \mathbf{F}_{\text{elec}} \cdot d\mathbf{s} = -\int q\mathbf{E} \cdot d\mathbf{s}$, the work corresponding to a $q = 1$ coulomb charge is simply $-\int \mathbf{E} \cdot d\mathbf{s}$.

Since it takes positive work to push a 1-coulomb test charge *against* the electric field, integrating from the negative to the positive plate ensures that V comes out positive. So here, we must integrate from the negative shell (at $r = c$) to the positive shell (at $r = b$). As my path of integration, I'll choose a straight radial line. That way, the field lines point parallel to my path, collapsing the dot product E.*ds* into the regular product *Edr*:

path
+q
−q

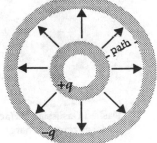

$$V = -\int_{\text{outer shell}}^{\text{inner shell}} \mathbf{E} \cdot d\mathbf{s}$$

$$= -\int_{r=c}^{r=b} E \cdot dr$$

$$= -\int_c^b \frac{q}{4\pi\varepsilon_0 r^2}\,dr \qquad \text{[using our expression for } E(r) \text{ from step 2]}$$

$$= \frac{q}{4\pi\varepsilon_0 r}\bigg|_c^b$$

$$= \frac{q}{4\pi\varepsilon_0 b} - \frac{q}{4\pi\varepsilon_0 c}.$$

That's the potential difference (voltage) between the two conducting spherical shells. It's the same potential difference generated by a point charge. Which makes sense, since the field between the two shells is the same field produced by a point charge.

By the way, if you accidentally obtain a negative voltage (by integrating from the positive to the negative plate), use the absolute value when calculating the capacitance below.

Step 4: Calculate the capacitance using $C = q/V$.

Using our result from step 3, we get

$$\begin{aligned} C &= \frac{q}{V} \\ &= \frac{q}{\frac{q}{4\pi\varepsilon_0 b} - \frac{q}{4\pi\varepsilon_0 c}} \\ &= \frac{1}{\frac{1}{4\pi\varepsilon_0 b} - \frac{1}{4\pi\varepsilon_0 c}} \\ &= \frac{4\pi\varepsilon_0 bc}{c-b}. \end{aligned}$$

The q's cancel, proving that it doesn't matter if you assigned the "wrong" charge to the shells. The ratio $C = q/V$ comes out the same no matter what.

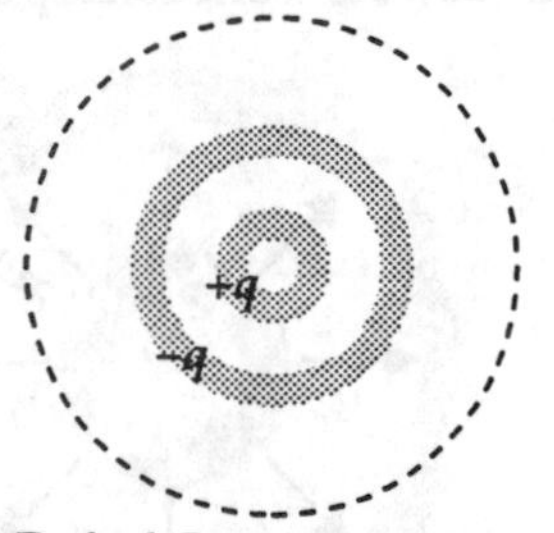

Dashed Gasussian surface encloses zero net charge.

(b) We're looking for the electric field at a point outside both spherical shells. Since the charge distribution is spherically symmetric, we can use Gauss' law, with a spherical Gaussian surface. By symmetry, the electric field is constant over the whole Gaussian surface, and therefore we can pull **E** outside the integral:

$$\oint \mathbf{E} \cdot d\mathbf{A} = \frac{q_{\text{encl}}}{\varepsilon_0}$$

$$E(4\pi r^2) = 0,$$

where in the second step, I noticed that the Gaussian surface encloses net charge $q_{\text{encl}} = 0$! The $+q$ on the inner shell "cancels" the $-q$ on the outer shell. So, according to Gauss' law, $E = 0$. The field vanishes *everywhere* outside the two shells.

We can understand this result in another way. As mentioned in part (a), a spherical shell of uniformly-distributed charge "behaves" like a point charge concentrated at its center, provided you're outside the shell. So, if you're outside both shells, they both behave like point charges at the center. These $+q$ and $-q$ "point charges" cancel each other out.

(c) To solve this, we must first calculate how much charge resides on the inside (small) shell and on the outside (big) shell. Then, we can figure out how those charges distribute themselves. For instance, does the small sphere's charge "live" on the inner surface or on the outer surface of that shell?

Subproblem 1: Find the total charge on each shell.

The battery applies a known potential difference, V_0, between the conducting spherical shells. Intuitively, the bigger the potential difference, the more charged up the shells get. Intuitively, by doubling the voltage V, you double the charge q on the shells. So, q and V are proportional: $q = CV$, where the proportionality constant C encodes the system's "willingness" to accept charge. As emphasized above, this proportionality constant is the capacitance, C. From the basic definition $C = q/V$, it follows that $q = CV$.

Using our part (a) answer for the capacitance, I get

$$\text{Charge on inside (small) sphere} = +q = +CV = +\frac{4\pi\varepsilon_0 bc}{c-b}V_0.$$

$$\text{Charge on outside (big) sphere} = -q = -CV = -\frac{4\pi\varepsilon_0 bc}{c-b}V_0.$$

Subproblem 2: Find the charge distribution.

From question 36-1, we know that charges settle on the *surface* of a conductor. That's because the charges rearrange themselves until the electric field inside the conductor vanishes, which happens only when the charges all migrate to the surface.

Let's focus first on the smaller spherical shell. We must think about whether the charge $q = CV_0$ coats the inner edge of that shell, the outer edge of that shell, or some combination.

To figure this out, remember that **the electric field inside a conducting material is zero.** (Otherwise, unbound electrons would keep sloshing around inside the conductor, as explained in question 36-1.) From this insight, we can conclude that no charge lives on the inner edge of the small sphere (at $r = a$). Here's why.

Physical impossibility

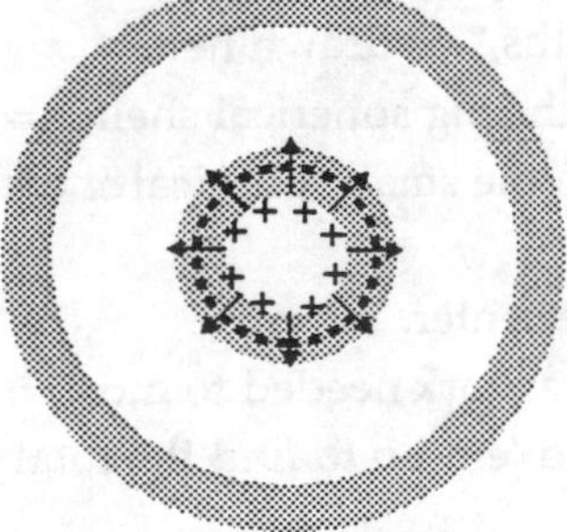

No electric field is "allowed" inside a conducting material.

Suppose, hypothetically, that some charge *does* reside on the inner surface of the small shell. Then a Gaussian surface inside the shell (between $r = a$ and $r = b$) encloses charge, as drawn here. Therefore, by spherical symmetry and Gauss' law, the electric field on that Gaussian surface is nonzero. But the electric field inside a conducting material is *always* zero. This contradiction proves that *no* charge resides on the inner surface of the small shell. In other words, the surface charge density at $r = a$ is $\sigma_a = 0$.

Therefore, all the charge on the small shell coats the outer surface, at r = b. The charge spreads out evenly over surface area $A = 4\pi r^2 = 4\pi b^2$. So, the surface charge density (charge per area) is

$$\sigma_b = \frac{q}{A} = \frac{\frac{4\pi\varepsilon_0 bc}{c-b}V_0}{4\pi b^2} = \frac{V_0\varepsilon_0 c}{b(c-b)}.$$

Now let's turn our attention to the outside (large) spherical shell, which carries overall charge $-q = -CV_0$. From the above reasoning, you might think all this charge lives on outer surface ($r = d$). But a careful application of the above reasoning proves that all the charge on the big

This "works"

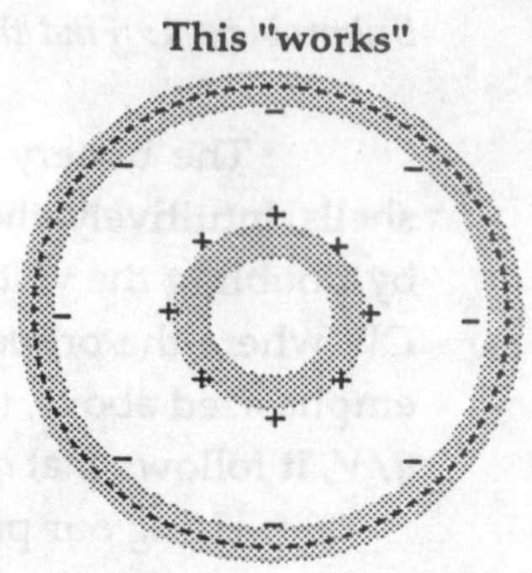

shell resides on the *inner* surface, at $r = c$. To see why, look at this picture. The two shells carry equal and opposite charge. And I've placed the big shell's charge entirely on the inner surface of that shell. Consider the Gaussian surface drawn inside the big shell, between $r = c$ and $r = d$. It encloses no net charge, because the positives and negatives cancel. Therefore, by spherical symmetry and Gauss' law with $q_{\text{encl}} = 0$, the electric field on this Gaussian surface vanishes. In other words, the electric field inside the big shell is zero—exactly as it must be, inside a conducting material. So, this picture is physically possible!

By contrast, if we moved some of the negative charges to the *outer* surface of the big shell, then fewer negative charges would remain on the inner surface. Therefore, my Gaussian surface would enclose more positive charge than negative charge. So, the electric field on the Gaussian surface would be nonzero. But the field inside a conducting material *must* vanish. This contradiction goes to show that the charges must arrange themselves as I've drawn, with all the negative charge density on the *inner* surface of the big shell. The surface charge density at the outer surface is $\sigma_d = 0$.

Summary: On both spherical shells, the charges arrange themselves so that no electric field exists inside the conducting material. The positive charge on the smaller shell coats the *outer* surface, at $r = b$. By contrast, the negative charge on the big shell coats the *inner* surface, at $r = c$.

The inner surface of the big shell has radius c, and surface area $A = 4\pi r^2 = 4\pi c^2$. Therefore,

$$\sigma_c = \frac{-q}{A} = \frac{\frac{4\pi\varepsilon_0 bc}{c-b}V_0}{4\pi c^2} = \frac{V_0\varepsilon_0 b}{c(c-b)}.$$

Because both shells carry the same charge, the surface charge *density* is higher on the small shell, since that charge has less room to spread out. Our mathematical answers confirm this.

(d) Consider pushing the particle along a straight path from very far away to the center of the spheres. We can break this path into three "subpaths," as drawn here:

(1) From far away to the inner edge of the big spherical shell ($r = c$);

(2) From the big spherical shell ($r = c$) to the small spherical shell ($r = b$); and then

(3) From the small spherical shell to the center.

Let's calculate W_1, W_2, and W_3, the work needed to move the particle along each subpath. Then we can add 'em up to find the total work.

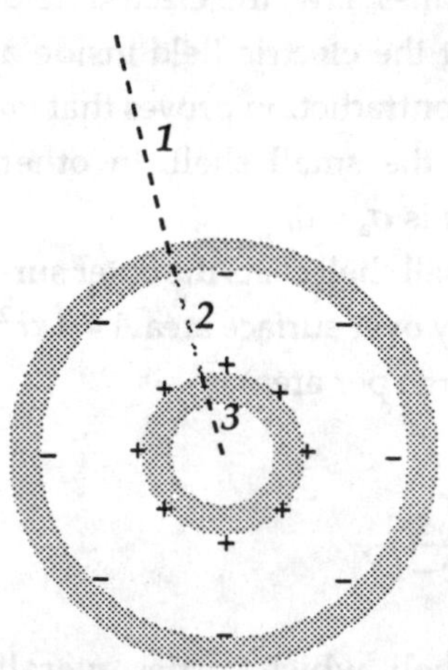

3-segment path of integration. Segment 1 actually extends to infinity.

Subpath 1: Very far away to the big spherical shell

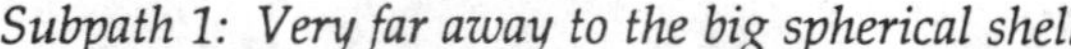

In part (b), we found that the electric field outside both shells vanishes, because the positive and negative shells "cancel." So, when pushing the particle along subpath 1, you're not "fighting" an electric field; the particle just "floats" along that path, with no effort. In other words, you perform no work: $W_1 = 0$. (This same reasoning applies inside the hole drilled into the big shell.)

Subpath 2: Between the conducting shells

The electric field between the conducting shells points radially outward, as we saw in part (a). Because you must "fight" this field, moving the particle along subpath 2 requires work.

To find W_2, we could start with our part (c) expression for the charge on the inner shell, calculate the corresponding electric field, and then evaluate

$$W_2 = -\int_{r=c}^{r=b} \mathbf{F}_{\text{elec}} \cdot d\mathbf{s} = \int_{r=c}^{r=b} q_1 \mathbf{E} \cdot d\mathbf{s} = -q_1 \int_{r=c}^{r=b} \mathbf{E} \cdot d\mathbf{s} = q_1 \Delta V,$$

where ΔV denotes the potential difference between the two shells. But we need not go to all this trouble, because the problem *gives* us the potential difference between the two shells. It's V_0. So, from the above equation, we immediately get

$$W_2 = q_1 V_0.$$

Should W_2 be positive? Yes, because you're pushing a positive charge *against* the electric field.

Subpath 3: Inside $r = b$

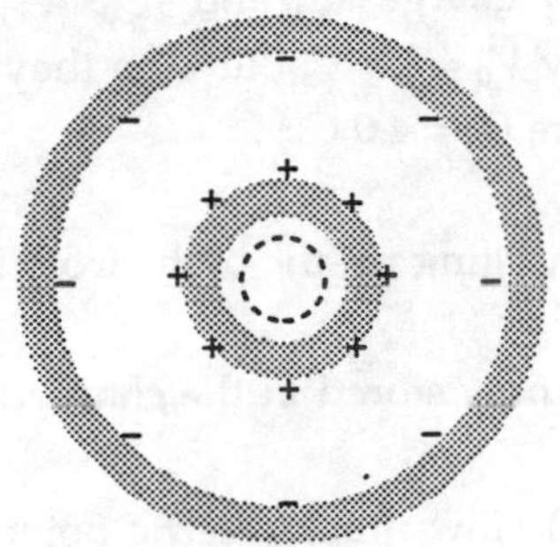

This Gaussian surface encloses no charge.

In the hollowed-out region between $r = a$ and $r = 0$, the electric field vanishes! To see why, draw a spherical Gaussian surface of radius $r < a$. This Gaussian surface encloses no charge: $q_{\text{encl}} = 0$. And by symmetry, we can pull the electric field outside the flux integral. So, according to Gauss' law,

$$\oint \mathbf{E} \cdot d\mathbf{A} = \frac{q_{\text{encl}}}{\varepsilon_0}$$

$$E(4\pi r^2) = 0,$$

and hence $E = 0$. Therefore, it takes no work to move the particle along subpath 3, fro the inner shell to the center: $W_3 = 0$.

OK, let's tie it all together. When pushing the particle to the center of the spheres from very far away, we needed to perform work *only* between the spheres. Everywhere else, the field vanished, and therefore the particle just floated along with no push needed. In summary,

$$W_{\text{total}} = W_1 + W_2 + W_3 = 0 + q_1 V_0 + 0 = q_1 V_0.$$

Brief digression: "Screening"

In general, if a conductor completely surrounds (encases) an empty region, the electric field in that region is zero—even if the conductor has no symmetry. See your textbook for a fuller explanation. For instance, the electric field inside an unopened can of soda is zero, even if you place a charged particle near the can. The metal doesn't "block" the electric field generated by the particle. Rather, the electrons in the metal rearrange themselves so as to eliminate the electric field inside the metal "skin" of the can. As a side effect, this rearrangement also cancels out the electric field inside the soda itself. Scientists and engineers use this "screening" effect to get rid of unwanted electric fields, by lining the walls, floor, and ceiling of a chamber with metal mesh.

End of digression

(e) The capacitance stays the same. At first glance, you might think the capacitance doubles when you double the voltage, because now the conducting shells hold twice as much charge. Well, the charge on the shells *does* double. But that's not because the capacitance changes. The system still holds the same charge *per volt of applied potential difference.* Since doubling the voltage causes the charge to double, the ratio of charge to voltage, q/V, stays the same. And that ratio *is* the capacitance.

Indeed, that's the whole point of "capacitance." By changing the potential difference between the two plates (shells), you can always adjust the charge on the plates. And by altering the charge on the plates, you tweak the potential difference (voltage). But no matter how you alter the charge or voltage, the *ratio* of charge to voltage ($C = q/V$) stays *constant*. In this way, the capacitance encodes how "willing" the system is to accept more charge. It depends only on the size and shape of the system.

QUESTION 36-4

Consider these two oddly-shaped metal blobs. Initially, they carry charge $+Q_0$ and $-Q_0$, respectively, where $Q_0 = 2.0$ C. The potential difference between them is initially $V_0 = 1.5$ V. But then they get hooked up to a battery that charges them up further, to $+Q_1$ and $-Q_1$, where $Q_1 = 4.0$ C.

At this point, the battery is disconnected.

(a) How much work did the battery do charging up the blobs from $\pm Q_0$ to $\pm Q_1$?

(b) How much electric potential energy is now stored in the charged-up blobs?

(c) (*Ask your instructor if you need to understand the topic of this problem*) How much electric potential energy is stored in the blobs *together with* the electric field created by the blobs? Hint: If you find yourself doing a lot of math, you've missed something.

(d) If the blobs were now moved closer together, with the battery still disconnected, would the electric potential energy stored in the system go up, go down, or stay the same? What about the capacitance? Explain your answers.

(e) Now the blobs are put back where they started. A nonconducting particle of charge $q = 3.0\ \mu$C is placed on the negative blob. How much work would it take to move that particle along a straight line to the positive blob?

ANSWER 36-4

This problem tests how well you understand work and energy in electrostatic systems. You'll have to distinguish between the work it takes to charge up a capacitor and the work it takes to "fight" the electric field created by the capacitor *once it's already charged up*.

(a) We can solve by thinking in terms of potential energy. As I've emphasized, work and potential energy are related by

$$W = \Delta U.$$

Intuitively, if I spend 50 joules of work lifting a block off the floor onto a table, then the block ends up with 50 more joules of potential energy than it started with. So, to find the work performed by the battery, we must figure out how much the system's potential energy *changes.*

In question 36-2d, I showed that the potential energy stored up in a capacitor is

$$U_{\text{capacitor}} = \frac{1}{2}qV, \text{ which can be rewritten } \frac{1}{2}CV^2 \text{ or } \frac{1}{2}\frac{q^2}{C}.$$

You can obtain the "rewritten" expressions from the original equation by using $C = q/V$. The unexpected factor of a half comes from the fact that, while the capacitor charges up, the *average* potential difference between the plates is half the *final* potential difference between the plates. Please see question 36-2d for a fuller explanation.

To use one of these formulas for $U_{\text{capacitor}}$, I need to know either the capacitance of this system, or the final voltage (when it has charge Q_1). Well, since capacitance is the charge *per volt*, and since the potential difference was V_0 when the charge was Q_0, we get $C = Q_0/V_0 = (2.0\text{ C})/(1.5\text{ V}) = 1.3$ farads. This ratio of charge to voltage stays the same when the system acquires more charge. So, the system has final potential energy

$$U_f = \frac{1}{2}\frac{Q_1^2}{C} = \frac{1}{2}\frac{Q_1^2}{Q_0/V_0}.$$

That's *not* the answer. It would be the answer if the battery were charging up the capacitor "from scratch." But the capacitor already has some potential energy before the battery does its work. In fact, its initial potential energy at charge Q_0 and voltage V_0 is

$$U_0 = \frac{1}{2}Q_0V_0.$$

Therefore, when "pushing the system" from U_0 to U_f, the battery does work

$$\begin{aligned} W = \Delta U &= U_f - U_0 \\ &= \frac{1}{2}\frac{Q_1^2}{Q_0/V_0} - \frac{1}{2}Q_0V_0 \\ &= \frac{1}{2}\frac{(4.0\text{ C})^2}{(2.0\text{ C})/(1.5\text{ V})} - \frac{1}{2}(2.0\text{ C})(1.5\text{ V}) \\ &= 6.0\text{ J} - 1.5\text{ J} \\ &= 4.5\text{ J}. \end{aligned}$$

Notice that the system ends up with four times its initial potential energy. By doubling the charge on the blobs (from 2.0 C to 4.0 C), you also double the potential difference between them (from 1.5 V to 3.0 V). Since both q and V double, the potential energy $\left(U = \frac{1}{2}qV\right)$ quadruples.

(b) As we saw in part (a), the system ends up with final potential energy

$$U_f = \frac{1}{2}\frac{Q_1^2}{C} = \frac{1}{2}\frac{Q_1^2}{Q_0/V_0} = \frac{1}{2}\frac{(4.0\text{C})^2}{(2.0\text{ C})/(1.5\text{ V})} = 6.0\text{ J}.$$

In other words, it takes 6 joules of work to charge up the system to Q_1. As shown above, the battery that charges it from Q_0 to Q_1 performs only 4.5 joules of this work. The other 1.5 joules were done by the battery that initially charged the capacitor from 0 to Q_0.

Let me briefly review where the formula $U_{capacitor} = \frac{1}{2}\frac{Q^2}{C}$ comes from. $U_{capacitor}$ is the work needed to charge the capacitor from 0 to Q. While transferring little bits of charge from one plate (blob) to the other, you must do work to fight the potential generated by the previously-moved charges. Specifically, when the capacitor is partially charged up, to charge q, the potential difference between the plates is q/C. Therefore, to transfer the next "bit" of charge, dq, requires work

$$dW = Vdq = \frac{q}{C}dq.$$

To find the total work needed to charge the capacitor from $q = 0$ to $q = Q$, add up the dW's.

$$W = \int_{q=0}^{q=Q} dW = \int_0^Q \frac{q}{C}dq = \frac{1}{2}($$

That's the potential energy stored in the fully-charged capacitor.

(c) This problem targets a common confusion associated with the concept of potential energy. Your textbook says that the electric potential energy stored in a capacitor is given by the expressions for $U_{capacitor}$ listed in part (b). Your textbook also asserts that energy is stored in the electric field, and derives a funky formula for the field's "potential energy per volume," $u = \frac{1}{2}\varepsilon_0 E^2$. So, you might get the impression that the total energy stored in the capacitor and in the electric field surrounding it is $U = \frac{1}{2}\frac{Q_1^2}{C} + \int \frac{1}{2}\varepsilon_0 E^2 d^3\mathbf{r}$. But this is wrong. To see why, we must dig deeper into the meaning of potential energy. We'll see that the answer is $U = \frac{1}{2}\frac{Q_1^2}{C}$ with no extra terms.

As usual, thinking about gravity can pump our intuitions. Consider the Moon orbiting the Earth. (For now, neglect the Sun and planets and all other heavenly bodies.) What's the potential energy of the moon? From last semester, you may know $U_{grav} = \frac{GMm}{r}$, where M and m denote the Earth's mass and Moon's mass. But then, what's the potential energy of the Earth (caused by the gravitational field of the Moon)? Again, the formula says $U_{grav} = \frac{GMm}{r}$.

Does this mean that the Earth and Moon together have potential energy $-2\frac{GMm}{r}$? No way. By definition, the potential energy of the Earth/Moon system is the work required to put those objects into their present configuration, a distance r apart. By integrating $F_{grav} = \frac{GMm}{r^2}$ over distance, from ∞ to r, you can confirm that the work equals $-\frac{GMm}{r}$. *That's the total potential energy of the Earth/Moon system.* The answer comes out the same no matter whether you think of the Moon as getting pushed toward a stationary Earth, or you think of the Earth as getting pushed toward a stationary Moon. In other words, we can picture the potential energy as belonging to the Moon, or as belonging to the Earth. Both ways of thinking work equally well; it's just a matter of bookkeeping. The only absolute "fact" is that the *total* potential energy of the Earth/Moon system is $-\frac{GMm}{r}$

(when we take $r = \infty$ as our "zero" point), because that's how much work it takes to place the Earth and Moon a distance r apart. You *cannot* think of the Earth and Moon as *both* possessing potential energy $-\frac{GMm}{r}$.

Similar reasoning applies to the capacitor. As we found in parts (a) and (b), the work needed

to charge up the capacitor plates for 0 to Q_1 is $U = \frac{1}{2}\frac{Q_1^2}{C} = 6.0$ J. This total potential energy is an absolute fact about the system, assuming we set $U = 0$ when the plates carry no charge. You may visualize these 6 joules of energy as "stored in" the charged plates, as we did in part (b). Alternatively, you may consider all 6 joules to be stored in the electric field generated by the charged plates. Or, you may think of the plates as storing half the energy (3 joules) and the electric field as storing half the energy (3 joules). It's your choice, as long as the *total* energy works out to be 6 joules.

So, you *can't* say the plates carry energy $U = \frac{1}{2}\frac{Q_1^2}{C}$ and the electric field carries *additional* energy $U = \int \frac{1}{2}\varepsilon_0 E^2 d^3\mathbf{r}$, because in that case the total electric potential energy comes out greater than 6 joules.

(d) I'll solve using two different methods. First, I'll think in terms of a gravitational analogy. ("Oh no, Andy, not again!") Then, I'll reason more directly about charge, voltage, and potential energy.

Method 1: Gravitational analogy

Since the two plates carry opposite charge, they attract. In other words, they "want" to fall towards each other, just as a dropped ball wants to fall towards the Earth. When a ball falls, it gains kinetic energy *at the expense of* potential energy. So, the potential energy decreases. Indeed, the ball loses potential energy even if we lower it to the ground gently, instead of letting it fall freely. In summary, when a ball moves in the direction that gravity pulls it, the ball loses potential energy.

Similarly, when the plates "fall" toward each other, the system loses electric potential energy. In other words, when the two plates move in the direction that the electric field pushes them—namely, towards each other—the potential energy decreases.

Method 2: Direct reasoning about electrostatic potential energy

Since $U_{\text{capacitor}} = \frac{1}{2}qV$, we can see whether the potential energy increases or decreases by figuring out what happens to the charge, and what happens to the potential difference between the plates, when the plates move closer together.

Since the battery remains disconnected, no charge can flow onto or off of the plates. No matter how we move them around, the plates retain charge $+Q_1$ and $-Q_1$, unless they touch. So, if U changes when we move the blobs closer together, it's *because* the potential difference between the plates changes, not because the charge on the plates changes.

Potential difference is the work it would take to move a 1-coulomb test charge from the negative blob to the positive blob. When the blobs are brought closer to together, they create a stronger (average) electric field between them. This tends to increase the work needed to push the test charge. But with the plates closer together, the test charge has less distance to travel. This decrease in distance *more than compensates* for the increase in force. It takes less work to move the test charge from one plate to the other, when the plates are closer together. In other words, the potential difference gets smaller when the plates get closer. Therefore, $U_{\text{capacitor}} = \frac{1}{2}qV$ also gets smaller when the plates get closer.

With these two oddly-shaped blobs, you might not have been able to guess that bringing the plates closer together decreases the potential difference. This general result pops out more readily when you consider a regular parallel-plate capacitor. As your textbook shows, the electric field strength inside a parallel-plate capacitor hardly changes when the plates are brought closer together. So, you can easily see that decreasing the plate separation, s, decreases the potential difference between the plates, $V = -\int \mathbf{E} \cdot d\mathbf{s}$.

OK, by two different methods, I've shown that bringing the plates closer together decreases the potential energy, U. But what about the capacitance? Well, potential energy and capacitance are inversely proportional: $U = \frac{Q_1^2}{2C}$. Therefore, since U decreases (and the charge stays the same), the capacitance must increase.

Let me make this result more physical. Capacitance tells us how much charge the system is willing to hold *per volt* of potential difference. As just argued, when we move the plates closer together, Q stays the same, but the potential difference V decreases. So, after moving closer together, the plates hold more charge *per volt*, because the voltage has decreased. In symbols, $C = q/V$ increases. Intuitively, with the plates closer together, transferring additional charge from one plate to the other becomes easier.

(e) Given everything we've done, you might be confused about whether to use $W = \Delta U = \frac{1}{2}qV$, or $W = \Delta U = qV$. I'll try to clear things up.

Here, we're not transferring charge from one plate to the other. The charge on the plates remains steady at $+Q_1$ and $-Q_1$. Therefore, the potential difference between the plates stays constant at $V_1 = 3.0$ volts. So, we're simply moving a charge q across a *fixed* potential difference V_1. Therefore,

$$W = \Delta U = qV_1 = (3.0 \times 10^{-6}\text{ C})(3.0\text{ V}) = 9.0 \times 10^{-6}\text{ J}.$$

By contrast, when you charge a capacitor, the potential difference between the plates keeps changing, as more and more charge builds up on the plates. For instance, as the system charges up from 0 to Q_1, the average potential difference between the plates is only $V_1/2$. That's why charging a capacitor requires work $W = \frac{1}{2}qV$, instead of a "full" qV. Intuitively, most of the tiny "bits" of charge that transfer from one plate to the other do not "fight" against the final potential difference, V_1. Since most of those bits of charge get transferred when the plates are only partially charged, most of them "fight" a potential difference smaller than V_1.

By the way, the potential created by the point charge itself cannot play a role in this problem. The electric field (and corresponding electric potential) generated by an object cannot push the object itself, any more than you can lift yourself off the ground. The particle "feels" only the electric fields created by the blobs. That's why I used V_1, the potential arising the blobs' electric field.

QUESTION 36-5

A long, thin conducting cylindrical rod of radius R and length l is placed inside a conducting cylindrical shell (tube) of radius $10R$ and length l. These two cylinders are concentric, as drawn here. In the side view, I colored in the rod, but drew only the outline of the cylindrical shell. The shell is basically a very long soda can, with the top and bottom cut out. My drawing fails to depict the fact that the rod and shell are much longer than they are wide: $l \gg 10R$.

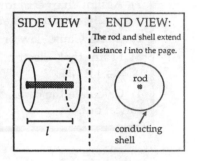

SIDE VIEW | END VIEW: The rod and shell extend distance l into the page.

l | rod | conducting shell

One terminal of a battery is hooked up to the rod, while the other terminal is connected to the cylindrical shell. As a result, the rod acquires an approximately uniform linear charge density $+\lambda_0$. The shell acquires equal and opposite linear charge density.

Express your answers in terms of R, l, and λ_0.

(a) (*Typical exam question*) What must the voltage of the battery be, to ensure that the rod acquires linear charge density λ_0? Derive any formulas you use.

(b) How many joules of chemical energy does the battery "use up" while charging up this system?

Part (c)

END VIEW: The light gray dielectric sheath envelops the rod.

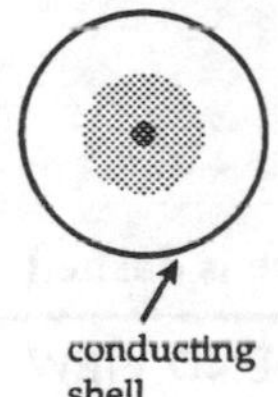

(c) Now consider a nonconducting dielectric sheath, which is a thick cylindrical shell of length l, inner radius R, and outer radius $5R$. Inside $r = R$, the sheath is hollow. The sheath has dielectric constant κ. *With the battery still connected to the rod and the conducting shell*, the sheath is slid over the rod, so that the sheath completely envelops the rod.

When the sheath is inserted, does the charge on the rod increase, decrease, or stay the same, as compared to its part (a) value? Justify your answer qualitatively, without heavy math.

(d) When the sheath is inserted, does the potential difference between the rod and the shell increase, decrease, or stay the same? Justify your answer.

ANSWER 36-5

(a) The potential difference between the two plates (the rod and the shell) must equal the voltage of the battery used to charge them up. For instance, if we charge the system with a 1.5-volt battery, then the rod acquires a potential that's exactly 1.5 volts higher than the shell's potential. A battery *enforces* a potential difference between the objects to which it's attached. So, to find the battery voltage that gives the rod and shell linear charge density $+\lambda_0$ and $-\lambda_0$, we must find the potential difference between those two "plates" when they carry that charge.

So, we're given the charge q (or at least, a closely related quantity), and we need to find the potential difference V. Therefore, you can conceptualize this as a capacitance problem. To solve for the capacitance, you'd find the potential difference V generated by the charge q. Then you'd calculate $C = q/V$. Here, although we don't care about C, we still need to find the potential difference generated by the charge on the rods. So, we need to complete the first three steps of the usual capacitance-finding strategy.

Capacitance-finding strategy:

1) Assign charge $\pm q$ to your plates. Here, we need not "pretend" to know the charge, because we really do. The rod and shell have linear charge density $+\lambda_0$ and $-\lambda_0$.
2) Using Gauss' law or other techniques, calculate the electric field between your two plates.
3) From that electric field, calculate the potential difference between the plates, using $V = -\int \mathbf{E} \cdot d\mathbf{s}$.
4) Calculate the capacitance using $C = q / V$. The arbitrary charge q you chose in step 1 should cancel out.

Let's do it.

Capacitance-finding strategy step 1: Assign charge to the plates.

As just mentioned, the rod carries linear charge density (charge per length) λ_0, while the cylindrical shell carries an equally large negative charge density.

Capacitance-finding step 2: Calculate the electric field between the plates.

We can approximate the rod and shell as infinitely long. This approximation introduces enough symmetry that Gauss' law allows us to solve for the field. Since I've applied Gauss' law to a charged rod in previous chapter, let me give an abbreviated presentation here:

Gauss' law steps 1 and 2: Draw in field lines, and draw a "good" Gaussian surface.

Since we need the electric field between the rod and the shell, that's where I'll draw my Gaussian surface. The field lines flow radially outward, as shown in my end-view drawing. By symmetry, the electric field has the same strength at any two points equidistant from the rod. Therefore, if we surround the rod with a "Gaussian cylinder," the electric field has constant strength over the entire side (tube) of that cylinder. I've given my Gaussian surface arbitrary radius r and length x. Notice that r must be between R and $10R$.

Gaussian surface is dashed

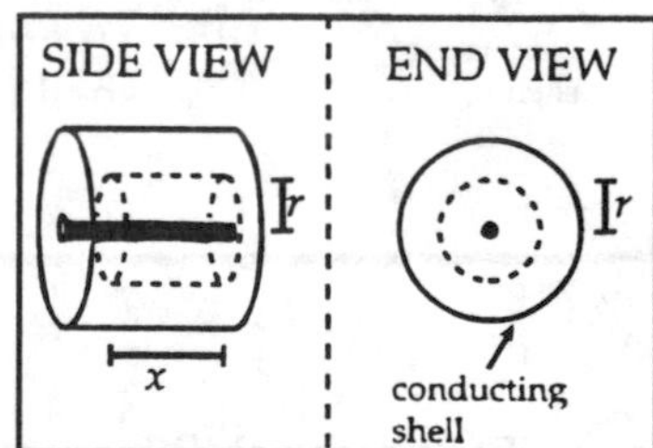

Gauss' law step 3: Use Gauss' law, $\int \mathbf{E} \cdot d\mathbf{A} = q_{\text{encl}} / \varepsilon_0$, to find the electric field.

Since E is constant over the whole side of the Gaussian cylinder, we can pull it outside the flux integral. Also, field lines only flow through the side of the Gaussian cylinder, not the ends. (Field lines "skim along" the ends, without actually piercing *through* those faces.) So,

$$\int \mathbf{E} \cdot d\mathbf{A} = \oint_{\text{side of Gaussian cylinder}} E\,dA = EA_{\text{side of Gaussian cylinder}} = E(2\pi r x),$$

since the side (tube) of a cylinder has surface area A = circumference $\times$ length.

Now let's work on the other side of Gauss' law. My Gaussian surface doesn't enclose any of the cylindrical shell. But it "chops off" a length x of the rod, which has linear charge density λ_0. Therefore,

$$q_{\text{encl}} = \frac{\text{charge}}{\text{length}} \times (\text{length of rod enclosed}) = \lambda_0 x.$$

Put all this together to get

$$\oint \mathbf{E} \cdot dA = \frac{q_{\text{encl}}}{\varepsilon_0}$$

$$E(2\pi rx) = \frac{\lambda_0}{2\pi\varepsilon_0 r}.$$

Cancel the x's and isolate E to get

$$E = \frac{\lambda_0}{2\pi\varepsilon_0 r},$$

where r is the distance from the central axis of the rod.

Now that we've calculated the electric field between the rod and shell, we can proceed to the next step of the capacitance-finding strategy.

Capacitance-finding strategy step 3: From the electric field, calculate the potential difference.

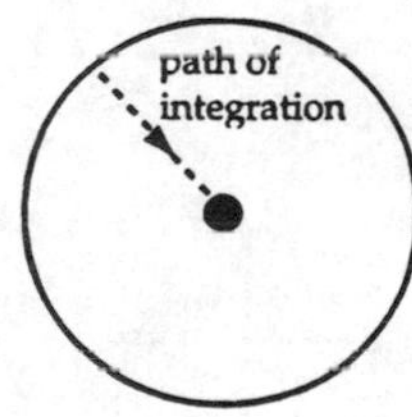

To find the potential difference between the rod and the shell, we must integrate the electric field over a path from the negative to the positive plate. I'll use a straight radial path from the shell ($r = 10R$) to the edge of the rod ($r = R$). Since the infinitesimal distance vector $d\mathbf{r}$ points radially outward, as does the electric field $\mathbf{E}$, the dot product $\mathbf{E}{\cdot}d\mathbf{r}$ reduces to the regular product, Edr. Therefore,

$$V = -\int_{\text{negative plate}}^{\text{positive plate}} \mathbf{E} \cdot d\mathbf{s} = \int_{r=10R}^{r=R} E dr$$

$$= -\int_{10R}^{R} \frac{\lambda_0}{2\pi\varepsilon_0} dr$$

$$= -\frac{\lambda_0}{2\pi\varepsilon_0} \ln r \Big|_{10R}^{R}$$

$$= -\frac{\lambda_0}{2\pi\varepsilon_0} [\ln R - \ln 10R]$$

$$= \frac{\lambda_0}{2\pi\varepsilon_0} [\ln 10R - \ln R]$$

$$= \frac{\lambda_0}{2\pi\varepsilon_0} \ln \frac{10R}{R}$$

$$= \frac{\lambda_0}{2\pi\varepsilon_0} \ln 10.$$

That's the potential difference between the rod and shell when they carry linear density λ_0, i.e., the battery voltage needed to give the rod and shell those charges. We need not complete step 4 of the capacitance-finding strategy, because we don't want the capacitance.

(b) Since energy is conserved, the chemical energy lost by the battery—i.e., the work done by the battery—converts into the electric potential energy of the rod and shell. For instance, if the battery loses 5 joules of chemical energy, then the capacitor ends up with 5 joules of potential energy. So, we can determine the chemical energy lost by the battery simply by figuring out the potential energy stored in the charged-up capacitor.

In question 36-2d, by treating the charge-up process as the successive transfer of tiny "bits" of charge from one plate to the other, we found that

$$U_{\text{capacitor}} = \frac{1}{2}QV,$$

where Q is the "final" charge and V is the final voltage (potential difference). Well, from part (a), we know the final potential difference between the plates. It's $V = \frac{\lambda_0}{2\pi\varepsilon_0}\ln 10$. And we can also figure out the total charge Q on the rod. It's

$$Q = \frac{\text{charge}}{\text{length}} \times (\text{length of rod}) = \lambda_0 l.$$

So,

$$\begin{aligned}\text{Chemical energy lost by battery} = U_{\text{capacitor}} &= \frac{1}{2}QV,\\ &= \frac{1}{2}(\lambda_0 l)\left(\frac{\lambda_0}{2\pi\varepsilon_0}\ln 10\right)\\ &= \frac{\lambda_0^2 l}{4\pi\varepsilon_0}\ln 10.\end{aligned}$$

(c) When the capacitor gets disturbed, by moving the plates or by inserting a dielectric, two things can happen. The potential difference can stay the same while the charges "adjust" themselves. Or, the charge on each plate can stay the same, while the potential difference changes. (Or the charge and voltage can both adjust.) Always begin this kind of problem by figuring out if the charge or voltage stays fixed. Once you know what *doesn't* change, it's easier to figure out what *does* change, and how.

In this case, as the dielectric gets inserted, the plates (rod and shell) remain connected to the battery. A battery "enforces" a fixed potential difference between the plates, equal to the battery's voltage. For instance, a 1.5-volt battery maintains the potential difference between the plates at 1.5 volts, no matter whether you insert dielectric, move the plates around, or watch glitzy Hollywood movies with lots of car chases. The battery will transfer charge from one plate to the other, if needed, to ensure that the potential difference stays fixed.

We just answered part (d)! Because the battery enforces a fixed potential difference, inserting the dielectric can't change it. V keeps its value from part (b).

So, if anything changes, it's the charge on the plates. To figure out whether the charge increases or decreases, we must think physically about what a dielectric does. A dielectric *dampens* electric fields. Specifically, the polarized dipoles in the dielectric generate their own electric field that partially cancels the external electric field. See your textbook for the details.

Given all this, I can explain why the charge on the rod (and shell) must *increase* after the dielectric gets inserted. I'll use what mathematicians call a "proof by contradiction." Specifically, I'll "assume" that the charge on the rod does *not* increase, and show that this assumption leads to an incorrect conclusion. Here's the argument:

Suppose the charge on the rod does not increase. Then the dielectric dampens the field between R and $5R$. Therefore, when we insert the dielectric, the average field strength between the rod and the shell decreases. But if the average field decreases, so does the potential difference between the rod and the shell, $V = \int \mathbf{E} \cdot d\mathbf{s}$. This conclusion contradicts the fact that the battery enforces a *fixed* potential difference. Therefore, my supposition that the charge doesn't increase must be wrong.

Let me re-explain this more intuitively. Since the battery enforces a fixed potential difference V, and since $V = -\int \mathbf{E} \cdot d\mathbf{s}$, the average electric field strength between the rod and shell must stay the same, when we insert the dielectric. Therefore, *even though* the dielectric dampens the field, the rod must somehow generate as big an (average) electric field as it did before. For this reason, the charge on the rod must increase. Intuitively, the charge on the plates must increase exactly enough to compensate for the field-dampening caused by the dielectric. Since dielectrics dampen electric fields, *more* charge is needed to generate the *same* old field.

(d) We've already answered part (d). As long as the battery remains connected to the plates, it enforces a fixed potential difference between them, transferring charge from one plate to other (if necessary) to maintain that fixed V.

QUESTION 36-6

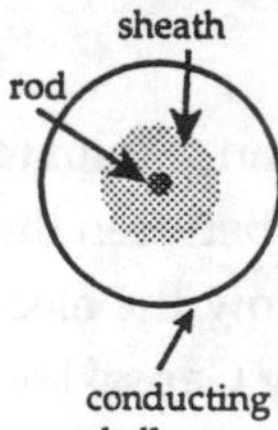

(Challenge problem. Check with your instructor to see if you need to be able to solve problems of this type and difficulty)

I'll now ask additional questions about the rod, shell, and dielectric sheath from question 36-5. Recall that the rod has radius R and length l. The cylindrical shell has radius $10R$ and length l. The dielectric sheath has radius $5R$, length l, and dielectric constant κ. You may freely use any results we obtained in question 36-5.

(a) Calculate the new capacitance of the rod-and-shell system, with the dielectric sheath inserted. Hint: For a given charge on the rod, the electric field *outside* the sheath is the same as it would have been if the sheath were missing. You can show this using Gauss' law.

(b) If we consider this system's energy to be "stored" in the electric field, what percentage of the total energy is stored by the field inside the dielectric sheath? Hint: the electric field drops to zero outside the conducting shell. You can solve this even if you didn't finish part (a).

(c) The sheath and rod are frictionless, and are aligned horizontally. Suppose the sheath is given a light tap, so that it slides rightward along the rod, while the rod and cylindrical shell are held still. Will the sheath slide all the way off the rod? Or will it slow down, turn around, and slide leftward, back to where it started? Justify your answer. You need not use heavy math.

(d) Calculate how much work is required to pull the sheath completely off the rod. Hint: Think about the potential energy of the overall system with the sheath in place, and the potential energy of the overall system with the sheath missing. If you didn't obtain an answer to part (a), call the capacitance C' and solve this problem in terms of C' and the given constants.

ANSWER 36-6

(a) The capacitance specifies how much charge the system is willing to carry, per volt of potential difference. In question 36-5c, we found that inserting the dielectric causes the plates to become *more* charged up, for fixed voltage. So, the dielectric increases the capacitance.

You may have heard that a dielectric increases capacitance by κ. That's true only if the dielectric fills the *whole* space between the capacitor plates, i.e., only if it weakens the electric field *everywhere* by a factor of κ. So, we can't jump to the answer. Instead, we must apply the capacitance-finding strategy summarized in question 36-5a.

Capacitance-finding strategy step 1: Assign arbitrary charge q to the plates.

This strategy involves "pretending" the plates hold known charge $+q$ and $-q$, and figuring out the potential difference V produced by that charge. Then we can calculate the capacitance using $C = q/V$. In this case, you might be tempted to "reverse" the usual strategy, by starting with the known V and calculating the unknown q on the plates, instead of starting with "pretend-known" q and calculating the corresponding V. But it's much easier to "pretend" you know q and then calculate the corresponding V. Using the "wrong" q won't mess you up, because when you divide q by V to obtain the capacitance, the q cancels out.

So, I'll assign arbitrary charge q to the rod and $-q$ to the shell. Hence, the rod carries linear charge density (charge per length) $\lambda = q/l$.

Capacitance-finding step 2: Find the electric field produced by the charge.

We need to know **E** between the rod and shell, so that in step 3 we can evaluate $V = -\int \mathbf{E} \cdot d\mathbf{r}$ The path of integration from shell to rod passes through the empty space between the dielectric and the shell, and also goes through the dielectric. Therefore, we need to know the electric field both inside and outside the dielectric. We can find those fields separately, using Gauss' law both times.

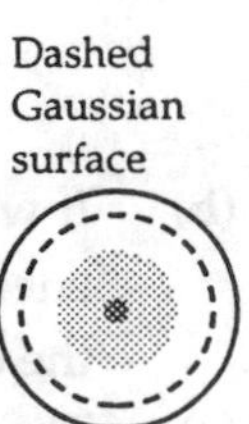

Let's start by finding the electric field in the empty space outside the dielectric, between the shell and the rod. The dashed Gaussian cylinder pictured here has radius $r > 5R$. What charge does it enclose? Well, the dielectric sheath carries no *net* charge. Due to polarization, the outer and inner surfaces of that sheath become coated with positive and negative charge, respectively. (See your textbook for details.) But those two coatings contain equal and opposite charge. Therefore, the net charge enclosed by my Gaussian cylinder arises entirely from the rod. In other words, q_{encl} is the same *as if* the dielectric sheath were missing. For this reason, the field outside the dielectric sheath is the same as if the sheath were absent. We already calculated that field in question 36-5a. It's $E = \frac{\lambda}{2\pi\varepsilon_0 r}$.

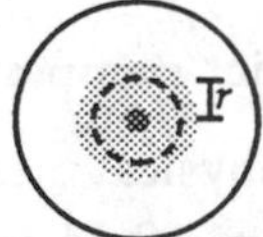

So far, we've found the field outside the dielectric. We still need the field inside the dielectric. To find it, I'll invoke Gauss' law, with my Gaussian surface *inside* the dielectric, as drawn here. This Gaussian cylinder has radius r, and extends distance x into the page.

Is q_{encl} simply the q on the rod (or rather, the part of the rod enclosed by my Gaussian surface)? No, because of the polarized surface charge density on the surfaces of the dielectric sheath. In "response" to the positive rod, a negative surface charge density gets induced on the inside surface of the sheath, at $r = R$. A positive surface charge density gets induced on the outside surface of the sheath, at $r = 5R$. But my Gaussian surface encloses only the negative surface charge, not the positive one. So, my Gaussian surface encloses the charge on the rod, *plus* the negative surface charge just discussed. When you work out the size of that negative charge, and then simplify things with algebra, the following modified version of Gauss' law pops out:

Modified Gauss' law $$\oint \mathbf{E} \cdot dA = \frac{q_{\text{encl free}}}{\kappa\varepsilon_0},$$

where $q_{\text{encl free}}$ refers to the "free" charges, i.e., the charges *other* than those induced in the dielectric. So here, $q_{\text{encl free}}$ means the charge on the rod. The κ factor "takes into account" the induced surface charge on the dielectric. I can't emphasize enough that this modified Gauss' law is *not* a new law. Regular Gauss' law applies to *all* situations, including dielectrics. But when you apply regular Gauss' law to dielectrics, this modified version emerges.

OK, let's simplify both sides of this equation in the usual way. Because the field points radially outward, the dot product $\mathbf{E} \cdot d\mathbf{A}$ reduces to a regular product. Also, by symmetry, the field strength is constant over the whole side (tube) of my Gaussian cylinder. Hence, we can pull E outside the flux integral:

$$\oint \mathbf{E} \cdot dA = E \int_{\text{side of Gaussian cylinder}} dA = EA_{\text{side of Gaussian cylinder}} = E(2\pi r x)$$

Now I'll address the other side of the modified Gauss' law. Since the Gaussian surface encloses length x of the rod, which has linear charge density $\lambda = q/l$, we get

$$q_{\text{encl free}} = \frac{\text{charge}}{\text{length}} \times (\text{length of rod enclosed}) = \lambda x.$$

Put all this information into modified Gauss' law to get

$$\oint \mathbf{E} \cdot d\mathbf{A} = \frac{q_{\text{encl free}}}{\kappa\varepsilon_0}$$

$$E(2\pi r^2) = \frac{\lambda x}{\kappa\varepsilon_0}.$$

Isolate E to get

$$E = \frac{\lambda}{2\pi\kappa\varepsilon_0 r}.$$

Compare this to the field outside the dielectric material, $E = \frac{\lambda}{2\pi\varepsilon_0 r}$. The dielectric "dampens" the field by a factor of κ, the dielectric constant. This result gives us insight into the physical meaning of κ. Roughly speaking, it's the factor by which the dielectric dampens the electric field from its "natural" strength.

At this point, we've found the electric field between the rod and shell, both inside and outside the dielectric, when the rod has linear charge density λ (and the shell has equal and opposite charge). Now we can calculate the potential difference between the shell and the rod.

Capacitance-finding strategy step 3: From the electric field, calculate the potential difference.

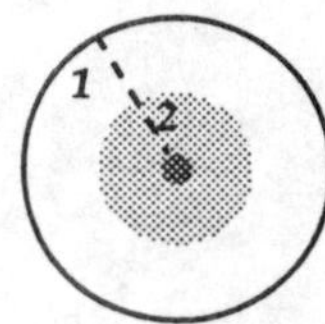

As in question 36-5a, we can integrate $\mathbf{E} \cdot d\mathbf{r}$ along a radial path from the shell to the rod. But now the path consists of two subpaths. Subpath 1, which is entirely outside the dielectric, goes from the shell ($r = 10R$) to the outer edge of the dielectric ($r = 5R$). Subpath 2, which is entirely inside the dielectric material, goes from the outer edge of the dielectric ($r = 5R$) to the edge of the rod ($r = R$). So, the total potential difference between the shell and the rod is

$$V = -\int_{\text{whole path}} \mathbf{E} \cdot d\mathbf{r}$$

$$= -\left(\int_{\text{subpath 1}} \mathbf{E} \cdot d\mathbf{r} + \int_{\text{subpath 2}} \mathbf{E} \cdot d\mathbf{r}\right)$$

$$= -\left(\int_{10R}^{5R} E_{\text{outside dielectric}}\,dr + \int_{5R}^{R} E_{\text{inside dielectric}}\,dr\right)$$

$$= -\left(\int_{10R}^{5R} \frac{\lambda}{2\pi\varepsilon_0 r}\,dr + \int_{5R}^{R} \frac{\lambda}{2\pi\varepsilon_0 r}\,dr\right).$$

Evaluate these integrals to get

$$V = \frac{\lambda}{2\pi\varepsilon_0}\left\{\ln r\Big|_{10R}^{5R} + \frac{1}{\kappa}\ln r\Big|_{5R}^{R}\right\}$$

$$= \frac{\lambda}{2\pi\varepsilon_0}\left\{\ln 10R - \ln 5R + \frac{1}{\kappa}(\ln 5R - \ln R)\right\}$$

$$= \frac{\lambda}{2\pi\varepsilon_0}\left\{\ln\frac{10R}{5R} + \frac{1}{\kappa}\ln\frac{5R}{R}\right\}$$

$$= \frac{\lambda}{2\pi\varepsilon_0}\left\{\ln 2 + \frac{1}{\kappa}\ln 5\right\}.$$

Capacitance-finding strategy step 4: Calculate the capacitance using C = q/V.

I've been working in terms of the rod's linear density, $\lambda = q/l$. So, $q = \lambda l$. Hence,

$$C' = \frac{q}{V} = \frac{\lambda l}{\frac{\lambda}{2\pi\varepsilon_0}\left(\ln 2 + \frac{1}{\kappa}\ln 5\right)} = \frac{2\pi\varepsilon_0 l}{\ln 2 + \frac{1}{\kappa}\ln 5},$$

which is higher than the system's capacitance without the dielectric sheath in place. This confirms our earlier conclusion that inserting the dielectric makes the system more "willing" to hold charge, by dampening the electric field. Notice that I used C' to denote the capacitance of the sheathed system.

(b) If we consider all the electric potential energy to be stored in the field itself, not in the capacitor plates, then the *energy density* , i.e., the energy per unit volume, is

$$u = \frac{1}{2}\varepsilon_0 E^2.$$

To find the energy "stored" in a volume of space, integrate the charge density over that volume. So, we can first find $U_{\text{in dielectric}}$ by integrating u over the region of space occupied by the sheath. Similarly, we can calculate $U_{\text{outside dielectric}}$ by integrating u over the empty space between the dielectric and the shell. Then we can evaluate

$$\text{fraction of energy sotored by field in dielectric} = \frac{U_{\text{in dielectric}}}{U_{\text{total}}} = \frac{U_{\text{in dielectric}}}{U_{\text{in dielectric}} + U_{\text{outside dielectric}}}.$$

(Since the field drops to zero outside the shell, as you can prove using Gauss' law, we need not worry about that region of space.)

Subproblem 1: Calculate $U_{\text{in dielectric}}$.

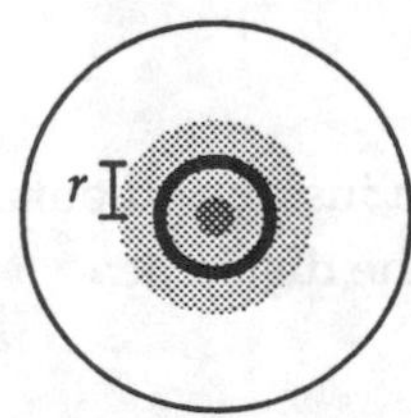

The black infinitesimal volume element extends distance l into the page.

We must evaluate $U_{\text{in dielectric}} = \int_{\text{dielectric}} u\,dV$, where dV is an infinitesimal volume element, not an infinitesimal potential. To find a useful expression for dV, think of the dielectric slab as consisting of concentric cylindrical shells. A given shell has length l (into the page), thickness dr, and arbitrary radius r. I've broken the slab into cylindrical shells because, within a given shell, the electric field has constant strength (by cylindrical symmetry).

What's the infinitesimal volume dV of this shell? Well, as the drawing shows, the top of the shell has area dA = length ↔ width = $2\pi r dr$. So, dV = height × area = $l dA = 2\pi l r dr$.

Using this expression for dV, and remembering that the dielectric extends from $r = R$ to $r = 5R$, we get

$$
\begin{aligned}
U_{\text{in dielectric}} &= \int_{\text{dielectric}} u\,dV \\
&= \int_R^{5R} u(2\pi l r dr) \\
&= \int_R^{5R} \frac{1}{2}\varepsilon_0 E^2 (2\pi l r dr) && \left[\text{since } u = \frac{1}{2}\varepsilon_0 E^2\right] \\
&= \int_R^{5R} \frac{1}{2}\varepsilon_0 \left(\frac{\lambda}{2\pi\kappa\varepsilon_0 r}\right)^2 (2\pi l r dr) && \left[\text{since } E_{\text{in dielectric}} = \frac{\lambda}{2\pi\kappa\varepsilon_0 r}\right] \\
&= \frac{\lambda^2 l}{4\pi\kappa^2\varepsilon_0}\int_R^{5R}\frac{1}{r}dr \\
&= \ldots \text{ evaluate integral} \ldots \\
&= \frac{\lambda^2 l}{4\pi\kappa^2\varepsilon_0}\ln 5.
\end{aligned}
$$

Subproblem 2: Calculate $U_{\text{outside dielectric}}$.

Divide the space between the dielectric and the conducting shell into (imaginary) cylindrical shells of radius r and length l. As we saw above, each shell has infinitesimal volume $dV = 2\pi l r dr$. So,

$$
\begin{aligned}
U_{\text{outside dielectric}} &= \int_{\text{outside dielectric}} u\,dV \\
&= \int_{5R}^{10R} \frac{1}{2}\varepsilon_0 E^2 (2\pi l r dr) \\
&= \int_{5R}^{10R} \frac{1}{2}\varepsilon_0 \left(\frac{\lambda}{2\pi\kappa\varepsilon_0 r}\right)(2\pi l r dr) && \left[\text{since } E_{\text{outside dielectric}} = \frac{\lambda}{2\pi\varepsilon_0 r}\right] \\
&= \frac{\lambda^2 l}{4\pi\varepsilon_0}\int_{5R}^{10R}\frac{1}{r}dr \\
&= \frac{\lambda^2 l}{4\pi\varepsilon_0}\ln 2.
\end{aligned}
$$

Now that we've calculated the electric potential energy stored both inside and outside the dielectric, we can calculate what fraction of the total energy "lives" inside the dielectric:

$$
\begin{aligned}
\frac{U_{\text{in dielectric}}}{U_{\text{total}}} &= \frac{U_{\text{in dielectric}}}{U_{\text{in dielectric}} + U_{\text{outside dielectric}}} \\
&= \frac{\dfrac{\lambda^2 l}{4\pi\kappa^2\varepsilon_0}\ln 5}{\dfrac{\lambda^2 l}{4\pi\kappa^2\varepsilon_0}\ln 5 + \dfrac{\lambda^2 l}{4\pi\varepsilon_0}\ln 2} \\
&= \frac{\ln 5}{\ln 5 + \kappa^2 \ln 2}.
\end{aligned}
$$

(c) In this problem, we must figure out whether the force on the dielectric sheath tries to push it off the rod (rightward) or back onto the rod (leftward). The force, involving a complex interaction between induced charge in the dielectric and the "fringe" fields at the edge of the rod, is difficult to sort out. Fortunately, we can *sidestep* these messy details. To figure out the direction of this force, we need to think *only* about the system's potential energy. Here's why.

A system "wants" to lower its potential energy. In other words, a field (such as a gravitational or electric field) always pushes an object so as to *lower* its potential energy. For instance, when you drop a ball, gravity pulls it downward, decreasing the ball's potential energy.

Given this insight, we can address the problem at hand. If the electric potential energy decreases as the sheath slides off the rod, then the force on the sheath pushes it off the rod. By contrast, if the potential energy is minimized when the sheath fully envelops the rod, then the electric force tries to push the sheath back into that position. So, we need to figure out whether removing the sheath leads to higher or lower potential energy.

Intuitively speaking, the more charged up the system becomes (for fixed potential difference), the more energy it stores. This intuition corresponds to the formula $U_{\text{capacitor}} = qV/2$. And in question 36-5c, we already figured out that inserting the dielectric *increases* the charge on the plates. Let me briefly review why. The potential difference—and hence, the average electric field between the plates—stays the same no matter what, because the battery enforces a fixed potential difference. But the dielectric "dampens" the electric field. Therefore, the rod and shell must become more charged up, to "compensate" for the field-dampening effect of the dielectric.

In summary, inserting the dielectric sheath causes the plates to become more charged up, even though the potential difference between them stays the same. So, the sheathed system has higher potential energy. Therefore, the sheath "wants" to slide off the rod, in order to lower the potential energy. In other words, the electric force pushes the sheath off the rod (unless the sheath is "perfectly centered"). The rightward-sliding sheath experiences a rightward force that accelerates it off the rod.

Before moving on, let me point out that, if the battery had been disconnected, then the sheath would have "wanted" to stay *on* the rod. See if you can explain why. (Hint: With the battery disconnected, the charge on each plate stays fixed, and hence, inserting the dielectric decreases the electric field between the plates.) Furthermore, ask your instructor whether you need to be able to calculate the magnitude of these forces, something we didn't do here.

(d) As just shown, when the battery stays connected, the sheath "wants" to slide off the rod. So, it takes *negative* work to pull the sheath off the rod. In other words, the sheath can do work on whoever removes the sheath from the rod, by pushing on his hand.

To calculate the work, keep in mind the relationship between work and potential energy, $W = \Delta U$. For instance, suppose it takes –5 joules of work to pull the sheath off the rod. Then, the potential energy with the sheath removed is 5 joules less than the potential energy with the sheath fully inserted. This example suggests a strategy. We can figure out the system's potential energy both before and after the sheath gets removed. The work needed to remove the sheath is simply the difference,

$$W_{\text{to remove sheath}} = \Delta U = U_{\text{sheath off}} - U_{\text{sheath inserted}}.$$

Fortunately, in earlier problems, we compiled enough information to write down these two potential energies. Remember, a capacitor carries potential energy

$$u = \frac{1}{2}qV, \qquad \text{which can be rewritten as } \frac{1}{2}\frac{q^2}{C} \text{ or } \frac{1}{2}CV^2.$$

Using this formula, let's first calculate $U_{\text{sheath off}}$. Way back in 36-5a, we found that when the unsheathed rod has linear charge density λ_0, the potential difference between the rod and the shell is $V = \frac{\lambda}{2\pi\varepsilon_0}\ln 10$. Since linear charge density is charge per length ($\lambda = q/l$), the unsheathed rod carries charge $q = \lambda_0 l$. So, the unsheathed system has potential energy

$$U_{\text{sheath off}} = \frac{1}{2}qV = \frac{1}{2}(\lambda_0 l)\left(\frac{\lambda_0}{2\pi\varepsilon_0}\ln 10\right) = \frac{\lambda_0^2 l}{4\pi\varepsilon_0}\ln 10.$$

Now let's figure out the system's potential energy with the sheath inserted. We don't know the charge on the rod. But since the same battery stays connected the whole time, the potential difference stays the same as it was before, $V = \frac{\lambda}{2\pi\varepsilon_0}\ln 10$. Furthermore, in part (a) above, we figured out the capacitance of the sheathed system: $C' = \frac{2\pi\varepsilon_0 l}{\ln 2 + \frac{1}{\kappa}\ln 5}$. So, the sheathed system carries potential energy

$$U_{\text{sheath fully inserted}} = \frac{1}{2}CV^2 = \frac{1}{2}\,\frac{2\pi\varepsilon_0 l}{\ln 2 + \frac{1}{\kappa}\ln 5}\left(\frac{\lambda_0}{2\pi\varepsilon_0}\ln 10\right) = \frac{\lambda_0^2 l}{4\pi\varepsilon_0}\frac{(\ln 10)^2}{\ln 2 + \frac{1}{\kappa}\ln 5}.$$

Therefore, the work needed to remove the sheath from the rod is

$$\begin{aligned} W_{\text{to remove sheath}} &= U_{\text{sheath off}} - U_{\text{sheath fully inserted}}. \\ &= \frac{\lambda_0^2 l}{4\pi\varepsilon_0}\ln 10 - \frac{\lambda_0^2 l}{4\pi\varepsilon_0}\frac{(\ln 10)^2}{\ln 2 + \frac{1}{\kappa}\ln 5} \\ &= \left(\frac{\lambda_0^2 l}{4\pi\varepsilon_0}\ln 10\right)\left(1 - \frac{\ln 10}{\ln 2 + \frac{1}{\kappa}\ln 5}\right). \end{aligned}$$

By plugging in any κ greater than one, you can confirm that this work is negative, as expected. Remember, the sheath "wants" to slide off the rod.

37 CHAPTER

Currents and Resistance in Materials

QUESTION 37-1

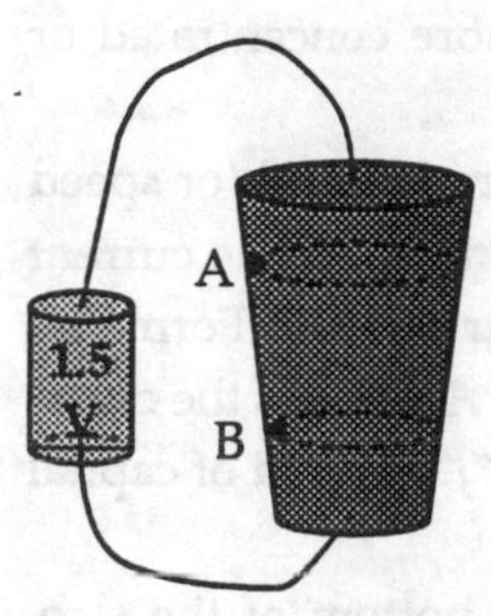

Consider a tapered slab of aluminum, as drawn here. By "circle A" and "circle B," I mean the imaginary dashed circles through points A and B.

The ends of the slab are connected, by essentially resistanceless wires, to opposite terminals of a 1.5-volt battery.

(a) Which electric current is bigger: the one through circle A, or the one through circle B? Or are they equally big? Explain your answer.

(b) Is the electric current *density* greater at the center of circle A or the center of circle B? Or are they equally big? Again, explain your answer.

(c) The current through circle A is 0.30 amps. What can you tell me about the resistive properties of the slab? Can you find its resistivity? Its resistance? Calculate what you can, and explain why you *can't* calculate the other things.

(d) Compare the resistivity of the slab near point A to its resistivity near point B. At which point, if either, is the resistivity greater? Hint: Don't get caught up in formulas. Think about what resistivity *means*.

(e) (*Challenge problem*) Where is the electric field bigger; at the center of circle A or the center of circle B? Or is it equally big at both points? Explain your answer.

ANSWER 37-1

(a) Parts (a) and (b) require a conceptual understanding of the difference between current and current density. Intuitively, you expect the current through circles A and B to differ in some way. As we'll see below, that intuition applies to the current density, but not the current.

Current is the charge per second passing a given point. For instance, the current at point B is the charge per second passing through circle B.

At first glance, you might want to say more current flows through circle A, because that circle is bigger. But the current through both circles *must* be the same, for the following reason. Any charge flowing through circle A also passes through circle B. Otherwise, charges would "build up" in the slab between those circles, something which cannot happen.

Let me elaborate upon this argument. Suppose, hypothetically, that 10 coulombs per second flow downward through circle A, but only 8 coulombs per second flow downward through circle B. Consider the segment of the slab between circles A and B. Ten coulombs per second enter that segment, while only 8 coulombs per second leave. Therefore, 2 coulombs per second of "excess" charge build up in the slab. This *can't happen* in a steady-flow circuit; if charges build up even temporarily, their mutual repulsion causes them to fly apart. For this reason, if 10 coulombs per second flow through circle A, then 10 coulombs per second must flow through circle B.

(b) Since the same charge per second flows through circles A and B, the charge flowing through circle B must be either be more concentrated (tightly packed), or more fast. To see why, imagine a crowd of people walking through a tapering tunnel. Suppose 5 people per second enter the wide end of the tunnel. Then, 5 people per second must exit the narrow end of the tunnel, or else bodies would start piling up in the tunnel, something that can't happen (for long!). In order for 5 people per second to exit the tunnel, they must either squeeze closer together or walk faster. Similarly, in order for the current through circle B to equal the current through circle A, the charges must become more concentrated or more fast near circle B.

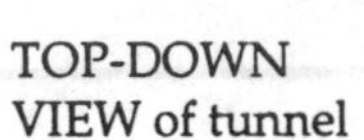

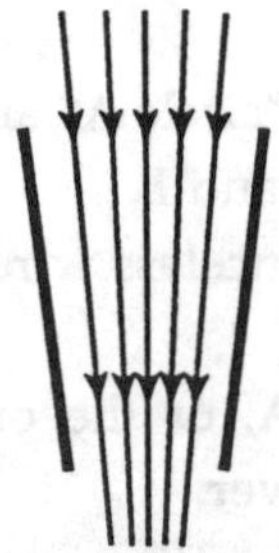

People walk through as indicated by the arrows.

Current density captures this intuitive idea of the concentration and/or speed of charges. For instance, if circle B has half the area of circle A, and the same current flows through each circle, then the current is twice as dense near point B. Formally, current density is the current *per cross-sectional area*: $J = I/A$, where A denotes the cross-sectional area and I denotes the current. Some books use "i" and "j" instead of capital letters to denote current and current density.

In summary, the current density increases towards the bottom of the slab, because the same current passes through a smaller and smaller area. The current den sity is higher at circle B.

By the way, starting with $J = I/A$, you can derive the formula $J = rv_d$, where r is the charge density of the mobile (conduction) electrons, and v_d is the drift velocity of those charges. See your textbook for a derivation. This formula confirms our "tunnel intuition" that current density depends on the concentration and speed of the flowing charges.

Here's a broader point. Although the "Currents and resistance in materials" chapter of your textbook may seem like a confusing barrage of formulas, the concepts and equations actually make sense, if you think about the underlying physical meaning.

(c) This problem forces us to clarify the difference between resistance and resistivity. Both concepts tell us, roughly speaking, how "resistant" an object is to having current pumped through it. But resistance refers to an *extrinsic* property of the whole object, while resistivity refers to an *intrinsic* property of the material comprising the object.

To illustrate the intrinsic/extrinsic distinction in a friendlier context, consider the "heaviness" of water. Picture two identical buckets of water, one of them half full, the other completely full. Which water is heavier? Well, in one sense, the water in the full bucket is heavier, because there's more of it. But in another sense, the water in the full bucket is no more "intrinsically heavy" than the water in the half-full bucket. Both blobs of water have the same density, i.e., the same mass *per volume*. So, in the extrinsic sense, the full bucket of water is heavier. But in the intrinsic sense, both blobs of water have equal "heaviness." From this example, we see that weight is an extrinsic property. It depends not just on the *type* of material, but also on the *amount*. By contrast, mass density is an intrinsic property, depending *only* on the type of material, not on the amount.

Similarly, resistivity is an intrinsic property of a material. It does not depend on the size or shape of the slab. We could look up r_{aluminum}, the resistivity of aluminum, in a table.

By contrast, resistance is extrinsic, like the weight of a bucket of water. It depends not just on the type of material, but also on the amount and configuration of material. (Resistance, unlike weight, depends on size *and* shape.) An aluminum slab is more resistant to having current pumped through it when it's long and thin, and more conducive to current when it's short and wide—just like a wider tunnel lets people "flow" through more easily.

Given all this, you might think we lack sufficient information to find the slab's resistance, since we don't know its exact dimensions. But here, we can infer the resistance using Ohm's law, usually written $V = IR$. The physical meaning of this law becomes clearer, however, if we write it

$$I = \frac{V}{R}. \qquad \textbf{Ohm's law}$$

Intuitively, the potential difference across the slab corresponds to a "hill" down which the charges "roll." A bigger V corresponds to a higher, steeper hill. At a rough intuitive level, that's why the current is proportional to V.

In this hill analogy, R corresponds to the "rockiness" of the hillside. As the charges roll, their progress gets slowed by collisions with rocks. Higher rockiness (resistance) leads to more collisions, and hence a slower flow. (Inside a conducting wire, atomic ions are the "rocks" against which the flowing electrons collide.) So, it makes sense that current is inversely proportional to resistance.

Here's a crucial point: In Ohm's law, V is the "cause" and I is the "effect." The potential difference *makes* current flow, and the resistance moderates the size of the current. Don't misinterpret "$V = IR$" as meaning that a voltage causes a resistance, or that a current times a resistance defines a voltage.

In any case, the battery's voltage—and hence, the potential drop across the slab—is $V = 1.5$ volts. And the current is $I = 0.30$ amps. Solve Ohm's law for R to get

$$R = \frac{V}{I} = \frac{1.5\text{V}}{0.30\text{A}} = 5.0\Omega,$$

where Ω stands for "ohms."

(d) You might be tempted to say the resistivity increases where the slab gets thinner, because the current becomes more "constricted." Well, current does indeed have more trouble flowing through the narrower section of the slab. So, the tapered slab allows less current to pass through it than an non-tapered cylinder would. In other words, the tapering affects an *extrinsic* property of the slab, namely its overall resistance. But resistivity is an *intrinsic* property, depending only on the material. Since the slab consists of aluminum at both A and B, those two points share the same resistivity.

If this seems weird, it's because we rarely distinguish between intrinsic and extrinsic quantities in everyday thinking. But here, the distinction becomes crucial. Remember, the full bucket is heavier not because that water is intrinsically denser, but simply because there's more of it. Similarly, the tapering near point B causes an increased resistance *not* because the aluminum near point B is more intrinsically resistive, but simply because a thinner slab constricts the current more, no matter what the slab is make of.

(e) First, I'll argue intuitively that the electric field at circle B must be larger, in order to push current through the more "constricted" part of the slab. Then, I'll show how the microscopic version of Ohm's law formalizes the intuitive argument.

As we saw in parts (a) and (b), the same current must flow through circles A and B, or else charge would illegally build up in the slab. Therefore, the current *density* increases at the narrow part of the slab, near circle B. As it turns out, the increased current density stems almost entirely from the increased speed of those charges, not from an increased concentration of charges.

So, the charges flowing through circle B move *faster* than the charges flowing through circle A. At both circles, however, the charges undergo collisions with atomic ions ("rocks"), collisions that slow the charges down. Therefore, since the charges move faster at circle B, the electric force pushing them through circle B must be greater than the electric force pushing them through circle A. Hence, the electric field at circle B must be greater than the field at circle A.

This intuitive reasoning gets formalized in the "microscopic" version of Ohm's law:

$$E = J\rho, \qquad \textbf{Ohm's law, microscopic version}$$

where ρ denotes resistivity. Intuitively, the electric field causes current density to flow. The stronger the field, the faster the charges move, i.e., the higher the current density J. But the material's resistivity, which corresponds to the intrinsic "rockiness," moderates the current density.

In this problem, points A and B share the same resistivity, because the whole slab consists of aluminum. But the current density is bigger at B, because the current is more constricted. Therefore, according to microscopic Ohm's law, the electric field must be bigger at B. We reached this same conclusion above using intuitive reasoning.

Notice how the microscopic Ohm's law corresponds to the regular version, $V = IR$. E corresponds to V, J corresponds to I, and r corresponds to R. Using these correspondences, you can derive the microscopic version of Ohm's law from the regular version, or vice versa. See your textbook for details.

QUESTION 37-2

A parallel-plate capacitor consists of conducting square plates of side-length $s = 0.50$ m, a distance $D = 0.010$ m apart. It's charged up using a $V_0 = 15$ volt car battery. As a result, the top plate acquires positive charge, and the bottom plate acquires equal negative charge. The battery is then disconnected.

Slab inserted between capacitor plates at $t = 0$, after the charged-up plates have been disconnected from the battery

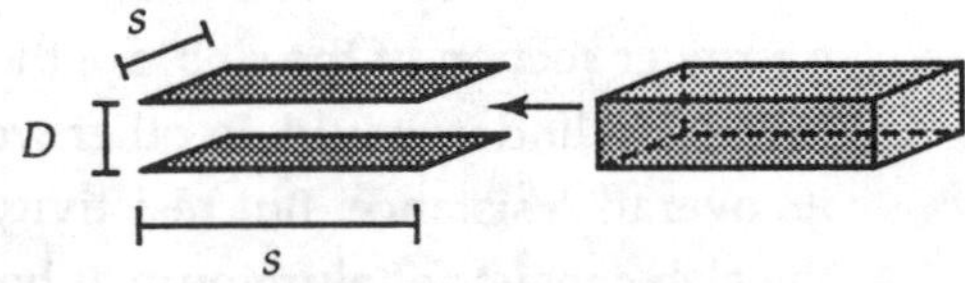

Next, a silicon slab of length s, width s, and height D is very quickly inserted. The slab, which has resistivity $\rho = 2500\ \Omega\cdot\text{m}$, completely fills the space between the plates. Let $t = 0$ denote the moment the slab was inserted.

(a) Immediately after $t = 0$, what is the current through the slab?

(b) Immediately after $t = 0$, what is the current density in the slab?

(c) Qualitatively, what happens to the charge on the plates as time passes? Does the charge on the top plate increase, decrease, or stay the same? Explain your answer, and sketch a rough, qualitative graph of Q vs. t.

(d) Same as part (c), but now think about the current through the slab. Does it increase, decrease, or stay the same? Again, sketch a rough qualitative graph.

(e) At $t = 0$, what is the electrical power generated in the slab?

(f) (*Very hard*) Assuming all the power gets "devoted" to creating heat, how much total heat dissipates in the slab during this discharge? Hint: You can solve this with minimal math. Feel free to use any textbook equations you want, without deriving them.

ANSWER 37-2

(a) Immediately after $t = 0$, the potential difference between the two plates is still $V_0 = 15$ V, the potential difference supplied by the car battery. So, that's the initial potential difference across the slab. We're looking for the initial current, call it I_0. To solve for I_0 using Ohm's law, $I_0 = V_0/R$, we need to know the resistance of the slab, when current flows through it vertically. Fortunately, we know the slab's resistivity, as well as its size and shape. Therefore, we can calculate the resistance using $R = \rho\frac{L}{A}$. But first, let me explain this formula intuitively.

The charges flowing through the slab are analogous to people walking through a tunnel. The wider the tunnel, the more people can flow through, other things being equal. In other words, the tunnel's resistance to "people-current" is inversely proportional to the area. Intuitively, that's why resistance is inversely proportional to the cross-sectional area A through which current flows.

What about a longer vs. shorter tunnel? Here, the analogy gets more subtle. After solving this problem, I'll take a few paragraphs to explain why a longer tunnel has more resistance, other things being equal. For now, let me take this result as given, and move on.

Finally, since resistivity corresponds to the concentration of rocks or potholes in the tunnel, the overall resistance should be proportional to resistivity.

In summary, based on the tunnel analogy, we expect resistance to be proportional to resistivity and length, but inversely proportional to area:

$$R = \rho\frac{L}{A},$$

where ρ is resistivity, L is length, and A is cross-sectional area. This formula is correct.

Current flows vertically through the dark-lined square of area s^2.

By contrast, if current flowed rightward, the rectangle drawn here would show the cross-sectional area.

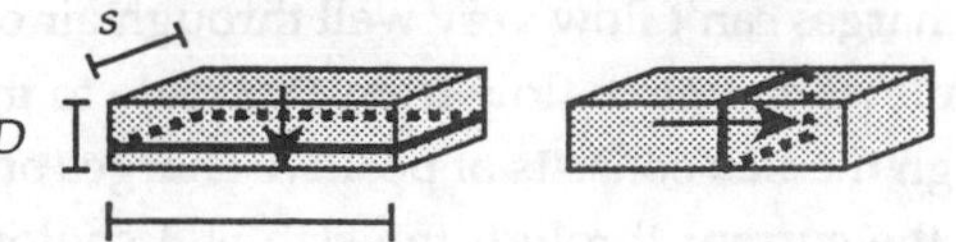

Here, we must figure out whether the length is s or D, and whether the cross-sectional area is sD or s^2. Picture the flowing charges. They travel vertically. So, they flow through length $L = D$, the distance from the top to the bottom of the slab. Furthermore, as this diagram indicates, the charges flow through cross-sectional area $A = s^2$. So,

$$R = \rho\frac{V}{A}\ \rho\frac{D}{s^2} = (2500\ \Omega\cdot\text{m})\frac{0.010\text{m}}{(0.50\,\text{m})^2} = 100\ \Omega\,.$$

Substitute this resistance into Ohm's law to get

$$I_0 = \frac{V_0}{R} = \frac{15\text{V}}{100\Omega} = 0.15\text{A}\,.$$

Theoretical clarification: Why is resistance proportional to the resistor's length?

Charges flow through a slab or wire because the electric field in the slab pushes them. Here, the charged capacitor plates create this field. In other problems, a battery creates the field. But in all cases, the potential difference relates to the electric field by

$$V = \int \mathbf{E} \cdot d\mathbf{s} = El,$$

where l is the distance between the capacitor or the length of the wire, i.e., the distance across which the charges must flow. For simplicity, I assumed that the field inside the slab or wire is uniform. This same reasoning works for messy fields, too. But for now, let's use $V = El$, and hence $E = V/l$. Here's the point: For a *given* potential difference V across a slab, if you increase the slab length l, you decrease the electric field—that is, you decrease the force pushing the charges through the slab. So, other things being equal, a longer slab allows less current to flow through it. That's why resistance is proportional to slab length.

Let me repeat this argument in terms of the tunnel analogy. A fixed potential difference corresponds to a fixed height (altitude) difference between the two ends of the tunnel. For instance, maybe the end of the tunnel is 20 meters below the entrance. In that case, the shorter the tunnel, the steeper it is. For example, a 40-meter tunnel that must descend 20 vertical meters is very steep. Cars inside this tunnel would "flow" extremely quickly, pulled by the steep incline. By contrast, a 500-meter tunnel that drops by 20 vertical meters isn't steep at all. Cars would roll down the slope more gently. So, for a fixed height difference between the two ends of the tunnel, a shorter tunnel allows "car-current" to flow through more quickly. In this sense, a longer tunnel puts up more "resistance" to car-current. Similarly, a longer slab or wire puts up more resistance to electrical current.

(b) Current density specifies the *concentration* of the current. It's the current *per cross-sectional area*. From part (a), the vertically-moving charges flow through cross-sectional area $A = s^2$. Therefore, immediately after $t = 0$, the current density is

$$J_0 = \frac{I_0}{A} = \frac{I_0}{s^2} = \frac{0.15\text{A}}{(0.50\text{m})^2} = 0.60 \text{ amps per square meter.}$$

(c) To solve (c) and (d), we must think physically about what's going on; formulas alone won't cut it! Let's follow the usual textbook convention of assuming that current consists of positive charges, even though electrons do the actual flowing.

The positive charges on the top plate feel attracted to the negative charges on the bottom plate. In other words, the charged plates create a downward electric field that pushes positive charges toward the bottom plate. But despite this force, the charges can't flow very well through air or empty space. By contrast, the semiconducting silicon slab allows charge to flow from one plate to the other.

Here's my point. The current flowing through the slab consists of positive charges on the top plate flowing to the bottom plate. For instance, if the current through the slab is 3 coulombs per second, that's just another way of saying that 3 coulombs per second flow off the top plate and onto the bottom plate. The current, I, is simply the rate at which charge flows off the top plate. So, if Q

denotes the charge on the top plate, then $I = -dQ/dt$. The minus sign indicates that the current is positive when the positive plate *loses* charge, i.e., when positive charges flow *off* the upper plate.

So, as time passes, the charge on the positive plate decreases. Simultaneously, as those positive charges flow onto the negative plate, the negative plate also "loses" charge. For instance, the charge on that plate might change from –5 coulombs to –4 coulombs to –3 coulombs . . . . In summary, the charge on the plates starts off high but gradually goes down to zero, as graphed here. In later chapters, we'll se that Q decreases exponentially. But for now, I just wanted to understand that Q decreases. The Q vs. t graph must reflect this.

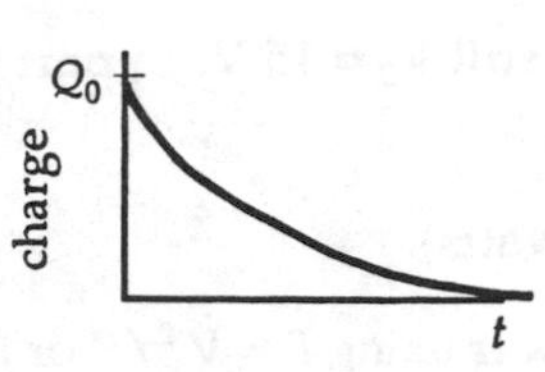

(d) Now let's figure out whether the current through the slab increases with time, decreases with time, or stays constant. You can't assume that I decreases just because Q decreases. For instance, if Q steadily decreases from its initial value to zero, then $I = -dQ/dt$ is constant. As it turns out, I does decrease as time passes. But to reach this conclusion, we must use more subtle reasoning.

By Ohm's law, the current through the slab is proportional to the potential difference between the top and bottom plate: $I = V/R$. This potential difference arises from the electric field between the plates, which in turn arises from the charge on the plates. As we saw in part (c), the charge on the plates decreases. Therefore, the potential difference between the plates decreases. For this reason, the current through the slab, $I = V/R$, decreases with time.

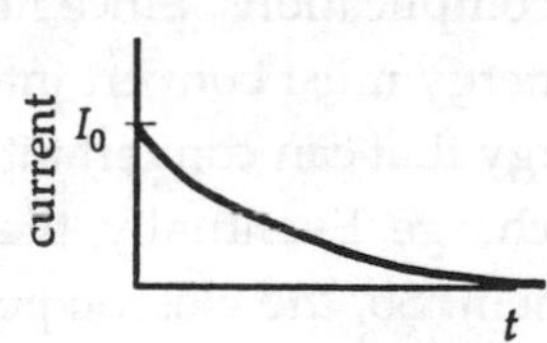

Later, you'll see that the current, like the charge on the plates, "decays" exponentially. But in this problem, I just wanted you to realize why I decreases instead of staying constant.

(e) We can solve using the electrical power formula $P = VI$. Before chugging away, let me show where that formula comes from.

By definition, power is the rate at which energy gets transferred. Here, the potential energy stored in the capacitor converts into the heat dissipated in the slab. So, the power is the rate at which, the potential energy decreases: $P = -dU/dt$. Recall that a capacitor carries potential energy $U = \frac{1}{2}qV$, which we can rewrite as $U = \frac{1}{2}\frac{q^2}{C}$. So,

$$\begin{aligned} P &= -\frac{dU}{dt} = -\frac{d}{dt}\left(\frac{1}{2}\frac{q^2}{C}\right) \\ &= -\frac{1}{2C}\left(2q\frac{dq}{dt}\right) \\ &= -\frac{q}{C}\frac{dq}{dt} \\ &= -V\frac{dq}{dt} \qquad \text{[since } C = q/V\text{]} \\ &= VI, \end{aligned}$$

where in the last step I used the relationship between the charge on the plates and the current in the slab, $I = -dq/dt$.

By the way, since $V = IR$, you can rewrite this equation for electrical power as

$$P = VI = \frac{V^2}{R} = I^2R.$$

Immediately after $t = 0$, the potential difference between the plates is still $V_0 = 15$ V. In part (a), we calculated the initial current to be $I_0 = 0.15$ A. So, the initial power is

$$P = VI = (15\text{ V})(0.15\text{ A}) = 2.2 \text{ joules per second (i.e., watts)}.$$

That's the initial rate at which heat gets produced. You get the same answer using $P = V^2/R$ or $P = I^2R$, if you substitute in the value we obtained for R in part (a), namely $R = 100$ W.

(f) Here, the "power" tells us the rate at which heat gets produced: $P = d(\text{Heat})/dt$. So, you could calculate the total heat by summing up the little "bits" of heat created during each microsecond, i.e., by integrating over the power:

$$\text{Total heat energy generated} = \int d(\text{Heat}) = \int P dt.$$

Unfortunately, we don't know the power as a function of time. We know only the initial power, immediately after $t = 0$. To obtain $P(t)$, we'd need to figure out the current and voltage as a function of time.

Fortunately, energy conservation allows us to sidestep these complications. Since heat is a form of energy, it can't just appear out of nowhere. Another kind of energy must convert into heat. Here, the electric potential energy stored provides a "reservoir" of energy that can convert into heat. As current flows through the slab (creating heat), the plates lose their charge. Eventually, the plates end up with no charge, and hence, no potential difference between them. So, the electric potential energy of the system drops to zero. Therefore, **all of the system's initial potential energy converted into heat**. For instance, if the capacitor carried 10 joules of potential energy, then 10 joules of heat gets produced, and the capacitor ends up with no potential energy.

This insight gives us a straightforward way to solve. Just figure out the total potential energy of the system before it "discharged." This lost potential energy equals the dissipated heat.

As we've seen before, a capacitor has potential energy

$$U_{\text{capacitor}} = \frac{1}{2}QV = \frac{1}{2}CV^2 = \frac{Q^2}{2C}.$$

We know the capacitor's initial potential difference, $V_0 = 15$ V. So, to calculate $U_{\text{capacitor}}$, we just need to know the system's initial charge, or its capacitance. Well, in Chapter 36, I demonstrated how to figure out a capacitance. For practice, you should derive the capacitance of a parallel-plate capacitor, and check your answer against the textbook's result, $C = \varepsilon_0 \frac{A}{D}$. Since your textbook supplies a good derivation, I won't derive that formula myself. In any case, since the plates have area $A = s^2$, we get

$$U_{\text{capacitor}} = \frac{1}{2}CV^2 = \frac{1}{2}\left(\varepsilon_0 \frac{s^2}{D}\right)V_0^2$$

$$= \frac{1}{2}(8.85 \times 10^{-12}\text{ C/N}\cdot\text{m}^2)\frac{(0.50\text{m})^2}{0.10\text{m}}(15\text{ V})^2$$

$$= 2.5 \times 10^{-9}\text{ joules}.$$

That's the initial potential energy stored in the capacitor, i.e., the potential energy lost by the system as it discharges. All this lost potential energy turns into heat.

You may be surprised that so little heat gets created, given our part (e) conclusion that the system initially produces heat at a rate of 2.2 joules per second. The paradox vanishes when you realize that the system discharges in a few billionths of a second.

QUESTION 37-3

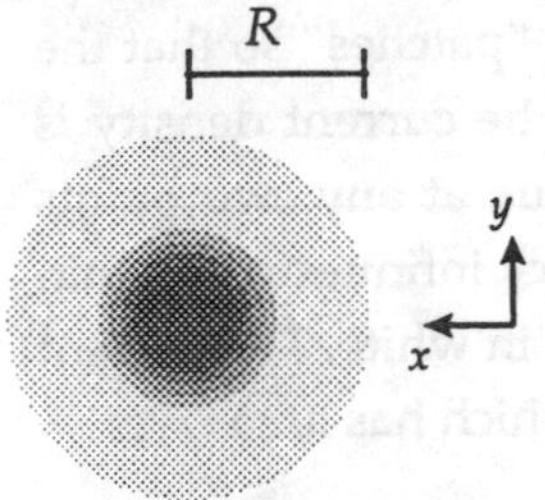

Beam is centered on the z-axis, which "points" into the page.

Consider a cylindrical beam of ions, flowing in the z-direction (into the page in this drawing). The drawing shows only a cross-section of the beam, which actually extends into and out of the page. The beam has radius R, centered on the z-axis. Crucially, the current density is higher near the middle of the beam, and lower near the edges, as indicated here with darker and lighter grays. If $r = \sqrt{x^2 + y^2}$ denotes the distance away from the z-axis (the center of the beam), then the beam's current density is

$$J = b - \frac{br}{R} \quad \text{between } r = 0 \text{ and } r = R.$$

$$= 0 \quad \text{for } r = R.$$

(a) What are the SI units of the constant b?

(b) What is the total current of this beam?

(c) (*Hard; break it into "subproblems"*) Suppose this ion beam is aimed at the pendulum bob drawn here. The pendulum has length l, and the bob has mass m. The beam stays on for time t_1. All the ions in the beam hit the pendulum bob and stick to it.

After the beam is turned off, a uniform leftward electric field of strength E_0 is turned on. As a result, the pendulum swings leftward, eventually settling at some angle. When the pendulum settles, what angle does the string make with the vertical? Answer in terms of the given quantities (m, l, b, R, t_1, E_0), and any universal constants you need. Neglect the ions' mass.

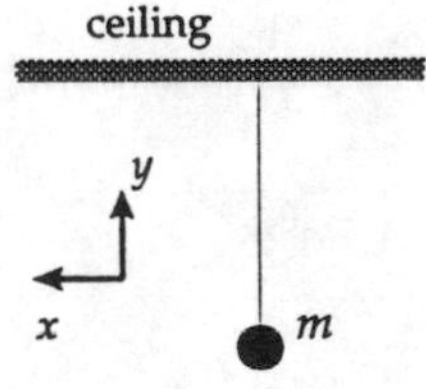

ANSWER 37-3

(a) Since

$$J = b + (\text{other term}),$$

the constant b must be a current density. Otherwise the units on the two sides of this equation would disagree. In fact, b is the current density in the middle of the beam, at $r = 0$. Since current density is current per area, the SI units of b are amperes per square meter. An ampere ("amp") is a coulomb per second.

(b) A common mistake is to multiply the current density by the beam's area:

Incorrect equation $I = JA = \left(b - \frac{br}{R}\right)\pi R^2.$

This works if the current density is constant. But it's not. J decreases near the edges of the beam. To use $I = JA$, we'd need to plug in the *average* current density of the beam, which we don't know.

Instead, we must use calculus. Think of the cross section of the beam as consisting of tiny "patches" of area dA. The infinitesimal "bit" of current through a given patch is $dI = JdA$. Since the current density varies, the different patches all contain different current. But by adding up (i.e., integrate over) those bits of current, we can obtain the total current.

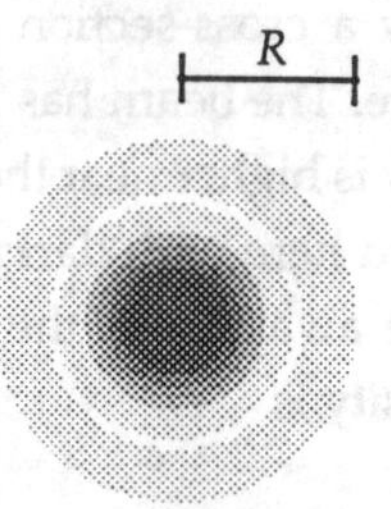

White area patch

When setting up this kind of integral, pick your "patches" so that the current density is constant *within* a given patch. Here, the current density is radially symmetric. In other words, J has the same value at any two points equidistant from the center. So, as my area patches, I'll pick infinitesimally thin rings or arbitrary radius r and "width" dr, as drawn here in white. The current density is constant over the entire area of a given ring, which has area

$$dA = \text{length} \times \text{width} = (2\pi r)dr.$$

Since $J = b - \frac{br}{R}$, the infinitesimal current through that infinitesimal ring is

$$dI = JdA = \left(b - \frac{br}{R}\right)2\pi r dr.$$

The cross section of the beam is "made of" these rings, starting at $r = 0$ and ending at $r = R$. Calculate the total current by adding up the contributions from each ring:

$$\begin{aligned}
I &= \int dI = \int JdA \\
&= \int_0^R \left(b - \frac{br}{R}\right)2\pi r dr \\
&= \int_0^R 2\pi b\left(r - \frac{r^2}{R}\right)dr \\
&= 2\pi b\left(\frac{r^2}{2} - \frac{r^3}{R}\right)\Bigg|_0^R \\
&= 2\pi b\left[\frac{R^2}{2} - \frac{r^3}{3R}\right] \\
&= \frac{\pi b R^2}{3}.
\end{aligned}$$

Since b has units of amps per square meter, and R has units of meters, this expression has units of amps, which is what we want.

(c) To formulate a strategy, let's think about what's happening. A beam of known current dumps ions onto the pendulum, for a known amount of time. Since current is charge per time, we can calculate the total charge that ends up on the bob. Next, an electric field pushes the bob leftward. Intuitively, the angle through it swings depends on how highly it's charged. By considering the electric force and the other forces on the bob, we can calculate that angle.

In summary, we can divide this problem into two subproblems. First, calculate the total charge on the bob. Then, use old-fashioned force reasoning—specifically, statics reasoning—to find at what angle the pendulum settles.

Subproblem 1: From the current, find the charge transferred to the pendulum bob.

Since current is charge per time ($I = \Delta q/\Delta t$), the bob receives a total charge $q = I\Delta t$. For instance, if a steady current of 2 coulombs per second flows for 5 seconds, then 10 coulombs gets dumped onto the bob. In symbols,

$$q = I\Delta t = \frac{\pi b R^2}{3} t_1,$$

where I used out part (a) answer for the current.

Subproblem 2: Force problem to calculate the swing angle.

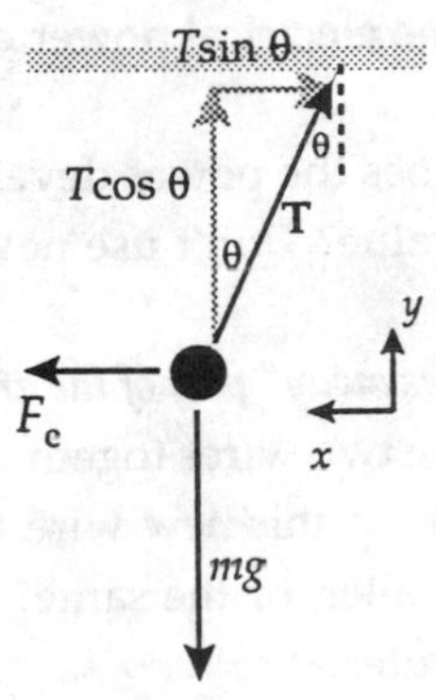

We want the angle at which the pendulum "settles," in response to a known electric field. So, we can use forces. Since the bob *settles* (comes to rest), it's no longer accelerating. Therefore, the net force, in both the x- and y-direction, must be zero. (You could also solve by setting the net torque equal to zero.)

In this kind of problem, always start with a force diagram. Tension pulls the bob in the direction that the string is stretched. The electric field, and hence the electric force on the positively charged bob, points leftward. As we saw way back in Chapter 33, $F_e = qE_0$. So, applying Newton's 2nd law in the horizontal and vertical directions, remembering that the settled bob doesn't accelerate, gives us

$$\sum F_x = ma_x$$
$$qE_0 - T\sin\theta = 0. \quad (1)$$

$$\sum F_y = ma_y$$
$$T\cos 0 - mg = 0. \quad (2)$$

We solved for q, the charge on the pendulum bob, in subproblem 1 above. So, we now have two equations in two unknowns, T and q. It's just a matter of algebra to solve for q. We're done with the "physics thinking."

Algebra starts here. Rewrite Eqs. (1) and (2) as

$$T\sin 0 = qE_0 \quad (1)$$
$$T\cos 0 = mg. \quad (2)$$

Now divide Eq. (1) by Eq. (2). The T's cancel, and $\frac{\sin\theta}{\cos\theta} = \tan\theta$. So, we get $\tan\theta = \frac{qE_0}{mg}$, and hence

$$\theta = \tan^{-1}\frac{qE_0}{mg} = \tan^{-1}\frac{\pi bR^2 t_1 E_0}{3mg},$$

where I've substituted in the expression for q obtained in subproblem 1, namely $q = \frac{\pi bR^2}{3}t_1$.

QUESTION 37-4

Pure aluminum is "malleable," easily pounded into different shapes. I have a small aluminum cube, with side-length s = 0.0050 meters (about 0.2 inches). I want $I = 0.10$ amps to flow through this aluminum chunk, when I connect opposite "ends" of it to opposite terminals of a $V = 120$ volt power source. Unfortunately, if I hook up the power source to the cube, too much current flows through.

(a) Assuming I have a hammer, what can I do to make sure only 0.10 amps flow through the chunk, when it's connected to the power source? Be as quantitative as possible. You may look up any formulas and constants you need in the textbook.

(b) [*You can answer parts (b) through (d), even if you didn't complete (a)*] What's the electrical power developed in the modified chunk, when hooked up to the power source?

(c) If we now cool down the aluminum substantially, using liquid nitrogen, does the power developed in the chunk go up, go down, or stay the same compared to its part (b) value? Don't use new formulas; explain your answer qualitatively.

(d) *Don't read on until you've tried part (a), because the wording of this question "gives away" part of the answer.* Suppose you take the wire from part (a), and cut it in half, and then tape the two wires together side by side, to make a single wire that's twice as thick and half as long. Hook up this new wire to the power source. Compared to your part (b) answer, is the power bigger, smaller, or the same? If it's bigger or smaller, by what factor? See if you can answer in under two minutes.

ANSWER 37-4

(a) We want the aluminum chunk to put up sufficient resistance that only 0.10 amps flow through it, when 120 volts get applied. From Ohm's law, $V = IR$, we can immediately calculate what the chunk's resistance must be:

$$R = \frac{V}{I} = \frac{120\ V}{0.10\ \text{A}} = 1200\ \Omega.$$

But aluminum has low resistivity. From the table in the textbook, $\rho = 2.8 \times 10^{-8}\ \Omega\cdot\text{m}$. Somehow, we must "re-shape" this naturally conductive chunk so as to give it high resistance.

Instead of jumping straight to a formula, let's think about resistance in terms of the tunnel analogy discussed earlier. A tunnel's "resistivity" corresponds to the prevalence of potholes or rocks. But the tunnel's overall resistance to car-current depends not just on resistivity, but also on area and length. The wider the tunnel, the more cars can flow through. And for less intuitive reasons, a longer tunnel has higher resistance than a shorter tunnel, for a fixed potential (height) difference between the entrance and the exit. So, the overall resistance is proportional to length, but inversely proportional to area:

$$R = \rho \frac{L}{A}. \tag{1}$$

This formula correctly relates resistance to resistivity. To achieve high resistance despite the chunk's low resistivity ($\rho = 2.8 \times 10^{-8}\ \Omega \cdot \text{m}$), we must pound it into a long, thin wire. The wire can be cylindrical, block-shaped, or whatever. But it must have a big enough length, and a small enough cross-sectional area, so that $R = 1200\ \Omega$. How long must the wire be?

Well, since we know R and r, we can solve Eq. (1) to obtain the *ratio* of L to A. The ratio turns out to be $4.3 \times 10^{10}\ \text{m}^{-1}$. So for instance, to obtain a rod with the correct resistance, we could make its cross-sectional area $A = 1$ square meter and its length $L = 4.3 \times 10^{10}$ meters. Or we could make its cross-sectional area $A = 2$ square meters and its length $L = 8.6 \times 10^{10}$ meters. Both of those (huge!) rods have the correct ratio of L to A, i.e., the correct resistance. But needless to say, we can't forge either of those rods from the given small chunk of aluminum, because it's too small. Pounding on the chunk changes its shape, but not its volume.

This insight gives us the extra information needed to solve the problem. As we pound on the aluminum chunk, changing its length and area, the *volume* stays constant. That's because we don't change the total amount of metal, or its mass density. Before getting pounded, the volume is s^3. After getting re-shaped into a wire of length L and area A, it has volume LA. Therefore, since the volume doesn't change,

$$s^3 = LA. \tag{2}$$

We now have two equations in the two unknowns, L and A. Intuitively, Eq. (1) tells us the ratio of L to A, while Eq. (2) determines the overall size of the wire.

Now I'll complete the algebra. From Eq. (2), $A = s^3/L$. Solve Eq. (1) for L, and substitute in this expression for A, to get

$$L = \frac{RA}{\rho} = \frac{R(s^3/L)}{\rho}.$$

Multiply through by L to get $L^2 = Rs^3/\rho$, and take the square root, to get

$$L = \sqrt{\frac{Rs^3}{\rho}} = \sqrt{\frac{(1200\ \Omega)(0.0050\ \text{m})^3}{2.8 \times 10^{-8}\ \Omega \cdot \text{m}}} = 73 \text{ meters!}$$

Solve Eq. (2) for area, and substitute in this length, to get

$$A = \frac{s^3}{L} = \frac{(0.0050\ \text{m})^3}{73\ \text{m}} = 1.7 \times 10^{-10}\ \text{m}^2,$$

about the cross-sectional area of a human hair.

In summary, we must "stretch" the aluminum into a hair-thin wire about a soccer field long, in order for its resistance to reach the desired 1200 ohms.

(b) As derived in question 37-2e, the power dissipated in a wire or slab—i.e., the rate at which heat gets generated in a wire or slab—is

$$P = VI.$$

So, even if you didn't get part (a), you can find the power:

$$P = VI = (120 \text{ V})(0.10 \text{ A}) = 12 \text{ W},$$

where "W" stands for watts. You get the same answer using V^2/R or I^2R, with $R = 1200\ \Omega$. Remember, since $V = IR$, you can rewrite $P = VI$ as $P = V^2/R$ or $P = I^2R$.

(c) A conductor's resistivity goes up when the temperature increases, and goes down when the temperature decreases. I'll explain why in a minute. But first, let me finish this problem. Since the wire's resistivity goes down, while its length and cross-sectional area stay the same, the resistance goes down. Therefore, the current goes up. Hence, the power ($P = VI$) also **increases.** (If you used $P = I^2R$, you had to realize that the increase in I^2 more than compensates for the decrease in R.)

Why is resistivity greater at higher temperatures? The "free-electron" model described in your textbook gives us a way to figure out the answer. In a conductor, the unbound valence electrons that carry the current behave like a gas, in some ways. As they bop around, the electrons collide with atomic nuclei in the metal. Consider a single electron. It moves around wildly, bouncing off nuclei like a pinball. Each collision "resets" (randomizes) the electron's velocity. But between collisions, the electron doesn't travel in a straight line. The electric field in the wire (created by the battery or power source) nudges the electron down the wire. The electric field slightly "curves" the electron's otherwise-random motion, so that the electron tends on average to drift down the wire.

The pinball analogy clarifies this. Imagine a pinball machine with lots of bumpers. Each time the ball hits a bumper, its direction of motion "randomizes." But between collisions, the ball doesn't travel in a straight line. Its motion "curves" toward the bottom of the machine.

I'll now argue that, when the electron bounces around more slowly and therefore undergoes fewer collisions per second, it drifts down the wire more quickly. Imagine two pinball machines. In one, the ball zips so quickly that it hits a bumper four times per second. In the other machine, the ball moves slower and hits a bumper only once a second. In which machine will the ball drift to the bottom faster? The one in which the ball moves slower. The less frequently the ball's velocity gets randomized, the more time gravity has to curve its path down towards the bottom. Ironically, the pinball's "drift velocity" (toward the bottom of the machine) is greater when its total (random) velocity is smaller.

Similarly, the electron in the wire drifts down the wire more quickly when it undergoes fewer collisions—i.e., when its random motion is slower. And free electrons, like all "gasses," move slower at lower temperature.

In summary, lowering the temperature decreases the electron's overall speed, thereby decreasing the number of collisions per second, thereby *increasing* its drift velocity down the wire. In other words, lowering the temperature decreases the resistivity.

(d) This new wire has half the length, but twice the cross-sectional area, of the old wire we made in part (a). If the wire were a tunnel, it would now be shorter and wider. Intuitively, current should flow through more easily.

The formula $R = \rho \frac{L}{A}$ confirms this. By halving L and doubling A, we make R one fourth as big as it used to be. Since the resistance goes down by a factor of four, while the voltage stays the same, the current *increases* by a factor of four. Therefore, from $P = VI$, the power quadruples, too. You can reach the same conclusion by looking at $P = V^2/R$ or $P = I^2R$.

38 CHAPTER

DC Circuits

QUESTION 38-1

Consider the following circuit. The batteries supply voltages ("EMFs") $\mathcal{E}_1 = 1.5$ V and $\mathcal{E}_2 = 1.0$ V, respectively, and have essentially no internal resistance. The resistors have resistance $R_1 = 1.0\ \Omega$, $R_2 = 2.0\ \Omega$, and $R_3 = 3.0\ \Omega$. I've labeled in some points, *a* through *f*, for future reference.

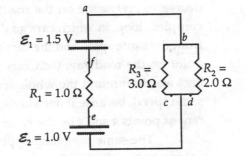

(a) At which point (if either) is the current greater, *a* or *e*? Answer and explain without doing detailed calculations.

(b) At which point (if either) is the current greater, *a* or *c*? Again, justify your answer without detailed calculations.

(c) Find the current through R_1.

(d) Find the potential drop across resistor 1.

(e) In 30 seconds or fewer, predict whether the potential difference between points *b* and *d* is greater than, less than, or the same as the potential difference between points *b* and *c*. After taking your guess, justify your answer.

(f) Suppose resistor 3 were cut, so that no current could flow through it. Would the current through R_1 go up, go down, or stay the same? After taking an intuitive guess and justifying it, calculate the new current through R_1, and compare it to the current before R_3 was cut.

ANSWER 38-1

(a) This "trick question" unveils a common misconception about circuits. Many students think that the current increases upon flowing through a battery, or decreases upon flowing through a resistor. But actually, current neither increases nor decreases when it flows through a circuit element. The same current flows past both points. The following "roadway" analogy explains why.

Think of the circuit as a roadway. The roadway, like the electrical circuit, forms a complete "loop"; cars eventually get back to where they started. The resistanceless wire segments correspond to a level stretches of road with no potholes. Resistors correspond to downhill segments of road filled with potholes—downhill, because charges lose potential energy when flowing through a resistor, just as a car loses potential energy when rolling down a hill. The potholes are so dense that cars

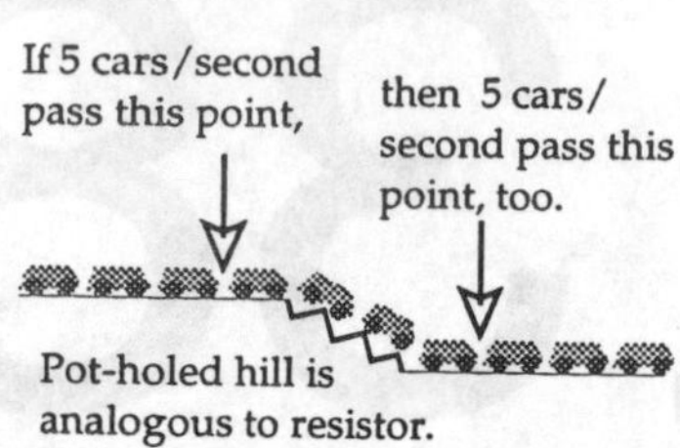
CARS FLOW RIGHTWARD
(downhill)
If 5 cars/second pass this point,
then 5 cars/ second pass this point, too.
Pot-holed hill is analogous to resistor.

flow down the hill at constant speed. A battery, which raises the potential energy of the charges flowing through it, corresponds to a giant "automobile elevator" that lifts cars from a lower to a higher altitude segment of road. In this analogy, think of the roadway as being packed with cars, bumper to bumper, so that they can't possibly get any more concentrated.

Now suppose 5 cars per second flow past a point on the smooth flat road. The cars then flow down a "resistor" (hill). How many cars per second flow past a point near the bottom of the hill? The answer must be 5 cars per second. Here's why. If only 4 cars per second pass that second point, then one car per second "builds up" on the hill. But when current flows steadily, charges (or cars) can't build up. In summary, the car current doesn't change when it passes through a resistor.

Don't get me wrong: raising the resistance does indeed decrease the current. But the current decreases *everywhere* on the roadway, not just on the hill. Remember, the circuit corresponds to a complete loop, in which cars go down a few hills, ride a few car elevators, and then start over again from the same spot. And the cars are bumper to bumper. Therefore, if a car gets slowed down at any point on the roadway, then cars get slowed down *everywhere*, because the traffic jam "propagates backwards" around the whole loop. The cars at all different points on the roadway still share the same speed, because if one slows down, they all slow down. So, doubling R_1 would reduce the current at points a and e by the same amount.

The same reasoning applies to the battery (car elevator). If 5 cars per second enter the elevator ("battery"), then 5 cars per second must leave the elevator (battery). Otherwise, cars—I mean charges—would build up in, or miraculously disappear from, the battery. This can't happen. By using a stronger battery, we increase the current *everywhere* in the circuit. If one car gets faster, they all get faster.

This all goes to show that the current is the same at *any* two points in the circuit, except where the circuit "forks" into two separate paths.

(b) Assume the current flows counterclockwise. As just noted, the same current flows past points a, b, e, and f. But between points c and d, the current branches. In the roadway analogy, think of that area as an intersection. Cars flow into the intersection from the one road below it, and then flow out of the intersection along two roads, which lead through c and d. The cars must obey the following law.

Intersection rule (also called the "node rule"): The number of cars per second flowing *into* the intersection must equal the cars per second flowing *out of* the intersection.

To see why, suppose hypothetically that 5 cars per second enter the intersection but only 4 cars per second leave. Then one car per second "builds up" in the intersection. This kind of build-up can't happen in steady-flow circuits. Or, suppose 5 cars per second enter the intersection but 6 per second leave. Then where did that "extra" car come from? The intersection rule formalizes the requirement that cars neither build up nor mysteriously appear.

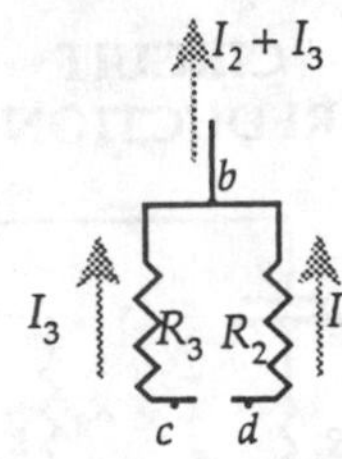

Similarly, the current flowing into intersection b must equal the current flowing out of that intersection. Let I_2 and I_3 denote the currents flowing through R_2 and R_3. Then $I_2 + I_3$ flows out the intersection through b. By contrast, only I_3 flows through point c. So, the current is higher at b. Roughly speaking, traffic merges at b, and therefore more traffic flows through that point than flows through either of its "feeder" roads (i.e., the roads containing c and d). Since the current through b equals the current through a, as we saw above, it follows that more current flows through a than flows through c.

(c) Following the standard textbook convention, I denote the positive terminal of a battery by the longer line. A battery tries to make current flow around the circuit from its positive to its negative terminal. So, battery 1 (with EMF $\mathcal{E}_1$) tries to make current flow counterclockwise around the circuit, while battery 2 tries to push current clockwise. Because battery 1 is stronger, it "wins"; the current flows counterclockwise.

As emphasized in part (a) using the roadway analogy, the current through resistor 1 depends not just on that resistance, but also on the other resistances. That's because, if a car gets slowed down anywhere in the circuit, then all the cars slow down. So, we need to find the "net resistance" of the whole circuit, as well as the net EMF (voltage). The we can substitute those net quantities into Ohm's law, $V = IR$, to solve for the current.

To find the net voltage and net resistance of simple circuits use "circuit reduction."

Circuit reduction

In circuit reduction, you "simplify" the circuit one step at a time, by focusing on two circuit elements and "combining" them into one. Let me show what I mean. Start with R_2 and R_3, the two resistors in parallel. "In parallel" means that a charge traveling around the circuit flows through one of those resistors *or* the other, but not both. As the textbook shows, two in-parallel resistors behave exactly like a single resistor with "equivalent resistance" given by

In parallel

$$\frac{1}{R_{eq}} = \frac{1}{R_2} + \frac{1}{R_3}$$

$$= \frac{1}{2\,\Omega} + \frac{1}{3\,\Omega}$$

$$= \frac{5}{6\,\Omega},$$

and hence $R_{eq} = \frac{6}{5}\,\Omega = 1.2\,\Omega$. *This is less than either R_2 or R_3 individually*. But that makes sense, in terms of the roadway analogy. Imagine a one-lane roadway branching into two lanes. Now there's room for more cars. Consequently, the two lanes taken together are less "resistant" to traffic flow, as compared to either lane individually.

At this stage in our circuit reduction, the simplified circuit consists entirely of resistors and batteries in series. Two circuit elements are "in series" if a charge flowing through one *must* also flow through the other. As intuitively expected, the equivalent resistance of two in-series resistors is simply their sum:

In-series $$R_{eq'} = R_1 + R_{eq\text{ from above}}$$
$$= 1.0\Omega + 1.2\Omega$$
$$= 2.2\Omega.$$

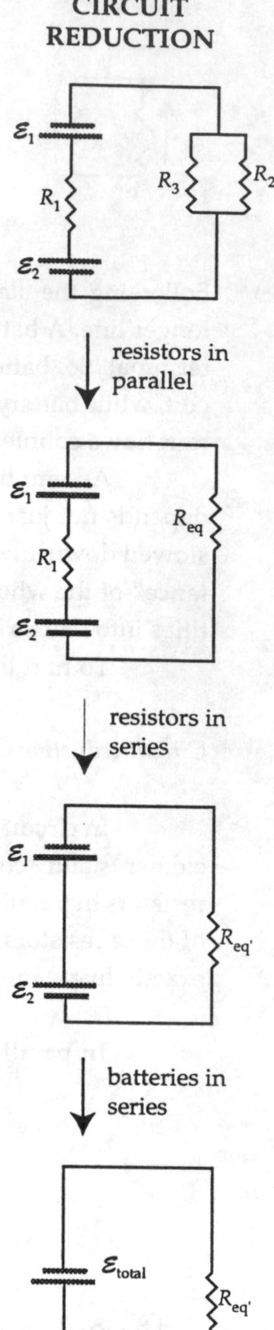

Now we just need to "reduce" the two batteries into one. Well, $\mathcal{E}_1$ tries to push current counterclockwise, while $\mathcal{E}_2$ tries to push current clockwise. If we call counterclockwise the positive direction, then

$$\mathcal{E}_{total} = \mathcal{E}_1 - \mathcal{E}_2$$
$$= 1.5\,V - 1.0\,V$$
$$= 0.5\,V.$$

Given these three steps of circuit reduction, we see that the overall circuit behaves *as if* it contained a single battery of voltage $\mathcal{E}_{total} = 0.5$ V, and a single resistor of resistance $R_{eq'} = 2.2\ \Omega$. From Ohm's law, the total current is therefore

$$I_{total} = \frac{\mathcal{E}_{total}}{R_{eq'}} = \frac{0.5\,V}{2.2\Omega} = 0.23\,A.$$

That's the current through resistor 1 and through both batteries.

(d) I see two ways of solving this. But they give different answers! On the one hand, according to your textbook, the potential across a resistor changes by $V = -IR$. Hence,

$$V_{ef} = -IR_1 = -(0.23\ A)(1.0\ \Omega) = -0.23\text{ volts},$$

where the minus sign indicates that the potential *decreases*. On the other hand, as we saw above, the circuit has "net voltage $V_{total} = \mathcal{E}_1 - \mathcal{E}_2 = 0.50$ volts. So, shouldn't V_{ef} be 0.50 volts, instead of 0.23 volts?

To understand why 0.23 volts is the correct potential drop, think in terms of the roadway analogy. Batteries are like car elevators, and resistors are like hills. In an electrical circuit, charges usually gain potential when flowing through a battery, and lose potential when flowing through resistors. Similarly, a car gains height when riding up a car elevator, and loses height when rolling down a hill. Since the roadway (like the circuit) forms a complete loop, the car eventually returns to where it started. Therefore, if the elevators lift the car 500 feet, then the car must *lose* 500 feet of height when traveling down the hills. For instance, if the roadway includes two hills, the car might descend 230 feet on one hill and 270 feet on the other hill. So, the car need not lose all 500 feet on the first hill. In general, the height lost by the car on *all* the hills combined must equal the height it gains in the elevators. Otherwise, the car can't get back to where it started.

Analogously, when charge flows around a complete circuit, and therefore returns to where it started, the potential it gains in the batteries must equal the potential it loses in the resistors. Here, the batteries cause a net potential gain of 0.50 volts. So, when flowing through resistors, the charge must lose 0.50 volts. But here's the crucial point. Just like the car went down two hills, each charge flowing around this circuit passes through *two* resistors: either R_1 and R_2, or R_1 and R_3. For this reason, the potential need not drop by the full 0.50 volts across resistor 1 alone. In fact, as we saw above, the potential drop across R_1 is only $V_{ef} = 0.23$ volts. Therefore, the potential drop across R_2 or R_3 must be 0.27 volts, to ensure that the combined potential drop equals 0.50 volts.

(e) Nothing except resistanceless wire lies between points *c* and *d*. In the roadway analogy, this corresponds to flat, smooth pavement. Just as a car has the same potential energy at any two points on a flat road, the potential is the same at any two points separated by nothing but resistanceless wire. So, $V_c = V_d$.

It follows that $V_c - V_b = V_d - V_b$. In words, the potential difference between *b* and *d* equals the potential difference between *b* and *c*. So, the potential drops by the same amount across R_2 and R_3. But hey, we reached that same conclusion at the end of part (d), by thinking in terms of the roadway analogy.

By the way, since $V = IR$, and since the potential drops by the same amount across R_2 and R_3, it follows that $I_2R_2 = I_3R_3$. Apparently, when current "branches" through in-parallel resistors, it splits up so as to equalize the potential drop across each different branch.

(f) In the roadway analogy, cutting R_3 corresponds to closing one of the lanes, in this case "lane *c*." Intuitively, this lane-closing slows down traffic in the rest of the circuit. So, the current through R_1 decreases.

Let me repeat this argument in more technical terms. Other things being equal, more current can flow through R_2 and R_3 combined than can flow through R_2 alone. In other words, the in-parallel equivalent resistance of R_2 and R_3 is *less* than the individual resistance, R_2.

Now I'll calculate the new current. With R_3 snipped, "lane *c*" becomes a dead end that plays no role in the circuit. My diagram here reflects that fact. Let's reduce this new circuit. Since the resistors are in series, $R_{\text{total}} = R_1 + R_2$. So, the new current through the whole circuit—and hence, through R_1—is

CIRCUIT REDUCTION

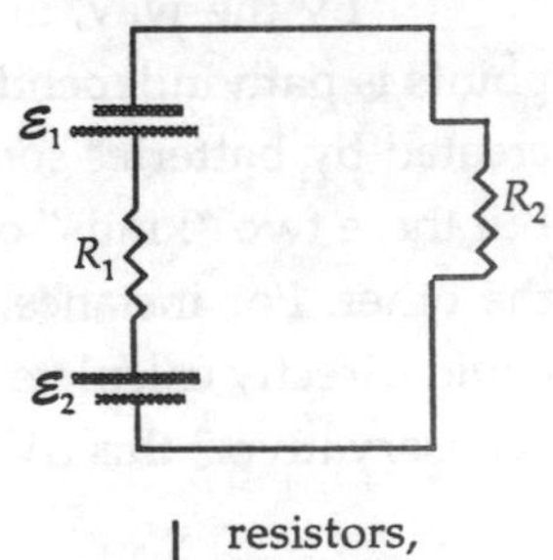

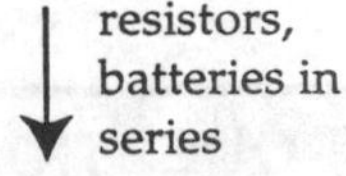

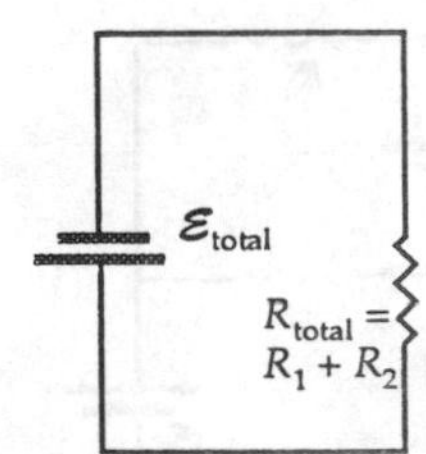

$$I = \frac{\mathcal{E}_{\text{total}}}{R_{\text{total}}} = \frac{\mathcal{E}_{\text{total}}}{R_1 + R_2} = \frac{0.5\,\text{V}}{1.0\,\Omega + 2.0\,\Omega} = 0.17\,\text{A},$$

compared to the 0.23 amps we got in part (c). The current goes down, as intuitively predicted.

QUESTION 38-2

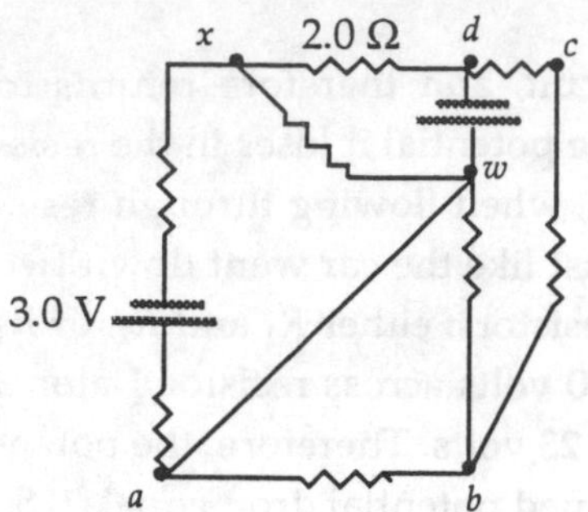

Consider this complicated circuit. Except for the one labeled battery and the one labeled resistor, we know none of the other EMFs and resistances.

Using my potentiometer, I measure the potential drop from point *a* to *b*, the potential drop from *b* to *c*, and the potential drop from *c* to *d*. Adding up these values, I find that the total potential drop from *a* to *d* along path *abcd*. It's 0.40 volts.

Consider the potential drop from *a* to *d* along path *awxd*. Is it greater than, less than, or equal to 0.40 volts? Is it possible to tell? Explain your answer.

ANSWER 38-2

You can answer this *without* finding the current through all the relevant resistors. The answer pops out of a crucial physical insight: **The potential difference between two points is the same, no matter what path you take.** Consequently, the potential difference between *a* and *d* is 0.40 volts along *any* path, including *awxd*.

Let me explain why, in terms of the roadway analogy. Suppose a car can travel from point *a* to point *d* along several different roads. Does the car's change in gravitational potential energy between *a* and *d* depend on which path it takes? Well, the car's potential energy depends on the altitude (height): $U_{grav} = mgh$. And while moving from *a* to *d*, the car gains or loses the *same* amount of altitude, no matter what path it takes. For instance, if you drive from the top of a 10000-foot mountain to sea level, then your altitude changes by 10000 feet, no matter what road you take. Similarly, a charge going from *a* to *d* gains or loses the same amount of potential energy, no matter what path it takes.

By the way, in Chapter 35, I emphasized *ad nauseam* that the potential difference between two points is path independent. Remember the contour map analogy? Conceptually, the potential differences created by batteries somehow "seem different" from the potential differences created by static charges. But these two "kinds" of potentials are really the same thing. Anything that's true about one is true about the other. For instance, batteries create electric fields in the wires and resistors. Given those fields, we could directly calculate the potential difference between *a* and *d* using $\Delta V = \int_a^d \mathbf{E} \cdot d\mathbf{s}$ Since electric fields are "conservative," this ΔV comes out the same no matter what path you integrate over.

QUESTION 38-3

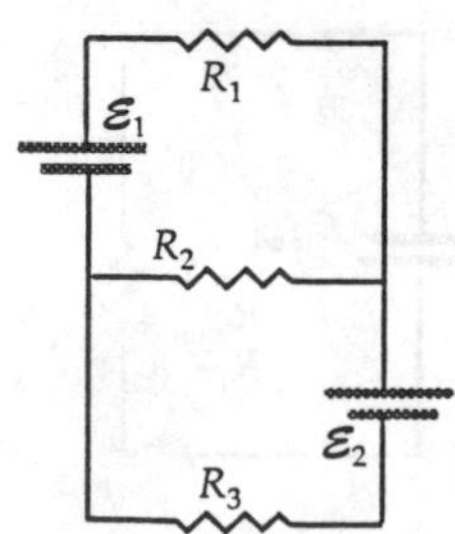

In this circuit, the batteries have voltages $\mathcal{E}_1 = 2.0$ V and $\mathcal{E}_2 = 4.0$ V. Two of the resistors have fixed resistances: $R_2 = 5.0\ \Omega$ and $R_3 = 8.0\ \Omega$. But R_1 is a variable resistor. You can manually adjust it.

(a) Suppose R_1 is set to 3.0 Ω. What's the current through R_1?

(b) Kelly wants to use this circuit to heat up some oil, by immersing R_1. Kelly positions the circuit so that no circuit element except R_1 is in the oil.

At what resistance should Kelly set R_1 so that the oil heats up as quickly as possible? Don't worry about completing the math; focus on setting up the relevant equation or equations.

ANSWER 38-3

(a) You might be tempted to treat the circuit as a bunch of batteries and resistors in series and in parallel. Unfortunately, you can't use circuit reduction here, for the following reason. In circuit reduction, we "combine" the two batteries into one, and pretend that a unique total current flows through the "composite" battery (of voltage $\mathcal{E}_1 + \mathcal{E}_2$). This game makes sense only if the same current flows through battery 1 and battery 2, as in question 38-1. But here, the two batteries live on different loops of circuit. The current through battery 1 might not equal the current through battery 2. Therefore, we can't "pretend" that a *single* current flows through a combined battery. Circuit reduction can't work. Instead, we must use Kirchhoff's laws.

Rather than listing Kirchhoff's laws ahead of time, I'll invoke them as needed, and summarize them afterwards. They make more sense once you've seen them in action.

To solve a multi-loop circuit, *divide it into individual loops and treat each loop separately*. On your diagram, clearly label the loops you'll consider. By "loop," I mean any closed path, i.e., any path that ends up where it started.

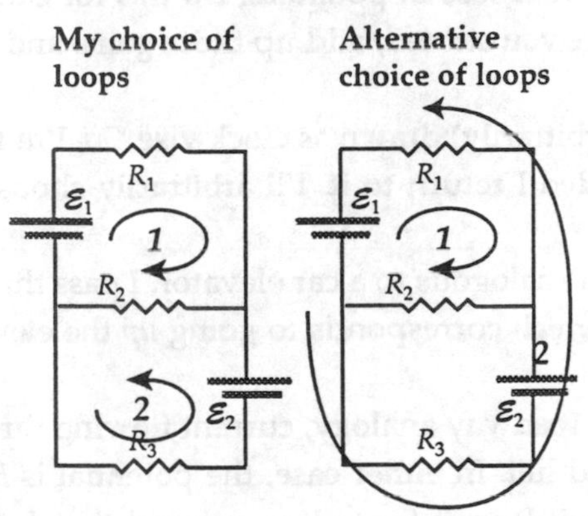

Now for some good news. There's no "incorrect" choice of loops. For instance, we can think of this circuit as consisting of a "top loop" and "bottom loop." I'll make that choice. But we could equally well divide the circuit into a "top loop" and "whole-circuit loop," as drawn here. (The whole-circuit loop goes around the perimeter of the circuit, avoiding R_2.) Or we could pick the bottom loop and the whole-circuit loop. Kirchhoff's laws work equally well for any of these choices. Furthermore, you may arbitrarily choose the orientation, clockwise or counterclockwise, of each loop. To simplify the reasoning, I try to select loop orientations that follow the direction in which current flows. But the same final answer pops out no matter what.

After picking your loops, label the current in all segments of the circuit. From the roadway analogy, remember that current doesn't change upon passing through a battery or resistor. It changes only when the wire (road) branches at an intersection. So here, the current through $\mathcal{E}_1$ equals the current through R_1. I'll call that current I_1. Similarly, the same current flows through $\mathcal{E}_2$ and R_3. Call it I_3. Finally, I'll let I_2 denote the current through R_2.

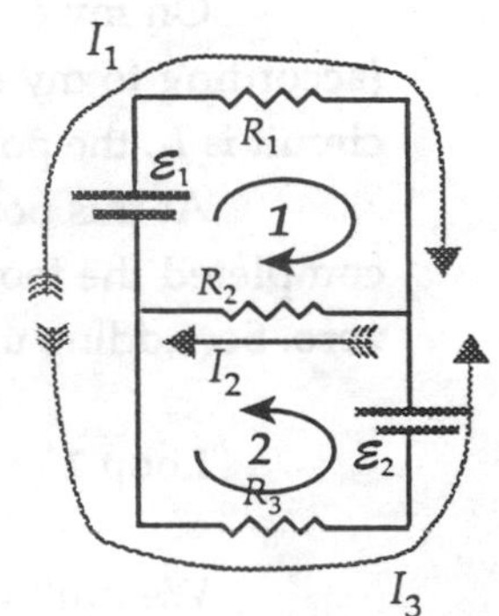

On your diagram, always "guess" the direction of each current, as I've done here. Fortunately, a wrong guess can't hurt you. For instance, if I_2 actually flows rightward, then I'd end up with a negative value for I_2. The minus sign would tell me that the current flows "backwards," as compared to my guess.

At this stage, we're tooled up to use Kirchhoff's laws. We already met Kirchhoff's 1st law, the "intersection rule," in question 38-1: The current entering an intersection equals the current leaving an intersection. Focus your attention on the intersection to the right of R_2. According to my diagram, I_1 and I_3 flow *into* that intersection, while I_2 flows *out*. Therefore,

Intersection rule $$I_1 + I_3 = I_2. \tag{0}$$

You reach the same conclusion by considering the intersection to the left of R_2, since I_2 flows into that intersection, while I_1 and I_3 flow out. So here, Kirchhoff's 1st law spits out only one equation.

To generate more equations about this circuit, use Kirchhoff's 2nd law: The total change in potential around a loop is zero. But that's nothing new! Remember, "potential" is proportional to potential energy. In the roadway analogy, imagine driving around a complete loop, so that you *end up back where you started*. Then your final potential energy *is* your initial potential energy. In other words, your potential energy *changes* by $\Delta U = 0$. If you gained potential energy during one part of the drive, you lost it somewhere else. Similarly, if a charge goes around a whole loop and ends up back where it started, then its final potential *is* its initial potential. In other words, the change in potential is $\Delta V = 0$.

This abstract fact about potential allows us to write useful equations, one equation for each loop. Here's the idea. Pretend you're walking around the loop. As you pass through a circuit element (resistor, battery, or capacitor), write down the gain or loss in potential. Do this for each circuit element you encounter. When you get back to where you started, add up those gains and loses in potential, and set the sum equal to zero.

Let's apply this strategy to loop 1, which I've (arbitrarily) drawn as clockwise. So, I'm walking clockwise. My starting point doesn't matter, provided I return to it. I'll arbitrarily choose the spot right under battery 1.

On my walk, I first encounter battery 1, which is analogous to a car elevator. I pass through the battery from the negative to the positive terminal, which corresponds to going *up* the elevator. So, the potential *rises* by $\mathcal{E}_1$.

Continuing the walk, I next encounter R_1. In the roadway analogy, current flowing through a resistor corresponds to cars flowing down a pot-holed hill. In either case, the potential is *higher* where the current enters the resistor, and lower where it leaves. Since I've guessed that I_1 flows clockwise, I'm walking *with* the current, i.e., "down a hill." Therefore, I *lose* potential. From Ohm's law, my ΔV is $-I_1R_1$, where the minus sign indicates a potential *drop*.

On my clockwise walk around loop 1, I next reach R_2. Again, I'm walking *with* the current (according to my diagram). So once again, my potential drops. Since the current in this part of the circuit is I_2, the potential across the resistor changes by $-I_2R_2$.

At this point, I encounter no more circuit elements before returning to where I started. I've completed the loop. As emphasized above, the total change in potential around the whole loop is zero. So, adding up my ΔV's from the previous three paragraphs, I get

Loop 1 $$\mathcal{E}_1 - I_1R_1 - I_2R_2 = 0. \tag{1}$$

We can't yet solve for I_1, because this equation contains a second unknown, I_2. Eq. (0) above doesn't help, because it introduces a third unknown, I_3. We need another equation. To get it, consider loop 2, which I've also arbitrarily drawn as counterclockwise. I'll start my walk right below battery 2.

First, I pass through the battery, from the negative to the positive terminal. This changes my potential by $+\mathcal{E}_2$.

Next, I walk leftward through R_2. Since I'm walking *with* the current (according to my diagram), the potential change is $-I_2R_2$. (By contrast, walking *against* the current would correspond in the roadway analogy to walking up a hill, resulting in a *positive* ΔV.)

Lastly, I walk through R_3, again with the current. So, $\Delta V = -I_3R_3$.

Set the sum of these potential changes equal to zero, to get

Loop 2 $$\mathcal{E}_2 - I_2R_2 - I_3R_3 = 0. \qquad (2)$$

At this stage, we're done with the physical reasoning. We have three equations, one from Kirchhoff's 1st law (the intersection rule) and two from Kirchhoff's 2nd law (the "loop" equations). in this part of the problem, we know both voltages and all three resistances. So, with these three equations, we can solve for the three unknowns, I_1, I_2, and I_3.

Algebra starts here. From Eq. (0), $I_2 = I_1 + I_3$. Substitute that expression for I_2 into Eqs. (1) and (2) to get

$$\mathcal{E}_1 - I_1R_1 - (I_1+I_3)R_2 = 0, \qquad (*)$$

$$\mathcal{E}_2 - (I_1+I_3)R_2 - I_3R_3 = 0. \qquad (**)$$

Now solve Eq. (*) for I_3 in terms of I_1. This yields $I_3 = \dfrac{\mathcal{E}_1 - I_1(R_1 + R_2)}{R_2}$. Finally, substitute this expression for I_3 into Eq. (**), to get

$$\mathcal{E}_2 = \left[I_1 + \frac{\mathcal{E}_1 - I_1(R_1 + R_2)}{R_2}\right]R_2 - \frac{\mathcal{E}_1 - I_1(R_1 + R_2)}{R_2}R_3 = 0.$$

Now we must isolate I_1. Expanding out the terms gives

$$\mathcal{E}_2 - I_1R_2 - \mathcal{E}_1 + I_1\,R_1 + I_1\,R_2 - \frac{R_3}{R_2}\mathcal{E}_1 + I_1\frac{R_1R_3}{R_2}\,I_1R_3 = 0.$$

Cancel the I_1R_2 with the $-I_1R_2$, and group the remaining terms to get

$$I_1\left(R_1\frac{R_1R_3}{R_2} + R_3\right) = \mathcal{E}_1\left(1 + \frac{R_3}{R_2}\right) - \mathcal{E}_2,$$

and hence

$$I_1 = \frac{\mathcal{E}_1\left(1 + \dfrac{R_3}{R_2}\right) - \mathcal{E}_2}{R_1\left(1 + \dfrac{R_3}{R_2}\right) + R_3} = \frac{(2.0\,\text{V})\left(1 + \dfrac{8.0\,\Omega}{5.0\,\Omega}\right) - 4.0\,\text{V}}{(3.0\,\Omega)\left(1 + \dfrac{8.0\,\Omega}{5.0\,\Omega}\right) + 8.0\,\Omega} = +0.076\,\text{A}.$$

End of algebra.

The positive sign indicates that I correctly guessed the direction of I_1 on my diagram.

Before proceeding to part (b), let me summarize the strategy I used to approach the multi-loop circuit.

Multi-loop circuit strategy

1) Choose your loops. Then, "guess" the current direction in each part of the circuit, and give those currents names.

2) Apply Kirchhoff's 1st law, the "intersection rule," to relate the currents in the different parts of the circuit. *The current flowing into an intersection equals the current flowing out of that intersection.*

3) Apply Kirchhoff's 2nd law to each loop individually: *The total change in potential around a loop is zero.* Keep organized by "walking around the loop" and writing down the potential change as you pass through each circuit element. Using the car elevator and pot-holed hill analogies for batteries and resistors, we saw that

(a) If you walk "up" the battery (from negative to positive terminal), the potential change is $+\mathcal{E}$. If you walk "down" the battery (positive terminal to negative terminal), it's $-\mathcal{E}$.

(b) If you walk across a resistor *with* the current, then the potential change is $-IR$. If you walk *against* the current ("uphill"), the potential change is $+IR$.

(c) If you walk across a capacitor, . . . *rule to be filled in later . . .*

Steps 2 and 3 usually give you enough equations to solve for the unknowns.

(b) Kelly wants R_1 to generate heat energy at the highest possible rate. The rate at which electrical energy converts into heat is called the *power*. So, in this problem, we're *maximizing the power* in R_1.

In Chapter 37, we derived the power dissipated in a resistor. It's $P = VI$, where V denotes the potential drop across the resistor, *not* the voltage of the battery. Since $V = IR$, we can rewrite the power as

$$P = VI = (IR)I = I^2R.$$

So, in resistor 1, "$P = I_1^2R_1$". We're finding the value of R_1 such that "$P = I_1^2R_1$" reaches its highest possible value for this circuit.

Before launching into heavy mathematics, let's see if we can guess the answer. To maxi-mize $I_1^2R_1$, we could try making R_1 huge. But that might not work, because the large resistance keeps the current small, and hence $I_1^2R_1$ might stay small. Alternatively, we could make R_1 small, so that I_1 becomes large. But this trick might not maximize $I_1^2R_1$, because the *other* resistors in the circuit prevent I_1 from getting too big.

Apparently, intuition alone can't get us to the answer. We have no choice but to employ a strategy I introduced in Volume 1, question 4-7, and reviewed in this volume, question 33-4c.

Maximization/minimization strategy

(1) Figure out what variable you're maximizing or minimizing. Then figure out which variable you "alter" in order to achieve maximization/minimization. This parameter is called the "free variable."
(2) Write the variable you're maximizing as a function of the free variable. The equation may also contain constants, but it may not contain other *variables* besides the free variable and variable you're maximizing.
(3) Differentiate the variable you're maximizing/minimizing with respect to the free variable. Set the derivative equal to 0, and solve for the free variable.

I'll now implement this strategy.

Step 1: Figure out what's being maximized and what's the free (adjustable) variable.

Here, we're maximizing P. The quantity we can adjust is R_1. That's the free variable.

Step 2: Express the maximized variable as a function of the free variable.

We can't just write "$P = I_1^2 R_1$" and call it quits, because the equation may contain no variables besides the free variable and the variable we're maximizing. When you alter R_1, you also alter I_1, the current through that resistor. So, I_1 is a variable; you can't assume I_1 keeps its value from part (a). For this reason, we must eliminate I_1 from our equation for P.

Fortunately, in part (a), we already derived an equation for I_1 as a function of R_1:

$$I_1 = \frac{\mathcal{E}_1\left(1+\frac{R_3}{R_2}\right) - \mathcal{E}_2}{R_1\left(1+\frac{R_3}{R_2}\right) + R_3}.$$

Since "$P = I_1^2 R_1$",

$$P = \left[\frac{\mathcal{E}_1\left(1+\frac{R_3}{R_2}\right) - \mathcal{E}_2}{R_1\left(1+\frac{R_3}{R_2}\right) + R_3}\right]^2 R_1.$$

In this messy equation, the free variable R_1 is the *only* variable on the right-hand side. Everything else is a given, fixed quantity. So, we're done with step 2.

Step 3: Differentiate the maximized variable with respect to the free variable, and set it to 0.

Here, we must set $dP/dR_1 = 0$, and solve for R_1. Many instructors would award almost full credit on a test if you carried the problem this far and then wrote in words, "set $dP/dR_1 = 0$, and solve for R_1." Check your instructor's policy. For the record, I'll complete the calculus.

Calculus starts here. The derivative starts looking a heck of a lot easier if you keep in mind that everything on the right-hand side, except R_1, is a constant. Indeed, we can rewrite the equation for P as

$$P = \left(\frac{K_1}{K_2K_1 + K_3}\right)^2 R_1,$$

where I've defined the K constants as $K_1 = \mathcal{E}_1\left(1 + \frac{R_3}{R_2}\right) - \mathcal{E}_2$, $K_2 = 1 + \frac{R_3}{R_2}$, and $K_3 = R_3$.

Differentiate this mess by using the product rule, $(uv)' = uv' + vu'$, with $u = \left(\frac{K_1}{K_2R_1 + K_3}\right)^2$ and $v = R_1$. Here we go . . .

$$0 = \frac{dP}{dR_1} = \frac{d}{dR_1}\left[\left(\frac{K_1}{K_2R_1 + K_3}\right)^2 R_1\right]$$

$$= \left(\frac{K_1}{K_2R_1 + K_3}\right)^2 \frac{d}{dR_1}R_1 + R_1\frac{d}{dR_1}\left(\frac{K_1}{K_2R_1 + K_3}\right)^2$$

$$= \left(\frac{K_1}{K_2R_1 + K_3}\right)^2 + R_1K_1^2\left[-2\frac{1}{(K_2R_1 + K_3)^3}K_2\right].$$

Divide through by K_1^2, then multiply through by $(K_2R_1 + K_3)^3$, to get

$$0 = (K_2R_1 + K_3) - 2K_2R_1.$$

We can immediately solve this for R_1, and substitute back in my definitions of K_2 and K_3, to get

$$R_1 = \frac{K_3}{K_2} = \frac{R_3}{1 + \frac{R_3}{R_2}} = \frac{8.0\Omega}{1 + \frac{8.0\Omega}{5.0\Omega}} = 3.1\Omega.$$

End of calculus.

This is only 0.1 ohm higher than R_1's resistance in part (a). So, the power was already almost maximized. Unfortunately, we had no way of knowing that ahead of time.

QUESTION 38-4

In this circuit, S_1 and S_2 are switches. Current can flow through a switch only when it's "closed." I apologize for this confusing terminology.

The batteries have voltages $\mathcal{E}_1 = 2.0$ V and $\mathcal{E}_2 = 1.5$ V. Neither battery has internal resistance. For the resistors, $R_1 = 3.0\ \Omega$, $R_2 = 5.0\ \Omega$, and $R_3 = 0.50\ \Omega$. The resistance R_4 is unknown. The capacitor has capacitance $C = 3.0 \times 10^{-6}$ coulomb per volt, and is initially uncharged.

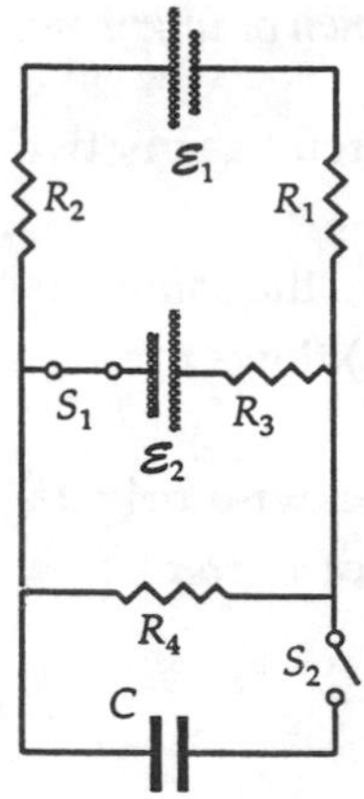

When S_1 is closed and S_2 is open, as drawn here, the current through R_4 is measured to be 0.10 amps, leftward.

(a) Solve for R_4. Hint: As far as current is concerned, a "dead end" in the circuit might as well not be there at all. Use this insight to simplify the circuit.

(b) Now S_1 is also opened, so that current flows through neither switch. What's the new current through R_4? Explain intuitively why it changed direction.

(c) [*Skip parts (c) through (e) if you don't need to deal with capacitors in circuits*] Now, with S_1 still left open, S_2 is closed at time $t = 0$. Immediately after $t = 0$, what's the current through the capacitor? Hint: the capacitor hasn't had time to get charged up by more than a few electron's worth of charge.

(d) *(Very hard)* What's the current through R_4 at time $t = 5$ seconds? Hint: Is the capacitor still getting charged up?

(e) At $t = 5$ s, what's the charge on the capacitor?

ANSWER 38-4

You can solve this standard multi-loop circuit problem using Kirchhoff's laws.

(a) In part (a), S_2 is open. Therefore, the segment of the circuit containing the capacitor is a "dead end." Current can't pass through that leg of the circuit. Therefore, as far as current is concerned, the circuit behaves as if the dead end were missing. Here, I've redrawn the circuit to reflect this. Of course, when S_2 gets closed, we'll need to "reinstate" the capacitor into our drawing and into our reasoning. But for now, it's gone.

To solve a multi-loop circuit, use the strategy introduced in the previous problem:

The "dead end" capacitor plays no role in the circuit.

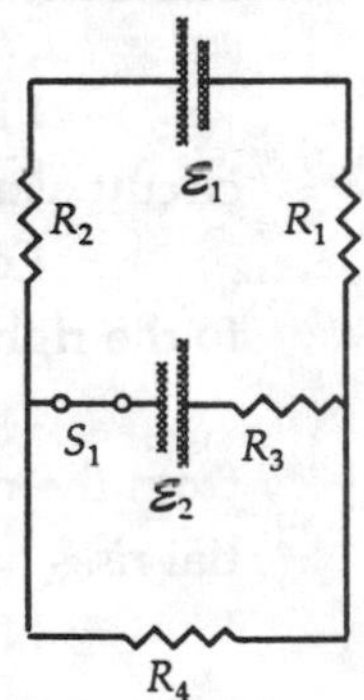

Multi-loop circuit strategy

1) Choose your loops. Also, label the currents, and guess their directions.

2) Apply Kirchhoff's 1st law: *The current flowing into an intersection equals the current flowing out of that intersection.*

3) Apply Kirchhoff's 2nd law to each loop individually: *The total change in potential around a loop is zero.* Keep organized by "walking around the loop" and writing down the potential change as you pass through each circuit element.

Usually, these steps give you enough equations to solve for whatever you want.

I'll now implement this strategy.

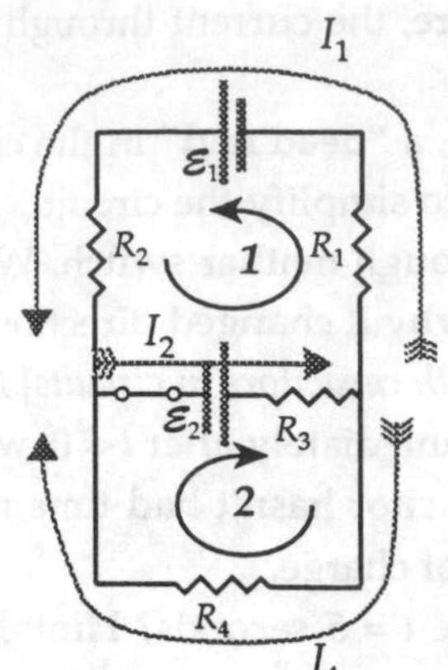

Step 1: Choose loops, and label the currents, whether they're known or unknown.

I've picked the top and bottom "squares" of the circuit as my two loops.

Because the current changes only at an intersection, the same current (I_1) flows through R_1, $\mathcal{E}_1$, and R_2. A *different* current (I_2) flows through the middle segment of the circuit containing $\mathcal{E}_2$ and R_3.

Notice that I chose the counterclockwise and clockwise orientations of loops 1 and 2 to match the (guessed) direction of current flow around each loop.

Step 2: Use Kirchhoff's 1st law, the Intersection rule.

Consider the intersection to the left of the switch. According to my diagram, I_1 and I_4 flow into that intersection, while I_2 flows out. So,

$$I_1 + I_4 = I_2. \tag{0}$$

You arrive at the same equation by examining the intersection to the right of R_3. So, Kirchhoff's 1st law gives us only one equation.

Step 3: Use Kirchhoff's 2nd law to write equations about each separate loop.

I'll pretend to walk around each loop, recording my charge in potential (ΔV) through each circuit element. These ΔV's sum up to zero.

Let's begin with loop 1, which I've drawn counterclockwise. I'll (arbitrarily) start my walk to the right of battery 1.

First, I encounter battery 1, which is analogous to a car elevator. I pass through the battery from the negative to the positive terminal. This corresponds to going *up* the elevator. So, the potential *rises* by $\mathcal{E}_1$.

Next, I encounter R_2. From the roadway analogy, current flowing through a resistor corresponds to cars flowing down a pot-holed hill. So, the potential is *higher* where the current enters the resistor, and lower where the current leaves the resistor. I'm walking *with* the current, "downhill." Hence, I *lose* potential. From Ohm's law, my potential changes by $-I_1R_2$.

Next on loop 1, I pass through battery 2, again from the negative to the positive terminal. So, $\Delta V = \mathcal{E}_2$. Notice that the potential change across a battery depends only on the battery's EMF, not on the current.

Then, I walk through R_3, with the current I_2. So, $\Delta V = -I_2R_3$. Finally, I traverse R_1 with the current I_1, and hence $\Delta V = -I_1R_1$. Then I'm back where I started.

As noted above, the sum of all these potential changes must be zero:

Loop 1 $$\mathcal{E}_1 - I_1R_2 + \mathcal{E}_2 - I_2R_3 - I_1R_1 = 0. \tag{1}$$

By the way, had you walked around loop 1 clockwise instead of counterclockwise, the same equation would pop out, but with all the signs reversed. This "global" sign flip corresponds to multiplying the equation through by -1; so it's really the same equation.

So far, Eqs. (0) and (1) give us two equations in two unknowns I_1 and I_2. (The problem tells us I_4.) So, we can solve for both of those currents. Unfortunately, the problem asks for R_4. We need another equation. I'll get it from loop 2.

Let's walk clockwise, starting just to the left of battery 2. Since I walk "up" the battery, the potential change is $\mathcal{E}_2$. Next, I walk with the current through R_3, and hence $\Delta V = -I_2R_3$. By similar reasoning, across R_4, $\Delta V = -I_4R_4$. Continuing my walk, I complete the loop without traversing any more circuit elements.

Add up these ΔV's to get

Loop 2 $$\mathcal{E}_2 - I_2R_3 - I_4R_4 = 0. \tag{2}$$

At this point, we have three equations in the three unknowns (I_1, I_2, and R_4). It's just a matter of algebra to solve for R_4.

Algebra starts here. I'll first solve Eqs. (0) and (1) for I_2. Then, I'll substitute that I_2 into Eq. (2), and solve for R_4.

From Eq. (0), $I_1 = I_2 - I_4$. Substitute this into Eq. (1) to get

$$\mathcal{E}_1 - (I_2 - I_4)R_2 + \mathcal{E}_2 - I_2R_3 - (I_2 - I_4)R_1 = 0.$$

Isolating the I_2 terms gives us $-(R_2 + R_1 + R_3)I_2 = -(\mathcal{E}_1 + I_4R_2 + \mathcal{E}_2 + I_4R_1)$. Hence,

$$I_2 = \frac{\mathcal{E}_1 + \mathcal{E}_2 + I_4(R_1 + R_2)}{R_1 + R_2 + R_3} = \frac{2.0\,\text{V} + 1.5\,\text{V} + (0.10\,\text{A})(3.0\,\Omega + 5.0\,\Omega)}{3.0\,\Omega + 5.0\,\Omega + 0.50\,\Omega} = 0.506\,\text{A}.$$

Now solve Eq. (2) for R_4, and substitute in this value for I_2, to get

$$R_4 = \frac{\mathcal{E}_2 - I_2R_3}{I_4} = \frac{1.5\,\text{V} - (0.506\,\text{A})(0.50\,\Omega)}{0.10\,\text{A}} = 12\Omega.$$

End of algebra.

(b) With S_1 open, the segment containing $\mathcal{E}_2$ and R_3 becomes a dead end; current can't flow through that leg of the circuit. And since S_2 remains open, the capacitor is also a dead end. In summary, no current can flow through $\mathcal{E}_2$, R_3, or the capacitor. As far as current is concerned, those segments of the circuit might as well be erased. In the accompanying picture, I've done exactly that.

What remains is a simple one-loop circuit! We can solve it using our old rules about resistors in series. But let me use Kirchhoff's laws, to make a point. Since the loop contains no intersections, we can't apply the 1st law; the same current flows around the whole circuit. So, let's go right to the 2nd law.

If we walk counterclockwise, starting from the top right corner, then we first walk "up" the battery ($\Delta V = +\mathcal{E}_1$). Next we walk with the current through R_2 ($\Delta V = -IR_2$), then through R_4 ($\Delta V = -IR_4$), and finally through R_1 ($\Delta V = -IR_1$), at which point we're back to the starting point. Since the total change is potential is zero,

$$+\mathcal{E}_1 - IR_2 - IR_4 - IR_1 = 0 \qquad \textbf{[whole-circuit loop]},$$

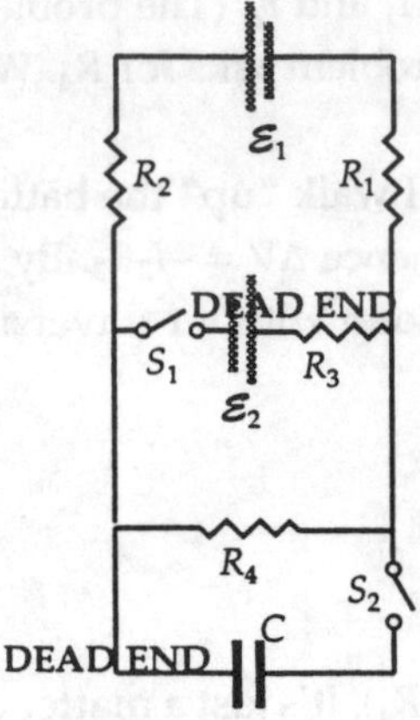

Equivalent to

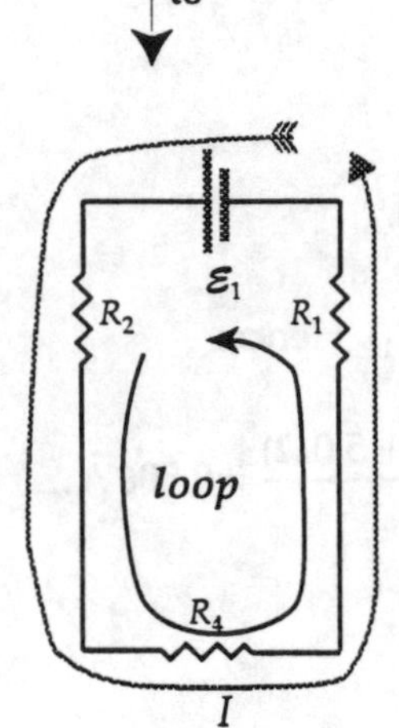

which we can rewrite as $+\mathcal{E}_1 = IR_{\text{total}}$, where $R_{\text{total}} = R_1 + R_2 + R_4$. We now see that the rule about adding resistors in series *comes from* Kirchhoff's 2nd law. Using similar reasoning, you can also derive R_{eq} for resistors in parallel. The circuit-reduction rules all come from Kirchhoff's laws. See your textbook for details.

Solving for I, and using our part (a) result that $R_4 = 12\ \Omega$, gives us

$$I = \frac{\mathcal{E}_1}{R_1 + R_2 + R_4} = \frac{2.0\ \text{V}}{3.0\Omega + 5.0\Omega + 12\Omega} = +0.10\ \text{A}.$$

The plus sign indicates that my diagram's guess was correct: current flows rightward through R_4. By contrast, in part (a), the current through R_4 flowed leftward. Why this reversal in direction? Because here we've "dead-ended" battery 2, the battery that forced current to flow clockwise around the bottom loop.

(c) Since S_1 stays open, the circuit leg containing $\mathcal{E}_2$ and R_3 remains a dead end, omitted from my drawing, But with S_2 closed, the capacitor is no longer a dead end. We've now got two loops. I_1 branches into I_4 and I_5, and then re-merges. We're solving for I_3.

At first glance, you might think no current can flow through a capacitor, because charges can't jump the gap between the two plates. Below, I'll show why current can flow "through" a capacitor, even though charges don't jump the gap. For now, let's forge ahead with Kirchhoff's 1st law. Looking at the intersection to the left of R_4, we see that I_1 flows in, while I_4 and I_5 flow out. So,

$$I_1 = I_4 + I_5. \tag{0}$$

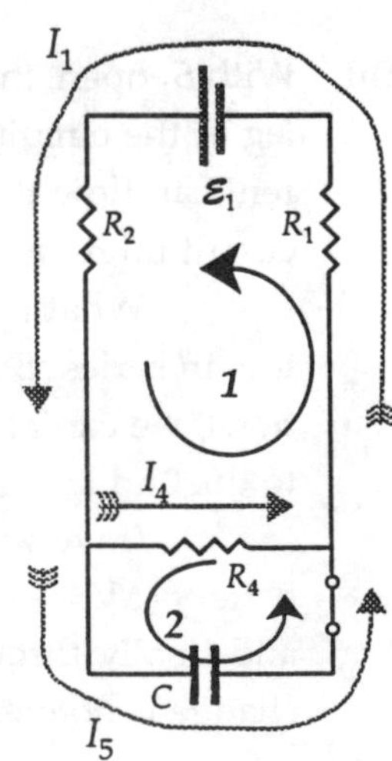

Without worrying about the capacitor, we can write an equation about loop 1, using Kirchhoff's 2nd law. As the loop diagram indicates, I'll walk counterclockwise. Notice that the current through R_4 is I_4, not I_1. So, by the usual reasoning,

Loop 1 $$+\mathcal{E}_1 - I_1R_2 - I_4R_4 - I_1R_1 = 0. \tag{1}$$

We have two equations in three unknowns (I_1, I_4, and I_5). To gather more information, consider loop 2, which includes the capacitor.

OK, it's time to explain how current flows through a capacitor, even though individual charges don't jump from one plate to the other. As positive charge flows onto the left plate, the charge accumulates on the plate. So, the current seems to be blocked. But while positive charges flow onto the left plate, an equal number of positive charges flow *off* the right plate. This leaves a deficit of positive charge, and hence a net neg-

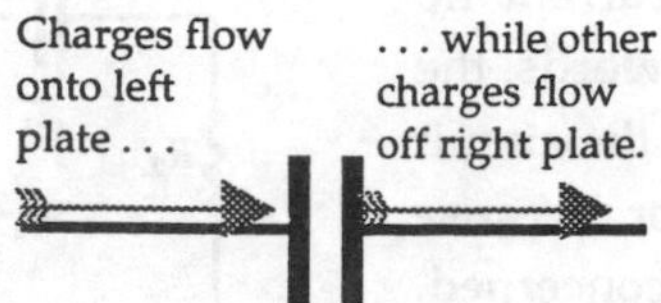
Charges flow
onto left
plate . . .
. . . while other
charges flow
off right plate.

ative charge, on the right plate. In this way, the plates acquire equal and opposite charge. As this diagram shows, on both sides of the capacitor, current flows, even though individual charges don't jump the gap. That's how current flows "through" a capacitor.

Given this insight, let's apply Kirchhoff's 2nd law to loop 2. To do so, we need to know the potential change across a capacitor. Well, by the definition of "capacitance," $C = q/V$. Therefore, the potential change across a capacitor is $V = q/C$, where q denotes the charge on the capacitor plates.

Immediately after the switch gets closed, hardly any charge has had time to flow onto the capacitor plates. So, the plates carry negligible charge: $q \approx 0$. Therefore, the potential change across the capacitor is $V = 0$. The usual loop-walk reasoning gives us

Loop 2, immediately after $t = 0$ $\qquad I_4R_4 + (0 \text{ for capacitor}) = 0. \qquad (2)$

I wrote "$+I_4R_4$" instead of "$-I_4R_4$" because, according to my loop 2 diagram, I'm walking against the current instead of with the current through R_4.

From Eq. (2), $I_4 = 0$. No current flows through R_4, immediately after $t = 0$. Here's why. The segment of the circuit containing the capacitor is resistanceless. Furthermore, since the capacitor remains uncharged at $t = 0$, the potential difference across it is still $\Delta V = 0$. So, at $t = 0$, the segment of the circuit containing the capacitor is "voltage-less" *and* resistanceless. Therefore, at $t = 0$, that segment behaves *as if* it were a resistanceless wire. This "resistanceless wire" is in parallel with R_4. In other words, when the current I_1 reaches the intersection to the left of R_4, it has a choice. The current can branch through R_4, through the capacitor, or some of each. But at $t = 0$, since the capacitor might as well be a resistanceless wire, *all* of I_1 branches through the capacitor, and none of it flows through R_4. In symbols, $I_5 = I_1$, while $I_4 = 0$. This is true only at $t = 0$, when the capacitor is still uncharged (and therefore voltage-less).

Given this simplification, we can quickly solve for I_3, the current through the capacitor. Starting with Eq. (1), set $I_4 = 0$ and $I_1 = I_3$ to get

$$\mathcal{E}_1 - I_3R_2 - 0 - I_3R_1 = 0,$$

and hence $I_3 = \dfrac{\mathcal{E}_1}{R_1 + R_2} = \dfrac{2.0\,\text{V}}{3.0\,\Omega + 5.0\,\Omega} = 0.25\,\text{A}.$

(d) Current flows through a capacitor *only* when it's getting charged up. This fact follows from my part (c) description of how current flows through, even though individual charges don't jump the gap between the plates. Positive charges flow onto the left plate and accumulate, while an equal number of positive charges flow off the other plate, leaving a net negative charge. Once the plates become fully charged, the left plate no longer "accepts" incoming charges, and the right plate no longer "spits out" charges. Therefore, all flow of charges—that is, all current—near the capacitor stops dead.

A typical capacitor charges up very quickly, usually in under a tenth of a second, and often much faster than that. Even without detailed calculations, we can safely assume that, by $t = 5$ s, the capacitor has become (essentially) fully charged. Therefore, current no longer flows through that segment of the circuit: $I_5 = 0$. In other words, the capacitor becomes a dead end. Consequently, *all* the current in the circuit flows through R_4. In this picture, I omit the charged-up capacitor, because a dead end might as well be missing, as far as current is concerned. Mathematically, erasing the dead end corresponds to setting $I_3 = 0$ in all our equations.

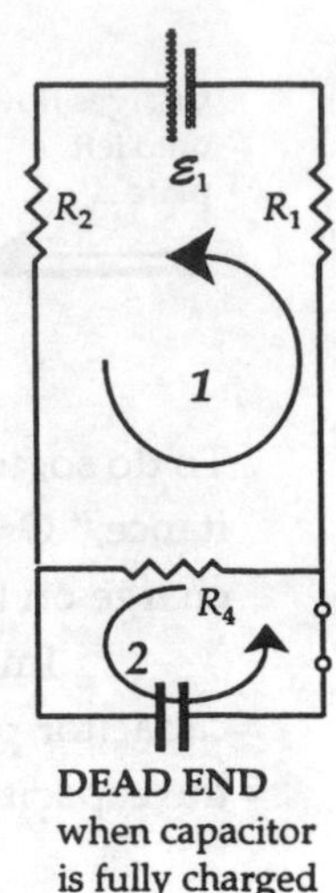

With the dead ends cut away, the circuit reduces to loop 1, the same loop we considered in part (b)! So, from part (b), we know the current through R_4. It's

$$I_4 = \frac{\mathcal{E}_1}{R_1 + R_2 + R_4} = \frac{2.0\,\text{V}}{3.0\,\Omega + 5.0\,\Omega + 18\,\Omega} = 0.77\,\text{A},$$

a result you can confirm by setting $I_1 = I_4$ in Eq. (1) above.

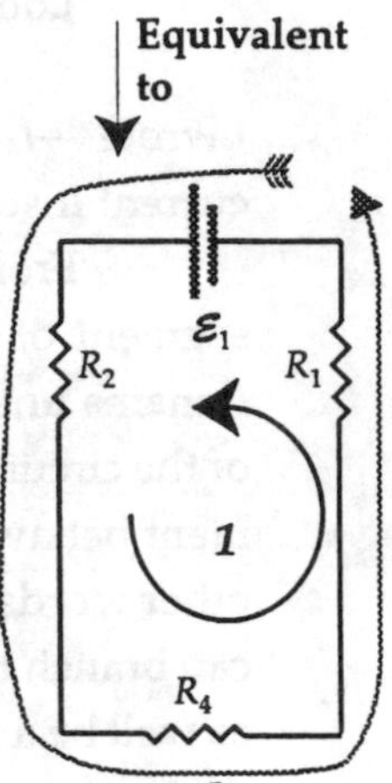

Let me summarize the main point of parts (c) and (d). When the capacitor is fully uncharged, it generates no potential difference, and therefore behaves like a resistanceless wire. But when it becomes fully charged, then current no longer flows through the capacitor, unless it discharges (which doesn't happen here). The capacitor becomes a dead end.

(e) Intuitively, the capacitor is "fully charged" when it generates a "backwards" potential difference that prevents additional current from flowing through it. To formalize this intuition, apply Kirchhoff's 2nd law to loop 2.

I'll walk counterclockwise, starting on right-hand side of R_4. Since loop 2 takes me leftward through that resistor, while I_4 flows rightward, I'm walking *against* the current. In the roadway analogy, this corresponds to climbing uphill. So, I *gain* potential energy, $\Delta V = +I_4 R_4$.

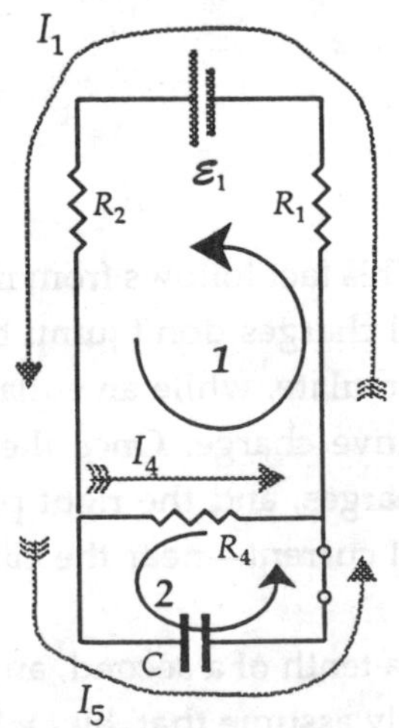

Next, I walk through the capacitor, from the left plate to the right plate. Since current has been flowing onto the left plate, it carries positive charge, while the right plate carries negative charge. So, my ΔV is the potential difference between the positive and negative plate. That ΔV must be negative, for the following reason. "Potential" is the potential energy of a 1-coulomb test charge. A 1-coulomb test charge "wants" to move from the positive to the negative plate, since the electric field between the plates pushes it in that direction. When an object "falls" in the direction that a gravitational or electric field pushes it, the object loses potential energy. So, when you walk through a capacitor from the positive to the negative plate, you're "falling" with the electric field, and therefore, your potential drops: $\Delta V = -Q/C$. Notice that I just used the definition of capacitance, $C = Q/V$.

Since the potential changes around loop 2 sum up to zero,

Loop 2 $+I_4R_4 - Q/C = 0.$

We can immediately solve for Q, since we know from part (d) that $I_4 = 0.10$ A, and we know from part (a) that $R_4 = 12\ \Omega$.

$$Q = I_4R_4C = (0.10\text{ A})(12\ \Omega)(3.0 \times 10^{-6}\text{ coulomb/volt}) = 3.6 \times 10^{-6}\text{ coulombs}.$$

That's the capacitor's charge, when fully charged up.

QUESTION 38-5

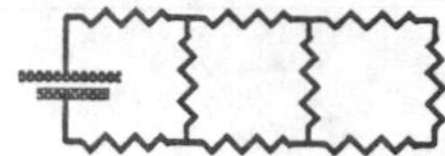

In this circuit, each resistor has resistance 1 Ω. The battery has voltage 1 volt. What current flows through the battery?

ANSWER 38-5

You could divide the circuit into three loops and use Kirchhoff's laws. This works fine. But here, we can "reduce" the circuit into a single loop. In other words, we can find the equivalent resistance R_{eq} of the whole circuit. From Ohm's law, current $i = V/R_{eq}$ flows through the battery, where $V = 1$ volt.

Many students wonder how they can tell when to use circuit reduction and when to use Kirchhoff's laws. As a rule of thumb, any circuit with two or more batteries on different loops demands that you use Kirchhoff's laws. Also, in some one-battery circuits, the resistors don't form a pattern of in-series and in-parallel connections. Again, such circuits require Kirchhoff's laws. I can't give you a simple to rule to describe which circuits are reducible. You need to get a feel for it by solving lots of problems. If unsure, use Kirchhoff's laws, because they are fundamental; they *always* work. By contrast, the in-series and in-parallel circuit-reduction formulas are *derived* from Kirchhoff's laws, and apply only in certain situations, as just discussed.

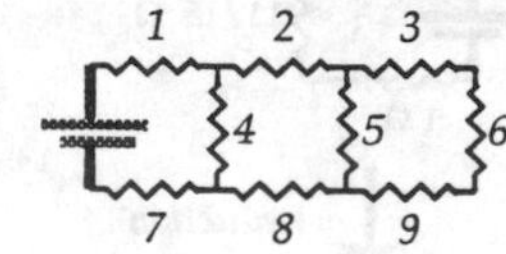

Before reducing the circuit correctly, let me point out some common mistakes. I'll refer to the resistors by numbers, as labeled here. These numbers are "names," not resistances. Assume current tends to flow clockwise.

Some students say that resistor 1 is "in series" with resistor 2, or that resistor 5 is in series with resistor 8. In a sense, this is true. But the "in-series" formula ($R_{eq} = R_1 + R_2$) applies only when *all* the current passing through one resistor also passes through the other. In this sense, resistors 1 and 2 are not in series, because some of the current through resistor 1 flows through resistor 4 instead of resistor 2. Similarly, 5 and 8 are not in series, because 8 gets "fed" by both 5 and 9.

A more common error is to say that resistor 2 is in parallel with resistor 4, or 3 is in parallel with 5. In a sense, this is right. But the "in-series" formula ($1/R_{eq} = 1/R_1 + 1/R_2$) applies to resistors R_j and R_k only when

(i) *all* current reaching those two resistors flows through one resistor *or* the other, but not both; and
(ii) the currents through R_j and R_k immediately "re-merge," without passing through any more resistors.

Resistors 2 and 4 satisfy condition (i), but not condition (ii). Pay particular attention to condition (ii). It's easy to forget. Notice that no two resistors in the entire circuit satisfy condition (ii).

Now that I've outlined the common mistakes, I'll reduce the circuit correctly, step by step. In the accompanying diagram, I always circle the resistors that I'm about to reduce.

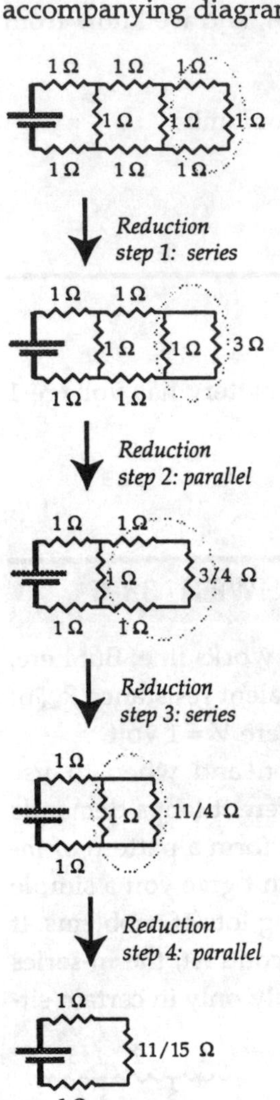

Reduction step 1: As noted above, the three resistors on the right-hand side of the circuit are in series. Their total resistance is $R_{tot} = 1\ \Omega + 1\ \Omega + 1\ \Omega = 3\ \Omega$. So, I've redrawn the circuit, erasing the three 1-ohm resistors and replacing them with a single 3-ohm resistor.

Reduction step 2: Now the "3 Ω" resistor is in parallel with resistor #5, which has resistance 1 Ω. To find the equivalent resistance of these two resistors, use

$$\frac{1}{R_{eq}} = \frac{1}{3\Omega} + \frac{1}{1\Omega} = \frac{4}{3\Omega},$$

and hence $R_{eq} = \frac{3}{4}\Omega$.

Reduction step 3: This "3/4 Ω" resistor is in series with two of the 1 Ω resistors, as indicated on the diagram. These three resistors have total resistance $R_{tot} = 3/4\ \Omega + 1\ \Omega + 1\ \Omega = 11/4\ \Omega$.

Reduction step 4: This "11/4 Ω" resistor is in parallel with resistor #4, as shown here. The equivalent resistance is given by

$$\frac{1}{R_{eq}} = \frac{1}{1\Omega} + \frac{1}{\frac{11}{4}\Omega} = \frac{1}{1\Omega} + \frac{4}{11\Omega} = \frac{15}{11\Omega},$$

and hence $R_{eq} = \frac{11}{15}\Omega$.

Reduction step 5: This "11/15 Ω" resistor is in series with two 1 Ω resistors, giving us a total resistance of

$$\frac{1}{R_{eq}} = \frac{11}{15}\Omega + 1\Omega + 1\Omega = \frac{41}{15}\Omega.$$

That's the equivalent resistance of the whole circuit. By Ohm's law, the current through the battery is

$$i = \frac{V}{R_{eq}} = \frac{1\ \text{volt}}{\frac{41\Omega}{15\Omega}} = \frac{15}{41}\ \text{amps} = 0.37\ \text{amps}.$$

QUESTION 38-6

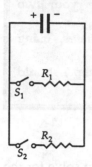

In this circuit, with both switches open, the capacitor is initially charged up to voltage V_0. In other words, the two capacitor plates have equal and opposite charges, which generate potential difference V_0 between the two plates.

The capacitor "plates" are actually thin cylindrical rods of length l and radius R. The distance between the centers of the rods, s, is much less than the length of the rods, l. Therefore, the rods "behave" as if infinitely long, to excellent approximation.

MAGNIFIED VIEW OF CAPACITOR

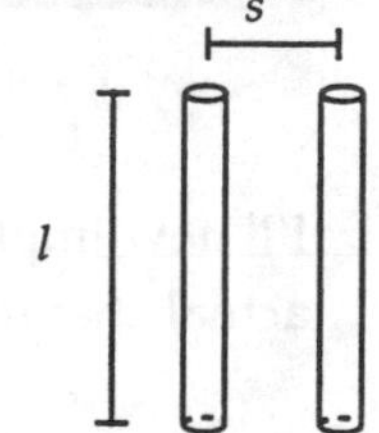

(a) *(Very hard, but you can answer the other parts of the problem without completing this part. You may want to skip this for now, and come back later.)* What charge does the positive capacitor plate carry, when the potential difference is V_0? Express your answer in terms of V_0, l, R, s, and any universal constants you need. Hint: Find the capacitance.

(b) To discharge the capacitor as quickly as possible, should I close switch 1 alone, switch 2 alone, or both switches? R_1 is greater than R_2. Justify your answer.

(c) Would squirting dielectric jelly into the space between the capacitor "plates" make the capacitor discharge more quickly or less quickly than it otherwise would? Explain your answer. The dielectric is inserted *after* the capacitor gets disconnected from the battery that charged it up.

(d) *(Read your textbook's section on RC circuits before attempting this part.)* The dielectric is removed. Suppose I close both switches at $t = 0$. Set up the equation or equations needed to solve for the current through R_1 at arbitrary time t. See if you can solve the equation(s). If you didn't finish part (a), just call the answer "Q_0" and solve in terms of Q_0 and the given constants (R_1, R_2, etc.).

ANSWER 38-6

(a) You might be tempted to start with the known potential difference V_0, and try to solve directly for Q_0, the initial charge on the plates. This turns out to be extremely difficult. Ironically, you can solve more easily using a "roundabout" strategy. Think of this as a standard capacitance problem. When finding the capacitance, you "pretend" to know the charge q on the plates, and then solve for the "unknown" potential difference V. That way, you can find the capacitance, $C = q/V$. This technique works, even if you use the "wrong" q, because the *ratio* q/V stays the same no matter charge the plates carry.

Once you know the capacitance, solve for Q_0 using $C = Q_0/V_0$, and hence $Q_0 = CV_0$.

So, the meat of the problem is figuring out the capacitance. Fortunately, we developed a strategy in Chapter 36.

Capacitance-finding strategy:

1) Assign equal and opposite charge q to your two plates.
2) Using Gauss' law or other techniques, calculate the electric field between your two plates.
3) From that electric field, calculate the potential difference between the plates, using $V = -\int \mathbf{E}\cdot d\mathbf{s}$.
4) Calculate the capacitance using $C = q/V$. The arbitrary charge q you chose in step 1 should cancel out.

I'll now implement this strategy for the two rods. Given the capacitance, we can then solve for the actual charge on the plates.

Capacitance-finding strategy step 1: Assign charges to the "plates."

I'll give the positive and negative rod charge q and $-q$, respectively. Since each rod has length l, the linear charge densities are $+\lambda$ and $-\lambda$, where $\lambda = q/l$. For the rest of this problem, I'll work in terms of these linear charge densities.

Capacitance-finding strategy step 2: Figure out electric field between the "plates."

Using Gauss' law, we've already calculated the electric field generated by a single "infinitely long" rod. But what about two rods? Unfortunately, no matter what Gaussian surface you draw around the two rods, you can't pull E outside the flux integral ($\int E\cdot dA$) in Gauss' law. Fortunately, the superposition principle saves us. According to this principle, the total electric field generated by two objects is simply the field generated by the first object, *plus* the field generated by the second object. So, we can *separately* figure out the fields created by the positive rod and the negative rod. Simply add those two fields to obtain the total electric field.

First, to find the field generated by the positive rod, I'll skim through the usual Gaussian reasoning. For a more detailed treatment, see question 34-3b.

Subproblem: Use Gauss' law to find the field produced by a single rod.

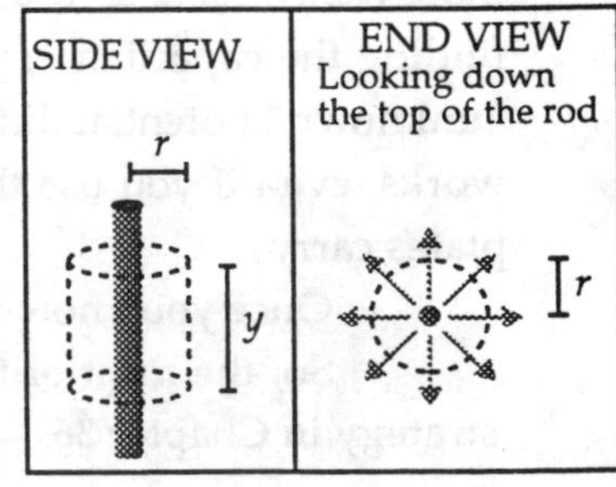

By cylindrical symmetry, if I enclose part of the rod with a cylindrical Gaussian surface, then the electric field has the same strength over the entire side (tube) of the Gaussian surface. Hence, we can pull E outside the flux integral:

$$\oint \mathbf{E}\cdot d\mathbf{A} = E\int_{\text{side of Gaussian cylinder}} dA = E(2\pi r y),$$

where I've used the surface area of a cylindrical tube of radius r and length y.

This Gaussian surface encloses length y of the rod, which has linear charge density λ, Therefore, $q_{\text{encl}} = \dfrac{\text{charge}}{\text{length}} \times (\text{length of rod enclosed}) = \lambda y$. So, according to Gauss' law,

$$\oint \mathbf{E} \cdot d\mathbf{A} = \frac{q_{\text{encl}}}{\varepsilon_0}$$

$$E(2\pi r y) = \frac{\lambda y}{\varepsilon_0},$$

and hence

$$E_{\text{single rod}} = \frac{\lambda}{2\pi\varepsilon_0 r}.$$

End of Gauss' law subproblem

The positive rod creates a field of this magnitude, pointing away from the rod. The negative rod produces a field of the same strength, but pointing towards the rod. We must add those two fields, to obtain the total field between the rods, $\mathbf{E}_{\text{total}}$. I'll do so below, while calculating the potential difference between the rods.

Capacitance-finding strategy step 3: Find the potential difference between the "plates."

The potential difference is $V = -\int \mathbf{E}_{\text{total}} \cdot d\mathbf{s}$, where the integral runs from the surface of the negative rod to the surface of the positive one (to ensure that the voltage comes out positive). I'll integrate along the straight path drawn below, from point a to point b.

At all points on this path, the electric field generated by the positive rod points rightward, away from that rod. The field produced by the negative rod *also* points rightward, toward the negative rod. Since both fields point rightward, we need not deal with vectors. The total electric field anywhere on the path points rightward, with strength

END-ON VIEW OF THE TWO RODS

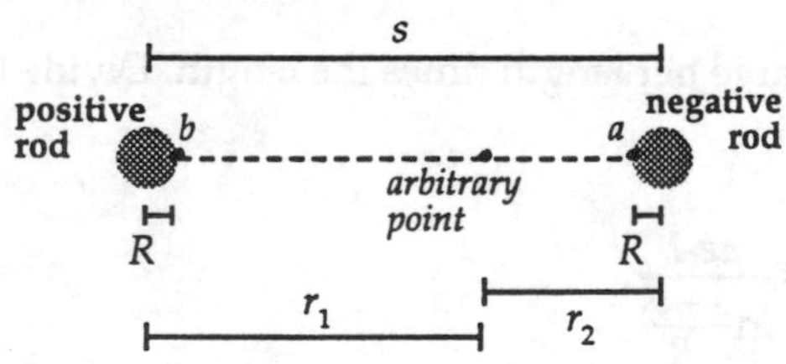

$$E = \frac{\lambda}{2\pi\varepsilon_0 r_1} + \frac{\lambda}{2\pi\varepsilon_0 r_2}.$$

Here, r_1 denotes the distance from the center of the positive rod, and r_2 denotes the distance from the center of the negative rod.

Since we're integrating from point a to point b, the path of integration starts at $r_2 = R$ and ends at $r_2 = s - R$. Please check this for yourself, using the diagram. So, I'll integrate over dr_2, using those values as my limits. Therefore, I must eliminate r_1 from the integral, by rewriting it in terms of r_2. That's not too hard. From the above diagram, $r_1 + r_2 = s$, and hence $r_1 = s - r_2$. So, the total electric field at an arbitrary point on the dashed line is

$$E = \frac{\lambda}{2\pi\varepsilon_0 (s - r_2)} + \frac{\lambda}{2\pi\varepsilon_0 r_2}.$$

Hence,

$$V = \int_a^b \mathbf{E} \cdot d\mathbf{s}$$

$$= \int_{r_2=R}^{r_2=s-R} \mathbf{E} \cdot d\mathbf{r}_2$$

$$= +\int_R^{s-R} E dr_2 \qquad [\mathbf{E} \cdot d\mathbf{r}_2 = -E dr_2, \text{ since } \mathbf{E} \text{ points rightward while } d\mathbf{r}_2 \text{ points leftward}]$$

$$= \int_R^{s-R} \left[\frac{\lambda}{2\pi\varepsilon_0 (s - r_2)} + \frac{\lambda}{2\pi\varepsilon_0 r_2} \right] dr_2$$

$$= \frac{\lambda}{2\pi\varepsilon_0} \left[-\ln(s - r_2) + \ln r_2 \right] \Big|_R^{s-R}$$

$$= \frac{\lambda}{2\pi\varepsilon_0} \left[-\ln(s - R) + \ln(s - R) \right] - \left[-\ln(s - R) + \ln s + R \right]$$

$$= \frac{\lambda}{2\pi\varepsilon_0} \left[-2\ln R + 2\ln(s - R) \right]$$

$$= \frac{\lambda}{\pi\varepsilon_0} \ln \frac{s - R}{R}.$$

Instead of re-expressing r_1 in terms of r_2, you could also have broken the integral into two "separate" integrals: $V = -\int_{s-R}^{R} \frac{\lambda}{2\pi\varepsilon_0 r_1} dr_1 + \int_R^{s-R} \frac{\lambda}{2\pi\varepsilon_0 r_2} dr$, which yields the same answer. The minus and plus signs in front of those integrals stem from the fact that $\mathbf{E} \cdot d\mathbf{r}_2 = -E dr_2$, since those two vectors point in opposite directions; while $\mathbf{E} \cdot d\mathbf{r}_1 = +E dr_1$, since both those vectors point rightward.

Capacitance-finding strategy step 4: Calculate capacitance using $C = q/V$.

The positive rod carries charge $q = \lambda l$, the charge per length times the length. Divide that by the potential difference we just found, to get

$$C = \frac{q}{V} = \frac{\lambda l}{\frac{\lambda}{\pi\varepsilon_0} \ln \frac{s - R}{R}} = \frac{\pi\varepsilon_0 l}{\ln \frac{s - R}{R}}.$$

End of capacitance-finding strategy

We just used a "hypothetical" charge to figure out the capacitance. Now we can use that C, along with the *actual* potential difference between the rods, to find the *actual* charge on the rods. Since capacitance is charge per volt ($C = q/V$), we get

$$Q_0 = CV_0 = \frac{\pi\varepsilon_0 l}{\ln \frac{s - R}{R}} V_0.$$

(b) The capacitor discharges most quickly if you close both switches, allowing current to flow around *both* loops of the circuit, as drawn here. When a capacitor discharges, positive charges leave the positive plate and flow around the circuit to the negative plate. This process continues until both plates have zero net charge. By making more "lanes" available to those charges, we enable more charges per second to flow, thereby speeding up the discharge.

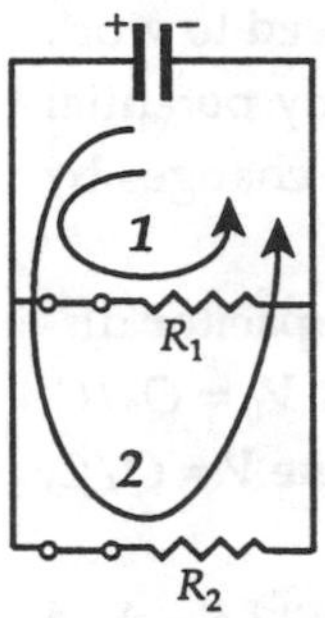

A more formal version of this argument, phrased in terms of equivalent resistance, leads to the same conclusion. R_1 and R_2 are in parallel. As we've seen before, the equivalent resistance of two in-parallel resistors is *less* than R_1 or R_2 individually, because the current can "spread out" over two paths. So, by making both resistors accessible, we lower the equivalent resistance, and thereby increase the current through the circuit. Since the current consists of positive charges flowing from the positive to the negative plate, increasing the current increases the rate at which the capacitor discharges.

(c) Intuitively, the rate at which the capacitor discharges depends on the charge, and also on the potential difference between the plates. So, let's begin by figuring out whether the dielectric changes the charge and/or the potential difference. Then, we can think about how those changes affect the rate of discharge.

Well, until we close a switch, the charge on each plate (rod) stays fixed, unable to adjust itself. A dielectric doesn't conduct; charges don't flow through it. For this reason, inserting the dielectric can't change Q_0 and $-Q_0$, the initial charge on the positive and negative plate.

By contrast, the dielectric *does* affect the potential difference between the plates. As explained in Chapter 36, a dielectric *dampens*(reduces) the electric field inside it. Therefore, since $V = -\int \mathbf{E} \cdot d\mathbf{s}$, the potential difference between the plates goes down. (Physically, by weakening the field, you decrease the work needed to push a test charge from one plate to the other.)

So far, we've found the physical effect of the dielectric. It reduces the potential difference between the plates, while leaving the initial charge unaffected. Now we must figure out how this reduced potential difference affects the rate of discharge.

Well, in this circuit, the potential difference across the capacitor "plays the role" of a battery. The potential difference is what drives current around the circuit. Therefore, reducing the V between the plates is just like turning down the voltage on a battery. In either case, the current decreases, in accordance with $V = iR$. In other words, less charge per second flows around the circuit. The capacitor discharges less quickly.

(d) As always, you can solve a circuit with Kirchhoff's laws. The presence of a capacitor and the absence of a battery does *not* alter our basic strategy.

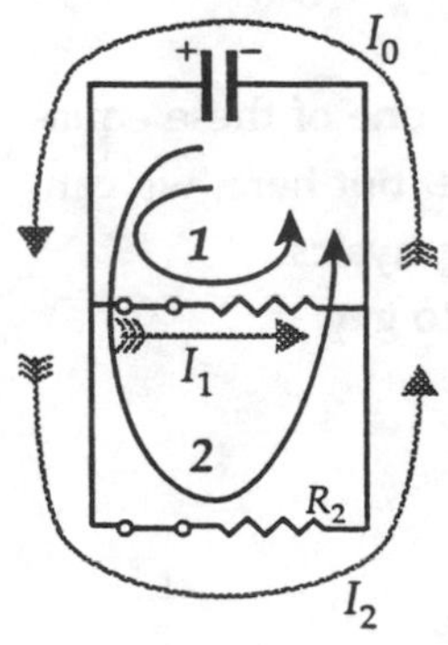

Begin by choosing loops, and labeling the currents in the different segments. As loop 1, I've chosen the top half of the circuit, from capacitor to R_1 to capacitor. As loop 2, I've chosen the "perimeter" of the circuit, from capacitor to R_2 to capacitor. As drawn here, the current I_0 through the capacitor splits into I_1 and I_2, the current through R_1 and R_2, respectively.

Now I'll apply Kirchhoff's 1st law, the intersection rule. Look at the intersection to the left of R_1. I_0 flows into that intersection, while I_1 and I_2 flow out. Therefore,

$$I_0 = I_1 + I_2. \qquad (0)$$

You get the same equation by considering the intersection to the right of R_1.

To gather more information, apply Kirchhoff's 2nd law to each loop individually. I'll walk counterclockwise around loop 1, starting from the upper right corner. First, I pass through the capacitor, from the negative to the positive plate. Since the electric field points from the positive to the negative plate, I'm walking against that field. If I were positively charged, I'd need to work hard to fight the field, just like a hiker needs to work hard to climb a mountain. So, my potential energy—and hence, my potential—increases. Therefore, since $C = Q/V$, my potential changes by $V = +Q/C$.

Here's a crucial point. Because the Q on the plates changes with time as the capacitor discharges, the potential difference changes as well. Sure, the potential difference starts at $V_0 = Q_0/C$. But later, when Q is less than Q_0, the potential is less than V_0. That's why you should write $V = Q/C$, where Q is a *variable*, instead of plugging in a specific value for Q.

Next on loop 1, I walk through R_1 with the current (i.e., "downhill"). So, $\Delta V = -I_1R_1$. And that's it for loop 1. Since the total potential change around a loop must equal zero,

Loop 1 $$\frac{Q}{C} - I_1R_1 = 0. \quad (1)$$

We can apply equivalent reasoning to loop 2, to get

Loop 2 $$\frac{Q}{C} - I_2R_2 = 0. \quad (2)$$

We found the capacitance in part (a). Still, we have three equations in *four* unknowns (I_0, I_1, I_2, and Q). Although we know the initial charge on the capacitor from part (a), we don't know $Q(t)$, the charge at arbitrary time. Therefore, except at $t = 0$, Q is unknown.

To generate another equation with Q in it, remember the relationship between charge and current. Here, the current "comes from" the charges flowing off the positive capacitor plate. For instance, if 3 coulombs per second flow off the positive plate and onto the negative plate, then the current "through" the capacitor is simply 3 coulombs per second. In general,

$$I_0 = -\frac{dQ}{dt}. \quad (3)$$

The minus sign indicates that I_0 is positive when the positive capacitor plate *loses* charge., i.e., when the capacitor discharges.

We now have four equations in four unknowns (Q, I_0, I_1, and I_2). Since one of these equations is differential instead of algebraic, there's no guarantee that we can solve. But here, we can. And the results deserve extra attention, because they encode lots of interesting physics.

Calculus and algebra starts here. Solve Eq. (1) and (2) for the currents to get

$$I_0 = -\frac{Q}{R_1C}$$

$$I_2 = -\frac{Q}{R_2C}.$$

Now add these two equations to get

$$I_1 + L_2 = \frac{Q}{C}\left(\frac{1}{R_1} + \frac{1}{R_2}\right)$$

$$I_0 = -\frac{Q}{R_{eq}C}.$$

In the last step, I invoked Eq. (0), $I_0 = I_1 + I_2$. As a convenient abbreviation, I also substituted in our old formula for the equivalent resistance of two resistors in parallel, $\frac{1}{R_{eq}} = \frac{1}{R_1} + \frac{1}{R_2}$.

OK, now for the cool part. From Eq. (3), $I_0 = -dQ/dt$. Substitute this expression for I_0 into the above equation, to get

$$-\frac{dQ}{dt} = \frac{Q}{R_{eq}C},$$

a standard first-order differential equation. We can solve for Q as a function of time, and then figure out the currents.

Address this differential equation by separation of variables: Multiply both sides by $-dt$, and divide both sides by Q, to get

$$\frac{dQ}{dt} = -\frac{1}{R_{eq}C}dt.$$

Now we can integrate both sides. The initial charge is Q_0, and the initial time is $t = 0$. We're looking for the charge Q at arbitrary later time t. So,

$$\int_{Q_0}^{Q} \frac{dQ}{Q} = \int_0^t \frac{1}{R_{eq}C}dt$$

$$\ln Q\Big|_{Q_0}^{Q} = -\frac{1}{R_{eq}C}t\Big|_0^t$$

$$\ln Q - \ln Q_0 = -\frac{t}{R_{eq}C}$$

Now for a cute maneuver. If $a = b$, then $e^a = e^b$. So, we can exponentiate both sides of the equation. Since logarithm is the inverse of the exponential, $e^{\ln a} = a$. Hence,

$$\frac{Q}{Q_0} = e^{-1/R_{eq}C}.$$

Multiply through by Q_0 to get, finally,

$$Q = Q_0 e^{-t/R_{eq}C}.$$

So, Q decreases exponentially as the capacitor discharges.

Now we can solve for I_1, using Eq. (1). Since $Q/C - I_1R_1 = 0$,

$$I_1 = \frac{Q}{R_1C} = \frac{Q_0}{R_1C}e^{-t/R_{eq}C},$$

where $C = Q_0/V_0$ is the capacitance found in part (a).

The current decreases exponentially, too! These results apply generally to discharging capacitors. The charge and current both decrease exponentially from their initial values. The decay factor is $e^{-t/\tau}$, where $\tau = RC$ is the decay "time constant." The bigger this constant, the longer the capacitor takes to discharge. Which makes perfect sense, physically. By increasing R, you make the circuit less "willing" to have charge flow through it, thereby slowing down the current. Similarly, the higher the capacitance, the more "willingly" the capacitor holds charge, and hence, the less forcefully it discharges.

39 CHAPTER

Magnetic Fields and Forces

QUESTION 39-1

A room is filled with a uniform magnetic field pointing straight up, from the floor to the ceiling. Let the z-direction be upward, the x-direction be eastward, and the y-direction be northward.

(a) A positively charged particle sits on the floor, initially motionless. What is the direction of the magnetic force on it, if any?

(b) Now the particle is dropped from a few meters off the floor. What path does it follow? In other words, does the particle curve right, curve left, go in a circle, follow a parabola, or what?

(c) Next, the particle is placed on the floor at arbitrary point A, and given an initial velocity in the x-direction. What path does it follow? Draw the path.

(d) (*Very hard, but cool*) Finally, the particle is held a few meters off the floor, and thrown in the x-direction. Describe and draw its path.

ANSWER 39-1

With magnetic fields and forces, people often get bogged down in the mathematics of cross products, without developing a *physical* understanding. This problem tests your qualitative understanding of magnetic forces.

(a) Part (a) highlights one of the crucial ways in which magnetic fields differ from electric and gravitational fields. Electric and gravitational fields don't "care" about an object's speed. For instance, gravity always exerts a downward force mg on a softball, no matter whether it's moving or sitting still. This does not mean a dropped softball follows the same trajectory as a tossed softball. They follow different trajectories, but *only* because they have different initial velocities, *not* because they experience different forces. Similarly, if an electric field $\mathbf{E}$ acts on a particle of charge q, the particle feels force $q\mathbf{E}$, whether or not it's moving.

The magnetic field, by contrast, acts only on *moving* charges. Specifically, the magnetic force on a charged particle is proportional to its speed:

$$\mathbf{F}_{\text{mag}} = q\mathbf{v} \times \mathbf{B}.$$

where $\mathbf{v}$ denotes the particle's velocity. Since the particle sits still ($\mathbf{v} = 0$), it experiences force $\mathbf{F}_{\text{mag}} = 0$.

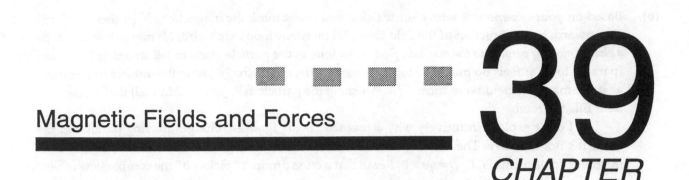

(b) Based on your experiences with electric fields, you might think the magnetic field pushes the particle upward, in the direction of the field lines. But magnetic fields are weird. *No magnetic force acts on a charge moving parallel to the magnetic field.* Therefore, as the particle starts to fall straight downward (parallel to **B**), it feels no magnetic force. Consequently, it falls freely under the influence of gravity, as if the magnetic field were turned off. Watching the particle fall, you couldn't tell that a magnetic field fills the room.

I can't explain intuitively why a magnetic field "ignores" charges moving parallel to the field. It's not intuitive. The best I can do is show how this strange physical fact follows from the magnetic force formula, $\mathbf{F}_{\text{mag}} = q\mathbf{v} \times \mathbf{B}$. Recall that a cross product "picks off" the components of vectors perpendicular to each other. For instance, according to the torque formula, $\tau = |\mathbf{r} \times \mathbf{F}| = F_{\perp} r$, only the component of a force *perpendicular* to the radius vector "contributes" to the torque. In general, a cross product has magnitude $|\mathbf{a} \times \mathbf{b}| = ab\sin\theta$, where θ denotes the angle between the two vectors. So, when the two vectors are parallel or "antiparallel" ($\theta = 0°$ or $180°$), the cross product equals zero.

For this reason, if the particle's velocity **v** is parallel (or antiparallel) to the magnetic field **B**, then $\mathbf{F}_{\text{mag}} = q\mathbf{v} \times \mathbf{B} = 0$. In other words, when a charged particle moves "along" a field line, the field can't act on the particle.

(c) Here, the particle moves perpendicular to the magnetic field. Therefore, it feels a magnetic force. To find the direction of this force, start with $\mathbf{F}_{\text{mag}} = q\mathbf{v} \times \mathbf{B}$. When you "cross" two vectors, the result is a new vector that's perpendicular to *each* of the vectors you're multiplying. Therefore, the magnetic force $\mathbf{F}_{\text{mag}}$ always points perpendicular to **v** (and **B**). In other words, the magnetic force always pushes a charged particle at right angles to its direction of motion. As the particle changes direction, the magnetic force changes direction as well, so that $\mathbf{F}_{\text{mag}}$ remains perpendicular to **v**.

This sounds vaguely familiar, from last semester. What kind of motion occurs when the force on an object always points perpendicular to the object's velocity? Circular motion! During circular motion, an object feels a "centripetal" (radial) force directed at right angles to its (tangential) motion. This radial force neither speeds up nor slows down the particle. It merely "nudges" the particle sideways, continuously changing its direction of motion, so that it traces a circle instead of a straight line. The magnetic field creates exactly this kind of "sideways" (centripetal) force. As a result, the charged particle traces out a circle.

To describe the particle's motion in more detail, we must decide whether that circle stays on the floor. Well, a cross product is perpendicular to both of the multiplied vectors. Hence, $\mathbf{F}_{\text{mag}}$ points perpendicular to **B**. Since **B** points vertically, in the z-direction, it follows that $\mathbf{F}_{\text{mag}}$ always points horizontally, along the floor (in the x-y plane). So, $\mathbf{F}_{\text{mag}}$ never pushes the particle up or down. It makes the particle travel in a circle on the floor, as drawn here. In these pictures, gray dots represent a magnetic field coming out of the page.

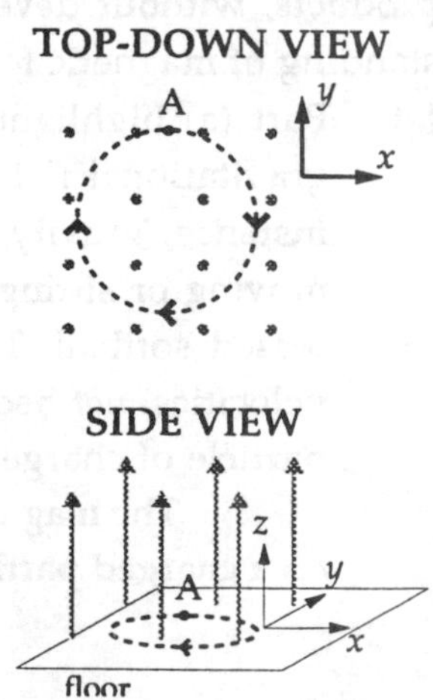

In summary, since $\mathbf{F}_{\text{mag}}$ always points perpendicular to the particle's direction of motion (**v**), the particle traces a circle. And since $\mathbf{F}_{\text{mag}}$ also points perpendicular to **B**, the circle is perpendicular to the magnetic field. Here, the z-directed field makes a charged particle circle in the x-y plane. But a y-directed field would make a charged particle circle in the x-z plane. And so on.

We still must decide whether the particle circulates clockwise or counterclockwise. To answer this, apply the *right-hand rule* to the cross product $\mathbf{v} \times \mathbf{B}$. First, point your right fingers in the direction of **v**. Initially, that's the x-direction. Then, curl your fingertips so that they point in the direction of **B**, twisting your wrist if necessary. In the top-down view, you fingertips should point out of the page. Whichever way your thumb now points, that's the direction of $\mathbf{v} \times \mathbf{B}$, and hence, the direction of $\mathbf{F}_{mag}$. Here, you thumb points "down the page" (in the top-down view). So, $\mathbf{F}_{mag}$ points in the $-y$ direction. In the top-down view, the particle gets deflected down the page. As a result, it circles clockwise instead of counterclockwise.

(d) To solve this, keep in mind the central lesson we learned last semester from studying projectile motion: *You can break an object's motion into "parts" (components), and treat those components separately*. For instance, when thrown sideways, a projectile keeps all its horizontal motion while accelerating vertically. The downward aspect of its motion doesn't affect the horizontal velocity, which stays constant. And the horizontal motion doesn't affect how much time the projectile takes to reach the floor. In brief, the horizontal and vertical motion are independent.

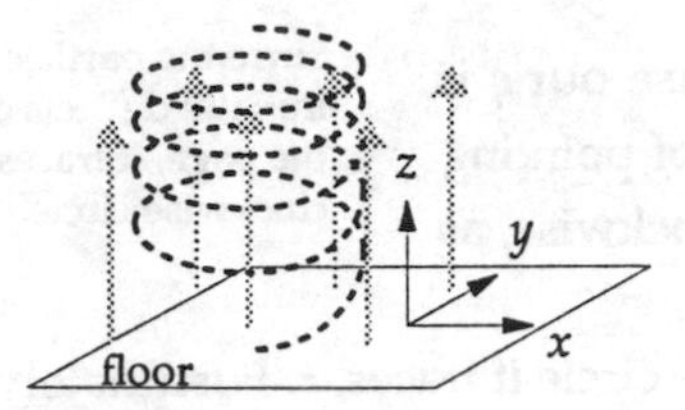

Similar reasoning applies here. In part (b), we saw that the magnetic field has no effect on the z-component of the particle's motion, since no magnetic force acts on a particle moving along a field line. A dropped particle would accelerate downward due to gravity, at $9.8\ \mathrm{m/s^2}$, *as if* the magnetic field didn't exist. Therefore, no matter what kind of complicated x-y motion the particle actually undergoes, its z-directed motion is nothing more than gravitational free fall, with acceleration $9.8\ \mathrm{m/s^2}$. This z-component of the motion is independent from the x-y components.

In part (c), we found that the x-y motion is circular. Therefore, as the particle falls, it traces out circles. The z-directed falling has no effect on the radius or period of these circles, since the x-y motion is independent from the z motion.

OK, let's tie this together. The particle's total motion combines free-fall in the z-direction with uniform circular motion in the x-y plane. This combination produces a helix. But unlike in a DNA strand, the spacing between the circles (i.e., the "pitch") gets bigger as the particle gets lower. That's because the particle *accelerates* downward, as opposed to falling at constant speed.

QUESTION 39-2

Top-down view
(from ceiling)

× × × × ×
× × × × ×
× ×• × × × (particle)
× × × × ×
× × × × ×

A frictionless particle of charge $Q = -2.0 \times 10^{-6}$ coulombs and mass $M = 3.0 \times 10^{-6}$ kg sits on the floor. The room is filled with a strong downward-pointing magnetic field of strength $B_0 = 0.10$ T. (In these drawings, gray x's indicate a magnetic field pointing into the page. The x's represent the tail feathers of arrows.)

(a) If I kick the particle rightward, will it trace out a clockwise circle or a counterclockwise circle, as seen from the top-down view?

(b) What initial speed must I give the particle so that it traces a circle of diameter 1.0 meter?

ANSWER 39-2

We can solve both parts by considering the force on the particle,

$$\mathbf{F}_{\text{mag}} = q\mathbf{v} \times \mathbf{B}.$$

(a) As discussed above, the particle traces a circle because the magnetic force always points perpendicular to the direction of motion, providing a centripetal acceleration. If the rightward-moving particle deflects "up the page" (in the above drawing), it will trace a counterclockwise circle. If it deflects "down the page," it will trace a clockwise circle. So,we need to find whether $\mathbf{F}_{\text{mag}}$ initially points up the page or down the page.

Since $\mathbf{F}_{\text{mag}} = q\mathbf{v} \times \mathbf{B}$, let's begin by finding the direction of $\mathbf{v} \times \mathbf{B}$, using the right-hand rule. Point your fingers in the direction of **v**, rightward. Then curl your fingertips in the direction of **B**, into the page. Your thumb now points up the page. That's the direction of $\mathbf{v} \times \mathbf{B}$.

But wait. That's *not* the direction of $\mathbf{F}_{\text{mag}} = q\mathbf{v} \times \mathbf{B}$, because our q is negative. The negative sign reverses the direction of $\mathbf{F}_{\text{mag}}$. Instead of pointing up the page, $\mathbf{F}_{\text{mag}}$ points down the page. The particle circulates clockwise, as drawn here.

particle

Since the particle initially deflects down the page, it traces a clockwise circle.

(b) To solve, we must relate the particle's speed, v, to the radius of the circle it traces, r. Fortunately, we can derive the relevant formula using circular-motion force reasoning from last semester.

When a particle travels in a circle, its radial acceleration (toward the center of the circle) is $a_c = v^2/r$. Hence, according to Newton's 2nd law, the particle must be feeling a net radial force $\Sigma F_{\text{radial}} = mv^2/r$. Since the magnetic field provides the radial force, we get

$$\sum F_{\text{radial}} = M\frac{v^2}{r}$$

$$|q\mathbf{v} \times \mathbf{B}| = M\frac{v^2}{r}$$

$$|QvB_0| = M\frac{v^2}{r}.$$

Let me clarify this derivation. The magnitude of the cross product reduces to a regular product, because **v** and **B** are perpendicular. (Remember, $|\mathbf{v} \times \mathbf{B}| = vB\sin\theta$; and here, $\theta = 90°$.) Also, since we found the direction of this force in part (a), I'm just worrying about the magnitude (absolute value). This derivation shows up again and again. Make sure you can do it.

Since we want the circle to have diameter 1.0 m, the radius must be $r = 0.50$ m. Solve the above equation for v to get

$$v = \frac{QB_0 r}{M} = \frac{(2.0 \times 10^{-6}\ \text{C})(0.10\ \text{T})(0.50\ \text{m})}{(3.0 \times 10^{-6}\ \text{kg})} = 0.033\ \text{m/s},$$

about 3 centimeters per second. That's very slow. If we kick the particle any harder, it will trace a bigger circle.

QUESTION 39-3

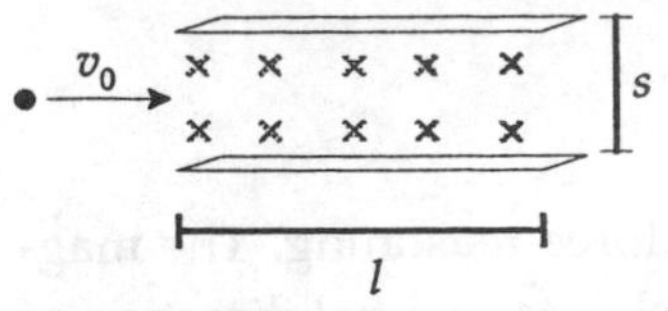

In outer space, where we can neglect gravity, a particle of charge $-Q$ and mass m gets shot rightward at initial speed v_0 between the two plates of a parallel-plate capacitor. The plates, fixed a distance s apart, have length l. The particle enters the capacitor midway between the plates, a distance $s/2$ from either one, as drawn here.

A uniform magnetic field of strength B_0 points into the page. This field exists only between the plates.

Express all answers in terms of the physical quantities listed above, and any universal constants you need. Initially, the capacitor is not charged up.

(a) The particle deflects and crashes into one of the plates. Does it hit the upper plate or the lower plate? And how far from the left end of that plate does the particle crash? Set everything up, even if you have trouble completing the math.

(b) (*This is a variant of a standard exam problem.*) Now the capacitor is connected to a power supply. As a result, one plate acquires positive charge, while the other acquires an equal negative charge. To what voltage (potential difference) must the power supply be set, to ensure that the particle travels in a straight line while between the plates, instead of deflecting? Hint: the electric field between the plates is approximately uniform.

(c) Suppose the power supply is set to a voltage a little below the one you found in part (b). Sketch the particle's trajectory, including its path after it exits the space between the two plates.

ANSWER 39-3

(a) In this part, the plates generate no electric field, because they carry no charge. Only a magnetic force acts on the particle.

Let's first figure out whether this force deflects the particle downward into the lower plate, or upward into the upper plate. Since $\mathbf{F}_{mag} = q\mathbf{v} \times \mathbf{B}$, I'll begin by applying the right-hand rule to $\mathbf{v} \times \mathbf{B}$. Point your fingers rightward to represent the velocity, then curl your fingertips into the page to represent the magnetic field. Your thumb now points upward. That's the direction of $\mathbf{v} \times \mathbf{B}$. Since $\mathbf{F}_{mag} = q\mathbf{v} \times \mathbf{B}$, a positive charge would deflect toward the upper plate. But this particle is negative. The minus sign reverses the direction of the force, indicating that the particle deflects toward the lower plate, not the upper one.

We need to figure out exactly where the particle crashes into the lower plate. As we've seen in previous problems, when a charged particle moves perpendicular to the magnetic field, it traces a circle. To figure out where that circle "hits" the lower plate, I'll first find the radius, r. Given the radius, we can then use pure math to calculate x, the distance from the left end of the plate to the crash point. Intuitively, if the radius is small, then the particle crashes very close to the left end of the plate. But if the radius is large, the particle travels a long way before crashing.

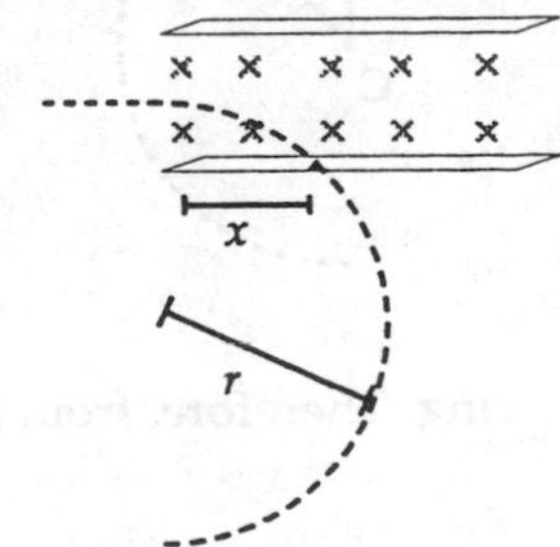

This drawing shows the entire path the particle would take if the bottom plate were removed, and if the magnetic field filled the space below the bottom plate. Of course, the particle crashes into the bottom plate, and never completes this semicircle. But this fictional con-

continuation of the particle's circular trajectory helps me solve the problem. Remember, I'll first find r, and then use it to calculate x.

Subproblem 1: Calculate the radius of the particle's trajectory.

To find r in terms of the given quantities, use circular-motion force reasoning. The magnetic field provides a radial force, i.e., a force perpendicular to the particle's tangential direction of motion. When traveling in a circle, a particle undergoes radial acceleration $a_c = v^2/r$. Therefore, from Newton's 2nd law,

$$\sum F_{\text{radial}} = ma_c$$

$$|q\mathbf{v} \times \mathbf{B}| = m\frac{v^2}{r}$$

$$Qv_0B_0 = m\frac{v_0^2}{r}, \tag{1}$$

where in the last step I used $|\mathbf{v} \times \mathbf{B}| = vB\sin\theta$, with $\theta = 90°$. Also, I wrote "Q" instead of "$-Q$," because I'm calculating the magnitude (absolute value) of the radial force. (We found the direction above.)

Solve for r, the radius of the circle:

$$r = \frac{mv_0}{QB_0}.$$

Notice that a bigger magnetic field produces a smaller r, i.e., a "tighter" circle.

Subproblem 2: Find x in terms of this r.

So far, we've found the radius of the particle's circular path. Given this r, we can figure out x using pure mathematics, with no more physical reasoning.

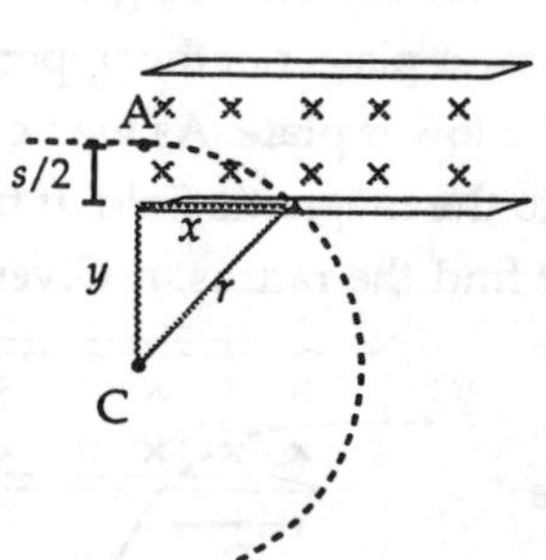

As usual, a good diagram takes us half way to the answer. C is the center of the circle, while A is the point at which the particle enters the space between the plates. Consider the (undrawn) vertical line segment from C to A. That line segment is a radius of the semicircle. Therefore, it has length r. But from the diagram, the distance from C to A is $y + s/2$. So, we have

$$y + s/2 = r,$$

and hence, $y = r - s/2$. That's the vertical leg of the triangle in my drawing. Therefore, from Pythagorean theorem ($x^2 + y^2 = r^2$), the horizontal leg has length

$$
\begin{aligned}
x &= \sqrt{r^2 - y^2} \\
&= \sqrt{r^2 - \left(r - \frac{s}{2}\right)^2} \\
&= \sqrt{r^2 - \left(r^2 - 2r\frac{s}{2} + \frac{s^2}{4}\right)} \\
&= \sqrt{sr - \frac{s^2}{4}} \\
&= \sqrt{s\frac{mv_0}{QB_0} - \frac{s^2}{4}},
\end{aligned}
$$

where in the last step, I used our expression for r from subproblem 1.

(b) Intuitively, the particle doesn't deflect if the electric force cancels the magnetic force. The plates must generate an electric field that pushes on the particle just hard enough to cancel the magnetic force. When that happens, the particle feels zero net force, and therefore floats in a straight line.

Given this insight, we can solve for the electric field. But the problem asks for the potential difference between the plates generated by the power supply. Fortunately, the electric field and the potential difference relate to each other in a simple way: $V = -\int_{\text{negative plate}}^{\text{positive plate}} \mathbf{E} \cdot d\mathbf{s}$. So, after finding the field between the plates, we can then find the corresponding potential difference.

In summary, we'll

1) figure out the electric field needed to cancel the magnetic force, and then
2) integrate over that field to find the corresponding potential difference between the plates.

The power supply must be set to that potential difference (voltage), in order to ensure that the plates produce the "correct" field.

Subproblem 1: Find the electric field between the plates.

To prevent the particle from deflecting, the electric field must exert a force that cancels the magnetic force:

$$
\begin{aligned}
F_{\text{elec}} &= F_{\text{mag}} \\
qE &= |q\mathbf{v} \times \mathbf{B}| \\
&= qvB_0,
\end{aligned}
$$

and hence, $E = v_0 B_0$.

Of course, this cancellation happens only if the electric and magnetic forces point in opposite directions. So, I'll now figure out the direction in which this electric field must point.

Well, as we found in part (a), the magnetic force initially pushes the particle downward. So, the electric force must push it upward. If the particle were positive, then an upward electric field would produce an upward force. But a negative particle gets pushed "backward" along electric field lines. Therefore, a *downward* electric field is needed to push the negative particle upward. Mathematically, this follows from $\mathbf{F}_{elec} = q\mathbf{E}$.

In summary, the electric force cancels the magnetic force if the electric field points downward, with strength $E = v_0B_0$. Don't put this formula on your cheat sheet. It's not general. It's the "answer" to a very specific problem.

Subproblem 2: Find the potential difference corresponding to this electric field.

Before continuing, let me briefly review where the formula $\Delta V = -\int \mathbf{E}\cdot d\mathbf{s}$ comes from. The potential difference is the work needed to "fight" the electric field while pushing a 1-coulomb test charge from one plate to the other (i.e., ΔU for a 1-coulomb charge). Since work $= -\int \mathbf{F}_{elec}\cdot d\mathbf{s} = -\int q\mathbf{E}\cdot d\mathbf{s}$, the work required to push a q = 1-coulomb charge is $-\int \mathbf{E}\cdot d\mathbf{s}$.

As we've seen before, the electric field inside a parallel-plate capacitor has approximately uniform strength, and points directly from the positive to the negative plate. Since we know that the field points downward, the upper plate must be positive, and the lower plate must be negative.

To evaluate $V = -\int_{\text{negative plate}}^{\text{positive plate}} \mathbf{E}\cdot d\mathbf{s}$, the potential difference between the plates, we must choose a path of integration. Integrating from the negative to the positive plate ensures that V comes out positive. I'll choose a straight line from the bottom to the top plate, the dashed line in this diagram Since $\mathbf{E}$ points in the opposite direction from $d\mathbf{s}$, the dot product $\mathbf{E}\cdot d\mathbf{s}$ reduces to a *negative* regular product. Also, since E has approximately constant strength between the plates, we can pull it outside the integral:

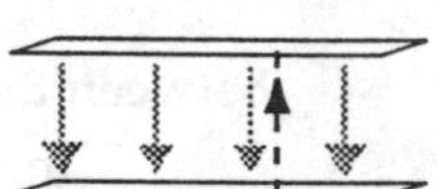

The dashed path of integration points oppositely to the electric field.

$$V = -\int_{\text{bottom plate}}^{\text{top plate}} \mathbf{E}\cdot d\mathbf{s}$$

$$= E\int_{\text{bottom plate}}^{\text{top plate}} ds$$

$$= Es,$$

since the path of integration has length s.

So, to ensure that the particle doesn't deflect, the power supply must generate a potential difference

$$V = Es = v_0B_0s,$$

where I've used my "answer" for E from subproblem 1, the field needed to ensure that the electric force cancels the magnetic force.

Before moving on to part (c), let me make a theoretical point. As demonstrated in this problem, when a particle feels both a magnetic field and an electric field, the total force is

$$\mathbf{F} = \mathbf{F}_{\text{elec}} + \mathbf{F}_{\text{mag}} = q\mathbf{E} + q\mathbf{v} \times \mathbf{B} = q(\mathbf{E} + \mathbf{v} \times \mathbf{B}).$$

Your textbook calls this the "Lorentz force law." It simply says that the total "electromagnetic" force is the sum of the electric and magnetic forces.

(c) If V is a little less than v_0B_0s, then the electric field is a little less than v_0B_0. In other words, the electric force is a little too weak to cancel the magnetic force. Therefore, the downward magnetic force slightly "overcomes" the upward electric force. The particle deflects downward, slightly. The particle does not follow a precisely circular arc while between the plates. That's because the electric force does not always point perpendicular to the particle's direction of motion, and hence, the net force is not "centripetal" (radial).

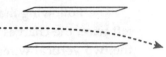

Once it leaves the space between the plates, the particle no longer feels any electric or magnetic forces. Since the net force vanishes, the particle follows a straight line, just like a hockey puck on frictionless ice or a rock in outer space.

QUESTION 39-4

REGULAR VIEW

Hanging straight down, before magnetic field turned on. The x-direction is into the page.

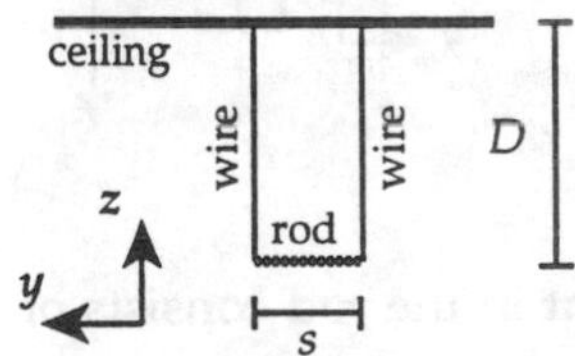

Consider a pendulum made of a cylindrical metal rod of mass $m = 0.0020$ kg, length $s = 0.20$ m, and resistance $R = 0.010\ \Omega$ The rod is connected at its ends to two stiff, resistanceless, massless wires, of length $D = 0.40$ m. Initially, the wires hang down from the ceiling, as drawn here.

Embedded inside the ceiling is a small battery (not pictured), of voltage $V = 1.5$ volts. The positive terminal touches the left wire, while the negative terminal touches the right wire, in the regular view picture.

I want to fill the room with a uniform magnetic field, designed to make the pendulum swing in the x-direction. Specifically, I want the pendulum to settle at a $\theta = 30°$ angle.

SIDE VIEW

End-on view of rod *after* it has swung. The y-direction is out of the page.

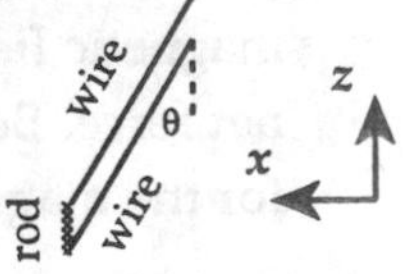

(a) In what direction should the magnetic field point, so that it pushes the rod in the x-direction as hard as possible? Please read the captions on these diagrams, to see what I mean by the x-direction.

(b) What must be the strength of this field?

ANSWER 39-4

As always, begin by visualizing what happens physically, *not* by plugging in formulas. Current flows around a circuit from the positive to the negative terminal of a battery. Therefore, this battery pushes counterclockwise current around the "circuit." Current flows down the left wire, rightward across the rod, then up the right wire, in the regular view.

An electric current consists of moving charges. Magnetic fields push on moving charges. Therefore, a magnetic field exerts a force on a current-carrying wire or rod. Starting with the basic magnetic force law, $\mathbf{F}_{\text{mag}} = q\mathbf{v} \times \mathbf{B}$, we'll derive the magnetic force acting on a current.

(a) For now, however, we can use purely qualitative reasoning. First, I'll show that the magnetic forces on the two wires point in opposite directions, and therefore cancel. Then, I'll figure out in which direction the field must point, to push the rod in the x-direction.

Following the textbook, I'll "pretend" that the current consists of positive charges. In the left wire, the charges travel downward. In the right wire, they travel upward. So, the charges in the two vertical wires have equal and opposite velocities. According to the magnetic force law, $\mathbf{F}_{\text{mag}} = q\mathbf{v} \times \mathbf{B}$, two charges with oppositely-directed velocities feel oppositely-directed forces. Therefore, the two wires get pushed in opposite directions. The magnetic forces on the two wires cancel out.

So, the pendulum swings because of the magnetic force on the *rod*. We must figure out the magnetic field direction that pushes the rod in the x-direction.

By applying the right-hand rule to $\mathbf{F}_{\text{mag}} = q\mathbf{v} \times \mathbf{B}$, you can find the answer. According to that rule, if you point your right fingers in the direction of $\mathbf{v}$, and then curl your fingertips in the direction of $\mathbf{B}$, your thumb then points in the direction of $\mathbf{F}_{\text{mag}}$. Here, the (positive) charges in the rod flow rightward, in the regular view. And $\mathbf{F}_{\text{mag}}$ points into the page (again in the regular view). So, if you point your fingers rightward (with $\mathbf{v}$), in which direction must you curl your fingertips (with $\mathbf{B}$) to make your thumb point into the page (with $\mathbf{F}_{\text{mag}}$)? By trying it out, you'll find that your curled fingertips must point downward, in the negative z-direction. That's the direction of $\mathbf{B}$. (Actually, any downward-tending vector in the y-z plane works; but the negative z-direction is best, because it gives the biggest force, and doesn't "twist" the pendulum.)

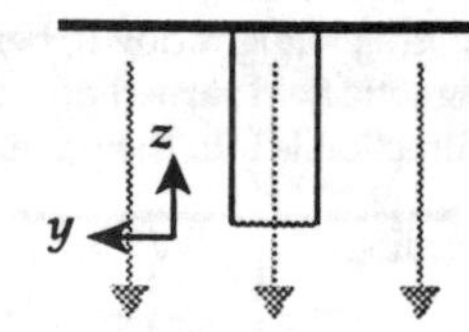

By the way, you reach this same conclusion by assuming the current in the rod consists of negative charges flowing leftward, instead of positive charges flowing rightward.

(b) When the pendulum settles at $\theta = 30°$, it's no longer accelerating. Therefore, the net force on it must vanish. Welcome to a classic force problem, specifically, a statics problem. We must calculate the magnetic field strength B needed to ensure that the pendulum, when displaced to $\theta = 30°$, feels no net force. Before drawing a force diagram and using Newton's 2nd law, I must derive the formula for the magnetic force acting on a current.

Derivation: The magnetic force on a current-carrying wire or rod

Consider the infinitesimal force on a tiny "piece" of the rod containing infinitesimal moving charge dq. Add up (integrate over) the tiny forces acting on all the different pieces, to obtain the total force on the rod. As a key step of the derivation, I'll recall that velocity is the rate of change of position. If we let $\mathbf{l}$ denote the position, then $\mathbf{v} = d\mathbf{l}/dt$.

OK, here goes. The infinitesimal force on a tiny piece of the rod containing charge dq is

$$\begin{aligned} d\mathbf{F}_{\text{mag}} &= dq\mathbf{v} \times \mathbf{B} \\ &= dq\frac{d\mathbf{l}}{dt} \times \mathbf{B} \qquad [\text{since } \mathbf{v} = d\mathbf{l}/dt] \\ &= dq\frac{d\mathbf{l}}{dt} \times \mathbf{B} \\ &= I\,d\mathbf{l} \times \mathbf{B}. \qquad [\text{since } v = dl/dt] \end{aligned}$$

You'll use this crucial formula throughout the course. Even if the magnetic field were non-uniform, you could integrate $Id\mathbf{l} \times \mathbf{B}$ to find the total force. Here, the integral is easy, for two reasons. First, the magnetic field is uniform, and therefore we can pull it outside the integral. Second, the magnetic field points perpendicular to the direction in which the rod "points." Therefore, the magnitude of the cross product $d\mathbf{l} \times \mathbf{B}$ reduces to $(dl)B$. Putting all this together, I get

$$\begin{aligned} F_{\text{mag}} &= \int dF_{\text{mag}} = \int I\,|d\mathbf{l} \times \mathbf{B}| \\ &= I\int (dl)B \\ &= IB\int dl \qquad [\text{since } B \text{ is constant}] \end{aligned}$$

where s is the rod's length.

End of derivation

So far, I've derived the formula for the magnetic force acting on this current-carrying rod. To solve for B, I'll now employ the standard force strategy we used dozens of times last semester. I could also solve by setting the net torque equal to zero. But forces alone get us all the way to the answer.

Whether you use forces or torques, start with a good free-body diagram. I'll use a side view, because otherwise the magnetic force would point into the page.

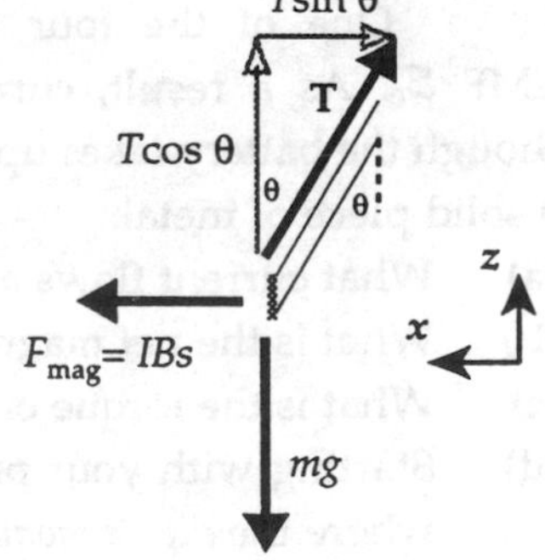

Each wire exerts a tension on the rod. By symmetry, those two tensions are equal. To simplify matters, let me combine those two tensions into a single force vector, **T**. What I'm calling "**T**" might be equivalent to "2**T**" on your force diagram.

As always, break up the force vectors along whatever (orthogonal) coordinate axes you want. "Tilted" axes work well here. But I'll use "regular" axes, so that **T** is the only vector I need to decompose into components.

Since the rod no longer accelerates when it settles, $a_x = a_z = 0$. Therefore, by Newton's 2nd law,

$$\sum F_x = ma_x$$

$$IBs - T\sin\theta = 0 \qquad (1)$$

$$\sum F_z = ma_z$$

$$T\cos\theta - mg = 0. \qquad (2)$$

We know $m = 0.0020$ kg, $s = 0.20$ m, $g = 9.8$ m/s², and $\theta = 30°$. We also know the current, from Ohm's law: $I = V/R = (1.5\text{ V})/(0.010\ \Omega) = 150$ A. So, Eqs. (1) and (2) contain only two unknowns, B and T. Only algebra remains.

Algebra starts here. Rewrite Eqs. (1) and (2) so that the tension terms appear alone:

$$T\sin\theta = IBs \qquad (1)$$

$$T\cos\theta = mg \qquad (2)$$

Now divide Eq. (1) by Eq. (2) to get $\tan\theta = \frac{IBs}{mg}$. Finally, isolate B and substitute in the numbers.

$$B = \frac{mg}{Is}\tan\theta = \frac{(0.0020\text{ kg})(9.8\text{ m/s}^2)}{(150\text{ A})(0.20\text{ m})}\tan 30° = 3.8\times10^{-4}\text{ T}.$$

End of algebra.

Notice that the pendulum length, D, makes no difference here.

QUESTION 39-5

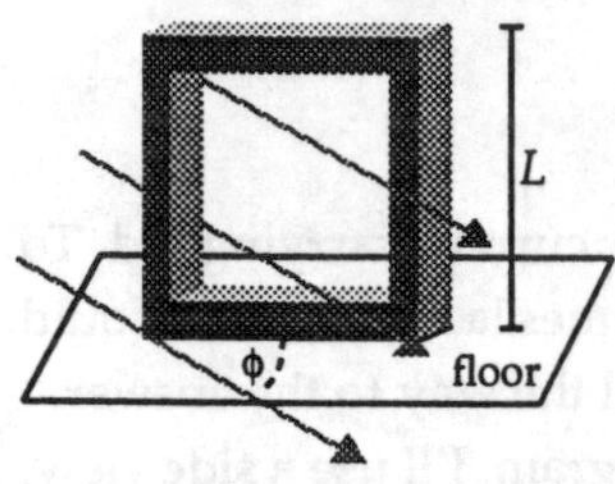

Consider a square metal "picture frame" sitting on the floor, as drawn here. Each side of the frame is a rectangular prism of mass M, length L, width s, and height s, where $L \gg s$. The metal has resistivity ρ_0.

A uniform magnetic field of strength B_0 fills the room. The field is parallel to the floor, and makes an angle ϕ with the plane of the frame.

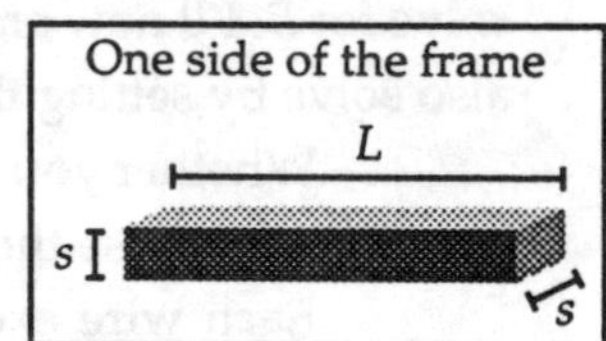

One of the four sides contains a small battery that supplies EMF $\mathcal{E}_0$. As a result, current flows clockwise around the frame. Even though the battery takes up space, you may treat each side of the square as a solid piece of metal.

(a) What current flows around the square?

(b) What is the net magnetic force on the square? Hint: Figure out the force on each of the four sides.

(c) What is the torque on the square. Hint: Take advantage of the four forces you found in part (b).

(d) Starting with your part (c) answer, show that the torque on the square can be written $\boldsymbol{\tau} = \boldsymbol{\mu}\times\mathbf{B}$, where the *dipole moment*, μ, is the current times the area vector: $\boldsymbol{\mu} = I\mathbf{A}$.

ANSWER 39-5

(a) Assuming we can find the square's resistance, Ohm's law ($V = IR$) will tell us the current.

In Chapter 37, using a tunnel analogy, I made it plausible that a slab has resistance

$$R = \rho\frac{l}{A},$$

where l denotes the length and A denotes the cross-sectional area. Intuitively, a wider tunnel allows more "car-current" to flow through, and therefore puts up less resistance. Also, if we assume a fixed height difference between the two ends of the tunnel, then a longer tunnel is less steep, and therefore pushes cars along less vigorously. Hence, a longer tunnel has higher resistance. For these intuitive reasons, a slab's electrical resistance is proportional to its length, but inversely proportional to its cross-sectional area.

Current flows through cross-sectional area $A = s^2$.

In each side, the current flows through length L, and through cross-sectional area $A = s^2$. Therefore, each side has resistance $R_{\text{side}} = \rho_0 \frac{L}{s^2}$. Since the four sides are in series, the total resistance is

$$R = 4R_{\text{side}} = 4\rho_0 \frac{L}{s^2}.$$

From Ohm's law, we immediately get

$$I = \frac{V}{R} = \frac{\mathcal{E}_0}{4\rho_0 \frac{L}{s^2}}.$$

(b) Following the "hint," I'll calculate the force on each of the four sides. We'll see that those forces cancel, leaving us with zero net force on the square.

In the previous problem, I derived the formula for the magnetic force on an infinitesimal piece of the slab. It's $d\mathbf{F}_{\text{mag}} = I d\mathbf{l} \times \mathbf{B}$, where $d\mathbf{l}$ is a tiny length of the slab. To find the force on an entire side, just integrate over the side. Since the magnetic field is uniform, we can pull it outside the integral:

Force on a side $$\mathbf{F}_{\text{mag}} = \int_{\text{one side}} I d\mathbf{l} \times \mathbf{B} = \left\{ I \int_{\text{one side}} d\mathbf{l} \right\} \times \mathbf{B} = I\mathbf{L} \times \mathbf{B},$$

since each side has length L. In this notation, the direction of $\mathbf{L}$ is the direction that current flows through the side.

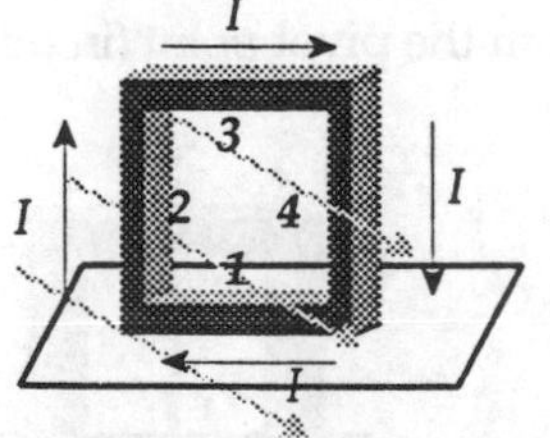

Let's start with sides 1 and 3, as labeled here. Both of those sides make the same angle, ϕ, with the magnetic field. Also, both sides have the same length, contain the same current, and "feel" the same magnetic field. Therefore, those two sides experience magnetic forces of the *same* strength:

$$F_1 = F_3 = I|\mathbf{L} \times \mathbf{B}| = ILB_0 \sin 9($$

However, those two forces point in opposite directions. They cancel. To see why, apply the right-hand rule to $\mathbf{F}_{\text{mag}} = I\mathbf{L} \times \mathbf{B}$. For side 1, $\mathbf{L}$ points leftward. So, point your fingers leftward. Now curl your fingertips so that they point in the direction of the magnetic field. Your thumb points straight up. That's the direction of $\mathbf{F}_1$, the magnetic force on side 1.

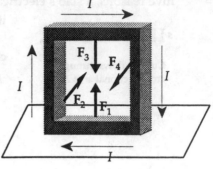

By similar reasoning applied to side 3, $\mathbf{F}_3$ points down. (Check it!) Because $\mathbf{F}_1$ and $\mathbf{F}_3$ have the same strength but point in opposite directions, they cancel out.

Do $\mathbf{F}_2$ and $\mathbf{F}_4$ also cancel? Yes. Since both those sides point vertically, while the magnetic field points horizontally (parallel to the floor), those sides make a 90° angle with the field. Therefore, both forces have strength

$$F_1 = F_3 = I|\mathbf{L} \times \mathbf{B}| = ILB_0 \sin 90^\circ = 1LB_0.$$

But once again, those two forces point in opposite directions. You can confirm this using the right-hand rule. Or, by letting $\mathbf{L}_2$ and $\mathbf{L}_4$ denote the length vectors for those sides, you can notice that $\mathbf{L}_4 = -\mathbf{L}_2$, and hence $\mathbf{F}_4 = I\mathbf{L}_4 \times \mathbf{B} = -I\mathbf{L}_2 \times \mathbf{B} = -\mathbf{F}_2$. Those two forces have the same magnitude, but point in opposite directions. Hence, they cancel.

In summary: As the diagram shows, the force on each side of the square gets canceled by its "partner" force on the opposite side. Therefore, no net force acts on the square. It does not slide across the floor. Instead, it stays in place. (More precisely, it *spins* in place, as we'll see below.)

(c) From looking at the above force diagram, you should see intuitively that $\mathbf{F}_2$ and $\mathbf{F}_4$ make the frame rotate clockwise around a vertical axis. In other words, a "clockwise" torque acts on the square.

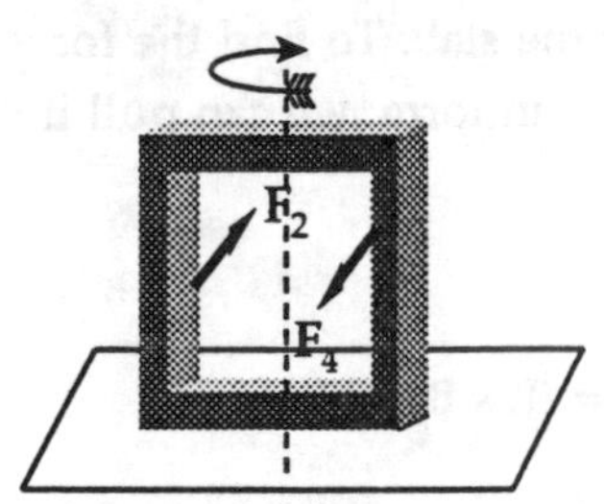

The frame rotates clockwise around the dashed axis of rotation.

We can figure out the total torque by summing up the individual torques on each side. Use the definition of torque,

$$\tau = |\mathbf{r} \times \mathbf{F}| = rF \sin\theta = F_{\perp} r,$$

which should look familiar from last semester. $F_{\perp}$ denotes the component of the force *perpendicular* to the relevant radius vector (moment arm), which goes from the pivot point (in this case, the center of mass) to the "contact point" where the force acts. Only this perpendicular component of the force "contributes" to the torques

For this reason, forces 1 and 3 exert no torque. For instance, look at $\mathbf{F}_1$. Its radius vector points from the center of the square to the midpoint of side 1. So, $\mathbf{F}_1$ points *along* this radius vector. Since $\mathbf{r}_1$ and $\mathbf{F}_1$ are parallel, that force generates no torques. The same goes for $\mathbf{F}_3$.

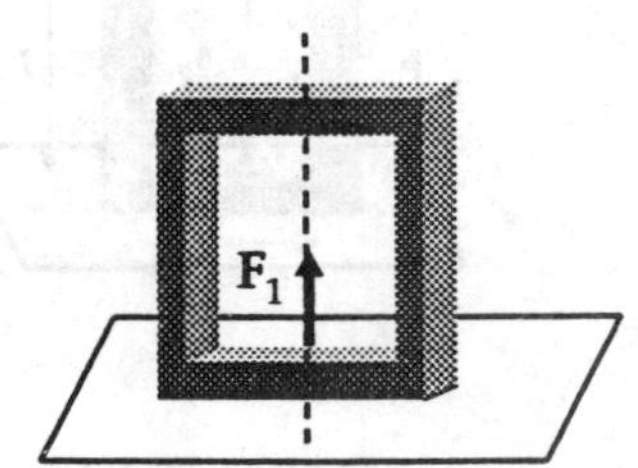

$\mathbf{F}_1$ exerts no torque, because it points toward the center of mass, i.e., the center of the square.

By contrast, forces 2 and 4 both generate torques. And those torques don't cancel, because they both make the frame rotate clockwise.

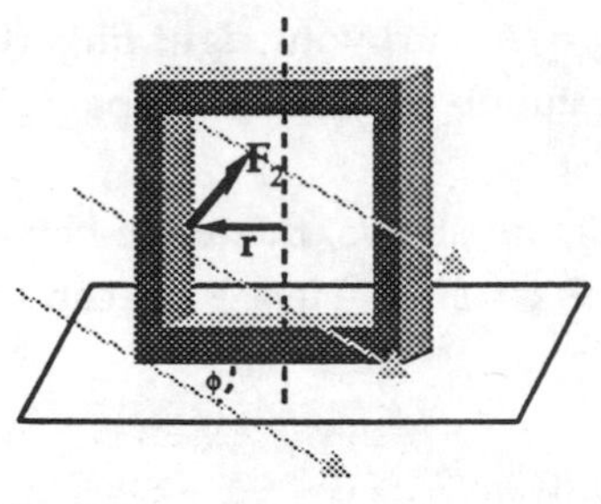

INSET (top-down view)

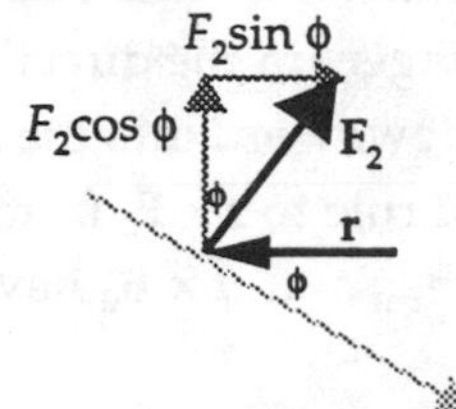

Let's begin by finding τ_2, the torque produced by $\mathbf{F}_2$. In this diagram, I'll drawn the relevant radius vector, from the center of mass to the contact point where $\mathbf{F}_2$ acts. I've also broken $\mathbf{F}_2$ into its components parallel and perpendicular to this radius vector. The "inset" diagram, drawn from a top-down vantage point, clarifies all this. Remember, $\mathbf{F}_2$ makes a 90° angle with the magnetic field (drawn in light gray), since a cross product always points at right angles to the "multiplied" vectors. And the radius vector is parallel to the bottom of the frame, which makes an angle ϕ with the magnetic field. So, as the diagram shows, $F_{2\perp} = F_2 \cos\phi$. Therefore,

$$\tau_2 = |\mathbf{r} \times \mathbf{F}_2| = rF_{2\perp} = rF_2 \cos\phi.$$

A common mistake is to write $\tau_2 = rF_2 \sin\phi$. In the formula $\tau = rF\sin\theta$, θ denotes the angle between $\mathbf{r}$ and $\mathbf{F}$. So, in this problem, $\theta = 90° - \phi$, and hence, $\sin\theta = \cos\phi$.

So far, we've figured out the torque on side 2. Let's rewrite this expression in terms of known quantities. As the diagram shows, r is half a side length: $r = L/2$. And in part (a), we found the force on side 2. It's $F_2 = ILB_0$. Putting all this together, I get

$$\tau_2 = rF_2 \cos\phi = \frac{L}{2} ILB_0 \cos\phi.$$

Using similar reasoning, you can confirm that $\mathbf{F}_4$ generates exactly the same torque. Since τ_2 and τ_4 both make the frame spin clockwise, they add together instead of canceling.

$$\begin{aligned}\tau_{net} &= \tau_2 + \tau_4 \\ &= 2\tau_2 \\ &= 2\frac{L}{2} ILB_0 \cos\phi \\ &= IL^2 B_0 \cos\phi,\end{aligned}$$

where I is the value we found in part (a).

(d) Since the square has area $A = L^2$, the magnetic moment has magnitude $\mu = IA = IL^2$. So, we can write the net torque in the form,

$$\tau_{net} = \mu B_0 \cos\phi.$$

We're getting close! To finish off this problem, we must think more carefully about the vectors. An area vector points in the direction that a surface "faces," i.e., perpendicular to the plane of the surface. For instance, your wall faces sideways. So, the square's area vector **A** points either into the page or out of the page.

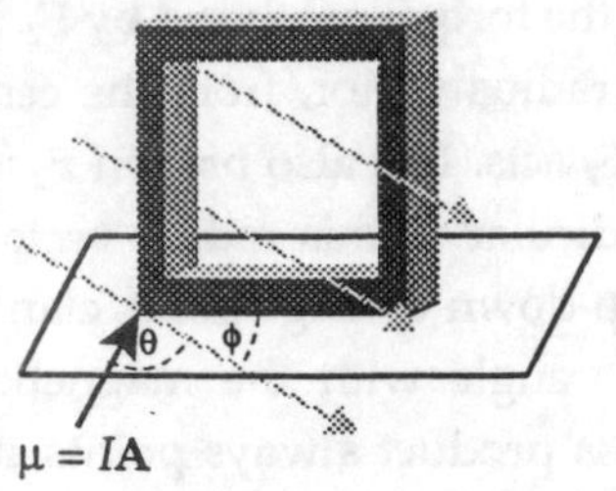

To find the direction of $\mu = IA$, curl your right fingertips in the direction of I, clockwise. Your thumb points *into* the page. That's the direction of μ, by definition.

Consequently, as this diagram shows, the angle between μ and $\mathbf{B}_0$ is $\theta = 90° - \phi$. Therefore, $\cos\phi = \sin\theta$. Hence, we can rewrite the net torque as

$$\tau_{net} = \mu B_0 \sin\theta.$$

This looks a lot like a cross product. Indeed, since $|\mu \times \mathbf{B}_0| = \mu B_0 \sin\theta$, we've shown that τ_{net} and $\mu \times \mathbf{B}_0$ have the same magnitude. We just need to show that $\mu \times \mathbf{B}_0$ and τ_{net} point in the same direction.

I'll begin by applying the right-hand to $\mu \times \mathbf{B}_0$. When I point my fingers in the direction of μ, and then curl my fingertips in the direction of $\mathbf{B}$, my thumb now points downward into the floor. That's direction of $\mu \times \mathbf{B}_0$. To find the direction of τ_{net}, apply the right-hand rule to $\mathbf{r} \times \mathbf{F}_2$ from part (c). You'll see that τ_{net} *also* points downward, into the floor. In summary, τ_{net} and $\mu \times \mathbf{B}_0$ have the same magnitude *and* point in the same direction. Therefore, $\tau_{net} = \mu \times \mathbf{B}_0$.

QUESTION 39-6

CYCLOTRON

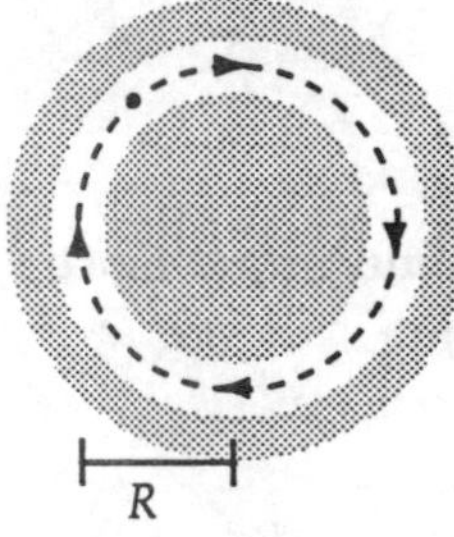

The particle travels clockwise along the dashed circle, inside a vacuum-filled tube (represented by the white space between the gray splotches).

A cyclotron is a circular tube in which charged particles get whipped around at high speeds by electric and magnetic fields.

Consider a particle of charge q and mass m inside a cyclotron of radius R. Neglect gravity.

(a) Suppose the particle is moving slowly. To speed it up, could the scientists use magnetic fields alone? Or would they also have to use electric fields? Explain.

(b) The scientists want the particle to circle the cyclotron with frequency f. That is, the particle should complete f circles per unit time. Assuming the particle acquired the appropriate speed in part (a), what magnetic field strength is needed to maintain this motion? The magnetic field points perpendicular to the plane of the circle traced by the particle. The particle's motion is nonrelativistic.

(c) Now the scientists feed that *same* charged particle into a smaller cyclotron, of radius $R/2$. They want the particle to circle the new cyclotron with the same frequency it had in part (b). Will they need to use a magnetic field bigger than, smaller than, or the same size as the field in part (b)? Explain your answer, intuitively and/or mathematically.

ANSWER 39-6

(a) A magnetic force *cannot* speed up or slow down a particle. A magnetic field changes a particle's direction of motion, but not its speed. Therefore, an electric field is needed to speed up the particle. Weird, huh? I'll now explain why the magnetic force doesn't "work."

To speed up an object, you must apply a force. But not just any force. At least a component of the force must be directed along the object's direction of motion. For instance, consider a hockey puck sliding rightward on frictionless ice. You can speed it up by applying a rightward or partially rightward "kick" (force). You can slow it down by applying a leftward or partially leftward force.

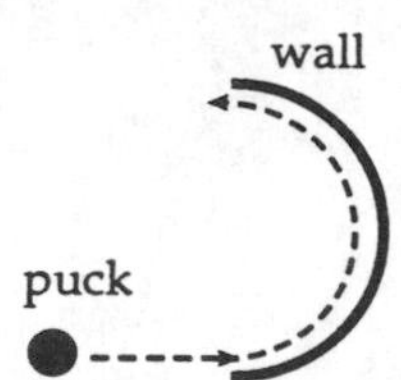

But what happens if you apply a continuous force perpendicular to the puck's motion? For instance, you could insert a semicircular wall that makes the puck move in a circle. The normal force exerted by the wall always points "radially" toward the center of the semicircle. In other words, the force is always perpendicular to the puck's tangential direction of motion. Such a force *cannot* speed up the puck. It can merely change the puck's direction of motion. A purely radial force produces *uniform* circular motion.

Here's another way to see why a force perpendicular to the direction of motion cannot speed up (or slow down) an object. According to the work-energy theorem,

$$W = \int \mathbf{F} \cdot d\mathbf{s} = \Delta K,$$

something must perform *work* on an object in order to change its kinetic energy, and hence, its speed. But as I emphasized in chapter 11, only the component of a force *parallel* to the direction of motion "contributes" work. The dot product in $W = \int \mathbf{F} \cdot d\mathbf{s}$ confirms this conclusion; when $\mathbf{F}$ points perpendicular to the direction of motion ($d\mathbf{s}$), the dot product vanishes, and hence, $W = 0$. So, a force perpendicular to the direction of motion cannot change an object's kinetic energy, and therefore cannot speed up or slow down the object.

OK, now for the punch line. The cross product in $\mathbf{F}_{\text{mag}} = q\mathbf{v} \times \mathbf{B}$ tells us that the magnetic force is perpendicular to $\mathbf{v}$. In other words, the magnetic force always points perpendicular to the direction of motion. Therefore, **a magnetic force alone can never speed up or slow down a charged particle. It can merely change the particle's direction of motion** (into a circle).

That's why particle physicists must use electric fields to speed up particles in cyclotrons. It's also why charged particles from outer space get deflected, but not slowed down, by the Earth's magnetic field.

(b) Intuitively, the faster the particle whips around the cyclotron, the stronger a magnetic field we must use to prevent the particle from flying into the outer wall of the cyclotron tube. With circular-motion reasoning, we can formalize this intuition, by finding the B needed to maintain the particle's circular motion at speed v. But wait; we don't know the particle's speed. Instead, we know its frequency. Fortunately, we can solve for v in terms of the given frequency, f.

In summary, to figure out the magnetic field needed to maintain the particle's motion at frequency f, I will

(i) find the particle's speed, v, in terms of the given frequency, f; and then
(ii) calculate the magnetic field, B, needed to keep the particle moving in a circle at that speed.

Subproblem 1: Kinematic reasoning to find the particle's speed, given its frequency

Intuitively, the more frequently the particle completes a full circle, the faster it must be moving. So, we expect v and f to be proportional. To derive the exact equation, consider the particle's period, T. The period is the time needed to complete one circle. By contrast, the frequency is the number of circles completed per time. So, $f = 1/T$, as we saw last semester.

In time T, the particle travels one full circumference, a distance $2\pi R$. Since the particle has constant speed, we get

$$v = \frac{\text{distance covered}}{\text{time}} = \frac{2\pi R}{T}.$$

Because $f = 1/T$, we can rewrite this equation in terms of frequency:

$$v = 2\pi Rf.$$

As intuitively expected, the speed is proportional to the frequency.

Now that we know how fast the particle whips around the cyclotron, we can figure out the magnetic field needed to maintain its circular motion at that speed.

Subproblem 2: Circular-motion reasoning to relate the particle's speed v to the magnetic field B

In general, a particle moving in a circle of radius r at speed v undergoes a radial (centripetal) acceleration $a_c = v^2/r$. Therefore, the net radial force needed to cause this acceleration is $\sum F_{\text{radial}} = mv^2/r$. Here, the magnetic field is what makes the particle move in a circle. In other words, the magnetic field supplies the radial force: $\sum F_{\text{radial}} = F_{\text{mag}} = |q\mathbf{v} \times \mathbf{B}|$. Putting all this together, I get

$$\begin{aligned}\sum F_{\text{radial}} &= ma_c \\ |q\mathbf{v} \times \mathbf{B}| &= \frac{mv^2}{R} \\ qvB &= \frac{mv^2}{R},\end{aligned}$$

where in the last step, I noticed that the magnetic field points perpendicular to the plane in which the particle moves, and hence $|q\mathbf{v} \times \mathbf{B}| = qvB\sin 90° = qvB$.

Solve for the field strength, and then substitute in our expression for v from subproblem 1, to get

$$B = \frac{mv}{qR} = \frac{m(2\pi Rf)}{qR} = \frac{2\pi fm}{q}.$$

That's the magnetic field needed to maintain the particle's circular motion at frequency f.

(c) I have conflicting intuitions. On the one hand, the particle now traces smaller circles than it did before, in the same amount of time. So, it's moving slower. Consequently, a smaller radial force is needed to maintain the particle's circular motion, other things being equal. This reasoning suggests that we can use a smaller magnetic field than the one in part (b).

On the other hand, since $\mathbf{F}_{\text{mag}} = q\mathbf{v} \times \mathbf{B}$, a slower particle feels a smaller magnetic force. To ensure that the force on the particle doesn't become *too* small, we might need to use a bigger magnetic field.

These two conflicting intuitions "cancel out." Our part (b) answer, $B = 2\pi fm/q$, does not depend on R. In words, the magnetic field needed to keep the particle circling the cyclotron with frequency f does *not* depend on the cyclotron's radius. For instance, the magnetic field needed to make a proton complete 100 small, slow circles per second is the *same* field needed to make the proton complete 100 big, fast circles per second.

Here's another way of framing this result. Solve $B = 2\pi fm/q$ for frequency:

$$f = \frac{qB}{2\pi m},$$

This is the so-called **cyclotron frequency**. It doesn't depend on the radius of the cyclotron. If you tell me the particle's mass and charge, and also the magnetic field strength, I can tell you the frequency with which the particle circles the cyclotron. This frequency is the same, no matter whether the particle traces big circles or small circles.

QUESTION 39-7

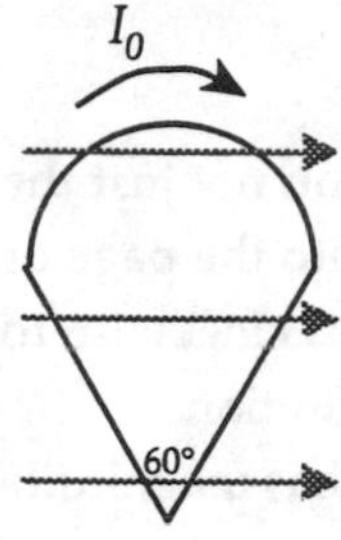

A funkily-shaped current loop consists of a semicircle of radius R, and two straight segments of length $2R$, with a 60° angle between them. A current I_0 flows clockwise. The loop is placed in a uniform magnetic field of strength B_0 pointing rightward, as drawn here. The magnetic field and the current loop both "live" in the plane of the page.

In this problem, you may use any formulas previously derived in this chapter. You need not rederive them.

(a) What's the net force on the current loop?

(b) What's the net torque on the current loop?

ANSWER 39-7

This "trick" problem tests how well you understand the generality of the results found in question 39-5. There, we considered a square current loop. By calculating the forces and torques on each side, we reached the following conclusions:

(i) The net magnetic force on a current loop in a uniform magnetic field is $\mathbf{F}_{\text{mag}} = 0$, because the forces on the different segments (sides) cancel out.

(ii) The net torque exerted on a current loop by a uniform magnetic field is

$$\tau_{\text{net}} = \boldsymbol{\mu} \times \mathbf{B},$$

where μ is the magnetic moment, $\boldsymbol{\mu} = I\mathbf{A}$.

These results are general. They apply to all current loops, not just rectangles. Therefore, you can solve this problem without suffering through tedious calculations involving $d\mathbf{F}_{\text{mag}} = I d\mathbf{l} \times \mathbf{B}$.

(a) As just noted, the total magnetic force on a current loop of *any* size and shape, when placed in a uniform magnetic field, is $\mathbf{F}_{mag} = 0$. The magnetic forces on the semicircle and on the two straight-edge sides cancel each other out.

(b) Since $\tau_{net} = \mu \times \mathbf{B}$, the torque depends on the current, and on the area of the loop. By thinking of the loop as a semicircle attached to an equilateral triangle, you can quickly calculate its area. From trig, the triangle's height is $2R\sin 60° = \sqrt{3}R$. So,

$$A = A_{triangle} + A_{semicircle}$$
$$= \frac{1}{2}(\text{base})(\text{height}) + \frac{1}{2}\pi R^2$$
$$= \frac{1}{2}(2R)(\sqrt{3}R) + \frac{1}{2}\pi R^2$$
$$= \frac{1}{2}(2\sqrt{3} + \pi)R^2.$$

By trig, the gray line has length $2R\sin 60°$.

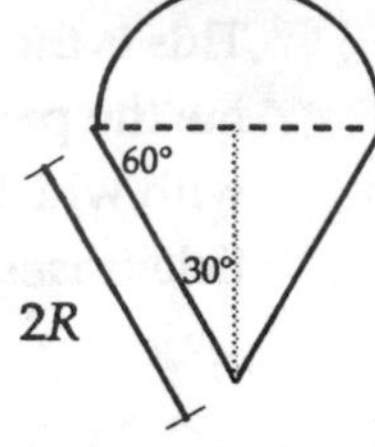

Therefore,

$$\mu = I_0 A = I_0 \frac{1}{2}(2\sqrt{3} + \pi)R^2.$$

Because $\tau_{net} = \mu \times \mathbf{B}$, we need to know the direction of this magnetic moment, not just the magnitude. Well, the loop lies in the plane of the page. Therefore, it "faces" either into the page or out of the page. But which? To find out, use the right-hand rule: Curl your fingertips clockwise to match the current. Your thumb points into the page. That's the direction of μ, by definition.

Since **B** points rightward, the angle between μ and **B** is 90°. Hence, the torque has magnitude

$$\tau_{net} = |\mu \times \mathbf{B}|$$
$$= \mu B_0 \sin 90°$$
$$= I_0 \frac{1}{2}(2\sqrt{3} + \pi)R^2 B_0.$$

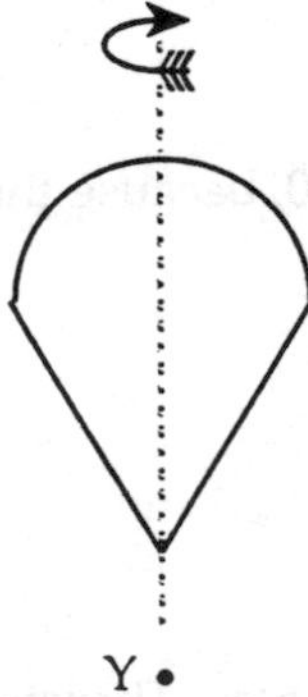

To find the direction of this torque, apply the right-hand rule to the cross-product $\mu \times \mathbf{B}$. Specifically, point your fingers into the page (with μ), then curl your fingertips to point along **B**. Your thumb now points down the page. That's the direction τ. Therefore, the object rotates in the direction your fingertips are curled, namely counterclockwise (as seen from point Y in the diagram). The dashed line is the axis of rotation.

Here's a more physical way to see which way the loop rotates. Apply the right-hand rule to the cross product in $d\mathbf{F}_{mag} = Id\mathbf{l} \times \mathbf{B}$. As you can confirm, the magnetic force on the left-hand side of the current loop points into the page. By contrast, the magnetic force on the right-hand side of the loop points out of the page. This combination of forces makes the loop rotate counterclockwise as seen from point Y.

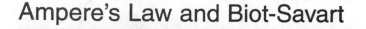

Ampere's Law and Biot-Savart

40

CHAPTER

QUESTION 40-1

(Before attempting this problem, please work through your textbook's sample problems that use the Biot-Savart law to calculate a magnetic field. This problem is too hard to serve as your first exposure.)

Take six segments of wire, each of length s and resistance R, and connect them into a hexagon. A tiny battery makes current flow clockwise around this circuit, by supplying voltage $\mathcal{E}_0$. The hexagon "lives" in the plane of this page.

(a) What's the magnetic field at the center of the hexagon? If you get stuck in the math, that's OK; but try to set everything up.

(b) A particle of charge Q is at the center of the hexagon, moving with speed v_0 into the page. What is the magnitude and direction of the magnetic force on that charge? Qualitatively describe the particle's trajectory in the long run.

(c) Now a particle of charge Q is at the center of the hexagon, moving rightward at speed v_0. Initially, what is the magnitude and direction of the force on the particle? For extra credit, qualitatively describe the particle's trajectory.

ANSWER 40-1

In this problem, you must use the Biot-Savart law, the magnetic equivalent of Coulomb's law from electrostatics. Recall that Coulomb's law specifies the electric field created by a tiny bit of charge (i.e., a point charge). Nonetheless, we found a way to apply Coulomb's law to rods and other "extended" charges. Specifically, we broke the continuous charge distribution into tiny "pieces"; calculated the infinitesimal electric field $d\mathbf{E}$ generated by a given piece; and then added up (integrated over) the $d\mathbf{E}$'s produced by all the pieces, to obtain the total electric field. (If this sounds unfamiliar, please look over question 33-3.) You can solve magnetism problems in a similar way. The Biot-Savart law specifies the infinitesimal magnetic field $d\mathbf{B}$ generated by a tiny piece of current-carrying wire. To obtain the total magnetic field, you can add up (integrate over) the $d\mathbf{B}$'s produced by all the infinitesimal pieces of wire.

Despite this conceptual similarity between Coulomb's and Biot-Savart's law, Biot-Savart is harder to use, because the magnetic field points perpendicular to the current.

Summary of Biot-Savart

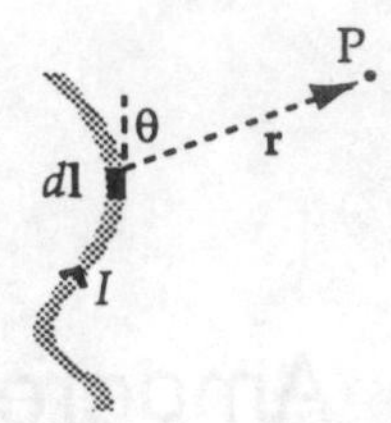

Before diving into this problem, let's review what the Biot-Savart law means, and how to use it. If you don't need a review, skip to part (a) below. I'll find the magnetic field generated at point P by the tiny blackened piece of wire. Current I flows up the gray wire. The blackened piece has length $d\mathbf{l}$, pointing up the page. In other words, $d\mathbf{l}$ points up the page.

First, draw the **r** vector, which starts at the blackened piece of wire and ends at the location where we're finding the field. Always label θ, the angle between $d\mathbf{l}$ and **r**.

Now for some important notation. Let $\hat{\mathbf{r}}$ denote a unit vector pointing in the direction as **r**. So, $\hat{\mathbf{r}}$ has length 1, while **r** has length r. Intuitively, $\hat{\mathbf{r}}$ encodes the direction of **r**, without telling you anything about its length. For instance, suppose **r** is 5 meters long and points northeast. Then $r = 5$ meters, and $\hat{\mathbf{r}}$ is a unit vector pointing northeast.

Given this notation, we can write the Biot-Savart law in a form that strongly resembles Coulomb's law. The infinitesimal magnetic field $d\mathbf{B}$ generated by the current I in that tiny length of wire, $d\mathbf{l}$, is

Biot-Savart $$d\mathbf{B} = \frac{\mu_0}{4\pi} I \frac{d\mathbf{l} \times \hat{\mathbf{r}}}{r^2}.$$

This agrees with the standard textbook version of the law, $d\mathbf{B} = \frac{\mu_0}{4\pi} I \frac{d\mathbf{l} \times \mathbf{r}}{r^3}$, because a vector equals its length "times" its direction: $\mathbf{r} = r\hat{\mathbf{r}}$.

My way of writing the law emphasizes that the magnetic field, like the electric field, falls off as $1/r^2$ (for point charges and for "bits" of current). Notice also that a wire produces a magnetic field proportional to its current, just like a particle produces an electric field proportional to its charge. Finally, the constant $\frac{\mu_0}{4\pi}$ plays the same role in "magnetostatics" that $\frac{1}{4\pi\varepsilon_0}$ plays in electrostatics. Please keep in mind the deep structural similarity between Coulomb's law and the Biot-Savart law.

Anyway, we can find the field at point P. To figure out the direction of the field ($d\mathbf{B}$), apply the right-hand rule to $d\mathbf{l} \times \hat{\mathbf{r}}$. First, point your fingers in the direction of $d\mathbf{l}$, "up the page." Then curl your fingertips to point in the direction of $\hat{\mathbf{r}}$, towards point P. Your thumb now points into the page. That's the direction of the $d\mathbf{B}$ generated by the blackened "piece" of current.

We can also figure out the strength (magnitude) of this field:

$$dB = \frac{\mu_0}{4\pi} I \frac{|d\mathbf{l} \times \hat{\mathbf{r}}|}{r^2} = \frac{\mu_0}{4\pi} I \frac{dl \sin\theta}{r^2},$$

where θ denotes the angle between $d\mathbf{l}$ and $\hat{\mathbf{r}}$. I labeled that θ in the above diagram. Notice that I used the general formula for the magnitude of a cross product: $|\mathbf{a} \times \mathbf{b}| = ab\sin\theta$. Since $\hat{\mathbf{r}}$ is a unit vector, it has length 1, and hence $|d\mathbf{l} \times \hat{\mathbf{r}}| = (dl)(1)\sin\theta = dl\sin\theta$.

End of Biot-Savart summary

Now that we've reviewed Biot-Savart, let's address the problem at hand.

(a) First, I'll find the direction of the magnetic field at the center. As just shown, a tiny piece of the hexagon generates a field pointing in the direction of $d\mathbf{l} \times \hat{\mathbf{r}}$. To find that direction, use the right-hand rule.

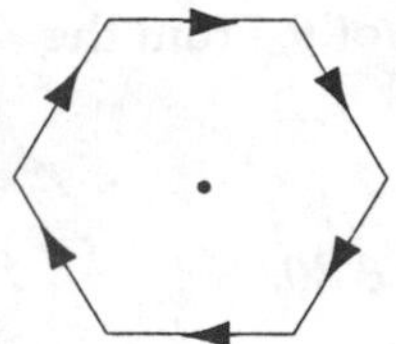

Specifically, you can choose an arbitrary piece of the hexagon, point your fingers in the direction of current flow ($d\mathbf{l}$), and then curl your fingertips in the direction of $\mathbf{r}$, towards the center. Your thumb points into the page. And this is true for *any* piece of the hexagon. So, each of the six sides generates an into-the-page field.

We can simplify things even further, using symmetry. The center of the hexagon sits the same distance from each side, along a perpendicular line through the midpoint of the side. Therefore, each side generates the same magnetic field at the center. And those six fields all point in the same direction.

Consequently, *the total magnetic field created by the whole hexagon is simply six times the field generated by a single side.* This simplification works only because of symmetry. A random point inside the hexagon would be located a different distance from each side. Therefore, each side would generate a different field at that point. But at the center, symmetry guarantees that each side "contributes equally" to the overall field.

In summary, we just need to find the field produced by *one* side of the hexagon, and then multiply it by six.

Finding the field generated by one side of the hexagon

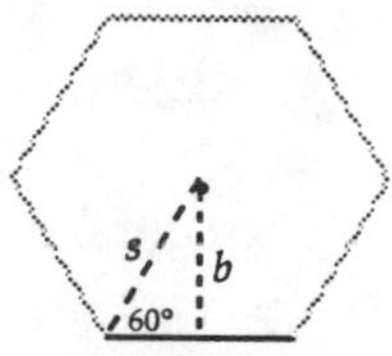

I'll work on the bottom side. How far is it from the center of the hexagon? Well, as you can show using geometry, the center is a distance s from each vertex, where s denotes the side length. Therefore, as this diagram indicates, the distance from the *midpoint* of a side to the center of the hexagon is $b = s\sin 60°$. So, I'll use "b" as an abbreviation for $s\sin 60°$.

Now I'll "zoom in" on the bottom side of the hexagon, and I'll blacken a little piece, of length dx. As mentioned above, you should draw the $\mathbf{r}$ vector connecting the blackened piece to the location at which you're finding the magnetic field, in this case the center of the hexagon. I've let x denote the distance from the blackened piece to the midpoint of the side. My strategy is to find the infinitesimal field contributed by this tiny piece of wire. Then, I'll add up the contributions from all the tiny pieces, by integrating.

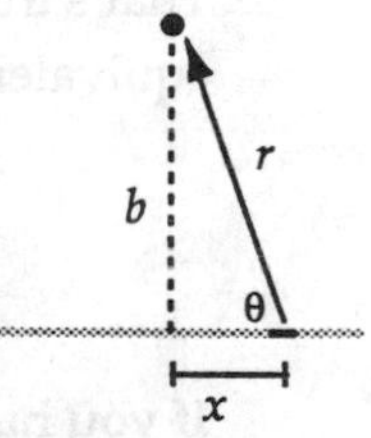

Bottom side of hexagon, with tiny piece blackened

From Biot-Savart, the infinitesimal field generated by the blackened piece of the wire has magnitude.

$$dB = \frac{\mu_0}{4\pi} I \frac{dl \sin\theta}{r^2}$$

$$= \frac{\mu_0}{4\pi} I \frac{dx}{x^2 + b^2} \sin\theta,$$

where in the second line I applied Pythagorean theorem to the above diagram, to get $r^2 = b^2 + x^2$. I also rewrote my infinitesimal length as "dx" instead of "dl."

To find the field generated by the entire bottom side, just integrate over the dB's contributed by all the "pieces." Before doing so, however, we must make sure that the only variable on the right-hand side is the variable we're integrating over. Unfortunately, the right-hand side contains two variables, x and θ. We must either express all the θ's in terms of x, or all the x's in terms of θ.

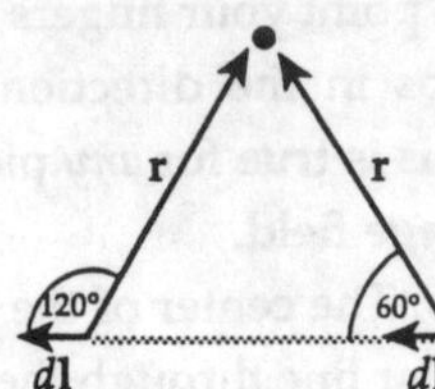

Because it simplifies the math, I'll express x in terms of θ. From the above diagram, $\tan\theta = b/x$. So,

$$x = \frac{b}{\tan\theta} = b\cot\theta, \qquad \text{and hence,} \qquad dx = b\csc^2\theta\, d\theta.$$

Furthermore, as this new diagram indicates, the angle between $d\mathbf{l}$ and $\mathbf{r}$ ranges from 60° to 120°. So,

$$B = \int dB = \frac{\mu_0}{4\pi} I \int \frac{dx}{x^2 + b^2} \sin\theta$$

$$= \frac{\mu_0 I}{4\pi} \int_{60°}^{120°} \frac{b\csc^2\theta\, d\theta}{b^2\cot^2\theta + b^2} \sin\theta$$

$$= \frac{\mu_0 I}{4\pi} \int_{60°}^{120°} \frac{b\csc^2\theta\, d\theta}{b^2\csc^2\theta} \sin\theta \qquad [\text{trig identity } \cot^2 + 1 = \csc^2]$$

$$= \frac{\mu_0 I}{4\pi b} \int_{60°}^{120°} \sin\theta\, d\theta$$

$$= \frac{\mu_0 I}{4\pi b} [-\cos 120° - (-\cos 60°)]$$

$$= \frac{\mu_0 I}{4\pi b}.$$

That's the magnetic field produced by one side of the hexagon. Therefore, since all six sides produce equivalent fields,

$$B_{\text{due to whole hexagon}} = 6B_{\text{due to one side}} = 6\frac{\mu_0 I}{4\pi b}.$$

If you had trouble evaluating the integral, don't worry about it. For now, focus on setting up these integrals correctly.

Before moving on to part (b), I must express this answer in terms of the given quantities. As shown above, $b = s\sin 60°$. We can also rewrite the current, I. Since the six sides of the hexagon are in series, and each side has resistance R, the hexagon has total resistance is $6R$. Therefore, from Ohm's law, $I = \frac{\mathcal{E}_0}{6R}$, Substitute these expressions into our answer for B, to get

$$B_{\text{due to whole hexagon}} = \frac{6\mu_0\mathcal{E}_0}{4\pi(s\sin 60°)(6R)} = \frac{\mu_0\mathcal{E}_0}{4\pi sR\sin 60°}$$

(b) Since the particle moves *along* a field line (parallel to **B**), it feels no magnetic force. Let me elaborate.

As shown in part (a), the field at the center of the hexagon points into the page. Because the particle moves into the page, **v** and **B** are parallel, and hence $F_{\text{mag}} = |q\mathbf{v} \times \mathbf{B}| = qvB\sin 0° = 0$. A particle moving parallel to the magnetic field feels no force. So, initially, the particle travels into the page, undeflected, along an imaginary axis through the center of the hexagon. Let me call that axis the "central axis."

Look at your textbook's drawing of the magnetic field produced by a symmetric current loop. By symmetry, the field at any location on the central axis points *along* that axis. So, as the particle travels along the central axis, it keeps moving parallel to the field. Therefore, it never feels a magnetic force. The particle continues moving in a straight line, blissfully undeflected.

(c) Here, the particle moves entirely perpendicular to the magnetic field. So,

$$F_{\text{mag}} = |q\mathbf{v} \times \mathbf{B}| = qvB \sin 90° = Qv_0B = Qv_0 \frac{\mu_0 \mathcal{E}_0}{4\pi sR \sin 60°},$$

where I've used our part (a) answer for B.

To find the direction of the force, apply the right-hand rule to $\mathbf{F}_{\text{mag}} = Q\mathbf{v}_0 \times \mathbf{B}$. Point your fingers in the direction of $\mathbf{v}_0$ (rightward), and curl your fingertips in the direction of $\mathbf{B}$ (into the page). Your thumb now points up the page. That's the direction of $\mathbf{F}_{\text{mag}}$, initially. The particle deflects toward the top side of hexagon, and stays in the plane of the page.

Extra credit problem. As the particle changes direction, the force on it also changes direction. Specifically, since $\mathbf{F}_{\text{mag}} = Q\mathbf{v} \times \mathbf{B}$, the force always points perpendicular to the direction of motion, and also perpendicular to the magnetic field. Because the field points perpendicular to the page, it follows that the magnetic force always stays in the plane of the page. For example, when the particle moves rightward, it deflects upward (in the plane of the page). When it moves upward (in the plane of the page), it deflects leftward. And so on. You can confirm these conclusions by applying the right-hand rule to the cross product in $\mathbf{F}_{\text{mag}} = Q\mathbf{v} \times \mathbf{B}$.

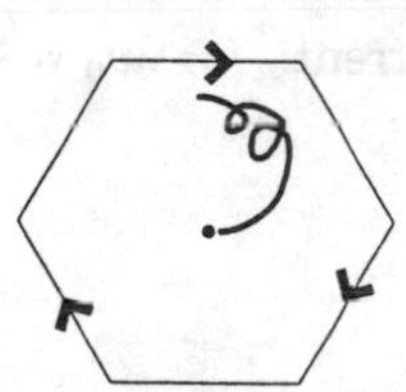

Solid curve is rough guess at particle's trajectory.

From the discussion so far, you might conclude that the particle traces counterclockwise circles. It *would* trace circles, if the magnetic field were uniform. But as the particle gets closer to some sides and farther from others, the magnetic field changes. As a result, the particle traces complicated loop-the-loops instead of perfect circles. To find the exact shape, we'd need to run a computer simulation.

QUESTION 40-2

(Don't try this problem until looking at your textbook's application of Ampere's law to a long wire.)

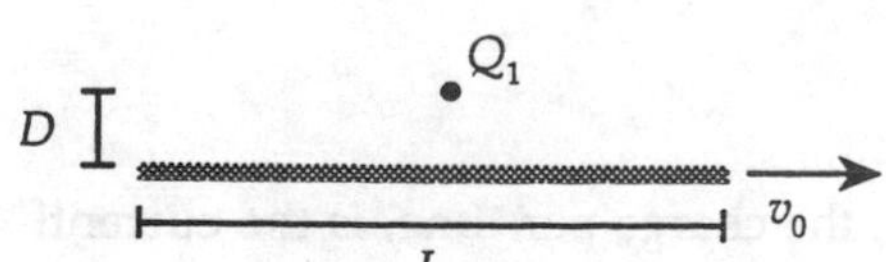

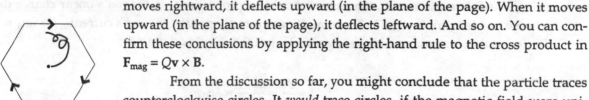

A very long rod (length L) has total positive charge Q_0, uniformly spread out. The rod moves right-ward with speed v_0. In this problem, you'll consider the forces acting on a positive particle of charge Q_1 located a small distance D away from the rod, where $D \ll L$.

Neglect gravity, and treat the rod as *very* long. Don't forget about the electrostatic repulsion between the two objects.

(a) Find the current due to the moving rod. Hint: Rewrite $\frac{dq}{dt}$ as $\frac{dq}{dx}\frac{dx}{dt}$.

What's the total force on Q_1 (magnitude and direction) if

(b) Q_1 **isn't moving?**

(c) Q_1 **is moving at speed v_1 into the page?**

(d) Q_1 **is moving at speed v_1 rightward?**

ANSWER 40-2

I've designed this problem to illustrate the similarities, and also the differences, between Gauss' law in electrostatics and Ampere's law in "magnetostatics." Although conceptually analogous, they differ in certain mathematical details that tend to confuse students.

(a) The moving rod constitutes an electric current, because charges are flowing. Current is the charge per time passing an imaginary point: $I = \frac{dq}{dt}$.

Let's try to guess the relevant formula intuitively. The faster the rod moves, the more charge per second passes a given point. So, we expect I to be proportional to v, the rod's speed. Also, the more densely charge is packed onto the rod, the more charge passes a given point, other things being equal. For this reason, the current should be proportional to the rod's linear charge density (charge per length), $\lambda = Q_0/L$. These intuitions correctly predict the rod's current: $I = \lambda v_0 = \frac{Q_0 v_0}{L}$. To derive this formula quickly yet rigorously, use the calculus chain rule:

$$I = \frac{dq}{dt} = \frac{dq}{dx}\frac{dx}{dt} = \lambda v_0,$$

where in the last step, I used the fact that that linear charge density is charge per length ($\lambda = dq/dx$), and velocity is distance per time ($v = dx/dt$).

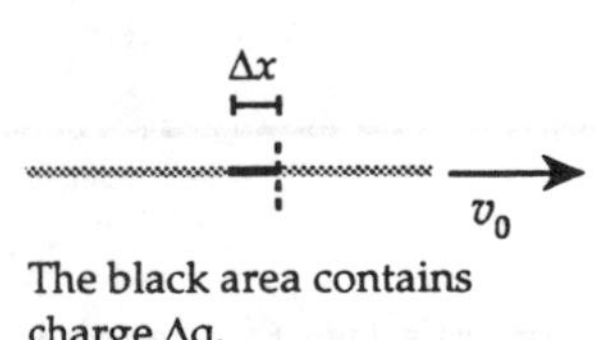

The black area contains charge Δq.

I'll now derive this formula one last time, in a more physical manner. As the rod travels rightward at speed v_0, how much charge Δq passes the imaginary dashed line in time Δt? Well, in time Δt, the *length* of rod passing the imaginary line is $\Delta x = v_0 \Delta t$. And since linear charge density is charge per length ($\lambda = \Delta q/\Delta x$), the charge contained in that length is

$$\Delta q = \lambda \Delta x = \lambda(v_0 \Delta t).$$

Divide this equation through by Δt to get $\frac{\Delta q}{\Delta t} = \lambda v_0$. But $\frac{\Delta q}{\Delta t}$, the charge per time, *is* the current! So, $I = \lambda v_0$.

(b) Since the charged particle doesn't move ($v = 0$), it *can't* feel a magnetic force. That's because the magnetic force on a charged particle is proportional to its velocity: $\mathbf{F}_{mag} = q\mathbf{v} \times \mathbf{B}$.

So, the particle feels an electrostatic repulsion, and nothing else. Since $\mathbf{F}_{elec} = q_1\mathbf{E}_{due\ to\ rod}$, the meat of this problem is finding the rod's electric field. As you know, Gauss' law enables you to calculate the field produced by an "infinite" rod.

Although we've used Gauss' law many times, I'd like to review the strategy, in order to illustrate the similarities and dissimilarities between Gauss' law and Ampere's law.

Gauss' law to calculate electric fields

1) Draw the electric field lines generated by the charge distribution.
2) Enclose (part of) the charge distribution with a Gaussian surface. The electric field should have constant strength over (at least part of) the Gaussian surface.
3) Use Gauss' law: $\int \mathbf{E}\cdot d\mathbf{A} = q_{encl}/\varepsilon_0$. If you can pull E outside the flux integral, then you can solve for it. *The whole point of choosing a symmetric Gaussian surface is to make sure the flux integral simplifies in this way.*

I'll now apply this strategy. Please read the next couple of pages, even if you're comfortable with Gauss' law, because I review some conceptual points relevant to Ampere's law.

Step 1: Draw the electric field lines.

Intuitively, the field lines spew radially outward from the rod. By symmetry, they radiate uniformly in all directions.

Step 2: Draw a "good" Gaussian surface.

By symmetry, the field has the same strength at any two points equidistant from the rod. Therefore, to ensure that E is constant over (part of) my Gaussian surface, I'll enclose the rod with a circle, as drawn in the end-on view. But wait; a circle can't be a Gaussian surface. Recall from the water version of Gauss' law that a Gaussian surface must be three-dimensional and closed. The Gaussian surface must extend into the page, forming a cylinder. I'll give the Gaussian surface arbitrary radius r. (Later, we'll set $r = D$.)

The overall Gaussian surface consists of a tube (i.e., the "side" of the Gaussian cylinder), and two ends.

Field lines pierce through dashed Gaussian surface.

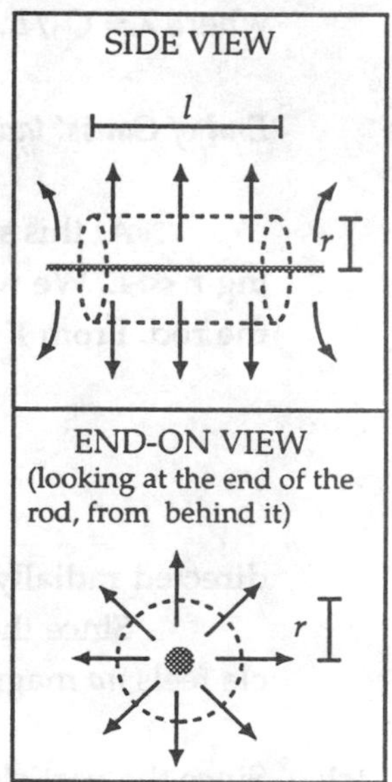

Step 3: Use Gauss' law to solve for the electric field.

By Gauss' law, the flux through the Gaussian surface is proportional to the charge enclosed:

$$\oint \mathbf{E}\cdot d\mathbf{A} = \frac{q_{encl}}{\varepsilon_0}.$$

My Gaussian surface encloses length l of the charged rod. Since the rod has total charge Q_0 uniformly spread over total length L, the linear charge density (charge per length) is $\lambda = Q_0/L$. Therefore, the Gaussian surface encloses charge

$$q_{\text{encl}} = \frac{\text{charge}}{\text{length}} \times (\text{length of rod enclosed by Gaussian surface}) = \lambda l = \frac{Q_0}{L} l.$$

And as we've seen before, field lines pierce only through the tube (side) of the Gaussian surface, not through the ends. Since the field lines pierce head-on through the tube of Gaussian surface, the dot product $\mathbf{E} \cdot d\mathbf{A}$ reduces to a regular product. Furthermore, as emphasized above, the electric field is constant over the entire tube. So, we can pull E outside the flux integral.

$$\oint \mathbf{E} \cdot d\mathbf{A} = \int_{\text{tube}} dA = EA_{\text{tube}} = E(2\pi r l),$$

where I've used the surface area of a cylindrical tube.

OK, now that we've simplified both sides of Gauss' law, we can solve for E:

$$\oint \mathbf{E} \cdot d\mathbf{A} = \frac{q_{\text{encl}}}{\varepsilon_0}$$

$$E(2\pi r l) = \frac{\lambda l}{\varepsilon_0},$$

and hence

$$E = \frac{\lambda}{2\pi\varepsilon_0 r},$$

where $\lambda = Q_0/L$.

End of Gauss' law strategy

At this stage, we've figured out the electric field an arbitrary distance r from the rod, assuming $r \ll L$. We want the force exerted by this field on a particle of charge Q_1, at distance $r = D$ from the rod. From $\mathbf{F} = q\mathbf{E}$, the force has magnitude

$$F_{\text{elec on particle}} = q_{\text{particle}} E = Q_1 \frac{\lambda}{2\pi\varepsilon_0 D} = \frac{Q_1 Q_0}{2\pi\varepsilon_0 DL}$$

directed radially away from the rod.

Since the particle isn't moving—at least, not until the electric force speeds it up—the particle feels *no* magnetic force. The electric force is the total force.

(c) Since the particle moves, it can experience a magnetic force. Does this mean the electric force "turns off?" No way! Two charged objects attract or repel whether or not they're moving. The electric force depends only on where the objects are located, *not* on how fast they're moving. Therefore, in parts (b) and (c), the particle feels the *same* electric force it felt in part (a). If the particle feels a magnetic force, it's *in addition to* that electrostatic repulsion.

Actually, as I'll now show below, the particle moves antiparallel to the magnetic field, and hence $F_{mag} = |Q_1\mathbf{v}_1 \times \mathbf{B}| = Q_1 v_1 B \sin 180° = 0$. To find the field direction, start with the Biot-Savart law,

$$d\mathbf{B} = \frac{\mu_0}{4\pi} I \frac{d\mathbf{l} \times \hat{\mathbf{r}}}{r^2},$$

and apply the right-hand rule to the cross product. Pick an arbitrary piece of the rod, and point your right fingers along dl, the direction of the current. So, your fingers point rightward. Now curl your fingertips so that they point toward Q_1. Your thumb now points out of the page. That's the direction of $d\mathbf{B}$. This conclusion holds no matter which piece of the rod you use. So, the magnetic field generated by the whole rod at Q_1 points out of the page. Since Q_1 moves antiparallel to this field (into the page), it feels no magnetic force. Therefore, as in part (b), the particle feels only an electrostatic force, $F_{elec} = q_1 E = Q_1 \dfrac{\lambda}{2\pi\varepsilon_0 D}$.

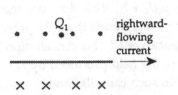

Q_1 rightward-flowing current

Gray dots denote a field out of the page, while crosses denote a field into the page.

For future reference, let's exploit Biot-Savart to find the field direction at points other than Q_1. Using the right-hand rule, please prove to yourself that the field below the wire points into the page. Indeed, by applying Biot-Savart to this end-on view of the rod, you can see that the field always points tangentially, not radially. The accompanying caption explains why. Furthermore, by symmetry, the magnetic field has the same strength at any two points equidistant from the rod. For these two reasons, the magnetic field lines circle the rod. This result, which will serve us well in part (d), applies to any long current-carrying "wire."

END-ON VIEW

The rod, and hence the current, flows into the page.

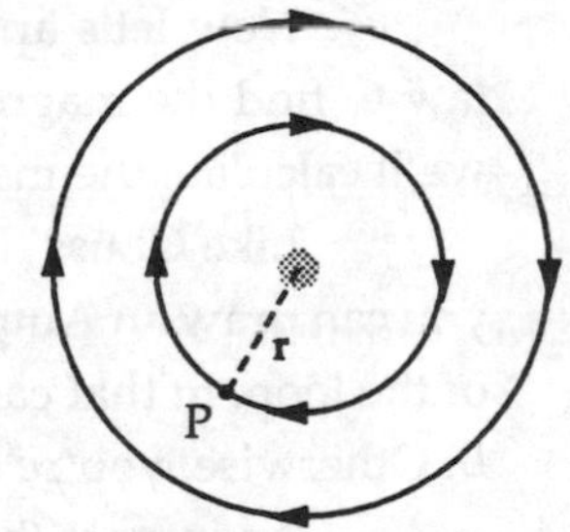

To find the field direction at P, point your fingers along dl (into the page), then curl your fingertips toward P. Your thumb, which indicates the field direction (B), points *perpendicular* to the dashed r vector. So, the field lines make circles around the rod.

(d) Finally, we'll use Ampere's law to calculate the magnetic field at Q_1. Given that field, we can find the magnetic force on Q_1. As noted in part (c), this magnetic force acts *in addition* to the electric force, which doesn't "switch off" simply because magnetism kicks in. The meat of this problem is figuring out the magnetic field.

Ampere's law: a conceptual review

Before applying Ampere's law to this particular problem, let me show you what it means in general, by making an analogy with Gauss' law. In Gaussian reasoning, you enclose some charge with a surface. The total electric flux coming out of the surface is proportional to the enclosed charge. Similarly, in Amperian reasoning, you "enclose" a current with an imaginary loop. An Amperian loop, unlike a Gaussian surface, does not "entrap" the whole wire. Instead, it's like a loop of string. It can be circular, rectangular, or any other shape.

Now focus your attention on the component of the magnetic field *along* the loop, at each point on the loop. You can integrate that field around the loop. This is not a two-dimensional flux integral over a surface. Rather, it's a one-dimensional line integral around a loop. According to Ampere's law, that integral is proportional to the current "enclosed" by the loop, i.e., the current passing through the loop:

Ampere's law $$\oint \mathbf{B} \cdot d\mathbf{s} = \mu_0 I_{\text{encl}},$$

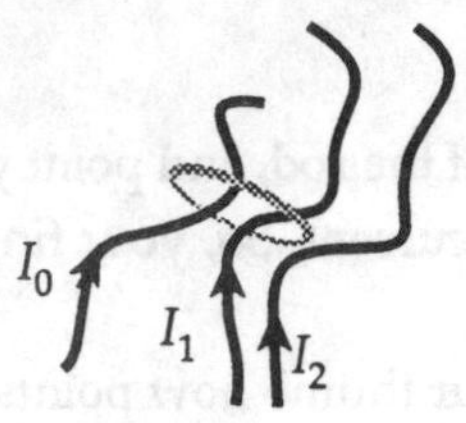

where ds is a little "piece" of the loop.

For instance, in this picture, the gray Amperian loop encloses only two of the three current-carrying wires. By Ampere's law, the line integral of the magnetic field around the loop equals $\mu_0(I_0 + I_1)$. This does not mean I_2 produces no magnetic field on the loop. It merely means the field produced by I_2, when integrated around the loop, "cancels out." This cancellation happens because, as you walk around the loop, the field produced by I_2 points "with" you part of the time and "against" you part of the time. No such cancellation occurs when the current flows *through* the loop.

By the way, the line integral in Ampere's law is just like the line integrals you used in Chapter 35 to calculate potential differences. The dot product picks out the component of the field pointing *along* the path of integration.

End of Ampere's law conceptual review

Now let's apply Ampere's law to the problem at hand. Remember, we're using Ampere's law to find the magnetic field near the moving rod, which carries current $I = \lambda v_0$. Given that field, we'll calculate the magnetic force on Q_1.

Like Gauss' law, Ampere's law is always *true*, but not always *useful*. It's useful only when you can draw an Amperian loop over which the magnetic field has constant strength, on at least part of the loop. In that case, you can pull B outside the line integral $\oint \mathbf{B} \cdot d\mathbf{s}$, and hence, you can solve for B. Otherwise, you're stuck using Biot-Savart.

As in part (b), we can treat the rod as infinitely long. So, it acts like an (effectively) infinite wire of current. Consequently, the magnetic field lines trace circles around the wire, as discussed in part (c). Therefore, by choosing a circular Amperian loop, we ensure that the field always points "along" the Amperian loop, reducing the dot product $\mathbf{B} \cdot d\mathbf{s}$ to a regular product. Even more important, since my Amperian loop consists of points equidistant from the rod, the magnetic field has the same strength over the entire loop. Consequently, the line integral in Ampere's law simplifies as follows:

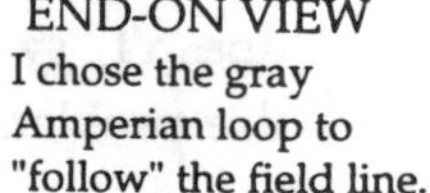
END-ON VIEW
I chose the gray Amperian loop to "follow" the field line.

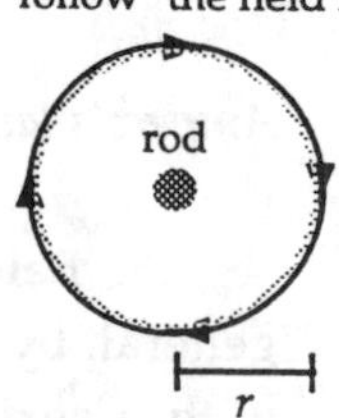

$$\oint \mathbf{B} \cdot d\mathbf{s} = B \oint ds = B(2\pi r),$$

where in the last step, I substituted in the length of a circumference, $s = 2\pi r$.

So, according to Ampere's law,

$$\oint \mathbf{B} \cdot d\mathbf{s} = \mu_0 I_{\text{encl}},$$

$$B(2\pi r) = \mu_0 I,$$

and hence

$$B = \frac{\mu_0 I}{2\pi r} = \frac{\mu_0 Q_0 v_0}{2\pi r L},$$

where in the last step I used my part (a) expression for the current. By the way, we just derived the textbook equation for the field generated by an infinite wire, $B = \frac{\mu_0 I}{2\pi r}$.

Now that we know the magnetic field at arbitrary distance r from the rod, we can find the magnetic force acting on Q_1. Since Q_1 sits a distance $r = D$ away from the rod, it feels magnetic field $B = \frac{\mu_0 Q_0 v_0}{2\pi D L}$. This field points out of the page, in the side view. Since the particle moves rightward, perpendicular to the field, the magnitude of the cross product $\mathbf{v} \times \mathbf{B}$ reduces to a regular product, and therefore

$$F_{\text{mag}} = |q\mathbf{v} \times \mathbf{B}| = qvB = Q_1 v_1 \frac{\mu_0 Q_0 v_0}{2\pi D L},$$

since the particle has velocity v_1.

To find the direction of this force, use the right-hand rule. Because $\mathbf{v}_1$ points rightward and $\mathbf{B}$ points out of the page, the cross product $\mathbf{F}_{\text{mag}} = q\mathbf{v} \times \mathbf{B}$ points down the page, towards the moving rod.

In summary, the magnetic force pulls the particle toward the rod, while the electric force pushes it away from the rod. Picking upward (away from the rod) as the positive direction, we get

$$\begin{aligned} F_{\text{total}} &= F_{\text{elec}} + F_{\text{mag}} \\ &= Q_1 \frac{Q_0}{L 2\pi \varepsilon_0 D} - Q_1 v_1 \frac{\mu_0 Q_0 v_0}{2\pi D L}, \text{upward}, \end{aligned}$$

where I substituted in our part (a) answer for the electric force, and I wrote out the linear charge density as $\lambda = Q_0/L$.

Digression for students interested in relativity theory. Otherwise, skip it.

Without numbers to plug in, you might think we can't tell whether this total force is positive or negative, i.e., whether the electric or the magnetic force "wins." But actually, we *can* tell. The electric force *has* to win, because we could always switch to a frame of reference in which the particle isn't moving and therefore experiences no magnetic force. To prove this result mathematically, you must know that light travels at speed $c = 1/\sqrt{\varepsilon_0 \mu_0}$, a fact you can derive from Maxwell's laws. Manipulate our above answer for F_{total} to get

$$\begin{aligned} F_{\text{total}} &= Q_1 \frac{Q_0}{L 2\pi \varepsilon_0 D} - Q_1 v_1 \frac{\mu_0 Q_0 v_0}{2\pi D L} \\ &= \frac{Q_0 Q_1}{2\pi \varepsilon_0 D L}(1 - v_0 v_1 \varepsilon_0 \mu_0) \\ &= \frac{Q_0 Q_1}{2\pi \varepsilon_0 D L}\left(1 - \frac{v_0 v_1}{c^2}\right). \end{aligned}$$

After learning relativity theory, you can explain this equation in terms of Lorentz contractions. Since the rod and particle move slower than the speed of light, $\frac{v_0 v_1}{c^2}$ is less than 1, and hence, F_{total} is positive. The electric force wins.

QUESTION 40-3

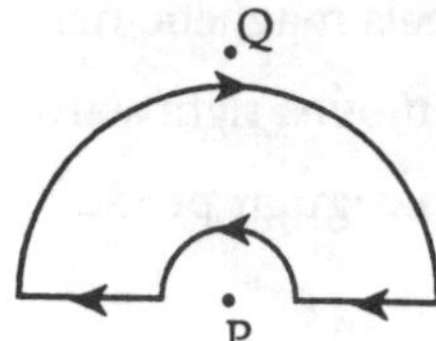

Current I_0 flows clockwise around this circuit, which consists of two semicircles and two straight segments. The inner semicircle has radius R, while the outer one has radius $3R$.

(a) What is the *direction* of the magnetic field at point Q? Justify your answer.

(b) What is the direction and magnitude of the magnetic field at point P, the center of the semicircles?

ANSWER 40-3

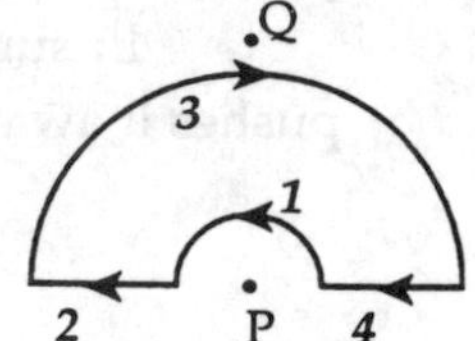

(a) The overall magnetic field is simply the sum of the fields produced by semicircle 1, wire segment 2, semicircle 3, and wire segment 4. According to the Biot-Savart law,

$$d\mathbf{B} = \frac{\mu_0}{4\pi} I \frac{d\mathbf{l} \times \hat{\mathbf{r}}}{r^2}.$$

the field strength falls off with the *square* of your distance from the wire. So, at Q, the field contributed by semicircle 3 probably "overwhelms" the fields generated by the other three segments.

To find the direction of the field generated by semicircle 3, apply the right-hand rule to Biot-Savart. According to that law, the field points in the direction of $d\mathbf{l} \times \mathbf{B}$. So, choose a piece of the semicircle near Q, and point your fingers in the direction of $d\mathbf{l}$, which is roughly rightward. Then curl your fingertips in the direction of $\hat{\mathbf{r}}$, i.e., towards Q. Your thumb points out of the page. That's the magnetic field direction.

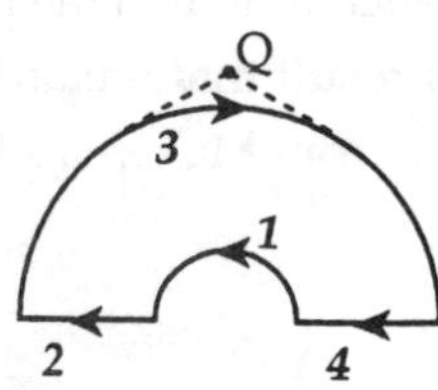

Interestingly, if you repeat this reasoning using a piece of semicircle 3 far from Q, the field points into the page, not out of the page. More precisely, in this picture, the part of semicircle 3 "inside" the dashed wedge generates a field at Q pointing out of the page; while the pieces of semicircle 3 "outside" the dashed wedge generate a field pointing into the page. Again, because the field strength decreases as $1/r^2$, the parts of semicircle 3 inside the dashed wedge probably "dominate." The overall field at Q points out of the page, most likely. But we don't know for sure.

(b) Unfortunately, no Amperian loop you could draw would allow you to pull B outside the Amperian line integral, $\oint \mathbf{B} \cdot d\mathbf{s}$. For this reason, Ampere's law can't get us to the answer. Even though Ampere's law is *true*, it's not calculationally *useful*. We have no choice but to invoke Biot-Savart,

$$d\mathbf{B} = \frac{\mu_0}{4\pi} I \frac{d\mathbf{l} \times \hat{\mathbf{r}}}{r^2}.$$

Fortunately, the superposition principle applies to all kinds of fields, not just electric ones. If $\mathbf{B}_1$ denotes the field created by semicircle 1, $\mathbf{B}_2$ denotes the field generated by wire segment 2, and so on, then the total field at point P is

$$\mathbf{B}_P = \mathbf{B}_1 + \mathbf{B}_2 + \mathbf{B}_3 + \mathbf{B}_4.$$

Using Biot-Savart, we can *separately* calculate each of these four fields.

Let's start with field at point P created by wire segment 2. It's zero. Here's why. From any infinitesimal piece of wire 2, the **r** vector from that piece to P goes rightward, as drawn here. But $d\mathbf{l}$ for that piece points leftward, in the direction of current flow. Because $\hat{\mathbf{r}}$ and $d\mathbf{l}$ are (anti)parallel, their cross product vanishes: $|d\mathbf{l} \times \hat{\mathbf{r}}| = dl \sin 180° = 0$. Therefore, according to Biot-Savart, wire segment 2 produces no magnetic field at point P.

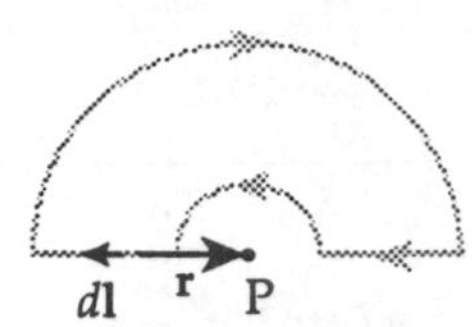

By equivalent reasoning, wire segment 4 also produces no magnetic field at point P.

Consequently, the only contributions to the field at point P come from the semicircles. To save work, I'll find the magnetic field produced by a semicircle of arbitrary radius r. Then I'll apply my general result to the specific semicircles in this problem.

Here's a strategy for keeping yourself organized while using Biot-Savart. I demonstrated this strategy in question 40-1, the hexagonal circuit.

Biot-Savart strategy for finding a magnetic field

1) Blacken a tiny piece of the wire (of length $d\mathbf{l}$), and draw in the vector **r** pointing from that blackened piece to the location at which you're finding the magnetic field.

2) Using Biot-Savart, calculate the magnitude and direction of the infinitesimal magnetic field $d\mathbf{B}$ generated by that piece.

3) Add up all those magnetic field contributions by integrating over the wire, using vector components if needed. (Fortunately, vectors won't be necessary here.)

Here we go . . .

Biot-Savart strategy step 1: Draw a good diagram, with **dl** *and* **r** *labeled.*

I've highlighted a tiny piece of the gray semicircle.

dl
r
P

Biot-Savart strategy step 2: Find **dB** *for that bit of wire.*

According to the Biot-Savart law, $d\mathbf{B}$ points in the direction of $d\mathbf{l} \times \hat{\mathbf{r}}$. As you can confirm with the right-hand rule, $d\mathbf{l} \times \hat{\mathbf{r}}$ points into the page at P, for *every* bit of wire in the semicircle. (Just point your fingers along the wire in the direction of current flow, and then curl your fingertips toward P. Your thumb points into the page.) Since all the $d\mathbf{B}$'s point in the same direction, we need not worry about vector components.

This infinitesimal field has magnitude

$$dB = \frac{\mu_0}{4\pi} I \frac{|d\mathbf{l} \times \hat{\mathbf{r}}|}{r^2} = \frac{\mu_0}{4\pi} I \frac{dl \sin\theta}{r^2},$$

where θ denotes the angle between $d\mathbf{l}$ and $\hat{\mathbf{r}}$. Since the current points tangential to the semicircle while $\hat{\mathbf{r}}$ points radially inward, $d\mathbf{l}$ and $\hat{\mathbf{r}}$ are perpendicular: $\theta = 90°$. Therefore, the magnitude of the cross product reduces to a regular product. So, our expression for dB reduces to

$$dB = \frac{\mu_0}{4\pi} I \frac{dl \sin 90°}{r^2} = \frac{\mu_0}{4\pi} I \frac{dl}{r^2}.$$

Biot-Savart strategy step 3: Add up the dB's by integrating over the wire.

Every piece of the semicircle is the same distance, r, from point P. So, r is a constant we can pull outside the integral:

$$\begin{aligned} B_{\text{semicircle}} &= \int \frac{\mu_0}{4\pi} I \frac{dl}{r^2} \\ &= \frac{\mu_0 I}{4\pi r^2} \int_0^{\pi r} dl \qquad \text{[the semicircle has length } \tfrac{1}{2}(2\pi r) = \pi r\text{]} \\ &= \frac{\mu_0 I}{4\pi r^2}(\pi r) \\ &= \frac{\mu_0 I}{4r}. \end{aligned}$$

The bigger the semicircle, the smaller a field it creates at the center.

Let's apply this result to semicircles 1 and 3. Since the current in semicircle 3 flows clockwise, $\mathbf{B}_3$ points into the page, as you can confirm by applying the right-hand rule to $d\mathbf{l} \times \hat{\mathbf{r}}$. And this field has strength $B_3 = \frac{\mu_0 I}{4(3R)}$. The current in semicircle 1 flows counterclockwise, and therefore produces an out-of-the-page field $B_1 = \frac{\mu_0 I}{4R}$. So, if we pick out-of-the-page as our positive direction, then the net field at point P is

$$\begin{aligned} \mathbf{B}_P &= \mathbf{B}_1 + \mathbf{B}_2 + \mathbf{B}_3 + \mathbf{B}_4 \\ &= \frac{\mu_0 I}{4R} + 0 - \frac{\mu_0 I}{4(3R)} + 0 \\ &= \frac{\mu_0 I}{6R}, \text{out of the page.} \end{aligned}$$

QUESTION 40-4

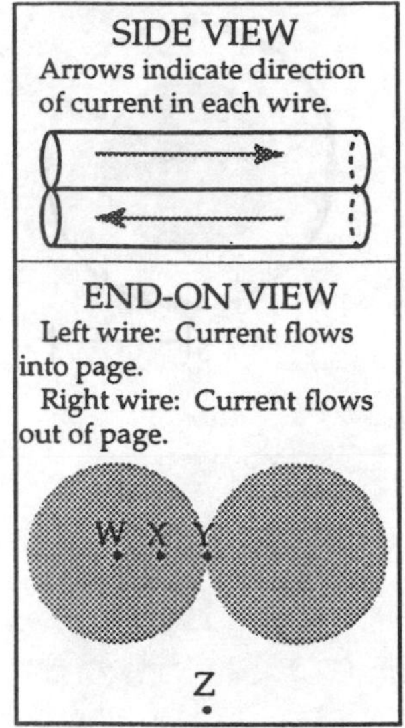

SIDE VIEW
Arrows indicate direction of current in each wire.
END-ON VIEW
Left wire: Current flows into page.
Right wire: Current flows out of page.
W X Y
Z

Two very long cylindrical wires of length L, each of radius R, run right next to each other, as pictured here. In the end view, the left wire carries current I^0 into the page, while the right wire carries equal current out of the page. In both wires, the current is uniformly distributed across the whole cross-sectional area.

In the side view, points W through Z are near the middle of the wires, not near the ends.

What is the strength and direction of the magnetic field at:

(a) Point W, the center of the left wire? Hint: Apply Ampere's law to each wire separately, and then use superposition.

(b) Point X, which is a distance $R/2$ from the center of the left wire, and distance $R/2$ from the left-hand edge of the right wire?

(c) Point Y, where the wires touch?

(d) Point Z, which is a distance $2R$ from the centers of both wires?

(d) What magnetic force does the right wire exert on the left wire?

Hint: For complicated reasons, the force is the same *as if* all the current in the left wire were concentrated at the central axis.

ANSWER 40-4

When using Ampere's law, $\int \mathbf{B} \cdot ds = \mu_0 I_{encl}$, your first instinct might be to draw an Amperian loop enclosing both wires. But no matter how cleverly you draw the loop, the magnetic field will vary in strength and direction at different points along the loop. For this reason, you can't pull B outside the line integral $\int \mathbf{B} \cdot ds$, and hence, you can't use Ampere's law to solve for B.

To avoid this roadblock, use the superposition principle. You can *separately* calculate the field due to each wire, and add them: $\mathbf{B}_{total} = \mathbf{B}_{due\ to\ left\ wire} + \mathbf{B}_{due\ to\ right\ wire}$. When calculating the field generated by the right wire, pretend that the left wire doesn't exist. And vice versa.

Things get a little complicated, because points W and X sit *inside* the left wire but *outside* the right wire. But hey, we can deal with this twist. First, using Ampere's law, I'll find a general expression for the magnetic field *outside* a wire. Then, again using Ampere's law, I'll find the magnetic field *inside* a cylindrical wire of uniform current density. With these two results, we can find the overall field at all four points.

Subproblem 1: Find the magnetic field outside a long wire carrying current I_0.

We want to invoke Ampere's law,

$$\oint \mathbf{B} \cdot ds = \mu_0 I_{encl}.$$

A long wire creates circular magnetic field lines. Therefore, if I enclose the wire with a circular Amperian loop, the magnetic field lines always point *along* the loop. Therefore, the dot product $\mathbf{B}\cdot d\mathbf{s}$ reduces to a regular product. Furthermore, since each point on this Amperian loop is equidistant from the wire, the magnetic field has the same strength over the entire loop, by symmetry. Therefore, we can pull B outside the Amperian line integral:

END-ON VIEW
Current flows into page.
Dashed Amperian loop "follows" the field line.

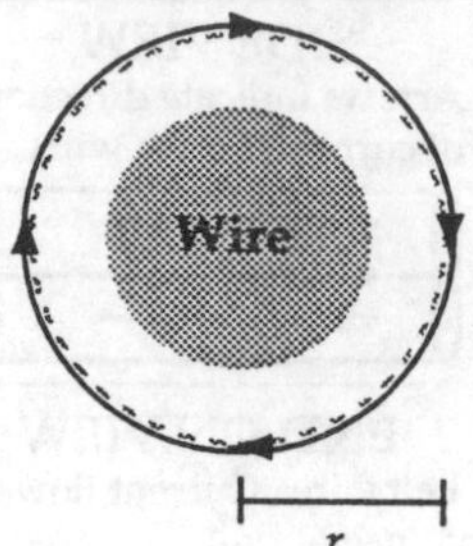

$$\oint \mathbf{B}\cdot d\mathbf{s} = \oint B\,ds = B\oint ds = B(2\pi r),$$

where r denotes the radius of the Amperian loop, *not* the radius of the wire. My Amperian loop has length $s = 2\pi r$, the circumference of the circle.

This loop encloses the whole wire, which carries current I_0. So, according to Ampere's law,

$$\oint \mathbf{B}\cdot d\mathbf{s} = \mu_0 I_{\text{encl}}$$

$$B(2\pi r) = \mu_0 I_0,$$

and hence

$$B = \frac{\mu_0 I_0}{2\pi r}.$$

That's the magnetic field *outside* a very long current-carrying wire. Since points W through Z all sit outside the right, this formula comes in handy. But W and X both sit *inside* the left wire. So, we *also* need to solve subproblem 2 . . .

Subproblem 2: Find the field inside a long wire of uniform current density, with total current I_0.

Dashed Amperian loop

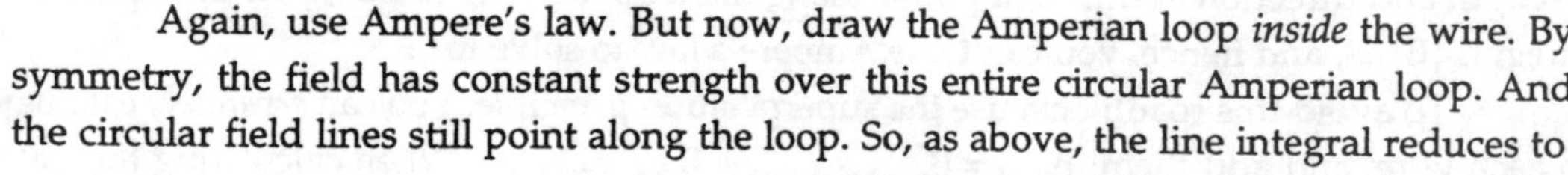

Again, use Ampere's law. But now, draw the Amperian loop *inside* the wire. By symmetry, the field has constant strength over this entire circular Amperian loop. And the circular field lines still point along the loop. So, as above, the line integral reduces to

$$\oint \mathbf{B}\cdot d\mathbf{s} = B\oint ds = B(2\pi r).$$

Now let's work on the other side of Ampere's law, $\mu_0 I_{\text{encl}}$. Unlike in subproblem 1, the Amperian loop does *not* enclose the entire current in the wire. It "captures" only the current inside the dashed line in my drawing. To figure out the enclosed current, think in terms of current density, or in terms of ratios. I'll demonstrate both methods, which are really two ways of saying the same thing.

Here's the quick and dirty ratio argument. The total current, I_0, is spread out over the entire cross-sectional area, $A_{\text{wire}} = \pi R^2$. But the Amperian loop encloses a smaller area, $A_{\text{encl}} = \pi r^2$. If the enclosed area is half the total area, then the enclosed current is half the total current. In general,

$$I_{\text{encl}} = I_0 \frac{A_{\text{encl}}}{A_{\text{wire}}} = I_0 \frac{\pi r^2}{\pi R^2} = I_0 \frac{r^2}{R^2}.$$

I'll now rederive this same expression by thinking in terms of current density, the current per cross-sectional area. Since the current density is uniform, $J = I_0/A_{wire} = I_0/(\pi r^2)$. The Amperian loop encloses area A_{encl}. Therefore, the loop encloses current

$$I_{encl} \frac{\text{current}}{\text{area}} \times (\text{area enclosed}) = JA_{encl} = \frac{I_0}{\pi R^2}\pi r^2 = I_0\frac{r^2}{R^2},$$

the same expression obtained using ratios.

Now we can invoke Ampere's law to find the magnetic field at arbitrary distance r from the center of the wire:

$$\oint \mathbf{B}\cdot d\mathbf{s} = \mu_0 I_{encl}$$

$$B(2\pi r) = \mu_0 I_0 \frac{r^2}{R^2},$$

and hence

$$B = \frac{\mu_0 I_0 r}{2\pi R^2}.$$

End of subproblem 2

So far, we've found the magnetic field both outside and inside a long wire of (uniformly distributed) current I_0. I'll summarize these results, letting B_{in} and B_{out} denote the field inside and outside the wire, respectively:

$$B_{in} = \frac{\mu_0 I_0 r}{2\pi R^2}$$

$$B_{out} = \frac{\mu_0 I_0}{2\pi r}.$$

In parts (a) through (d), I'll invoke these results repeatedly.

(a) Point W is inside the left wire and outside the right wire. But at the center of the left wire, where $r = 0$, B_{in} vanishes. Intuitively, the different "bits" of the left wire generate fields that cancel each other out at the center. So, the magnetic field at W stems entirely from the right wire. Since point W sits a distance $r = 2R$ from the center of the right wire,

$$B_W = B_{\text{out, due to right wire}} = \frac{\mu_0 I_0}{2\pi r} = \frac{\mu_0 I_0}{2\pi(2R)} = \frac{\mu_0 I_0}{4\pi R}.$$

To find the direction of this field, apply the right-hand rule to Biot-Savart, $d\mathbf{B} = \frac{\mu_0}{4\pi} I \frac{d\mathbf{l} \times \hat{\mathbf{r}}}{r^2}$. In the end-on view, look at the right wire. Point your fingers in the direction of current flow ($d\mathbf{l}$), out of the page. Then curl your fingertips in the direction of $\hat{\mathbf{r}}$, i.e., toward W. Your thumb now points down the page. That's the direction of the magnetic field generated by the right wire *at point W*.

Important digression: A right-hand rule shortcut. As you know, a long wire generates circular field lines. Before continuing to part (b), let me introduce a right-hand rule "shortcut" that quickly allows you to find whether those field lines circulate clockwise or counterclockwise around the wire. Simply point your right thumb in the direction of the current, and curl your fingertips. If they curl clockwise, the field lines flow clockwise. Similarly for counterclockwise.

Let's apply this shortcut to the right wire. When you point your thumb out of the page, your fingertips curl counterclockwise. At W, those counterclockwise field lines point down the page, confirming our conclusion from above. By playing around with some examples, you can prove that this right-hand rule shortcut always agrees with the "regular" right-hand rule.

(b) Point X is inside the left wire, and outside the right wire. So,

$$\mathbf{B}_X = \mathbf{B}_{\text{in, due to left wire}} + \mathbf{B}_{\text{out, due to right wire}}$$

Since these fields are vectors, we must figure out their directions before adding them. But that's easy, using the right-hand rule shortcut introduced at the end of part (a).

Since the current in the left wire flows into the page, point your thumb into the page, and curl your fingertips. They curl clockwise, indicating clockwise field lines. At point X, those clockwise field lines point down the page. Using the same shortcut, you can see that the right wire produces counterclockwise field lines, which *also* point down the page at X. So, at X, both wires produce a magnetic field pointing down the page.

END-ON VIEW

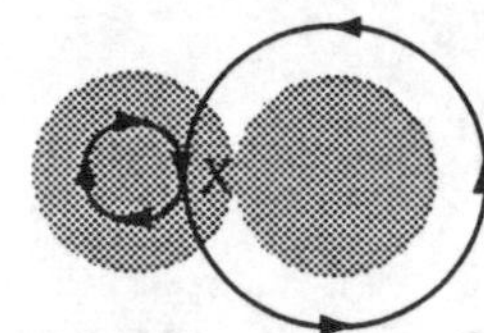

At point X, both wires generate fields pointing down the page.

Therefore, to find the total field at X, we can add the fields produced by the individual wires, without worrying about vector components. Point X is inside the left wire, a distance $r_{\text{left}} = R/2$ from the center; and out side the right wire, a distance $r_{\text{right}} = 3R/2$ from the center. Therefore

$$\begin{aligned}\mathbf{B}_X &= \mathbf{B}_{\text{in, due to left wire}} + \mathbf{B}_{\text{out, due to right wire}} \\ &= \frac{\mu_0 I_0 r_{\text{left}}}{2\pi R^2} + \frac{\mu_0 I_0}{2\pi r_{\text{right}}} \\ &= \frac{\mu_0 I_0 (R/2)}{2\pi R^2} + \frac{\mu_0 I_0}{2\pi(3R/2)} \\ &= \frac{\mu_0 I_0}{2\pi R}\left[\frac{1}{2} + \frac{2}{3}\right] \\ &= \frac{\mu_0 I_0}{2\pi R}\left(\frac{7}{6}\right).\end{aligned}$$

Notice that the right wire contributes more to the total field than the left wire does. Can you explain why, intuitively?

(c) For point Y, you might wonder whether to use the formula for B inside a wire or B outside a wire. Since Y sits on the boundary between "inside" and "outside," either formula should work. In other words, our expressions for B_{in} and B_{out} must "agree" at the edge of the wire, where $r = R$.

Fortunately, they *do* agree. Substituting $r = R$ into B_{in} and B_{out}, you get

$$B_{in} = B_{out} = \frac{\mu_0 I_0}{2\pi R} \qquad \text{at } r = R.$$

By the reasoning of part (b), both wires generate a down-the-page field at point Y. The total field has strength

$$\begin{aligned} B_Y &= B_{\text{due to left wire}} + B_{\text{due to right wire}} \\ &= \frac{\mu_0 I_0}{2\pi R} + \frac{\mu_0 I_0}{2\pi R} \\ &= \frac{\mu_0 I_0}{2\pi R}(2), \end{aligned}$$

stronger than the field at point X.

(d) Point Z sits the same distance, $r = 2R$, from both wires. Therefore, both wires generate the same field at that point:

$$B_{Z \text{ due to left wire}} = B_{Z \text{ due to right wire}} = \frac{\mu_0 I_0}{2\pi r} = \frac{\mu_0 I_0}{2\pi(2R)} = \frac{\mu_0 I_0}{4\pi R}$$

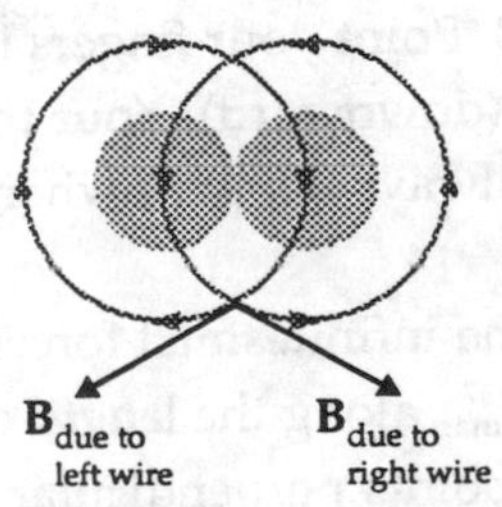

You can't just double this result to obtain the total field. At point Z, the fields generated by the left and right wires point in different directions, as this diagram shows. We must add the fields vectorially.

To sort out the vectors, I'll break these fields into their x- and y-components, as drawn here. The x-components cancel. So, the total field field points straight down, in the y-direction. Notice that both fields have the same y-component. Therefore, the total downward field is *twice* the downward field generated by the left wire or the right wire alone:

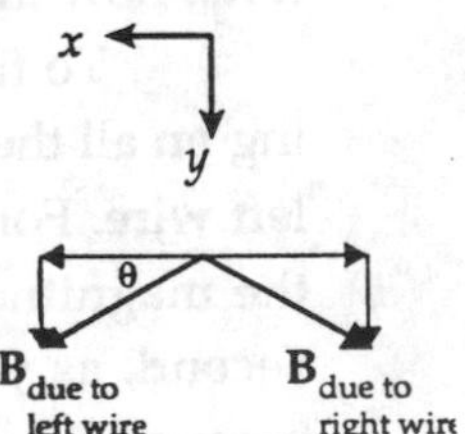

$$B_{total} + 2B_{y \text{ due to left wire}} = 2\frac{\mu_0 I_0}{4\pi R}\sin\theta.$$

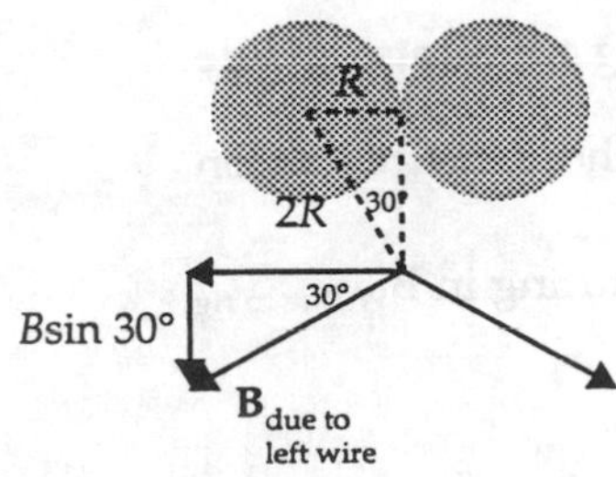

To complete the problem, we need to know θ in the above diagram.

To find that angle, look at the dashed triangle in this new drawing. The hypotenuse has length $2R$, while the shorter leg has length R. So, it's a 30°-60°-90° triangle. Since the dashed triangle is congruent to the vector triangle consisting of $\mathbf{B}_{\text{due to left wire}}$ and its components, $\theta = 30°$, too. Therefore, at Z, the total field points downward, with strength

$$B_Z = 2B_{y \text{ due to left wire}}$$

$$= 2\frac{\mu_0 I_0}{4\pi R}\sin 30°$$

$$= \frac{\mu_0 I_0}{4\pi R}.$$

(e) I lack the space to show mathematically why the hint "works." In general, you can't make "simplifications" like this. Here, we got lucky.

As we saw in part (a), the field generated by the right wire at the center of the left wire (point W) is

$$B_{W \text{ due to right wire}} = \frac{\mu_0 I_0}{4\pi R},$$

pointing downward. And from Chapter 39, we know that a magnetic field exerts a force

$$d\mathbf{F}_{\text{mag}} = I_0 d\mathbf{l} \times \mathbf{B}.$$

on an infinitesimal piece of wire of current I_0 and length $d\mathbf{l}$. In this equation, I_0 and $d\mathbf{l}$ refer to the wire *on which the force acts*, while **B** denotes the field generated by *other* wires. So here, $d\mathbf{l}$ refers to the left wire, and therefore points into the page. **B** refers to the field generated by the right wire, and therefore points down the page.

To find the direction of $d\mathbf{F}_{\text{mag}}$, apply the right-hand rule to $d\mathbf{l} \times \mathbf{B}$. Point your fingers in the direction of $d\mathbf{l}$ (into page), and curl your fingertips in the direction of **B** (downward). Your thumb points leftward. So, the right wire magnetically *repels* the left wire. This always happens when currents flow in opposite directions.

To find the magnitude of this repulsive force, we must add up the infinitesimal forces acting on all the pieces of the left wire. In other words, we must integrate dF_{mag} along the length of the left wire. Fortunately, the integral simplifies in two ways. First, since $d\mathbf{l}$ points perpendicular to **B**, the magnitude of the cross product reduces to a regular product: $|d\mathbf{l} \times \mathbf{B}| = (dl)B\sin 90° = (dl)B$. Second, as you "walk down" the central axis of the left wire, $\mathbf{B}_{\text{due to right wire}}$ stays constant. That's because $\mathbf{B}_{\text{due to right wire}}$ depends only on your distance from the central axis of the right wire. So, we can pull B outside the integral:

$$F_{\text{magnetic on left wire}} = \int dF_{\text{mag}} = \int I_0(dl)B$$

$$= I_0 B\int dl \qquad \text{[pulling constants outside}$$

$$= I_0 BL \qquad \text{[since the left wire has len}$$

$$= I_0\frac{\mu_0 I_0}{4\pi R}L \qquad \text{[substituting in } B_{W \text{ due to rig}}$$

$$= \frac{\mu_0 I_0^2 L}{4\pi R}.$$

We just derived a special case of the textbook formula for the magnetic force between two parallel or antiparallel currents. It's $F = \frac{\mu_0 I_0^2 L}{2\pi r}$, where r denotes the center-to-center distance between the two wires. Here, $r = 2R$.

QUESTION 40-5

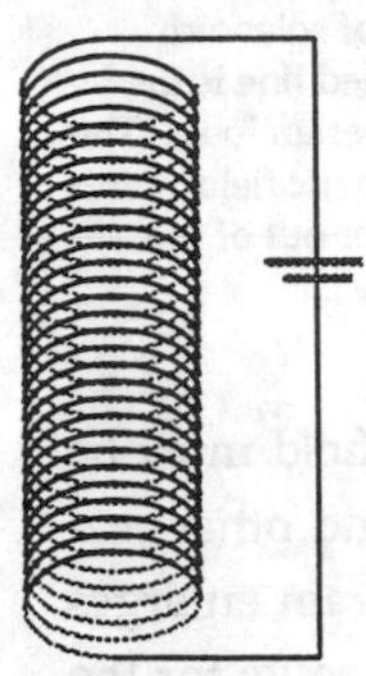

Consider a solenoid consisting of N coils, each with resistance R_0. (A solenoid is essentially an inflexible slinky.) The solenoid is hooked up to a battery of voltage V_0, as drawn here. The solenoid has length D and radius s, where $s \ll D$.

Throughout this problem, neglect the magnetic fields generated by the straight wires connecting the solenoid to the battery. Also, derive any solenoid-specific formulas from basic laws. Don't just plug in pre-derived solenoid formulas from your textbook.

(a) To good approximation, what is the magnetic field strength on the central axis of the solenoid, midway between the two ends of the solenoid?

(b) To good approximation, what is the magnetic field strength a distance $s/2$ from the central axis of the solenoid, midway between the two ends of the solenoid?

(c) Suppose you shoot a particle of charge q, mass m, and velocity v_0 down the central axis of the solenoid, from the top end of the solenoid towards the bottom end. This happens in outer space, where we can neglect gravity. Does the particle come out the other end? If not, why not? If so, at what speed?

ANSWER 40-5

Your textbook contains excellent diagrams of the magnetic field lines generated by a solenoid. Please study them carefully. You'll see that

(i) For solenoids that are much longer than they are wide, the field inside the solenoid is much stronger than the field outside the solenoid.

(ii) Near the ends of the solenoid, the field lines diverge. But at points deep inside the solenoid, the field lines point parallel to the central axis of the solenoid. And this field is approximately uniform, for reasons we'll explore below.

(a) To find the magnetic field on the central axis, you could reason as follows. First, using Biot-Savart, find the field generated by a single ring (coil) of current. A solenoid consists of "stacked" current rings. So, by summing up the fields contributed by each coil, you could calculate the overall field produced by the solenoid.

This technique works fine. But it's easier to use Ampere's law, the magnetic analog of Gauss' law. When a charge distribution displays sufficient symmetry, Gauss' law can help you find the electric field. Similarly, when the current distribution displays enough symmetry, Ampere's law gets you to the answer faster than Biot-Savart does.

The hardest part is choosing a "good" Amperian loop. Given your experiences with charged cylinders, you might draw the circular Amperian loop shown here. But this loop encloses no current. Therefore, by Ampere's law,

"BAD" CHOICE OF AMPERIAN LOOP

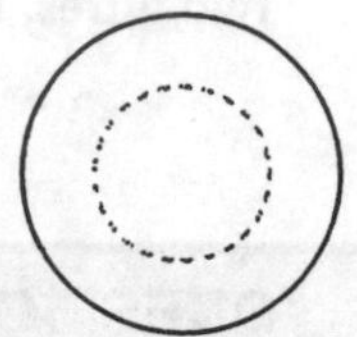

END-ON VIEW, looking down the end of solenoid. Dashed line is the Amperian loop. The magnetic field points into or out of the page.

Ampere's law $\oint \mathbf{B}\cdot d\mathbf{s} = \mu_0 I_{encl}$,

the line integral $\oint \mathbf{B}\cdot d\mathbf{s}$ equals 0. This is *not* because the magnetic field vanishes inside the solenoid. Rather, it's because the field points into or out of the page, in this drawing. Since the loop lies in the plane of the page, **B** is perpendicular to $d\mathbf{s}$, and hence $\mathbf{B}\cdot d\mathbf{s} = 0$. So, Ampere's law applied to this loop is *true*. But it's not *useful*; we can't use it to solve for the magnetic field.

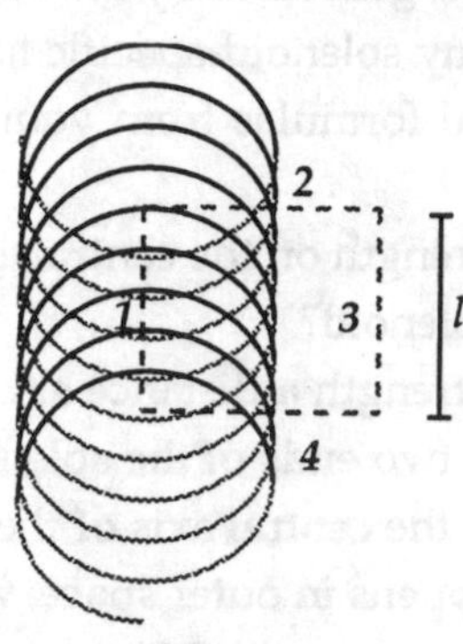

In order to enclose some current, i.e., to capture some coils, the Amperian loop must come out the side of the solenoid. Furthermore, to ensure that we can pull B outside the line integral $\oint \mathbf{B}\cdot d\mathbf{s}$, the magnetic field must be constant along (some sides of) the Amperian loop. For this and other reasons discussed below, I'll draw a rectangular loop. This diagram enlarges a section of the solenoid. By applying Ampere's law, we can solve for the field along side 1 of this loop, i.e., the field along the solenoid's central axis.

First, I'll find the enclosed current. As drawn here, the Amperian loop encloses five coils. But in general, we can calculate the enclosed current as follows. Let n denote the linear coil density. That's the number of coils per unit length. For instance, n might be 30 coils per inch, or 100 coils per meter. Since the solenoid contains N coils spread out over length D, the linear coil density is $n = N/D$. Therefore, the number of coils enclosed by my Amperian loop is

$$N_{encl} = \frac{\text{coils}}{\text{length}} \times (\text{length of solenoid enclosed}) = nl = \frac{N}{D}l,$$

where l denotes the length of the loop, as drawn here. For brevity's sake, let me keep writing "n" instead of "N/D." How does Nencl relate to the current through the loop?

Well, each coil enclosed by my Amperian loop "pierces" the loop. (To see this clearly, visualize the above drawing three-dimensionally, or build a model with a slinky and some string.) So, if I_0 denotes the current in the solenoid, then I_0 flows through my Amperian loop N_{encl} times. Each time the current goes around another coil, it passes through the Amperian loop *again*. Each of those "piercings" counts, in Ampere's law. For instance, in the above drawing, the current pierces the Amperian loop $N_{encl} = 5$ times, in which case the loop encloses current $5I0$. More generally, the Amperian loop encloses current

$$I_{encl} = N_{encl} I_0 = nlI_0.$$

At this stage, we have an expression for the enclosed current. I'll now simplify the other side of Ampere's law, namely the line integral $\oint \mathbf{B}\cdot d\mathbf{s}$. Since the Amperian loop consists of four sides, we can break up the integral:

$$\oint \mathbf{B}\cdot d\mathbf{s} = \int_1 \mathbf{B}\cdot d\mathbf{s} + \int_2 \mathbf{B}\cdot d\mathbf{s} + \int_3 \mathbf{B}\cdot d\mathbf{s} + \int_4 \mathbf{B}\cdot d\mathbf{s}.$$

Let me save side 1 for last, and start with side 2. As mentioned above, the magnetic field inside the solenoid points parallel to its central axis. But side 2 runs perpendicular to the central axis. Therefore, along side 2, $\mathbf{B}$ is perpendicular to $d\mathbf{s}$, and hence $\mathbf{B}\cdot d\mathbf{s} = 0$. This same conclusion applies to side 4. Since the parts of side 2 and side 4 outside the solenoid feel a negligible field, it follows that $\int_2 \mathbf{B}\cdot d\mathbf{s} = \int_4 \mathbf{B}\cdot d\mathbf{s} = 0$.

What about side 3? As just mentioned, the field outside a very long solenoid is negligible compared to the field inside the solenoid. So, $\int_3 \mathbf{B}\cdot d\mathbf{s} \approx 0$. We now see that only side 1 contributes to the Amperian line integral. Indeed, *I chose a rectangular Amperian loop so that things would simplify in this way.*

Since the magnetic field points along side 1, the dot product $\mathbf{B}\cdot d\mathbf{s}$ reduces to a regular product. Furthermore, because B has constant strength along side 1, we can pull it outside the line integral. So,

$$\begin{aligned}\oint \mathbf{B}\cdot d\mathbf{s} &= \int_1 \mathbf{B}\cdot d\mathbf{s} + \int_2 \mathbf{B}\cdot d\mathbf{s} + \int_3 \mathbf{B}\cdot d\mathbf{s} + \int_4 \mathbf{B}\cdot d\mathbf{s}.\\ &= B\int_1 ds + 0 + 0 + 0\\ &= Bl,\end{aligned}$$

where l denotes the arbitrary length of side 1.

Now that I've found the enclosed current and simplified the Amperian line integral, let's use Ampere's law to solve for B.

$$\begin{aligned}\oint \mathbf{B}\cdot d\mathbf{s} &= \mu_0 I_{\text{encl}},\\ Bl &= \mu_0 n l I_0,\end{aligned}$$

where n denotes the linear coil density. Cancel the l's to get

$$B = \mu_0 n I_0,$$

the textbook formula for the field inside a solenoid. Notice that the radius of the solenoid makes no difference, provided the solenoid is much longer than it is wide.

Here, $n = N/D$. To complete the problem, we must express I_0 in terms of the given constants. Since each coil has resistance R_0, and the solenoid consists of N coils in series, the total resistance is $R = NR_0$. Therefore, by Ohm's law, $I_0 = V_0/(NR_0)$. Therefore, the magnetic field inside the solenoid is

$$B = \mu_0 n I_0 = \mu_0 \frac{N}{D}\frac{V_0}{NR_0} = \frac{\mu_0 V_0}{DR_0}.$$

(b) In part (a), we found the field along the central axis of the solenoid. Does the field have this same strength *off* the central axis? Yes! To show why, I'll repeat the reasoning of part (a), looking for differences.

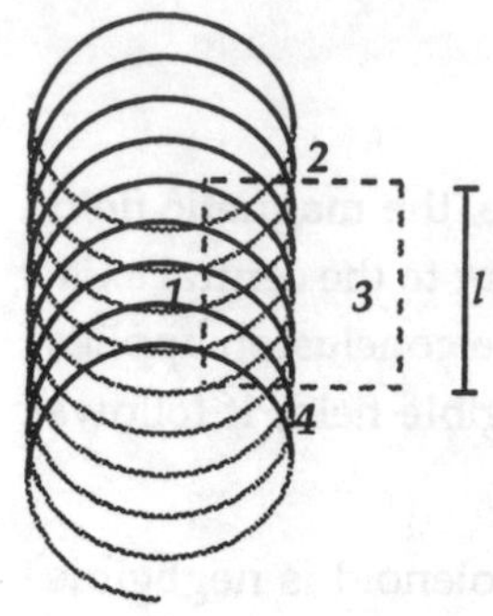

The new Amperian loop looks like the old one, except side 1 is now a distance $s/2$ from the central axis. Nonetheless, this new loop encloses the same current the old loop did. And for the same reasons as before, $\int \mathbf{B} \cdot d\mathbf{s} = 0$ along sides 2, 3, and 4. (The field points perpendicular to sides 2 and 4, and vanishes along side 3, to good approximation.) So, once again, only side 1 contributes to the Amperian integral. Since the field points parallel to side 1, and has constant strength along that side, $\int_1 \mathbf{B} \cdot d\mathbf{s} = B\int_1 ds = Bl$. In summary, all the reasoning of part (a) applies to part (b), with no modifications:

$$\oint \mathbf{B} \cdot d\mathbf{s} = \mu_0 I_{\text{encl}},$$

$$Bl = \mu_0 n l I_0,$$

and hence,

$$B = \mu_0 n I_0,$$

exactly as in part (a). Apparently, the field inside a solenoid stays the same, no matter how far from the central axis you go. Actually, the field gets *slightly* weaker near the sides. But for long, thin solenoids, this effect is negligible.

(c) As we saw in parts (a) and (b), the magnetic field inside a long solenoid has uniform strength and points parallel to the central axis. So, this particle gets fired *parallel* to the magnetic field. But a particle moving *along* a magnetic field line feels no magnetic force. Mathematically, that's because $\mathbf{F}_{\text{mag}} = q\mathbf{v} \times \mathbf{B}$. When $\mathbf{v}$ is parallel to $\mathbf{B}$, this cross product vanishes.

Consequently, the particle feels no forces as it floats through the solenoid. It might as well be floating in empty space. Therefore, it travels in a straight line, neither speeding up nor slowing down. The particle comes out the bottom of the solenoid with speed v_0.

QUESTION 40-6

A long thin solenoid of radius $R = 0.0100$ meters and length $L = 0.250$ meters consists of $N = 500$ coils. Point P is located on the central axis of the solenoid, at the center of the 250th coil. Current $I = 0.100$ A flows through the solenoid.

In this problem, derive all formulas from basic laws, even if you've derived them before. (You need to know this stuff cold.)

(a) What is the magnetic field at point P? Solve symbolically before plugging in numbers.

(b) (*Very hard; think conceptually.*) Now the 250th coil is removed and replaced with a twisting piece of wire that creates negligible magnetic field. Because this twisting wire connects the 249th to the 251st coil, the current through the rest of the solenoid remains unaffected. What is the new magnetic field at point P? Solve symbolically before plugging in numbers.

(c) By what fraction (or percentage) did the magnetic field decrease when we removed the 250th coil? Explain intuitively why this fraction works out to be greater than 1/500, despite the fact that only one of the 500 coils is missing.

ANSWER 40-6

(a) Inside a solenoid, the magnetic field is approximately uniform. It has the same strength at point P that it has elsewhere, provided you're not too near an end.

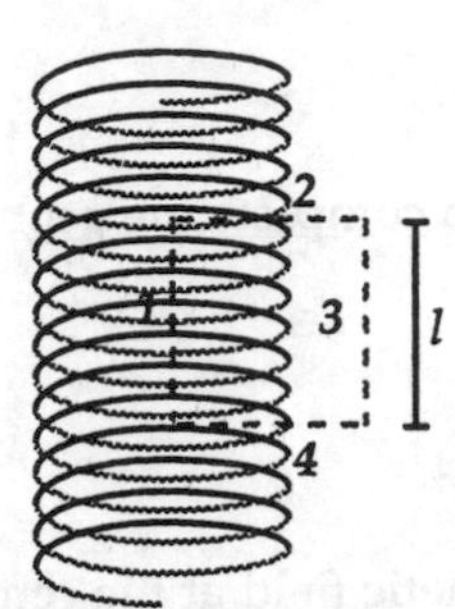

Since I just demonstrated how to find the magnetic field produced by a long solenoid, I'll race through the derivation here. We're solving for the magnetic field using Ampere's law,

$$\oint \mathbf{B}\cdot d\mathbf{s} = \mu_0 I_{\text{encl}}.$$

First I'll simplify the left-hand side. The field inside the solenoid points parallel to the central axis, while the field outside the solenoid is negligible. Therefore, with respect to my rectangular Amperian loop, **B** is perpendicular to $d\mathbf{s}$ along sides 2 and 4, and hence $\int_2 \mathbf{B}\cdot d\mathbf{s} = \int_4 \mathbf{B}\cdot d\mathbf{s} = 0$. Also, $\int_3 \mathbf{B}\cdot d\mathbf{s} = 0$, because the field (nearly) vanishes outside the solenoid. Only side 1 contributes to the overall line integral around the loop. Since the field points along that side, $\mathbf{B}\cdot d\mathbf{s}$ reduces to a regular product. And since B is uniform along that side, we can pull it outside the integral:

$$\oint_1 \mathbf{B}\cdot d\mathbf{s} = B\oint_1 ds = Bl.$$

Now I'll simplify the right-hand side of Ampere's law. My Amperian loop encloses the following number of coils:

$$N_{\text{encl}} = \frac{\text{coils}}{\text{length}} \times (\text{length enclosed}) = \frac{N}{L}l.$$

Therefore, the loop encloses current $I_{\text{encl}} = N_{\text{encl}}I = \frac{N}{L}lI$.

So, by Ampere's law,

$$\oint \mathbf{B}\cdot d\mathbf{s} = \mu_0 I_{\text{encl}},$$

$$Bl = \mu_0 \frac{N}{L}lI.$$

Cancel the l's and substitute in the numbers to get

$$B = \mu_0 \frac{N}{L} I = (1.26\times10^{-6}\,\text{T}\cdot\text{m/A})\left(\frac{500}{0.250\text{m}}\right)(0.100\text{ A}) = 2.52\times10^{-4}\,\text{T}.$$

(b) To solve this efficiently, use the infamous superposition principle. The field due to the solenoid with the 250th coil missing equals the field due to the whole solenoid, *minus* the field due to the 250th coil. Diagrammatically,

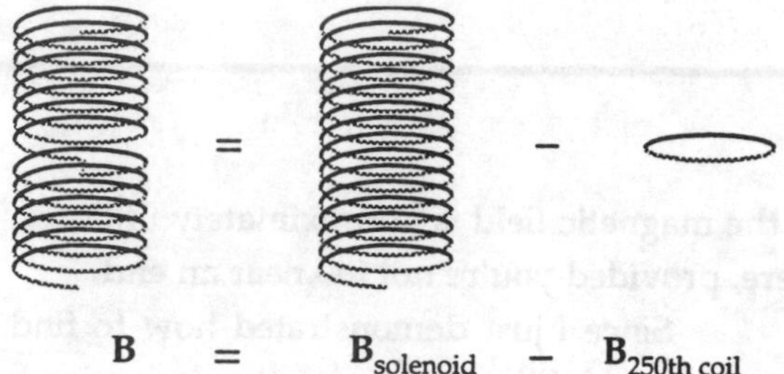

$$\mathbf{B} = \mathbf{B}_{\text{solenoid}} - \mathbf{B}_{\text{250th coil}}$$

In part (a), we found the field generated at point P by the whole solenoid. So, to complete the problem, we just need to find the field generated at point P by the 250th coil.

*Subproblem: Find **B** created by 250th coil at point P.*

Since point P sits at the center of the 250th coil, we're finding the magnetic field at the center of a circular current ring.

Unfortunately, no matter how cleverly you draw an Amperian loop through point P, the magnetic field is not constant over your whole loop. Therefore, you can't pull B outside the line integral $\oint \mathbf{B}\cdot d\mathbf{s}$. As a result, you can't solve for B using Ampere's law. We need another technique.

Well, when you couldn't use Gauss' law in electrostatics, you divided the charge distribution into tiny "pieces," used Coulomb's law to find the electric field produced by each piece, and then added up (integrated) these infinitesimal field contributions. Here, we must use the magnetic analog of this strategy:

Biot-Savart strategy for finding a magnetic field

1) Blacken a tiny piece of the wire, of length dl, and draw in the vector **r** pointing from that blackened piece to the location at which you're finding the magnetic field.
2) Using Biot-Savart, calculate the field, $d\mathbf{B}$, generated by that piece.
3) Add up all those magnetic field contributions by integrating.

Consider the magnetic field generated by this infinitesimal length of the current ring, $d\mathbf{l}$. According to Biot-Savart, that bit of wire generates a field

THE 250th COIL

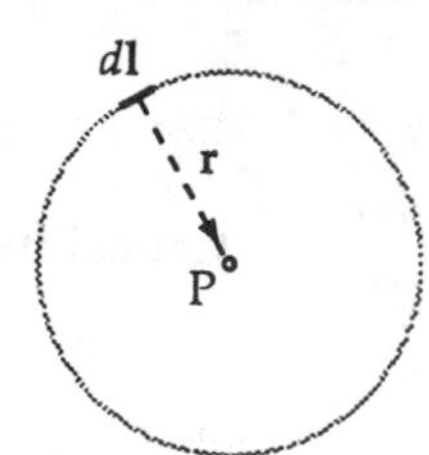

$$d\mathbf{B} = \frac{\mu_0}{4\pi} I \frac{d\mathbf{l} \times \hat{\mathbf{r}}}{r^2},$$

The "radius" vector **r**, indicated by the dashed line, runs from the blackened piece of current to P. Since $d\mathbf{l}$ points tangentially to the ring while $\hat{\mathbf{r}}$ points radially inward, $d\mathbf{l}$ and $\hat{\mathbf{r}}$ are perpendicular. Therefore, the magnitude of the cross product $\hat{\mathbf{r}}$ reduces to a regular product. And since the unit vector $\hat{\mathbf{r}}$ has length 1, we get $|d\mathbf{l} \times \hat{\mathbf{r}}| = dl\sin 90° = dl$.

Let's assume the current flows clockwise. By applying the right-hand rule $d\mathbf{l} \times \hat{\mathbf{r}}$, you can confirm that $d\mathbf{B}$ points *into* the page. This is true for *every* infinitesimal piece of the ring. Since all the $d\mathbf{B}$'s point in the same direction, we can add them up without worrying about vector components.

The integral is straightforward, because every piece of the ring is the same distance, $r = R$, from point P. Since r is constant, we can pull it outside the integral:

$$B = \int dB = \int_{\text{ring}} \frac{\mu_0}{4\pi} I \frac{dl}{r^2}$$

$$= \frac{\mu_0}{4\pi R^2} I \int_{\text{ring}} dl$$

$$= \frac{\mu_0}{4\pi R^2} I(2\pi R) \quad \text{[since the ring has circumference } 2\pi R\text{]}$$

$$= \frac{\mu_0 I}{2R}.$$

End of subproblem

We just found the magnetic field produced by a ring of current, at its center. So, that's the magnetic field generated by the 250th coil at point P: $B_{\text{250th coil}} = \frac{\mu_0 I}{2R}$. This field, like the solenoid's field, points along the central axis.

As noted above, the field created by the solenoid with the 250th coil missing is simply the field generated by the whole solenoid, minus the field produced by the 250th coil. Combining our part (a) expression for B_{solenoid} with this newly-derived expression for $B_{\text{250th coil}}$, we get

$$B = B_{\text{whole solenoid}} - B_{\text{250th coil}}$$

$$= \mu_0 \frac{N}{L} I - \frac{\mu_0 I}{2R}$$

$$= 2.52 \times 10^{-4}\,\text{T} - \frac{\left(1.26 \times 10^{-6}\,\frac{\text{T}\cdot\text{m}}{\text{A}}\right)(.100\text{ A})}{2(0.0100\text{ m})}$$

$$= 2.52 \times 10^{-4}\,\text{T} - 6.30 \times 10^{-6}\,\text{T}$$

$$= 2.46 \times 10^{-4}\,\text{T}.$$

(c) As just shown, removing the 250th coil decreases the field at point P by $\frac{\mu_0 I}{2R} = 6.30 \times 10^{-6}$ T from the original value of $M_o \frac{N}{L} I = 2.52 \times 10^{-4} T$. So, the fraction by which the magnetic field declines is

$$\frac{B_{\text{250th coil}}}{B_{\text{solenoid}}} = \frac{6.30 \times 10^{-6}\,\text{T}}{2.52 \times 10^{-4}\,\text{T}} = 0.025,$$

which is 2.5%.

At first glance, we might have expected each of the 500 coils in the solenoid to contribute equally to the total magnetic field at point P. In that case, removing the 250th coil would reduce the field at P by $1/500 = 0.002$, which is only 0.20%. So, the 250th coil contributes much more than its "fair share" to the total field at point P, more than 10 times its fair share. Intuitively, this makes sense. The 250th coil is right next to point P. All the other coils are farther from point P, and most of them are *much* farther away. That's why the 250th coil contributes disproportionately to the magnetic field at point P.

41 CHAPTER

Faraday's Law: Induction

QUESTION 41-1

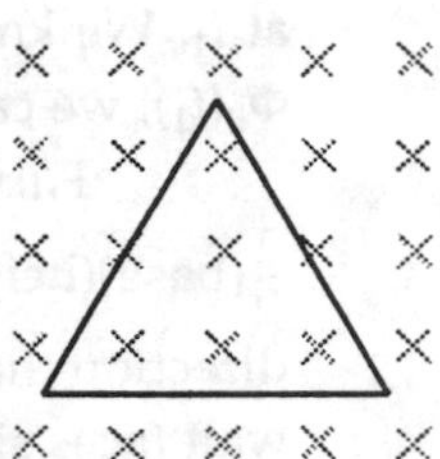

An equilateral triangle is constructed of three thin wires, each with length s and resistance R. The triangle sits inside a uniform magnetic field of strength B_0, pointing into the page. At time t_1, the magnetic field starts decreasing in strength at a steady rate, until it vanishes completely at time $3t_1$. After time $3t_1$, the magnetic field stays off.

What current flows around the triangle at time

(a) $.5t_1$?

(b) $2t_1$?

(c) $4t_1$?

(d) When current flows, does it flow clockwise or counterclockwise? Explain your reasoning.

ANSWER 41-1

According to the laws you learned previously, no current should ever flow around the triangle. Current flows in response to electric fields, corresponding to a voltage around the triangle. But here, it looks as if the triangle experiences only magnetic fields. And magnetic fields can't make current flow. Remember, magnetic fields deflect already-moving charges; but they can't get stationary charges moving in the first place.

The point of this chapter, however, is that electric fields *do* get created in the wires. According to *Faraday's law of induction*, electric fields get *induced* whenever a magnetic field *changes*.

(a) By explaining Faraday's law, I'll show you why **no current flows** at time $.5t_1$.

Let $\mathcal{E}_{\text{induced}}$ denote the induced voltage around the "current loop," in this case a triangle. (The word "loop" indicates a closed path, not a particular shape.) Many textbooks call this voltage an "electromotive force" (EMF), which is misleading, because $\mathcal{E}$ is a voltage, not a force.

Let Φ_B denote the magnetic flux through the current ring. Roughly speaking, "flux" specifies how many field lines pierce through the surface. (Remember the water streamline analogy from Chapter 34?) You calculate it by integrating the field strength over the surface area. So, for magnetic flux, $\Phi_B = \int \mathbf{B} \cdot d\mathbf{A}$, just as for electric flux, $\Phi_E = \int \mathbf{E} \cdot d\mathbf{A}$.

By Faraday's law, the induced voltage around the loop equals the *rate of change* of the magnetic flux through that loop:

Faraday's law: $$\mathcal{E}_{\text{induced}} = -\frac{d\Phi_B}{dt}.$$

I'll explain the physical meaning of the minus sign in part (d).

According to Faraday's law, an induced voltage kicks in only when the magnetic flux *changes*. So, before time t_1, no voltage gets induced, and therefore no current flows around the triangle. When it comes to inducing voltage and current, the size of the magnetic flux makes no difference. All that matters is whether the flux *changes*, and how quickly.

(b) At time $2t_1$, the magnetic field—and hence, the magnetic flux through the current loop—is changing. Therefore, by Faraday's law, a voltage gets induced. After calculating that voltage, we can use $V = iR$ to figure out the corresponding current.

First, we must find the rate of change of flux at time $2t_1$. To do so, I'll first find $\Phi_B(t_1)$, the flux at t_1. We know $\Phi_B(3t_1) = 0$, since the field vanishes at that time. Therefore, given $\Phi_B(t_1)$, we can figure out the rate of change of the flux between t_1 and $3t_1$.

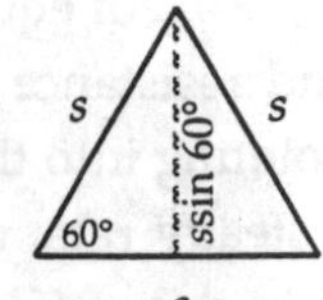

Flux at t_1. Since the triangle has height $s\sin 60°$, its area is $A = \frac{1}{2}(\text{base})(\text{height}) = \frac{1}{2}s(s\sin 60°) = \frac{1}{2}s^2 \sin 60°$. Recall that an area vector points in the direction that the surface "faces." For instance, your floor faces upward, and your wall faces sideways. Therefore, the magnetic flux through the triangle at time t_1 is

$$\Phi_B(t_1) = \int \mathbf{B}\cdot d\mathbf{A}$$

$$= \int B\,dA \qquad \text{[since triangle "faces" same direction as } \mathbf{B}]$$

$$= B\int dA \qquad \text{[since } B \text{ is uniform over whole area of triangle]}$$

$$= B_0 A$$

$$= B_0 \frac{1}{2}s^2 \sin 60°.$$

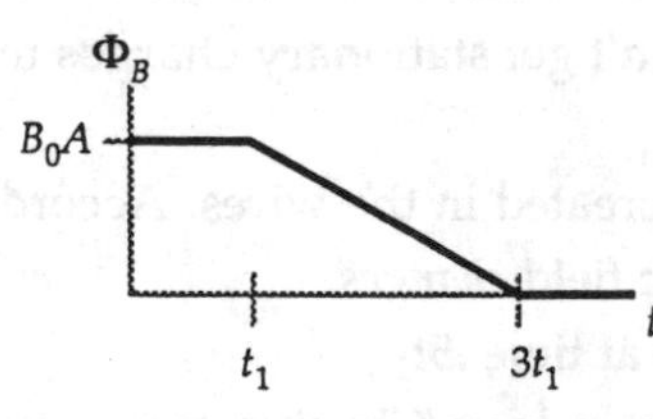

This flux decreases until hitting zero at time $3t_1$. Because it decreases at a constant rate, we can find the rate of change without using calculus. Just graph Φ_B vs. t. Between t_1 and $3t_1$, the graph is a negatively-sloped straight line. The slope—i.e., $\frac{d\Phi_B}{dt}$—is the same at all times between t_1 and $3t_1$. We can calculate that slope (derivative) using "rise over run":

$$\frac{d\Phi_B}{dt} \text{ at all times between } t_1 \text{ and } 3t_1 = \frac{\Delta\Phi_B}{\Delta t} \text{ between } t_1 \text{ and } 3t_1$$

$$= \frac{\Phi_B(3t_1) - \Phi_B(t_1)}{3t_1 - t_1}$$

$$= \frac{0 - \frac{1}{2}B_0 s^2 \sin 60°}{2t_1}$$

$$= -\frac{B_0 s^2 \sin 60°}{4t_1}.$$

The minus sign indicates that the flux decreases.

Faraday's law immediately tells us the induced voltage around the triangle at time $2t_1$:

$$\mathcal{E}_{\text{induced}} = -\frac{d\Phi_B}{dt}$$

$$= \frac{B_0 s^2 \sin 60°}{4t_1}.$$

Given that voltage we can quickly calculate the current. The triangle consists of three sides, each with resistance R. Since those three sides are in series, they have total resistance $R_{\text{total}} = 3R$. Therefore, by Ohm's law,

$$i_{\text{induced}} = \frac{\mathcal{E}_{\text{induced}}}{R_{\text{total}}} = \frac{B_0 s^2 \sin 60°}{4t_1(3R)} = \frac{B_0 s^2 \sin 60°}{12Rt_1}.$$

This current stays constant between t_1 and $3t_1$, because the magnetic flux through the current loop *changes* at a *constant rate*. See the above graph.

In part (d), we'll figure out whether this current flows clockwise or counterclockwise.

(c) After time $3t_1$, the magnetic flux stops changing. Therefore, according to Faraday's law, no more current gets induced, at any time past $3t_1$.

Here, many students get the right answer for the wrong reason. The induced current is zero, but *not* simply because the flux is zero. The induced current vanishes because the flux isn't *changing*. If the flux was temporarily zero, but was in the process of changing (from a positive to a negative value, for instance), then current would be induced, *even at the moment the flux hit zero.* Induced voltage and current only "care" about how fast the flux changes, not about the value of the flux.

(d) You can use either mathematical or physical reasoning to figure out the direction of the induced voltage (and current). I'll demonstrate both methods.

Mathematical method. Using your right hand, point your thumb in the direction that the magnetic flux *changes*, i.e., the direction of $d\Phi_B/dt$. In this case, the flux points into the page but decreases. So, the direction of $d\Phi_B/dt$ is *out* of the page. When you point your thumb out of the page, your fingertips curl counterclockwise. So, if $\mathcal{E}_{\text{induced}}$ equaled $+\frac{d\Phi_B}{dt}$, then the induced voltage (and current) would be counterclockwise. But since $\mathcal{E}_{\text{induced}}$ equals $-\frac{d\Phi_B}{dt}$, the voltage and current are clockwise.

Physical method. I recommend using the following reasoning, because it highlights an important *physical* insight about induced currents. According to Lenz' law, the induced current always acts so as to counteract the changing magnetic flux. Let me clarify this law by applying it to the problem at hand.

The triangle hates change. It wants the flux through it to stay the same, instead of decreasing. In other words, the triangle "wants" the into-the-page magnetic field through it to stay big, instead of decreasing. So, the induced current in the triangle generates its own magnetic field pointing into the page. In this way, the induced current tries to "counteract" decreasing field. This "induced" magnetic field can't prevent the overall magnetic field inside the triangle from decreasing. Nonetheless, this reasoning helps us figure out the direction of the induced current.

I just used Lenz' law to show that the induced current in the triangle flows so as to create an into-the-page field, at points inside the triangle. To finish this problem, we need to know whether a clockwise or a counterclockwise current produces this field. Well, using the right-hand rule shortcut introduced in Chapter 40, you can confirm that current must flow *clockwise* around the triangle to create an into-the-page field, at points inside the triangle. Picking any side of the triangle, point your thumb in the direction of current flow, assuming a clockwise current. Now curl your fingertips, which represent the field produced by the wire. At locations inside the triangle, your fingertips now point into the page. This proves that a clockwise current generates an into-the-page field. (The same result holds for all three sides of the triangle.)

QUESTION 41-2

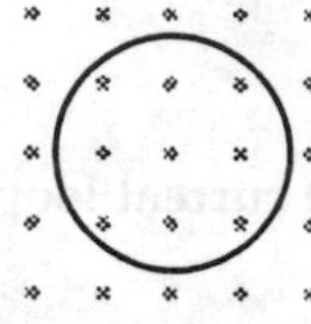

A circular brass ring, of radius $s = 0.10$ m and resistance $R_0 = 2.0\ W$, sits in a uniform magnetic field pointing out of the page. The following graph shows the strength of that field as a function of time. A negative value indicates that the field points into the page, instead of out of the page.

(a) At what time (or times) is the *largest* current induced in the ring?

(b) Between $t = 2$ s and $t = 3$ s, is the induced current clockwise or counterclockwise? Explain your reasoning.

(c) (*Very hard*) Between $t = 2$ s and $t = 3$ s, is the induced current increasing or decreasing in magnitude? Explain.

(d) What is the induced current at time $t = 4.7$ s?

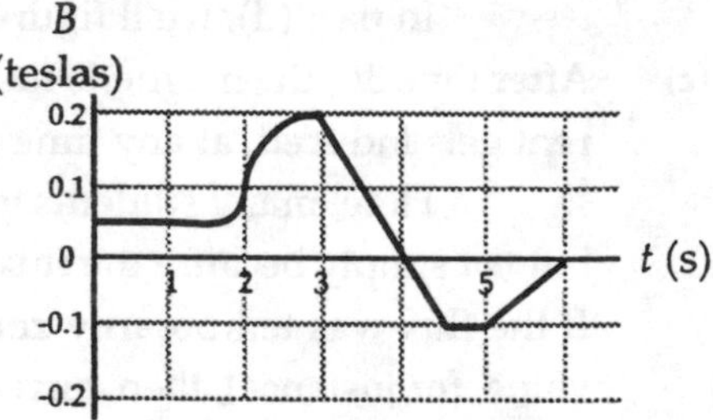

(e) What is the induced current at time $t = 4$ s?

(f) Between $t = 5$ s and $t = 6$ s, is the induced current increasing, decreasing, or constant in magnitude? Explain.

ANSWER 41-2

(a) You might be tempted to say $t = 3$ s, because that's when the flux through the ring becomes largest. But the induced voltage and current do not depend on the size of the flux. According to Faraday's law, the induced voltage is given by the *rate of change* of magnetic flux through the ring:

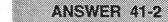

$$\mathcal{E}_{\text{induced}} = -\frac{d\Phi_B}{dt}.$$

The magnetic flux is proportional to the magnetic field, and also proportional to the area of the ring: $\Phi_B = \int \mathbf{B} \cdot d\mathbf{A} = BA_{\text{ring}}$. In this problem, the ring's area stays constant. So, the flux changes *because* the magnetic field changes. So, the rate of change of the flux, $d\Phi_B/dt$, is proportional to the rate of change of the magnetic field, dB/dt. In other words, Φ_B changes most rapidly when B changes most rapidly. Therefore, according to Faraday's law, the induced voltage and current are largest when the magnetic field changes most rapidly.

On a B vs. t graph, the rate of change of the magnetic field, dB/dt, corresponds to the **slope**. The magnetic field changes most rapidly wherever the slope (derivative) is biggest. Which appears to be at about $t = 2$ s.

(b) Between $t = 2$ s and $t = 3$ s, the magnetic field points out of the page, and *increases*. Don't get fooled by the fact that the slope decreases. This decreasing slope means that the magnetic field increases at a slower and slower rate. But the field keeps increasing, throughout the entire time interval between $t = 2$ s and $t = 3$ s.

According to Lenz' law, the induced current tries to counteract this change in the field. Since the out-of-the-page magnetic field is increasing, the ring "wants" to *decrease* the field, by creating its own magnetic field pointing *into* the page. In this way, the ring tries (unsuccessfully) to prevent the overall field from increasing.

As you can confirm using the right-hand rule shortcut, the induced current must flow clockwise in order to create an into-the-page magnetic field (at points inside the ring). Assuming a clockwise current, pick any segment of the ring, and point your thumb with the current. Now curl your fingertips, which represent the field lines created by that current. At locations inside the ring, your fingertips point into the page. So, a clockwise current creates an into-the-page field, at locations inside the ring.

Digression: Lenz' law and acceleration. Compare this result to answer 41-1d above. There, the induced current counteracted a magnetic field pointing *into* the page and *decreasing*. By contrast, in this problem, the induced current counteracts a magnetic field pointing *out* of the page and *increasing*. In both cases, the induced current produces an into-the-page field. Why? To prevent an into-the-page field from decreasing, you must crank up the field into the page. To prevent an out-of-the-page field from increasing, you must also generate an into-the-page field.

If this seems strange, imagine yourself trying to prevent a train from changing its velocity. If the train is moving forward and speeding up, you counteract the change in velocity by exerting a backward force. And if the train is moving backward and slowing down, you *also* counteract the change in velocity by exerting a backward force.

(c) Since the field and flux increase between $t = 2$ s and $t = 3$ s, you might be tempted to say the induced current increases as well. But according to Faraday's law, the induced voltage has nothing to do with the size of the magnetic flux. The induced voltage and current depend entirely on the rate at which the magnetic flux *changes*: $\mathcal{E}_{\text{induced}} = -d\Phi_B/dt$. As explained in part (b), the rate of change of the flux is proportional to the **slope** of the B vs. t graph. Between $t = 2$ and $t = 3$ s, the slope *decreases*. In other words, the magnetic flux increases, but at a slower and slower rate. So, the induced voltage (and current) gradually decrease during this time interval.

(d) At $t = 4.7$ s, the B vs. t graph has zero slope. At that moment, the magnetic flux through the ring is steady, not changing. Therefore, the induced voltage is zero. No current flows.

(e) Many people say that the induced current is zero, since the magnetic flux through the ring is zero at $t = 4$ s. But as I've emphasized *ad nauseam*, the induced voltage and current depend on the *rate of* *change* of the flux, not on its value. The flux through the ring is *changing* at $t = 4$ s, even though the flux happens to be zero at that moment. Therefore, a voltage gets induced.

To find the induced voltage and current, we can reason as follows. First, using the slope of the B vs. t graph, figure out $d\Phi_B/dt$, the rate at which the magnetic flux changes. Faraday's law then gives us the induced voltage around the ring. Finally, calculate the induced current using Ohm's law, $\mathcal{E}_{\text{induced}} = i_{\text{induced}}R$.

Let me spell out the first step of this strategy in more detail, before implementing it. The slope of the B vs. t tells us dB/dt, the rate of change of the magnetic field. The magnetic field and flux are related by

$$\Phi_B = \int \mathbf{B}\cdot d\mathbf{A}$$

$$= \int B dA \qquad \text{[since the ring "faces" same direction as } \mathbf{B}\text{]}$$

$$= B\int dA \qquad \text{[since } B \text{ is uniform over whole are of ring]}$$

$$= BA$$

$$= B\pi s^2.$$

So, once we know dB/dt, we can immediately find $d\Phi_B/dt$ using

$$\frac{d\Phi_B}{dt} = \frac{d(B\pi r s^2)}{dt} = \pi s^2 \frac{dB}{dt}.$$

By the way, if you're wondering why I pulled B outside the flux integral, even though B changes in time, please read the "mathematical clarification" below.

OK, let's implement the strategy outlined above. First, we must find dB/dt, the slope of the B vs. t graph, at $t = 4$ s. Between $t = 2$ s and $t = 4.5$ s, B decreases at a steady rate. Therefore, we can find the slope using rise over run. I'll use the points (3, 0.2) and (4, 0) to perform the calculation:

$$\frac{dB}{dt} \text{ at all times between } t = 3\text{ s and } t = 4.5\text{ s} = \frac{\Delta B}{\Delta t} \text{ during that time interval}$$

$$= \frac{B(4) - B(3)}{4\,\text{s} - 3\,\text{s}}$$

$$= \frac{0 - 0.2\,\text{tesla}}{1\,\text{s}}$$

$$= -0.2\ \text{tesla/s}.$$

So, as explained above, the flux changes at rate

$$\frac{d\Phi_B}{dt} = \frac{d(B\pi s^2)}{dt} = \pi s^2 \frac{dB}{dt}$$

$$= \pi(.10\,\text{m})^2(-0.2\,\text{tesla/s})$$

$$= -6.28 \times 10^{-3}\,\text{webers/s}.$$

By Faraday's law, this rate of change of flux tells us the induced voltage around the ring:

$$\mathcal{E}_{\text{induced}} = -\frac{d\Phi_B}{dt} = 6.28 \times 10^{-3} \text{ volts}.$$

Finally, Ohm's law gives us the induced current:

$$i_{\text{induced}} = \frac{\mathcal{E}_{\text{induced}}}{R_0} = \frac{6.28 \times 10^{-3}\text{ V}}{2.0\,\Omega} = 3.14 \times 10^{-3}\,\text{A},$$

about three milliamps.

Does this current flow clockwise or counterclockwise? Well, at $t = 4$ s, the magnetic flux through the ring is decreasing. Specifically, a positive (out-of-the-page) field is turning into a negative (into-the-page) field. To counteract this trend, the induced current "wants" to create its own field pointing in the positive (out-of-the-page) direction. Using the right-hand rule shortcut explained in question 41-4d, you can confirm that i_{induced} must flow counterclockwise to create an out-of-the-page field, at points inside the ring.

Crucial mathematical clarification

Mathematically minded readers might freak out over my derivation of the flux through the ring:

$$\Phi_B = \int \mathbf{B}\cdot d\mathbf{A} = \int B\,dA = B\int dA = \ldots$$

I pulled B outside the integral. That's valid only if B is constant. But here, B is emphatically *not* constant; it changes in time. So, what gave me the right to pull it outside the flux integral?

There are two different senses in which something can be "constant." It can be constant in space, or constant in time. For instance, consider the mass density (mass per volume) of a sofa. It's constant in the sense that it doesn't change with time. A sofa will be just as dense tomorrow as it is today. But the sofa's density is not constant over space. The legs are denser than the cushions. In summary, the sofa's density is temporally constant, but spatially varying. Roughly put, the density is constant in time, but non-uniform in space.

Now consider your body temperature. It's spatially uniform. In other words, at any given time, all the different parts of your body have the same temperature, to good approximation. But when you catch a fever, your body temperature is non-constant in time; it fluctuates up and down. So, body temperature is uniform in space but non-constant in time. When your left arm gets a fever, so does your right arm!

In this problem, the magnetic field is spatially uniform. Sure, B varies in time. But at any given moment, B on the left side of the ring equals B on the right side of the ring. Since $\int \mathbf{B}\cdot d\mathbf{A}$ is an integral over space, not an integral over time, we can pull B outside the integral. By contrast, we could *not* pull B outside $\int B\,dt$. Subtle, eh?

(f) You might want to say the induced current is changing, because Φ_B is changing. But as I emphasized in part (c), the induced current depends only on the rate of change of the magnetic flux. Between $t = 5$ s and $t = 6$ s, the slope of B vs. t is constant. The magnetic flux changes at a constant rate. In other words, $d\Phi_B/dt$ is constant. Therefore, the induced voltage and current stay constant during that interval.

QUESTION 41-3

(*Hard, but this is a classic exam problem. Come back to it several times before your next test.*) Minh holds a rectangular loop of wire, of length l, width w, and resistance R. She pushes it at constant speed v_0 toward a region containing a uniform into-the-page magnetic field, B_0. As the rectangle enters the magnetic field, Minh maintains its leftward motion at steady speed v_0.

The following questions refer to the moment at which the rectangle extends distance x into the magnetic field, as drawn here.

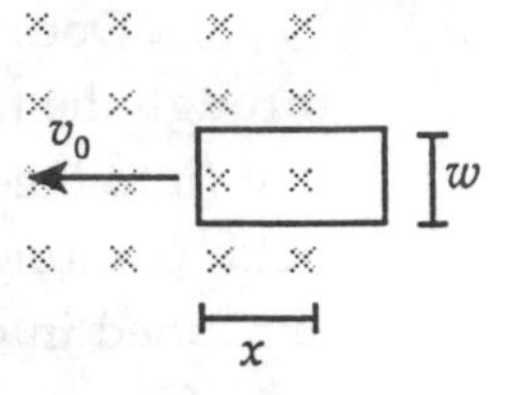

(a) What is the induced current in the rectangle?

(b) What magnetic force does the field exert on the rectangle? Give the magnitude and direction.

(c) What force is Minh exerting on the rectangle? Give the magnitude and direction.

(d) What is the electrical power dissipated in the rectangle?

(e) With what power is Minh pushing the rectangle?

(f) Which, if either, is bigger: Your answer to (d) or your answer to (e)? Explain why, even if you didn't solve parts (d) and (e) completely.

ANSWER 41-3

This problem requires you to synthesize concepts from several different chapters.

(a) According to Faraday's law, current gets induced in the rectangle because the flux through the rectangle changes: $\mathcal{E}_{\text{induced}} = -d\Phi_B/dt$. So, we must find the rate at which the flux changes.

In previous problems, the flux changed *because* the field got stronger or weaker. Here, the flux through the rectangle changes, *not* because the field strength increases or decreases, but because the rectangle moves into the field. As Minh pushes the rectangle leftward, the area of the rectangle "immersed" in the field gets bigger and bigger. So, more and more field lines can pierce through the rectangle. In other words, the flux increases.

Intuitively, the faster Minh pushes the rectangle, the more quickly the flux changes. So, we expect $d\Phi_B/dt$ to be proportional to v_0. To find $d\Phi_B/dt$ exactly, I'll first write an expression for Φ_B, the flux at an arbitrary moment. Then I'll differentiate that expression with respect to t.

Let $A_{\text{in field}}$ denote the area of the rectangle that's inside the magnetic field. Until the rectangle becomes fully immersed, $A_{\text{in field}}$ is less than $A_{\text{whole rectangle}}$. Indeed, from the above diagram,

$$A_{\text{in field}} = wx,$$

where x is the length of the rectangle inside the field. So, the flux through the rectangle is

$$\Phi_B = \int \mathbf{B}\cdot d\mathbf{A}$$
$$= B_0 A_{\text{in field}}$$
$$= B_0 wx.$$

In this expression, B_0 and w are constant in time, while x varies in time. This formalizes our earlier observation that the flux changes because the rectangle moves deeper into the field, *not* because the field fluctuates. So, when we differentiate with respect to time, B_0 and w get pulled out front:

$$\frac{d\Phi_B}{dt} = \frac{d(B_0 wx)}{dt}$$
$$= B_0 w \frac{dx}{dt}$$
$$= B_0 w v_0,$$

where in the last step, I used the definition of velocity as change in distance over change in time.

According to Faraday's law, this rate of change of flux gives us the induced voltage, $\mathcal{E}_{\text{induced}} = -d\Phi_B/dt = -B_0 w v_0$. And by Ohm's law, the corresponding induced current is

$$i_{\text{induced}} = \frac{\mathcal{E}_{\text{induced}}}{R} = -\frac{B_0 w v_0}{R}.$$

The minus sign reminds us that the induced current tries to counteract the changing flux. Here, the into-the-page flux increases as the rectangle enters the field. So, the induced current "wants" to prevent the flux from increasing, by creating an out-of-the-page magnetic field. Using the right-hand rule shortcut, you can confirm that the induced current must flow counterclockwise, in order to generate an out-of-the-page field at points inside the rectangle.

Below, I'll show you another way to apply Lenz' law, i.e., another way to figure out that the induced current flows counterclockwise.

(b) From part (a), we know the counterclockwise current flowing around the rectangle. So, we can calculate the force on each of the four sides, and then add them up.

Arrows show direction of the induced current.

Side 3 experiences no force, since it lies entirely outside the magnetic field. By contrast, sides 2 and 4—or more precisely, length x of those sides—get pushed by the magnetic field. But the forces on those two sides point in opposite directions. To see why, recall from Chapter 39 that the infinitesimal force acting on a tiny piece of wire is

$$d\mathbf{F} = i\, d\mathbf{l} \times \mathbf{B}.$$

Let's first find the direction of the force acting on side 4, by applying the right-hand rule to that cross product. Along side 4, $d\mathbf{l}$ points leftward, and $\mathbf{B}$ points into the page. So, point your fingers leftward, and then curl your fingertips into the page. Your thumb points down the page. That's the direction of the force on side 4. By similar reasoning, the force on side 2 points up the page.

Since sides 2 and 4 both carry the same current, both "feel" the same magnetic field, and both extend the same distance into the field, it follows from $d\mathbf{F} = i\, d\mathbf{l} \times \mathbf{B}$ that both sides experience equal magnetic forces. And because those forces point in opposite directions, they cancel. Therefore, the net magnetic force on the rectangle stems entirely from side 1.

To find the direction of $\mathbf{F}_1$, point your fingers down the page (with $d\mathbf{l}$), and curl your fingertips into the page (with $\mathbf{B}$). Your thumb points rightward. That's the direction of $\mathbf{F}_1$.

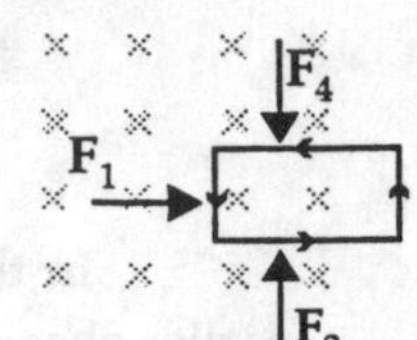

The magnetic forces on sides 2 and 4 cancel out.

To find the magnitude of this force, integrate $d\mathbf{F}$ over all the infinitesimal "pieces" of side 1. For all of those pieces, $d\mathbf{l}$ is perpendicular to $\mathbf{B}$, and hence the magnitude of the cross product reduces to a regular product. Furthermore, i and B are constant along side 1. So,

$$F_1 = \int_{\text{side 1}} i|d\mathbf{l} \times \mathbf{B}| = \int_{\text{side 1}} idlB = iB_0 \int_{\text{side 1}} dl = iB_0 w,$$

since side 1 has length w. Here, "i" denotes the induced current. In part (a), we found that $i_{\text{induced}} = B_0 w v_0 / R$. So, the rectangle feels a net magnetic force

$$F_1 = i_{\text{induced}} B_0 w = \left(\frac{B_0 w v_0}{R}\right) B_0 w = \frac{B_0^2 w^2 v_0}{R}.$$

This magnetic force points rightward, opposite to the direction in which Minh pushes. But that's no surprise. From Lenz' law, we *knew* ahead of time that the magnetic force would "resist" Minh's push. Remember, according to Lenz' law, the induced current tries to counteract the changing flux. Here, the flux changes because Minh pushes the rectangle deeper into the magnetic field. Therefore, the induced current flows so that the magnetic force on it will counteract Minh's force.

This realization gives us an alternate way to figure out whether the induced current flows clockwise or counterclockwise. Here, the current must flow down side 1 (instead of up side 1), to ensure that $\mathbf{F}_1$ opposes Minh's push.

(c) At first glance, you might think Minh needs to push harder than F_1, in order to keep the rectangle moving leftward. In other words, you might think the push force needs to overcome the magnetic force (F_1). But Minh pushes just hard enough to keep the rectangle moving at *constant* velocity. Therefore, the rectangle has no acceleration. Since $\mathbf{F}_{\text{net}} = m\mathbf{a}$, the net force on the rectangle must be zero. So, Minh's push is exactly big enough to cancel the magnetic force. Since $F_{\text{net}} = F_{\text{Minh}} - F_1 = 0$,

$$F_{\text{Minh}} = F_1 = \frac{B_0^2 w^2 v_0}{R}.$$

By Newton's 2nd law, a nonzero net force is needed to *initiate* an object's motion, but not to *maintain* an object's motion at constant velocity.

(d) Here, electric power dissipates as heat in the wires comprising the rectangle. From Chapter 38, recall that the power (energy per time) dissipated in any circuit element is $P = Vi$. Since the rectangle behaves as a resistor, $V = iR$ by Ohm's law. Therefore,

$$P = Vi = (iR)i = i^2 R.$$

We already calculated the induced current in part (a). It's $i_{\text{induced}} = B_0 w v_0/R$. So, the rectangle dissipates power

$$P = I^2 R = \left(\frac{B_0 w v_0}{R}\right)^2 R = \frac{B_0^2 w^2 v_0^2}{R}.$$

According to this formula, if Minh pushes the rectangle into the magnetic field twice as quickly, the dissipated power quadruples.

(e) I'll first solve this the "long" way. Then, in part (f), I'll show you a shortcut that stems from a deep physical insight.

Power is work (i.e., energy) per time. And work is force integrated over distance. So,

$$P = \frac{dW}{dt}$$

$$= \frac{\mathbf{F}\cdot d\mathbf{x}}{dt} \qquad \left[\text{since } W = \int \mathbf{F}\cdot d\mathbf{x}, \text{and hence } dW = \mathbf{F}\cdot d\mathbf{x}\right]$$

$$= \mathbf{F}\cdot\mathbf{v} \qquad [\text{since } \mathbf{v} = d\mathbf{x}/dt].$$

Because $\mathbf{F}_{\text{Minh}}$ and $\mathbf{v}_0$ both point leftward, the dot product reduces to a regular product:

$$P_{\text{Minh}} = F_{\text{Minh}} v_0 = \frac{B_0^2 w^2 v_0}{R} v_0 = \frac{B_0^2 w^2 v_0^2}{R},$$

where I used our part (c) answer for Minh's force.

(f) Parts (d) and (e) yield the same answer. And it's no coincidence. Energy conservation *demands* that those two answers agree. Here's why.

As Minh pushes the rectangle into the magnetic field, she does work on it. In other words, she gives the rectangle energy. This energy can't just disappear; it must convert into another form. Usually, when you push something, some or all of your work energy converts into kinetic energy, and the object speeds up. For instance, if you perform 5 joules of work pushing a frictionless ice block across the floor, it ends up with 5 joules of kinetic energy.

In this problem, none of Minh's work converts into kinetic energy, because the rectangle does not speed up. Nor does the rectangle gain potential energy. Therefore, *all* of her work must turn into heat or some other dissipative form of energy. For instance, if Minh performs 5 joules of work pushing the rectangle into the magnetic field, then 5 joules of heat must dissipate in the resistive wires. Symbolically,

$$W_{\text{Minh}} = Heat_{\text{dissipated in resistive wires}}.$$

Therefore, the *rate* at which Minh does work ($P_{\text{Minh}} = dW_{\text{Minh}}/dt$) must equal the *rate* at which heat dissipates in the wires, $P = d(Heat)/dt$.

We now see why the power exerted by Minh equals the power dissipated in the resistors. If you figure out one of these powers, you've automatically figured out the other.

QUESTION 41-4

OBLIQUE VIEW

Current flows up the wires, as indicated by black arrowheads. Above the cardboard plane, magnetic field lines point rightward. Below the plane, they point leftward.

Consider a huge square sheet of cardboard, of side length s, completely covered with N uniformly-spaced current-carrying wires, as drawn here. In this drawing, $N = 9$. But actually, N is huge, and the wires are closely spaced, just like in a solenoid. Each wire is connected to the same power source (not drawn here).

At points close to the cardboard, the magnetic field lines (drawn here in gray) point perpendicular to the wire, but parallel to the plane of the cardboard sheet.

A small circular ring of radius r and resistance R_0 is centered a distance D above the cardboard sheet, where $D \ll s$. The plane of this ring is perpendicular to the magnetic field lines. In this picture, the ring "faces" rightward. If you've learned about inductance, neglect the "self-inductance" of this ring.

The current in the wires is steadily increased from 0 to I_0 in time T. What current gets induced in the ring during this interval? Hint: Using Ampere's law, begin by figuring out the field produced when current I_0 flows in the wires. The key is to draw a "good" Amperian loop.

ANSWER 41-4

Start by thinking physically about what happens. As the current in the wires increases, the magnetic field increases. Consequently, the magnetic flux through the ring increases. Because the flux through the ring *changes*, a voltage (EMF) gets induced. This voltage makes current flow around the ring.

According to Faraday's law, $\mathcal{E}_{\text{induced}} = -d\Phi_B/dt$. So, we must find the rate of change of flux. To do so, we can figure out the initial and final flux, and work from there. And to find the flux, we need to know the magnetic field produced by the cardboard sheet of wires.

Let me spell out this strategy in more detail.

1. Find the magnetic field—and the corresponding magnetic flux through the ring—when the current is I_0.
2. Using the given information about how quickly the current changes from 0 to I_0, figure out the rate of change of the magnetic flux, $d\Phi_B/dt$.
3. Given this rate of change of flux, use Faraday's law to write down the induced voltage, and Ohm's law to find the corresponding induced current.

As we dive deep into details, please come up for air periodically, to see how each step fits into the overall strategy.

Subproblem 1: Find the magnetic field produced by this "infinite sheet" of wires.

Because the ring is very close to the cardboard plane of wires, we can approximate the cardboard plane as infinite. This approximation introduces sufficient symmetry that Ampere's law allows us to find the magnetic field.

According to Ampere's law, the line integral of the magnetic field around an imaginary "Amperian loop" is proportional to the current enclosed by that loop:

$$\oint \mathbf{B}\cdot d\mathbf{s} = \mu_0 I_{\text{encl}}.$$

This law, though always true, can't help us solve for the magnetic field unless we can pull B outside the line integral. In other words, B must be uniform along at least some legs of the Amperian loop.

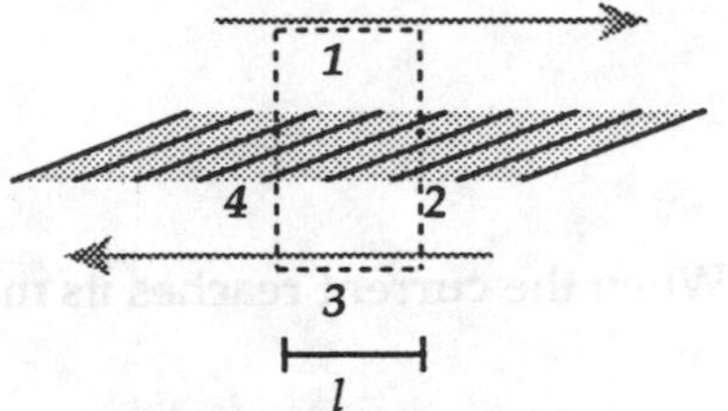

This dashed Amperian loop, as drawn here, encloses only 2 wires.

For this reason, I'll choose the rectangular loop drawn here. this rectangle extends equally far above and below the cardboard plane. The black dots indicate where sides 2 and 4 pierce through the cardboard. This rectangle is an imaginary Amperian loop, not a wire. We can integrate around the loop by integrating along all four sides:

$$\oint \mathbf{B}\cdot d\mathbf{s} = \int_1 \mathbf{B}\cdot d\mathbf{s} + \int_2 \mathbf{B}\cdot d\mathbf{s} + \int_3 \mathbf{B}\cdot d\mathbf{s} + \int_4 \mathbf{B}\cdot d\mathbf{s}.$$

Since sides 2 and 4 are perpendicular to the magnetic field, $\mathbf{B}\cdot d\mathbf{s} = 0$. Therefore, those two sides don't contribute to the line integral. By contrast, if I walk clockwise around the loop, then sides 1 and 3 both point *with* the field lines. So, $\mathbf{B}\cdot d\mathbf{s}$ reduces to a positive regular product. Because side 1 consists of points equidistant from the cardboard, the field strength along side 1 is uniform. The same goes for side 3. Furthermore, since I drew those sides the same distance from the cardboard, B along side 1 equals B along side 3. Therefore,

$$\begin{aligned}\oint \mathbf{B}\cdot d\mathbf{s} &= \int_1 \mathbf{B}\cdot d\mathbf{s} + \int_2 \mathbf{B}\cdot d\mathbf{s} + \int_3 \mathbf{B}\cdot d\mathbf{s} + \int_4 \mathbf{B}\cdot d\mathbf{s} \\ &= B\int_1 ds \;+\; 0 \;+\; B\int_3 ds \;+\; 0 \\ &= Bl \;+\; 0 \;+\; Bl \;+\; 0 \\ &= 2Bl,\end{aligned}$$

where I've given my Amperian loop arbitrary length l, where $l \le s$.

A common mistake is to think that $\int_1 \mathbf{B}\cdot d\mathbf{s}$ cancels $\int_3 \mathbf{B}\cdot d\mathbf{s}$, because the fields along those two sides point in opposite directions. But remember, taking a line integral corresponds to walking around the Amperian loop. Suppose you walk clockwise, starting from the upper left-hand corner. Along side 1, you're walking rightward, *with* the field. Therefore, the dot product $\mathbf{B}\cdot d\mathbf{s}$ is positive. Along side 3, you're walking leftward, again *with* the field. So, the dot product $\mathbf{B}\cdot d\mathbf{s}$ is again positive. The field switched direction, but so did you! For this reason, the contributions from sides 1 and 3 add up, instead of canceling.

Let me step back for a minute to get my bearings. We're using Ampere's law to find the magnetic field produced by the plane of wires, in order to find the flux through the ring. So far, we've drawn a well-chosen Amperian loop and simplified the Amperian line integral, $\oint \mathbf{B}\cdot d\mathbf{s}$. To finish finding B, we must figure out the current enclosed.

In the drawing, the Amperian loop encloses only $N_{\text{encl}} = 2$ wires. In general, the Amperian loop encloses length l of the cardboard plane. How many wires are contained in length l? Well, the cardboard plane contains N wires uniformly spread out over length s. So, the "linear wire density," i.e., the number of wires per unit length, is N/s. Therefore, length l of the plane contains this many wires:

$$N_{\text{encl}} = \frac{\text{wires}}{\text{length}} \times (\text{length of plane enclosed})$$

$$= \frac{N}{1} l.$$

When the current reaches its maximum value, each of those enclosed wires carries current I_0. Therefore,

$$I_{\text{encl}} = N_{\text{encl}} I_0$$

$$= \frac{NlI_0}{s}.$$

Given these results, we can apply Ampere's law:

$$\oint \mathbf{B} \cdot d\mathbf{s} = \mu_0 I_{\text{encl}}.$$

$$2Bl = \mu_0 \frac{NlI}{s}.$$

Cancel the l's and isolate B to get

$$B = \frac{\mu_0 N I_0}{2s}.$$

Notice that B does not depend on D, the distance from the cardboard plane. If you're sufficiently close to the plane so that it "seems" effectively infinite from your perspective ($D \ll s$), then the strength of the magnetic field does not decrease with distance. This makes sense, given your previous experiences with electrostatics. Recall that an infinite plane of charge generates a uniform electric field. We've found a magnetic analog of the infinite charged plane!

Given this field, we can now calculate the flux through the ring. Since the magnetic field is uniform, we can pull it outside the flux integral. Furthermore, since the ring "faces" the same direction in which the magnetic field points, the dot product $\mathbf{B} \cdot d\mathbf{A}$ reduces to a regular product. So,

$$\Phi_B = \int_{\text{ring}} \mathbf{B} \cdot d\mathbf{A} = B \int_{\text{ring}} dA = BA_{\text{ring}} = \left(\frac{\mu_0 N I_0}{2s}\right)(\pi r^2),$$

where in the last step I used our expression for B, along with $A_{\text{circle}} = \pi r^2$.

At this point, we've found $\Phi_B(I_0)$, the magnetic flux through the ring when the wires on the cardboard plane carry current I_0. Keep in mind that I_0 refers to the current in the wires, not the induced current in the ring. To find the induced voltage (and current) in the ring, we need to know the rate at which this flux *changes*. Then we can use Faraday's law.

Subproblem 2: Find the rate of change of this flux, due to the changing current.

The current in the wires increases at a steady rate from 0 to I_0, in time T. When $I = 0$, the wires generate no field, and hence the ring has no flux. So, the flux increases steadily from 0 to $\Phi_B(I_0)$ in time T. Therefore, during that interval, the flux changes at rate

$$\frac{d\Phi_B}{dt} = \frac{\Delta\Phi_B}{\Delta t} = \frac{\Phi_B(I_0) - 0}{T} = \frac{\mu_0 N I_0 \pi r^2}{2sT},$$

where in the last step, I used our expression for $\Phi_B(I_0)$ from subproblem 1 above.

By the way, you could also calculate $d\Phi_B/dt$ by using the chain rule, $\frac{d\Phi_B}{dt} = \frac{d\Phi_B}{dI}\frac{dI}{dt}$. I'll use that technique extensively in the next chapter. Notice that our answer for $d\Phi_B/dt$ is proportional to $dI/dt = I_0/T$, the rate of change of the current through the wires. The quicker the current in the wires increases, the quicker the flux through the ring changes.

Subproblem 3: Find the current induced in the ring.

Given $d\Phi_B/dt$, this is straightforward. By Faraday's law, the voltage induced in the ring is given by the rate of change of flux through the ring:

$$\begin{aligned}\mathcal{E}_{\text{induced}} &= -\frac{d\Phi_B}{dt} \\ &= -\frac{\mu_0 N I_0 \pi r^2}{2sT}.\end{aligned}$$

The minus sign reminds us that the induced current flows so as to counteract the changing flux through the ring. More on that in a minute. But first, let me calculate the induced current, using Ohm's law:

$$i_{\text{induced}} = \frac{\text{voltage}}{\text{resistance}} = \frac{\mathcal{E}_{\text{induced}}}{R_0} = \frac{\mu_0 N I_0 \pi r^2}{2sTR_0}.$$

OK, now I'll figure out whether that current flows clockwise or counterclockwise. According to Lenz' law, the induced current tries to "resist" any changes in the magnetic flux. Here, the rightward magnetic field through the ring increases, as the current in the wires increases. Therefore, to oppose this change, the induced current in the ring "wants" to create a field pointing leftward.

Clockwise induced current

Using the right-hand rule shortcut, you can confirm that the induced current must flow clockwise, in order to create a leftward field at points inside the ring, I'll briefly review how the shortcut works. Assuming the current flows clockwise, pick any piece of the ring, and point your right thumb in the direction of current flow. When you curl your fingertips, they represent the field lines generated by that piece. Here, when you curl your fingertips towards points inside the ring, your fingertips point leftward. This proves that a clockwise current around the ring produces a leftward field, at points inside the ring. A counterclockwise current would have generated a rightward field.

QUESTION 41-5

(*Very hard*) A platinum wire rectangle of length $l = 0.20$ m, width $w = 0.10$ m, and resistance $R = 1.0 \times 10^{-4}\ \Omega$ sits inside a uniform magnetic field of unknown strength. The field points out of the page.

Over time $T = 3.0$ s, the magnetic field steadily decreases to one third its initial value. During that 3-second interval, the current around the rectangle is observed to be $i = 0.50$ A.

What was the initial strength of the magnetic field? As always, solve symbolically before plugging in numbers.

ANSWER 41-5

Current gets induced because the magnetic flux through the rectangle changes. Usually in this kind of problem, you're given information about the changing magnetic field, and asked to calculate the induced current. Here, you're given the induced current, and asked to figure out something about the magnetic field. So, you can use the same concepts and formulas as always. You can even use the same old strategy, by "pretending" you know the initial field B_0, solving for the induced current in terms of B_0, and then inverting your equation to solve for B_0 in terms of i (instead of vice versa). Or, you can "work backwards," starting with the induced current. Let me demonstrate what I mean.

Since the induced current is $i = 0.50$ A, we can find the induced voltage (EMF) using Ohm's law:

Ohm's law

$$\mathcal{E}_{\text{induced}} = iR.$$

By Faraday's law, this induced voltage equals the rate of change of flux, $d\Phi_B/dt$. Here, since the field and flux change at a constant rate, we can replace the derivative $d\Phi_B/dt$ with $\Delta\Phi_B/\Delta t$. So, Faraday's law says

Faraday's law

$$\frac{\Delta\Phi_B}{\Delta t} = -\mathcal{E}_{\text{induced}}$$

$$\frac{\Delta\Phi_B}{T} = iR.$$

We could easily solve for $\Delta\Phi_B$ in terms of known physical quantities (i, R, and T). But we're trying to find the initial magnetic field strength B_0, not the change in flux. Fortunately, we can express $\Delta\Phi_B$ in terms of B_0. That is, we can relate the change in flux to the initial strength of the magnetic field. After doing so, we can immediately solve Faraday's law for B_0.

Subproblem: Relate the change in flux to the initial magnetic field strength.

The field starts at (unknown) value B_0 and ends up one third as strong, $B_f = B_0/3$. So, the magnetic field changes by two thirds of its initial strength:

$$\Delta B = B_f - B_0 = \frac{1}{3}B_0 - B_0 = -\frac{2}{3}B_0.$$

Since the field is uniform and pierces "head-on" through the rectangle, the magnetic flux is simply $\Phi_B = BA_{\text{rectangle}} = Blw$. Therefore, the flux *changes* by

$$\Delta\Phi_B = (\Delta B)lw$$
$$= \left(-\frac{2}{3}B_0\right)lw.$$

End of subproblem.

At this stage, I've related the change in flux to the initial magnetic field strength. Substitute this expression for $\Delta\Phi_B$ into the above Faraday's law equation to get

$$\frac{\Delta\Phi_B}{\Delta t} = -\mathcal{E}_{\text{induced}}$$
$$\frac{\Delta\Phi_B}{\Delta t} = -iR$$
$$\frac{(-\frac{2}{3}B_0)lw}{T} = -iR,$$

which we can solve for B_0 to get

$$B_0 = \frac{3iRT}{2lw} = \frac{3(0.50\text{ A})(1.0\times10^{-4}\,\Omega)(3.0\text{ s})}{2(0.20\text{ m})(0.10\text{ m})} = 0.011\text{ T}$$

As mentioned above, you could also solve for B_0 by working forward (instead of backward), provided you "drag along" the unknown B_0 through each step, until you relate it to the known induced current.

QUESTION 41-6

Coordinate axes

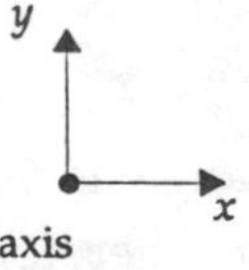

z-axis points out of page.

A complicated current distribution creates a magnetic field with z-component

$$B_z = \frac{C}{x^2},$$

where the x-direction rightward, the y-direction is up-the-page, and the z-direction is out of the page. Here, C is a constant. And yes, I meant to write "x^2," not "z^2," in the denominator.

A metal wire square of side-length s and resistance R is sitting in this magnetic field, with its lower left corner at $x = 2s$, $y = 0$. The square lies in the x-y plane (i.e., the plane of the page).

(a) What are the standard SI units of the constant C?

(b) What is the magnetic flux through the square? Hint: The magnetic field is *not* uniform over the entire square. Draw a picture.

(c) If the square gets pushed in the y-direction at constant speed v, what voltage gets induced in the square, immediately after the person starts pushing it? Hint: As you "walk" in the y-direction, does the magnetic field strength increase, decrease, or stay the same?

(d) (*Very hard*) If the square is pushed in the x-direction at constant speed v, what voltage gets induced in the square, immediately after the person starts pushing it?

(e) (*Hard*) Immediately after the person starts pushing the square in the x-direction at speed v, what magnetic force acts on the square? If you didn't solve part (d), call the answer "$\mathcal{E}$," and solve this problem in terms of $\mathcal{E}$.

(f) (*Extra credit for masochists and physics majors*) Assuming the person starts pushing the square at $t = 0$, write an expression for the current in the square as a function of time.

ANSWER 41-6

(a) $B = C/x^2$, where B is a magnetic field (units of tesla) and x is a length (units of meters). Solve for C to get $C = Bx^2$, which has units of tesla·meter². So, C has units of magnetic field times area. But field times area is flux! In SI units, a weber of magnetic flux is one tesla·meter². On this question, either "webers" or "tesla·meter²" earns full credit.

(b) To help yourself visualize what's happening, always draw a picture, especially if the problem doesn't provide one.

The magnetic field here points out of the page. Following the usual convention, I'll represent "out of the page" by arrowheads (as opposed to arrow tails). Arrowheads, viewed head on, look like dots.

The magnetic field strength falls off as $1/x^2$, where x is the distance along the x-axis. So, the field gets weaker as we move rightward. But it has no y-dependence. For instance, the field at $x = 2$, $y = 3$ has the same strength as the field at $x = 2$, $y = 100$. Put another way, if you walk "up the page," the field stays the same.

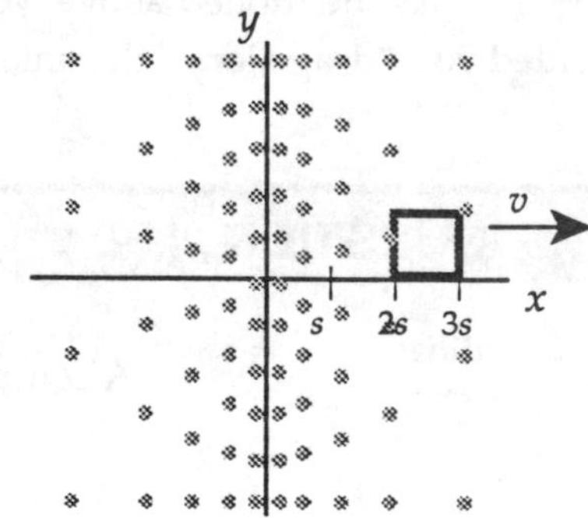

We can represent all this by drawing the "field dots" as getting less dense when we move rightward, but keeping the same density as we move up the page.

This picture highlights the following crucial insight: The field near the left side of the square ($x = 2s$) is stronger than the field near the right side ($x = 3s$). The field isn't uniform over the whole area of the square. Therefore, when calculating $\Phi_B = \int \mathbf{B} \cdot d\mathbf{A}$, we can't pull $\mathbf{B}$ outside the integral. We'll have to evaluate the integral more carefully.

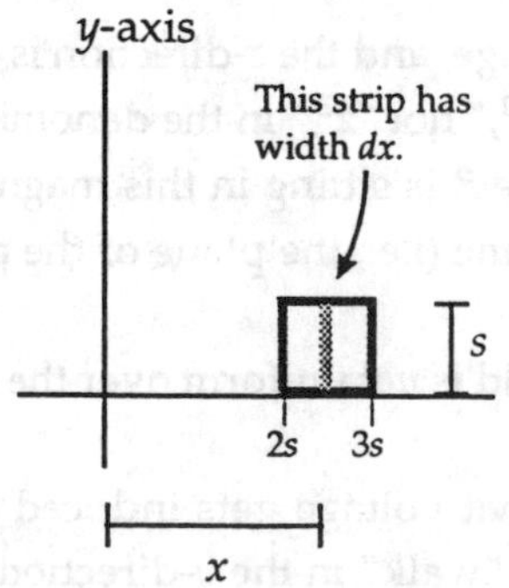

To do so, exploit the fact that the field doesn't change as you move in the y-direction. B is a function only of x, not of y. Therefore, within any thin vertical strip of the square, the field is uniform. This suggests an integration strategy. We can divide the square into infinitesimally thin vertical strips; find the infinitesimal magnetic flux through a given strip; and add up those flux contributions by integrating.

An arbitrary strip, of height s and width dx, has infinitesimal area $dA = s\,dx$. The z-directed field at all points in the strip is $B_z = C/x^2$. So, the infinitesimal flux through the strip is

$$d\Phi_B = B_z dA = \left(\frac{C}{x^2}\right)(s\,dx) = \frac{Cs}{x^2}dx.$$

(Only the z-component of the field "contributes" to the flux through a rectangle that faces the z-direction, as this one does. Mathematically, since $d\mathbf{A}$ points in the z-direction, the dot product in $d\Phi_B = \mathbf{B}\cdot d\mathbf{A}$ "picks out" the z-directed field.)

OK, now we can add up the $d\Phi_B$'s from all the strips, by integrating. The leftmost strip is at $x = 2s$, while the rightmost one is at $x = 3s$. So, those are the limits of integration. Crucially, you can't say the integral goes from 0 to s (or something like that), because the "x" in $B = C/x^2$ refers to the distance from the y-axis, not the distance from the left side of the square. See the above diagram.

Integrating over all the strips in the square gives

$$\begin{aligned}\Phi_B = \int d\Phi_B &= \int_{\text{square}} B\,dA \\ &= \int_{2s}^{3s} \frac{Cs}{x^2}\,dx \\ &= Cs\left(-\frac{1}{x}\right)\Bigg|_{2s}^{3s} \\ &= Cs\left(-\frac{1}{3s}+\frac{1}{2s}\right) \\ &= C\left(\frac{1}{2}-\frac{1}{3}\right) \\ &= \frac{C}{6}.\end{aligned}$$

That's the flux through the square, when its left edge is at $x = 2s$.

By the way, since C has units of flux (i.e., webers), we knew that Φ_B would work out to be some multiple or fraction of C.

(c) As the above diagram tries to show, when the square moves up the page (in the y-direction), the magnetic field through it doesn't get stronger or weaker. The field stays the same strength. Mathematically, that's because B depends only on x, not on y.

Therefore, as the square moves up the page, the magnetic flux through it stays the same. The *change* in flux is $d\Phi_B/dt = 0$. So, according to Faraday's law, no voltage gets induced.

This answer reminds us that, no matter how big or small the flux through the square happens to be, voltage gets induced only when that flux *changes*.

(d) As the square moves rightward, the magnetic field through it gets weaker and weaker. Therefore, the flux through the square decreases. Since the flux changes, voltage gets induced. Since $\mathcal{E}_{\text{induced}} = -d\Phi_B/dt$, we must find the rate at which the flux changes, immediately after the person starts pushing.

Well, in part (a), we found the initial flux through the square: $\Phi_B = C/6$. So, by differentiating that expression with respect to t, we should obtain the rate of change of flux, $d\Phi_B/dt$. But the derivative of a constant is zero: $\frac{d}{dt}\left(\frac{C}{6}\right) = 0$. Something just went wrong.

Digression: Where's the calculus error? To understand what went wrong, consider a car whose position (in meters) is given by $x = 2t^2$, where t in is seconds. How would we find its velocity at time $t = 3$ s? Well, its position at that time is obviously $x = 2(3)^2 = 18$ m. And $v = dx/dt$. So, the velocity at $t = 3$ s is given by $d(18)/dt = 0$, right? Wrong. My mistake was plugging in a specific value for time (namely, $t = 3$ s) *before* differentiating. When differentiating a physical quantity such as x, the quantity must be expressed in terms of variables (such as t), not in terms of specific values of that variable (such as 3 seconds). As you learned in calculus, always substitute in the specific value *after* differentiating. So here, you would differentiate x with respect to t to get $v = \frac{dx}{dt} = \frac{d(2t^2)}{dt} = 4t$. Only then would you plug in the specific value for t, such as $t = 3$ s.

End of digression.

Similarly, our part (b) expression for flux plugs in specific values for the square's position: the left side is at $x = 2s$, and the right side is at $x = 3s$. These specific values appear as limits of the integral. In order to differentiate Φ_B, we must express it as a general function of x, without plugging in specific values for x. Only *after* differentiating may we substitute in the specific values for the square's initial position.

To write Φ_B as a general function of position, let "X" denote the x-coordinate of the left-hand side of the square, as drawn here. The right-hand side is at X + s. Now repeat the reasoning of part (b), with "X" and "X+ s" as our limits of integration:

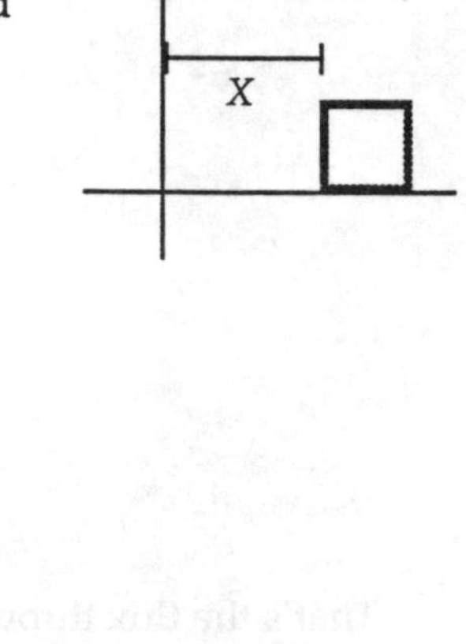

$$\Phi_B = \int d\Phi_B = \int_{\text{square}} B\,dA$$

$$= \int_X^{X+s} \frac{Cs}{x^2}\,dx$$

$$= Cs\left(-\frac{1}{x}\right)\Big|_X^{X+s}$$

$$= Cs\left(-\frac{1}{X+s} + \frac{1}{X}\right).$$

In part (b), we plugged in the specific value X = 2s. But here, I won't plug in that specific value until *after* differentiating Φ_B with respect to time. To differentiate, use the chain rule:

$$\frac{d\Phi_B}{dt} = \frac{d\Phi_B}{dX}\frac{dX}{dt} = Cs\left[\frac{1}{(X+s)^2}\frac{dX}{dt} - \frac{1}{X^2}\frac{dX}{dt}\right]$$

$$= Cs\left[\frac{1}{(X+s)^2} - \frac{1}{X^2}\right]\frac{dX}{dt}$$

$$= Cs\left[\frac{1}{(X+s)^2} - \frac{1}{X^2}\right]v.$$

In the last step, I used a crucial physical insight. Since X is the position of a side of the square, the rate at which X changes is the square's velocity: $v = dX/dt$.

At this stage, we have a *general* expression for the rate of change of flux through the square, as a function of the square's position. Here, we want the rate of change of flux when the left side of the square is at X = 2s. Substituting in X = 2s gives

$$\left.\frac{d\Phi_B}{dt}\right|_{2s} = Cs\left[\frac{1}{(2s+s)^2} - \frac{1}{(2s)^2}\right]v$$
$$= Cs\left[\frac{1}{9s} - \frac{1}{4s^2}\right]v$$
$$= \frac{5Cv}{36s}.$$

The minus sign indicates that the flux decreases as the square moves rightward.

By Faraday's law, $\mathcal{E}_{\text{induced}} = -d\Phi_B / dt = \frac{5Cv}{36s}$.

(e) I see two equally good ways of solving this. We can figure out the force directly. Or, we can make a clever argument using power and energy conservation. I'll demonstrate both methods.

Method 1: Force reasoning

First, I'll figure out the current flowing around the square. Then I'll calculate the magnetic force acting on each side. By summing up forces, I'll obtain the overall magnetic force.

By Ohm's law, the induced current (with the square at X = 2s) is

$$i = \frac{\mathcal{E}_{\text{induced}}}{R} = \frac{5Cv}{36sR}.$$

According to Lenz' law, this current tries to counteract the changing flux. As the square moves rightward, the out-of-the-page magnetic field through it gets smaller and smaller. To oppose this decreasing flux, the induced current creates a magnetic field pointing out of the page. Using the right-hand rule shortcut, you can confirm that current must flow counterclockwise around the square, to create an out-of-the-page magnetic field (at points inside the square).

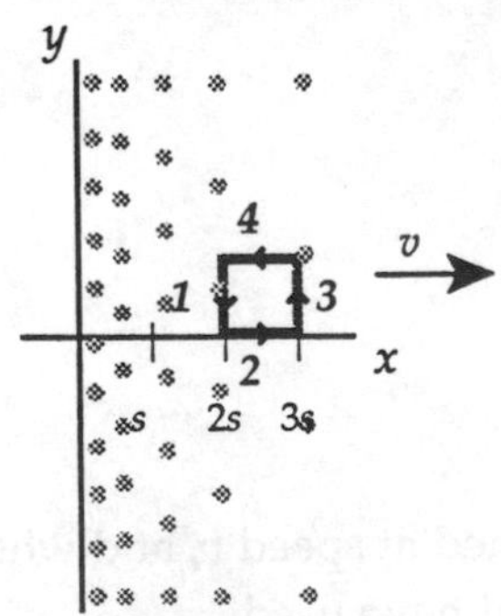

The induced current flows counterclockwise.

Now that we know the magnitude and direction of the current, we can calculate the magnetic force on each side of the square. As shown in Chapter 39, an infinitesimal piece of wire feels an infinitesimal magnetic force

$$d\mathbf{F} = i\ d\mathbf{l} \times \mathbf{B}.$$

Integrating $d\mathbf{F}$ over an entire side gives you the force on that side.

Let's start with side 1. To find the direction of the force on that side, use the right-hand. Point your fingers down the page with $id\mathbf{l}$, then curl your fingertips out of the page with **B**. Your thumb points leftward. That's the direction of $\mathbf{F}_1$, the force on side 1. By similar reason ing, $\mathbf{F}_3$ points rightward. But those forces *don't* cancel, because side 1 feels a bigger magnetic field than side 3 does.

To calculate F_1, integrate $d\mathbf{F}_1 = i\ d\mathbf{l} \times \mathbf{B}$ over all the "pieces" of side 1. The integral simplifies for two reasons. Since $d\mathbf{l}$ and **B** point at right angles, the magnitude of the cross product reduces to a regular product: $|d\mathbf{l} \times \mathbf{B}| = (dl)B$. And since all the pieces of side 1 have the same x-coordinate, namely $x = 2s$, the magnetic field ($B = C/x^2$) is uniform over side 1. Therefore, we can pull it outside the integral:

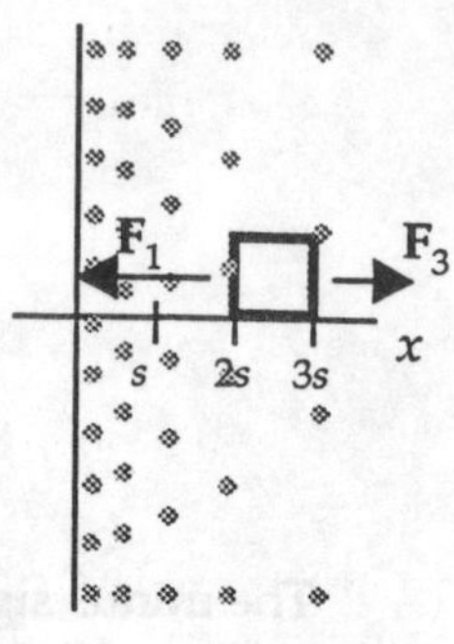

The magnetic force on side 1 "beats" the force on side 3.

$$F_1 = \int_1 i\,dlB = iB\int_1 dl = iBs = \left(\frac{5Cv}{36sR}\right)\left[\frac{C}{(2s)^2}\right]s,$$

where in the last step, I used our above expression for the induced current, along with the value of the magnetic field along side 1 (at $x = 2s$). This force points leftward.

For side 3, every step of reasoning is the same, except the field strength is $B = \frac{C}{(3s)^2}$ instead of $\frac{C}{(2s)^2}$. Therefore,

$$F_3 = \left(\frac{5Cv}{36sR}\right)\left[\frac{C}{(3s)^2}\right]s.$$

What about sides 2 and 4? The forces on those sides cancel. To see why, consider an arbitrary piece of side 2, colored gray in this drawing. Directly above it is a corresponding piece of side 4. Since those two pieces carry the same current, and since they feel the same magnetic field (because they're at the same x), those two pieces experience the same magnetic force: $dF_2 = dF_4$. But the currents in sides 2 and 4 flow in opposite directions. Therefore, the magnetic forces point in opposite directions. So, the forces on these two gray pieces cancel out. Since every little chunk of side 2 has a "partner" on side 4, the force on side 2 cancels the force on side 4.

For this reason, the overall magnetic force on the square points leftward, with value

$$\begin{aligned} F_{\text{mag}} &= F_1 - F_3 \\ &= \frac{5Cv}{36sR}\frac{C}{(2s)^2}s - \frac{5Cv}{36sR}\frac{C}{(3s)^2}s \\ &= \frac{5C^2v}{36s^2R}\left[\frac{1}{4} - \frac{1}{9}\right] \\ &= \left(\frac{5}{36}\right)^2\frac{C^2v}{s^2R}. \end{aligned}$$

I just solved for the magnetic force on the square when it's pushed at speed v, and when its left side is still at $x = 2s$. Let me *briefly* outline another method you could have used.

Method 2: Energy conservation and power

I'll rush through this reasoning. For a fuller presentation, see question 41-3f.

Because the square gets pushed at constant speed, it doesn't gain kinetic energy. Therefore, all the work done on the square by the person must convert into the heat dissipated as induced current flows around the square. In fact, energy conservation demands that the rate at which work gets done on the square equals the rate at which energy dissipates as heat. Since the rate of energy exchange is *power*, we can write this physical insight as

$$P_{push} = P_{elec},$$

where P_{push} denotes the power exerted by the person pushing the square, and P_{elec} denotes the electrical power, i.e., the rate of heat dissipation due to the current.

As we saw in question 41-3e, since the person's push force must cancel the magnetic force, $P_{push} = F_{push}v = F_{mag}v$. And as shown in Chapter 39, the electrical power dissipated in a resistor is $P_{elec} = Vi$. Here, the voltage comes from the induced EMF, $\mathcal{E}_{induced}$, which we found in part (d). So, we can now use energy conservation to find F_{mag}:

$$\begin{aligned} P_{push} &= P_{elec} \\ F_{mag}v &= \mathcal{E}_{induced}i \\ &= \left(\frac{5Cv}{36s}\right)\left(\frac{5Cv}{36sR}\right) \\ &= \left(\frac{5}{36}\right)^2 \frac{C^2v^2}{s^2R}. \end{aligned}$$

Divide through by v to get

$$F_{mag} = \left(\frac{5}{36}\right)^2 \frac{C^2v}{s^2R},$$

the same answer obtained above using method 1. Cool, huh?

(f) As shown in part (d),

$$\mathcal{E}_{induced} = Cs\left[\frac{1}{X^2} - \frac{1}{(X+s)^2}\right]v,$$

where X denotes the x-coordinate of the left side of the square. So,

$$i_{induced} = \frac{\mathcal{E}_{induced}}{R} = \frac{Cs}{R}\left[\frac{1}{X^2} - \frac{1}{(X+s)^2}\right]v.$$

That's the induced current as a function of position. But we want the induced current as a function of time. Therefore, we need to find the square's position as a function of time.

But that's not hard, because the square moves at constant velocity. Its acceleration is 0. And its initial position is $X = 2s$. So, a kinematic formula from last semester immediately gives us

$$\begin{aligned} X &= X_0 + v_0t + \frac{1}{2}at^2 \\ &= 2s + vt + 0. \end{aligned}$$

To obtain the induced current as a function of time, simply substitute this expression for $X(t)$ into the above equation for i_{induced}:

$$i_{\text{induced}} = \frac{Cs}{R}\left[\frac{1}{X^2} - \frac{1}{(X+s)^2}\right]v.$$

$$= \frac{Cs}{R}\left[\frac{1}{(2s+vt)^2} - \frac{1}{(3s+vt)^2}\right]v.$$

QUESTION 41-7

PENDULUM AT EQUILIBRIUM

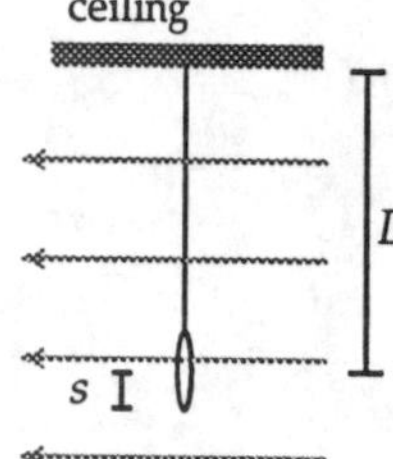

A pendulum consists of a small circular metal ring attached to a massless nonconducting rigid rod. The pendulum's length, from the top of the rod to the center of the ring, is L. The ring has mass M, resistance R, and radius s, where $s \ll L$.

The room is filled with a uniform horizontal magnetic field of strength B. When the pendulum hangs at equilibrium, the magnetic field points perpendicular to the plane of the ring (i.e., the field s pierce "head-on" through the ring).

Suppose the pendulum is displaced to *small* angle θ_0 and released from rest at time $t = 0$. This happens in a vacuum, where there's no air resistance.

(a) (*Extra credit*) If you covered simple harmonic motion in detail last semester, show that the pendulum's angle with the vertical, for the first couple of oscillations, is given approximately by

$$\theta = \theta_0 \cos \omega_{\text{osc}} t,$$

where the angular frequency of oscillation is $\omega_{\text{osc}} = \sqrt{g/L}$. If you never delved deeply into oscillatory motion, then take this result as given, and move on to part (b).

PENDULUM WHEN RELEASED

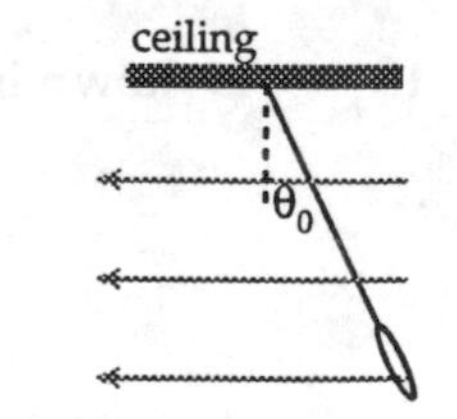

(b) Given this result from part (a), what is the induced current in the ring as a function of time, for the first couple of oscillations? Express you answer in terms of the given physical quantities (L, s, M, B, θ_0, g). Hint: Think about why the flux through the ring changes, as it swings.

(c) Sketch a rough graph of the pendulum's angle vs. time (θ vs. t) showing not just the first couple of oscillations, but also the pendulum's behavior in the long run. Explain in words why your graph looks like it does. Remember, there's no air resistance.

(d) How much total heat dissipates in the ring during its swinging? Hint: The pendulum's initial height above equilibrium is $H_0 = L(1 - \cos \theta_0)$. If you find yourself bogged down in math, you've missed the conceptual "trick."

ANSWER 41-7

(a) This is a classic simple harmonic motion problem. Instead of just writing down the relevant equation, let me review where it comes from. If you're not interested in oscillatory motion, skip to part (b).

FORCE DIAGRAM

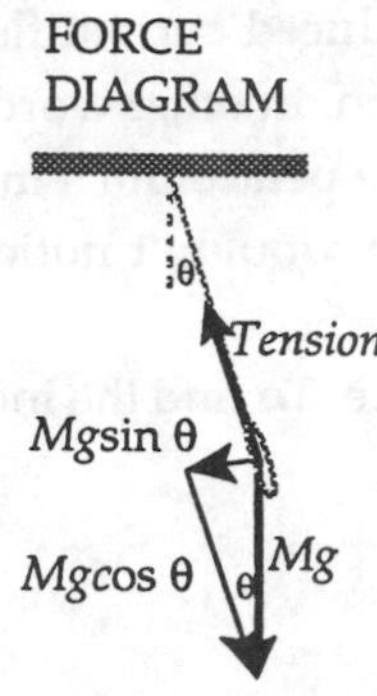

With a simple pendulum, we can work either in terms of force and acceleration, *or* in terms of torque and angular acceleration. Since we're finding θ as a function of time, let's use torque.

Start with a free-body diagram of the system when it's displaced an arbitrary distance or angle from equilibrium. The component of gravity parallel to the string, $Mg\cos\theta$, cancels the tension force. Therefore, the net force is $-Mg\sin\theta$, the other component of gravity. The minus sign indicates that the force pushes the pendulum back towards equilibrium. Because θ is small, $\sin\theta = \theta$. So,

$$F_{\text{net}} \approx -Mg\theta.$$

Recall that $\tau = |\mathbf{r} \times \mathbf{F}| = F_{\perp}r$, where **r** is a vector from the pivot point to the "contact point" where the force acts, and $F_{\perp}$ is the component of the force perpendicular to **r**. Here, the relevant force points entirely perpendicular to the radius vector (moment arm), which has length $r = L$. So,

$$\tau = F_{\perp}r = -MgL\theta.$$

(Those of you familiar with my "stuff" strategy can now see that stuff = MgL.)

According to Newton's 2nd law in rotational form, $\sum\tau = I\alpha$, where I is the rotational inertia and α is the angular acceleration. Since the ring is very small compared to the pendulum's length, we can treat it as a point mass, in which case $I = Mr^2 = ML^2$. Now that we know the torque and the rotational inertia, we could solve for α. But we're looking for θ, the pendulum's angle as a function of time. Fortunately, angular acceleration is the rate of change of angular velocity, which is the rate of change of angle: $\alpha = \frac{d\omega}{dt} = \frac{d^2\theta}{dt^2}$. So, from Newton's 2nd law, we get

$$\sum\tau = I\alpha$$

$$-MgL\theta = ML^2\frac{d^2\theta}{dt^2}.$$

Divide through by ML^2 to get

$$\frac{d^2\theta}{dt^2} = -\frac{g}{L}\theta.$$

This is a second-order differential equation. As you can confirm by double differentiating, the solution is sine or cosine:

$$\theta = A\cos\omega_{\text{osc}}t, \text{with } \omega_{\text{osc}} = \sqrt{g/L},$$

where A is the amplitude, the maximum angle reached by the pendulum during its oscillations. Here, $A = \theta_0$. Also, I used cosine instead of sine, and set the "phase constant" equal to 0, because the pendulum begins its motion at maximum displacement. In words, the pendulum swings back and forth in simple harmonic motion, as least for the first couple of oscillations.

Now actually, as we'll see in part (c), a magnetic force acts on the induced current flowing through the swinging ring. This magnetic force always resists the ring's motion. In other words, the magnetic force on the induced current acts as a resistive force, dampening the pendulum's motion. Consequently, the pendulum's oscillations get smaller and smaller. But you wouldn't notice this effect during the first few oscillations, unless the magnetic field is huge.

(b) Ohm's law tells us that $i_{\text{induced}} = \mathcal{E}_{\text{induced}}/R$, where R denotes the ring's resistance. To find the induced voltage, use Faraday's law:

$$\mathcal{E}_{\text{induced}} = -\frac{d\Phi_B}{dt}.$$

We need to figure out the rate at which the flux through the ring changes.

At first glance, you might think the flux doesn't change, because the magnetic field is constant, and the ring's area is constant. But flux also depends on the *angle* between the ring and the field. Let me review the intuition behind this.

Roughly speaking, the flux through a surface (such as a ring) specifies how many field lines pierce through the surface. As this drawing shows, fewer field lines pass through the surface when it's tilted with respect to the field. In the extreme case, where the surface "faces" perpendicular to the field, *no* field lines pierce through it ($F = 0$). For instance, no flux pierces through the third ring in the drawing. In general

For each ring, I've drawn the "area vector." These three rings have different fluxes, because they make different angles with the field.

$$\Phi = \int (\textbf{Field})\cdot d\textbf{A} = \int (\text{Field})(dA)\cos\theta,$$

where θ denotes the angle between the field and the "area vector." The area vector indicates the direction in which the surface *faces*. (Your floor faces upward, and your wall faces sideways.) So, when the surfaces faces the same direction that the field lines point ($\theta = 0°$), the flux is largest, since $\cos 0° = 1$. When the surface faces perpendicular to the field lines ($\theta = 90°$), the flux is zero, since $\cos 90° = 0$. These intuitions, and the accompanying equation, apply to *any* field, including magnetic ones.

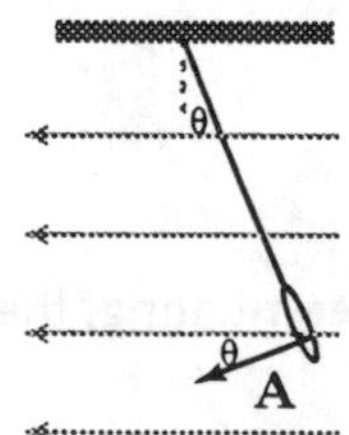

As this diagram shows, the angle that the ring's area vector makes with the magnetic field is the *same* angle that the pendulum makes with the vertical. When the ring swings, this angle changes. As a result, the flux changes.

To find the rate of change of flux, I'll use our part (a) result to write the flux as a function of time. Then, I'll differentiate with respect to time.

When the pendulum is at arbitrary angle θ,

$$\begin{aligned}\Phi_B &= \int_{\text{ring}} \textbf{B}\cdot d\textbf{A} \\ &= \int_{\text{ring}} B\,dA\cos\theta \\ &= B\cos\theta \int_{\text{ring}} dA,\end{aligned}$$

since B is uniform, and so is θ. You may wonder why we can pull θ outside the integral, given that it changes with time. But the flux integral is an integral over space (specifically, over the area of the ring), not over time. We're finding the flux at a particular "frozen" moment in time. At that frozen moment, every bit of the ring's area makes the same angle with the magnetic field. That's why we can pull θ outside the integral.

Let me finish evaluating the integral.

$$\begin{aligned}\Phi_B &= B\cos\theta \int_{\text{ring}} dA \\ &= (B\cos\theta)A_{\text{ring}} \\ &= B\pi s^2 \cos\theta.\end{aligned}$$

Before differentiating Φ_B with respect to time, we must rewrite Φ_B as a function of time. To do so, use our part (a) expression for θ:

$$\begin{aligned}\Phi_B &= B\pi s^2 \cos\theta \\ &= B\pi s^2 \cos\left(\theta_0 \cos\sqrt{g/L}\,t\right).\end{aligned}$$

Differentiate this using the chain rule. By Faraday's law,

$$\begin{aligned}\mathcal{E}_{\text{induced}} &= -\frac{d\Phi_B}{dt} \\ &= -B\pi s^2\left[-\sin\left(\theta_0 \cos\sqrt{g/L}\,t\right)\right]\cdot\left(-\theta_0 \sin\sqrt{g/L}\,t\right)\cdot\sqrt{g/L}.\end{aligned}$$

If you had trouble with the tricky derivative, that's OK Just make sure you can set everything up properly.

Now that we know the induced voltage, simply divide through by resistance to get the induced current:

$$i_{\text{induced}} = \frac{\mathcal{E}_{\text{induced}}}{R} = -\frac{B\pi s^2\theta_0}{R}\sqrt{\frac{g}{L}}\left[\sin\left(\theta_0 \cos\sqrt{g/L}\,t\right)\right]\cdot\left(\sin\sqrt{g/L}\,t\right)$$

(c) A regular pendulum, swinging back and forth with no friction or air resistance, would keep oscillating at the same amplitude forever. The oscillations would never get smaller. Therefore, θ vs. t would be a simple cosine curve.

By contrast, if air resistance or some other drag force opposes the pendulum's motion, then the oscillations get smaller and smaller. Nonetheless, even as the oscillations "decay," the frequency of oscillation doesn't change. The pendulum keeps swinging back and forth the same number of times per minute.

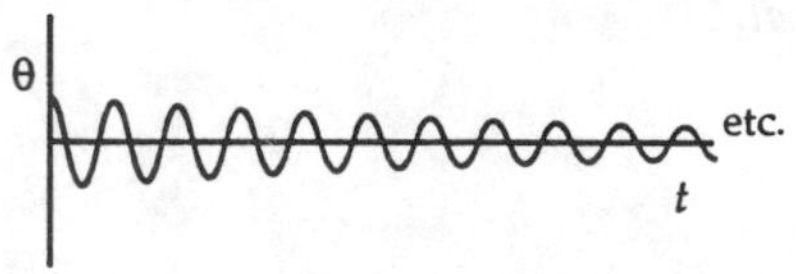

The magnetic force on the induced current acts as a drag force. Let me explain why. By Lenz' law, the induced current opposes changes in flux. Here, the flux changes because the pendulum's angle changes as it swings. Therefore, the induced current always flows so that the magnetic force opposes the pendulum's motion. Consequently, the amplitude of the oscillations decreases with time. We haven't calculated the exact mathematical function describing this decay. But the problem doesn't require us to do so. A rough sketch is fine.

I've oversimplified things here. Actually, the magnetic field exerts only a torque, not a force, on the ring. But as a result of the ring's interaction with the rod, the overall pendulum behaves as described. Besides, even without thinking about forces and torques, I *knew* the oscillations would get smaller. When induced current flows around the ring, heat gets generated. This heat energy must come "at the expense" of the pendulum's mechanical energy, as discussed below.

(d) Before demonstrating the short way to find the total heat dissipated, let me outline the long, tedious way.

The long way. Since power is energy per time ($P = dE/dt$), it follows that $E = \int P dt$. We can find the dissipated energy by integrating over the dissipated power. For a resistor, $P = Vi = i^2R$. So, given i as a function of t, we can square that current, multiply by R, and then integrate from $t = 0$ to $t = \infty$. Unfortunately, in part (b), we found $i(t)$ during the first few oscillations, *not* $i(t)$ at arbitrary later time.

The short way. Energy is conserved. It can't just disappear, even though the pendulum gradually slows down and eventually stops. The lost kinetic and potential energy all turns into heat (and other dissipative forms of energy).

I'll now flesh out this reasoning. When the pendulum gets released from initial angle θ_0, it carries potential energy. ($K_0 = 0$, since the pendulum gets released from rest.) But after swinging for a long time, the pendulum eventually ends up at rest, hanging straight down. At that point, it has no potential *or* kinetic energy. Therefore, the initial potential energy converts into another form—the heat dissipated in the ring as induced current flows. Since no other form of energy plays a role, energy conservation says that *all* the initial potential energy converts into heat. So, to find the total heat, we just need to calculate the initial potential energy.

As this diagram indicates, $H_0 + L\cos\theta_0$ equals the length of the pendulum when it's hanging straight down. But the pendulum always has length L. So,

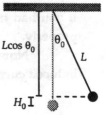

Lcos θ_0
θ_0
L
H_0

$$H_0 + L\cos\theta_0 = L.$$

Solve for H_0 to get the pendulum's initial height above equilibrium,

$$H_0 = L(1 - \cos\theta_0).$$

Therefore, the pendulum has initial potential energy $U_0 = MgH_0 = MgL(1 - \cos\theta_0)$. Energy conservation then gives us

$$Energy_0 = Energy_f$$

$$K_0 + U_0 = K_f + U_f + Heat_f$$

$$K_0 + MgL(1 - \cos\theta_0) = 0 + 0 + Heat_f.$$

So, without further ado, $Heat_f = MgL(1 - \cos\theta_0)$.

Magnetism In Matter

QUESTION 42-1

(Before attempting this problem, please read your textbook's sections on diamagnetism, paramagnetism, and ferromagnetism. Though very hard, this question can help you learn these concepts more deeply.) Here's a detective game. I've got three samples of material, call them A, B, and C. I know that one of the samples is diamagnetic, one is paramagnetic, and one is ferromagnetic (though currently unmagnetized). But I don't know which sample is which.

To sort things out, I shove samples A and B into a strong external magnetic field for a while. I then pull those samples out of the field, and hang them next to each other by thin threads. Nothing happens; they neither attract nor repel, except for their minuscule gravitational attraction. Next, I dangle sample B near the fringes of that external field. The sample swings towards the field, as if getting sucked in. Amazingly, from these two experiments, it's possible to tell which sample is which. For all three samples!

Now I place all three samples into the external magnetic field. Inside which sample is the magnetic field the strongest? The second strongest? The least strong?

ANSWER 42-1

Before solving this puzzle, let me review the basics of diamagnetism, paramagnetism, and ferromagnetism.

PARAMAGNETIC MATERIAL

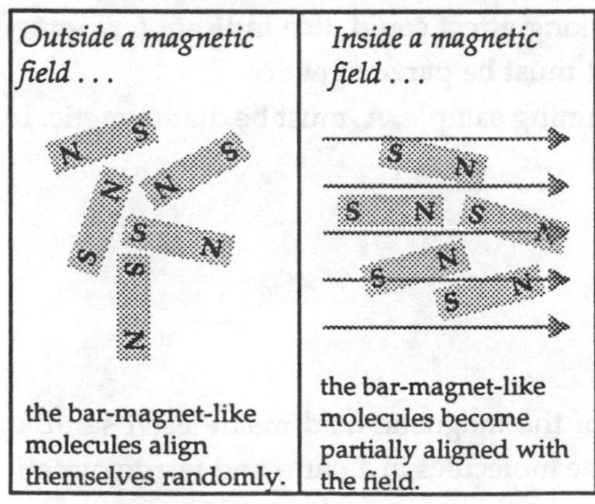

Outside a magnetic field . . .
Inside a magnetic field . . .
N S
S N
S N
N S
S N
S N S N
S N
S N
the bar-magnet-like molecules align themselves randomly.
the bar-magnet-like molecules become partially aligned with the field.

Each molecule of a paramagnetic substance has a *magnetic moment*. In other words, the molecules behave like tiny bar magnets. When immersed in an external magnetic field, those tiny bar magnets partially align themselves with the external field.

This alignment leads to two related effects. First, the aligned bar magnets are attracted towards the stronger part of the external field. So, paramagnetic material gets "sucked" towards other magnets. Second, the aligned bar magnets generate their own magnetic field, which reinforces the external magnetic field. Therefore, the total magnetic field inside the paramagnetic substance is *larger* than the external magnetic field. Briefly put, magnetic fields are amplified inside a paramagnetic substance. By contrast, when the paramagnetic material is removed from the external field, those tiny bar magnets once again become unaligned (randomized). For this reason, the paramagnetic effects discussed in this paragraph occur *only* while the material is inside an external magnetic field.

Ferromagnetic substances are similar, but with two differences—one major and one minor. The minor difference is that ferromagnetic substances are more extreme. When immersed in an external field, a ferromagnetic substance amplifies the field much more than a paramagnetic material does. But the major difference between paramagnetic and ferromagnetic materials is "memory." Once the tiny bar magnets in the ferromagnetic substance become aligned by an external field, they *stay aligned*, even when removed from the external field. In other words, the ferromagnetic substance stays "magnetized," while the paramagnetic material loses its magnetization when removed from the external field.

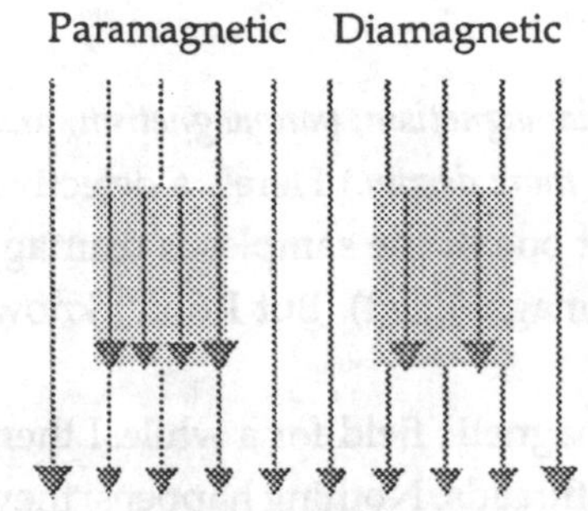

The magnetic field is amplified inside the paramagnetic material, but "dampened" inside the diamagnetic one. This diagram fails to show how the magnetic moments also affect the field outside the blocks of material.

The molecules of a diamagnetic material have no magnetic moment. But a tiny magnetic moment gets induced when the material is placed inside an external field. As explained in your textbook, the induced magnetic moment is aligned *oppositely* to the external field lines. As a result, the diamagnetic material gets "sucked" towards the *weaker* part of an external field, and away from the stronger part. Therefore, it gets repelled from other magnets. Also, this oppositional alignment of the induced magnetic moments leads to a dampening of the magnetic field inside it.

Now that we've reviewed the differences between dia-, para-, and ferromagnetism, let's solve the puzzle. I'll first consider the experiment involving samples A and B. From this "negative-result" experiment, we can conclude that neither sample A nor sample B is ferromagnetic. Here's how I know. Suppose, hypothetically, that sample A were ferromagnetic. Then it would have retained its magnetization after leaving the external field. In other words, sample A would have become a permanent magnet. That magnet would have attracted sample B (if B were paramagnetic), or would have repelled sample B (if B were diamagnetic). But in reality, sample A neither attracted nor repelled B. This goes to show that A must not be ferromagnetic. Equivalent reasoning shows that B can't be ferromagnetic, either. Since neither A nor B is ferromagnetic, C must be the ferromagnetic sample.

From the second experiment, we learn that B is paramagnetic; it gets attracted ("sucked in") to the stronger part of the external magnetic field. In general, that sucking effect could also indicate ferromagnetism. But above, we deduced that B is not ferromagnetic. So, it must be paramagnetic.

Since C is ferromagnetic and B is paramagnetic, the remaining sample, A, must be diamagnetic. In conclusion,

A = diamagnetic,
B = paramagnetic,
C = ferromagnetic.

We can now use these conclusions to "rank" the size of the magnetic field inside each sample, when it's immersed in an external field. Recall from above that the molecules in a para- and ferromagnetic substance behave like tiny bar magnets. These bar magnets align themselves with the external field lines, thereby *enhancing* (amplifying) the external field. So, inside samples B and C, the magnetic field is stronger than the external field. But the ferromagnetic material is more extreme. It amplifies the field more strongly than the paramagnetic substance does. Therefore, ferromagnetic substance C has the strongest field inside it. Paramagnetic substance B has the second strongest field inside it. And bringing up the rear is diamagnetic substance A. Remember, diamagnetic materials "dampen" (as opposed to amplify) the external field. The field inside sample A is weaker than the external field.

Please revisit this problem when studying for an exam, because it "tests" many different facets of dia-, para-, and ferromagnetism.

QUESTION 42-2

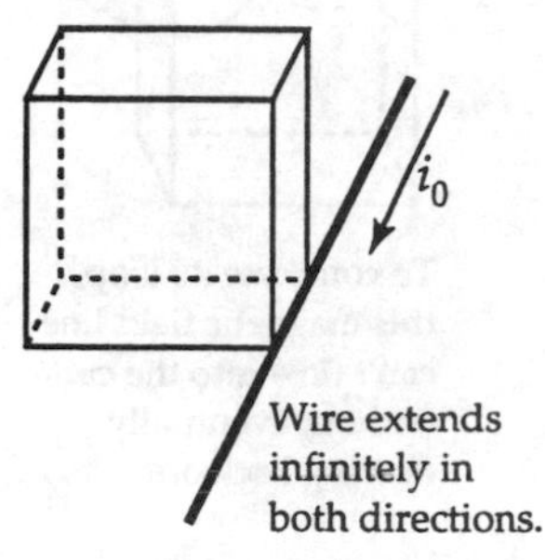

(*Check with your instructor to see if you need to understand this topic. Many instructors skip it.*) Consider a cardboard cubical box of side-length s. Along its bottom-right edge runs an infinitely long wire carrying current i_0, as drawn here.

(a) What is the net magnetic flux through the cube? This net flux includes the flux through all six sides of the cube. Recall that flux flowing *into* the cube counts as negative, while flux flowing out of the cube counts as positive.

(b) Would your answer to part (a) be bigger, smaller, or the same if the cubical box were made of iron instead of cardboard?

(c) Now the box is again made of cardboard, and a big magnet is placed inside it. What effect does this have on the net flux through the cube? Does the flux get bigger, get smaller, or stay the same?

ANSWER 42-2

(a) You could solve by using Ampere's law to find the magnetic field at each point on a given side of the cube, and then integrating $\int \mathbf{B} \cdot d\mathbf{A}$ over all six sides. Yecch.

By contrast, if you understand Gauss' law for magnetism, then you can just write down the answer. According to that law, the net magnetic flux through *any* closed (Gaussian) surface is zero. It doesn't matter whether the surface is real or imaginary, cardboard or iron. And it doesn't doesn't matter whether magnets are placed inside of it or outside of it. Let me explain why.

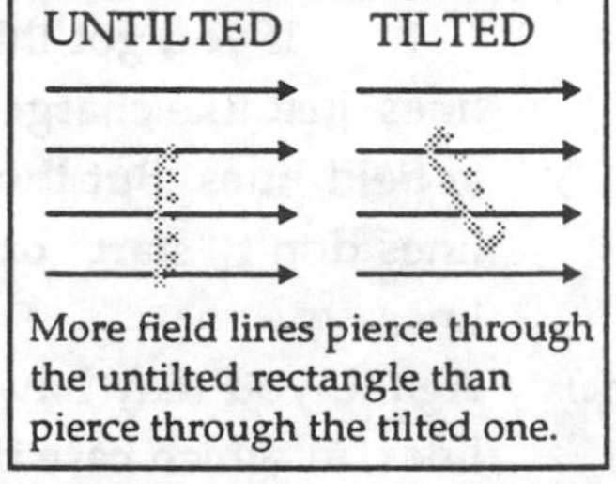

More field lines pierce through the untilted rectangle than pierce through the tilted one.

Roughly speaking, the "flux" through a surface specifies how many field lines pierce through it. That's why tilting a surface changes the flux, as drawn here. I could be talking about electric, magnetic, or even gravitational field lines—the concept is the same.

For closed (Gaussian) surfaces, the flux is positive when field lines flow out of the surface, and negative when they flow into the surface. For instance, when a positive charge sits inside a cube,

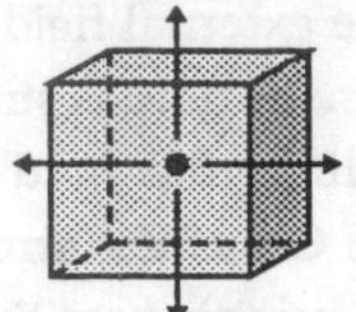

Since field lines flow out, flux is positive.

the net flux through the cube is positive, because the electric field lines flow out of the surface *and don't flow back in.* The positive charge "spews out" field lines, just like a sprinkler spews out water.

That's how electric field lines differ from magnetic field lines. Electric field lines have sources and sinks. They flow out of positive charges and into negative charges. By contrast, *magnetic field lines don't have sources. They don't start or end anywhere. Instead, they form complete loops.* A magnetic field line, no matter how funkily shaped, always "meets up" with itself. (A magnetic monopole would be a source of magnetic field lines; but as of 1997, monopoles seem not to exist.)

Because a magnetic field line forms a complete loop, any field line flowing into a cube must also flow back out. Otherwise, the field line could never "meet up" with itself. For instance, in this drawing, the field line enters the top of the cube, and leaves the side.

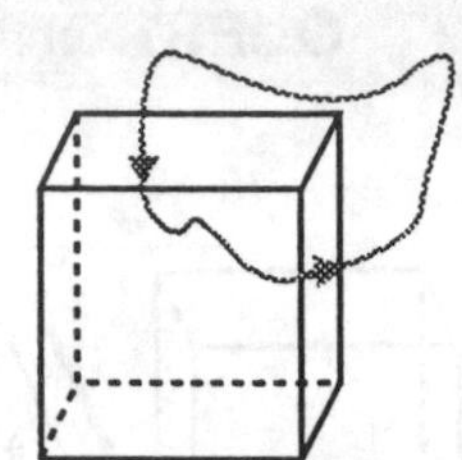

To complete its loop, this magnetic field line can't flow into the cube without eventually flowing back out.

Since any field line flowing into a closed surface must also flow back out, and vice versa, **the net magnetic flux through any Gaussian surface is zero.** That's Gauss' law for magnetism. The answer to this problem must be zero. Intuitively, a field line contributes positive flux when it flows out of the cube, and negative flux when it flows back in. These positive and negative contributions cancel.

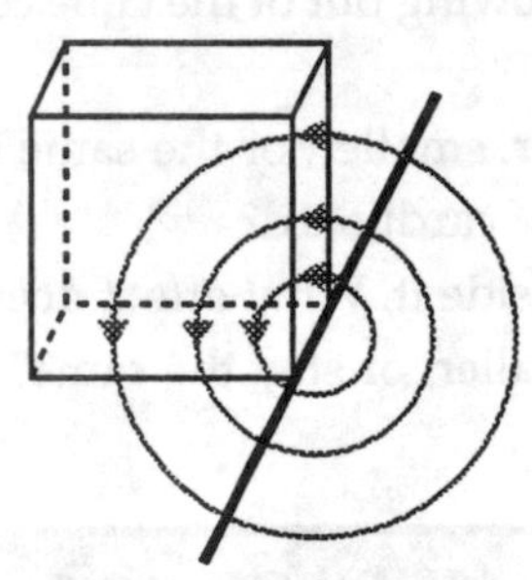

Let's see how this general law plays out here. An infinite wire generates circular magnetic field lines. These field lines flow into the right-hand side of the cube, and out the bottom, as drawn here. (Some of them flow into the top and out the left-hand side.) Of course, many field lines totally "miss" the cube, and therefore don't contribute to the magnetic flux. But any field line flowing into the cube also flows back out. Therefore, the net magnetic flux is zero.

(b) Constructing the cube out of iron (instead of cardboard) would make the field lines more numerous and more complicated. But the reasoning of part (a) still applies. Since magnetic field lines form complete loops, any field line flowing into the cube also flows back out. Therefore, the net magnetic flux is still zero.

If you got this wrong, you were probably thinking of iron as a "source" of magnetic field lines, just like charges are a source of electric field lines. Well, it's true that iron generates magnetic field lines. But those field lines form complete loops that go through the iron atoms. The field lines don't "start" or "stop" in the iron. In this way, magnetic field lines differ from electric field lines.

(c) Again, you may have thought that the magnet inside the box serves as a source of magnetic field lines, in which case the net magnetic flux is positive. But as I emphasized in part (b), the field lines created by the magnet form a complete loop. They come back to where they started. Therefore, any field line that flows out of the cube eventually flows back in. The field lines generated by the magnet are like boomerangs; they come back, completing a loop. So, once again, the net magnetic flux is zero.

QUESTION 42-3

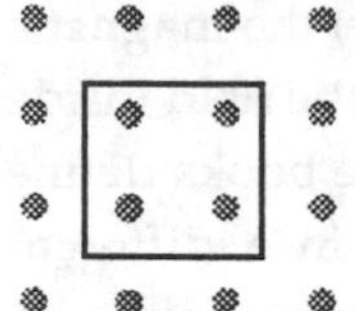

A giant solenoid (not pictured) creates a uniform magnetic field B_0, pointing out of the page. A metal square, of side-length s, sits in the plane of the page. When the magnetic field is steadily reduced to 0 over time T, $i = 2.00$ amps of induced current flow around the square. This all happens in a vacuum.

The experiment is now repeated: The external magnetic field is again reduced from B_0 to 0 over time T. But this time, the square was immersed in a mystery liquid. Observation reveals that the induced current was 2.01 amps, instead of 2.00 amps.

(a) Is the mystery liquid paramagnetic or diamagnetic? Explain your answer.

(b) (*Very hard*) What is the magnetic *permeability* of this liquid? (Some books denote permeability by μ. Others denote it by κ_m.)

ANSWER 42-3

(a) To answer this, we must think about why current gets induced. A *changing* magnetic flux through the square generates an induced EMF (voltage). This induced voltage makes current flow around the square, in accordance with $\mathcal{E}_{\text{induced}} = iR$. The bigger the induced voltage, the bigger the induced current.

So, the induced voltage must have been bigger when the square was immersed in mystery liquid. How could this happen? Well, according to Faraday's law, the induced voltage is proportional to the rate of change of the magnetic flux through the square. Since the flux changes at a steady rate, we can write Faraday's law as

$$\mathcal{E}_{\text{induced}} = -\frac{\Delta\Phi_B}{\Delta t}$$

$$= -\frac{\Phi_{B\text{ final}} - \Phi_{B\text{ initial}}}{T}.$$

Since the magnetic field ends up at 0, this expression simplifies to

$$\mathcal{E}_{\text{induced}} = -\frac{0 - \Phi_{B\text{ initial}}}{T}$$

$$= \frac{\Phi_{B\text{ initial}}}{T}.$$

From this equation, we can reach the following conclusion: In order to increase the induced voltage (and current), the mystery liquid must somehow increase the initial magnetic flux through the square. With the mystery liquid present, the initial magnetic field through the square is *bigger* than it was with the liquid absent. So, the liquid *amplifies* the magnetic field inside of it. In other words, the liquid must be **paramagnetic**.

Let me clarify this reasoning. With the liquid missing, the initial magnetic field through the square is B_0. But with the liquid present, the field through the square is slightly bigger: $B = \mu B_0$, where μ is slightly greater than 1. In both cases, the field decreases to 0 in time T. So, when the field starts off bigger, it must decrease at a slightly faster rate, thereby inducing a slightly larger voltage and current.

(b) I'll denote permeability by μ. Permeability tells you how much stronger (or weaker) the magnetic field is inside a substance, as compared to the external field. Let B denote the magnetic field inside the liquid, and B_0 denote the external magnetic field created by the solenoid. Some books define the permeability by $\mu = B/B_0$, in which case $B = \mu B_0$. Other books, working in a different system of units, define $\mu = \mu_0(B/B_0)$, where μ_0 denotes the permeability of the vacuum. Make sure you know your book's (or your instructor's!) definition. In either case, however, permeability specifies how much the paramagnetic (or diamagnetic) material amplifies (or weakens) the field inside of it.

To solve this problem, we need an exact relationship between the permeability and the induced current. Actually, all we need is a ratio. As we'll see below, the induced current is directly proportional to the permeability. Therefore, the ratio of the permeability with and without the liquid equals the ratio of the induced currents with and without the liquid:

$$\frac{i_{\text{with liquid}}}{i_{\text{in vacuum}}} = \frac{\mu_{\text{liquid}}}{\mu_{\text{vacuum}}}$$

$$\frac{2.01 \text{ amps}}{2.00 \text{ amps}} = \frac{\mu_{\text{liquid}}}{1},$$

which we can immediately solve for μ liquid to get 1.005.

OK, I pulled that ratio out of thin air. Let me show where it comes from. As we saw in part (a), since $\Phi_{B\text{ final}} = 0$,

$$\mathcal{E}_{\text{induced}} = \frac{\Phi_{B\text{ initial}}}{T}$$

$$iR = \frac{Bs^2}{T},$$

where in the second line, I invoked Ohm's law ($V = iR$), along with the fact that flux equals field times area, when the field is uniform and points perpendicular to the plane of the surface.

The initial field through the square is $B = \mu B_0$, where μ is the permeability of the material in which the square is immersed. So,

$$iR = \frac{\mu B_0 s^2}{T}.$$

For a vacuum, $\mu = 1$ or μ_0, depending on units. For the liquid, it equals μ_{liquid}. This equation shows that the induced current is proportional to the permeability. Therefore, my above "ratio reasoning" was valid. Inserting the liquid increases i by 0.5%, from 2.00 to 2.01 amps. Therefore, the liquid's permeability must be 0.5% more than the vacuum's permeability—1.005 instead of 1.00, or $1.005\mu_0$ instead of μ_0.

QUESTION 42-4

A cube of uniformly magnetized material of side-length $s = 0.050$ m has magnetization $M = 2000$ amperes per meter, when it's sitting on my desk. I now hold the cube in a uniform, upward-pointing magnetic field of strength 0.0040 teslas, with the cube's magnetization vector also pointing upward. The magnetization remains at $M = 2000$ A/m.

(a) From what I've told you, is the cube paramagnetic, diamagnetic, or ferromagnetic? Answer without checking any tables.

(b) When the cube is inside the external magnetic field, what is the magnetic field inside the cube?

(c) (*Hard, but a good exam question*) When the cube is inside the external field, how much work would it take to flip the cube over?

ANSWER 42-4

(a) Even *before* I put the cube into the external magnetic field, the cube is magnetized. Therefore, it must be ferromagnetic. Remember, paramagnetic and diamagnetic materials lose their magnetization as soon as they get removed from an external field. Microscopically speaking, the bar-magnet-like molecules in a paramagnetic material align themselves with an external field, but become randomly aligned when the substance leaves the external field. By contrast, a ferromagnetic substance "remembers" its magnetization; the molecular bar magnets stay aligned, even when removed from the external field.

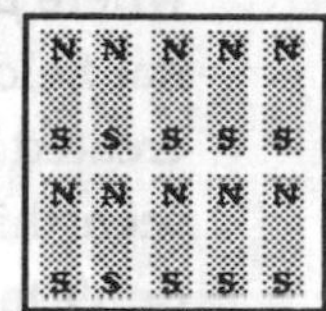

The molecular bar magnets in a ferromagnet are permanently aligned.

(b) The total magnetic field inside the magnetized cube comes from two contributions. First, the external magnetic field exists both inside and outside the cube, with strength $B_0 = 0.0040$ teslas. Second, the cube's magnetization generates its own magnetic field, of strength $\mu_0 M$, where μ_0 denotes the permeability of the vacuum. This makes sense, because M specifies the density and strength of the bar-magnet-like molecules in the cube. The stronger and denser these tiny bar magnets are, the bigger a magnetic field they create. This field points in the direction of the magnetization.

In summary, the total magnetic field inside the cube points upward, with strength

$$\begin{aligned} B &= \text{external field} + \text{field generated by the magnetization} \\ &= B_0 + \mu_0 M \\ &= 0.0040 \text{ testlas} + (4\pi \times 10^{-7} \text{ tesla} \cdot \text{meter/amp})(2000 \text{ amps/meter}) \\ &= 0.0040 \text{ teslas} + 0.0025 \text{ teslas} \\ &= 0.0065 \text{ teslas.} \end{aligned}$$

(c) To solve this, you need to know two things. First, magnetization specifies the strength and density of the tiny bar-magnet-like molecules in the cube. More precisely, M is the dipole moment density, by which I mean the dipole moment per volume: $M = \mu/V$, where μ denotes the total dipole moment and V denotes the volume. (Here, μ does not mean permeability, and V does not mean voltage.) So, we can easily solve for the total dipole moment of the cube:

$$\begin{aligned}\mu &= MV \\ &= Ms^3 \\ &= (2000\ \text{A/m})(0.050\ \text{m})^3 \\ &= 0.25\ \text{A}\cdot\text{m}^2.\end{aligned}$$

Second, you need to know the potential energy of a magnetic dipole inside an external magnetic field. It's

$$U = -\mu\cdot\mathbf{B}_{\text{ext}} = -\mu B_{\text{ext}}\cos\theta,$$

where μ is the magnetic dipole moment, and q is the angle between the dipole vector and the external field vector. Crucially, this potential energy depends only on the *external* field, not on the field created by the dipole itself. This makes sense, because the dipole can't exert a force on itself. (You've seen this before. When calculating the potential energy of a point charge in an electric field, you don't take into account the field created by the point charge itself. You include only the "external" electric field, i.e., the field created by *other* charges.)

To see why this potential energy formula helps us, recall the relationship between work and potential energy. If I spend 5 joules of work lifting a book onto a table, then it ends up with 5 more joules of potential energy than it started with. In general, the work you do on an object equals the object's change in potential energy, if begins and ends at rest: $W_{\text{by you}} = \Delta U$. So, the work needed to flip the cube is simply the change in the cube's potential energy.

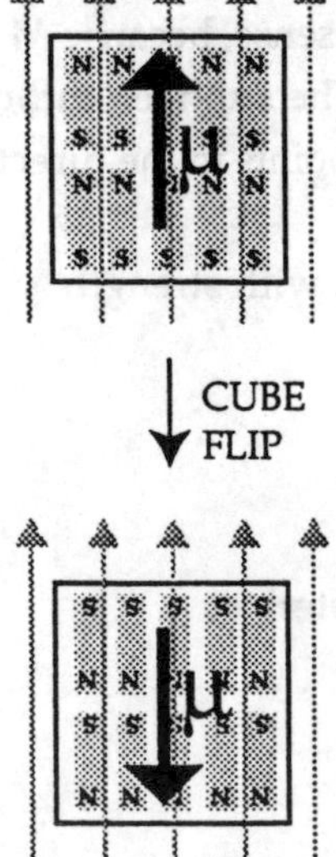

Here, the cube's dipole moment initially points upward, parallel to the magnetic field. But when we flip the cube, so that its dipole moment points downward, the angle between μ and $\mathbf{B}_{\text{ext}}$ becomes q = 180°. Since $B_{\text{ext}} = 0.040$ tesla, we get

$$\begin{aligned}W = \Delta U = U_{\text{after flip}} - U_{\text{before flip}} \\ &= -\mu B_{\text{ext}}\cos 180° - (-\mu B_{\text{ext}}\cos 0°) \\ &= \mu B_{\text{ext}} + \mu B_{\text{ext}} \qquad [\text{since } \cos 180° = -1, \text{ and } \cos 0° = 1] \\ &= 2\mu B_{\text{ext}} \\ &= 2(0.25\ \text{A}\cdot\text{m}^2)(0.040\ \text{tesla}) \\ &= 0.020\ \text{joules}.\end{aligned}$$

It's easy to mix up the minus signs. Let's check to see if a positive answer makes intuitive sense. Well, the cube's dipole moment "wants" to be aligned with the external field. When it's aligned, the potential energy is as low as possible, $U = -\mu B_{ext}\cos 0° = -\mu B_{ext}$. Flipping the cube raises its "magnetic potential energy," just as lifting a book raises its gravitational potential energy. In either case, you must *do* work. So, it makes sense that the work comes out positive.

To gain more physical insight into dipole moments, think about a small current loop, which has dipole moment $\mu = i\mathbf{A}$, where **A** denotes the area vector. As we saw in question 39-5 and 39-7, a current loop in a uniform magnetic field feels a torque that tries to align the loop with the field, so that **B** and μ point in the same direction. To tilt the current loop, you must fight that torque.

QUESTION 42-5

First, I take a sample of paramagnetic material, dunk it in liquid nitrogen (using tongs), and then hold it in a strong magnetic field. It takes 0.10 joules of work to pull the sample out of the magnetic field.

Second, I take that same sample, dunk it in boiling water, and then again hold it in the same strong magnetic field. Now, it takes *less* than 0.10 joules of work to pull the sample out of the field.

Why?

Hint: Think about how temperature affects the bar-magnet-like molecules comprising the paramagnetic substance.

ANSWER 42-5

I'll first discuss why it takes work to pull a paramagnetic material out of an external magnetic field. As we'll see, the more magnetized the sample becomes (i.e., the bigger its dipole moment), the more work is needed to remove it from the field. Then, I'll explain why a cold sample becomes more magnetized than a hot sample, other things being equal. Because the cold sample becomes more magnetized, and a more magnetized sample requires more work to remove it from the external field, the experimental results make sense.

Why it takes work to remove the paramagnetic sample from the external magnetic field

The molecules of paramagnetic substance behave like tiny bar magnets. When outside an external field, these bar magnets align themselves randomly, "canceling out" each others' effects. But when placed in an external magnetic field, the tiny bar magnets become partially aligned with the external field. As a result, the overall sample acquires a net magnetic dipole moment, aligned with the external field.

As we saw in the previous problem, the potential energy of a magnetic dipole is $U = -\mu \cdot \mathbf{B}_{ext}$. Therefore, when inside the field with μ parallel to $\mathbf{B}_{ext}$, the sample has potential energy $U_0 = -\mu B_{ext}$. But when removed from the external field, the sample no longer "feels" a field ($B = 0$). In addition, the molecular dipole moments become randomized, leaving the sample with net dipole moment $\mu = 0$. For these two reasons, the sample's "final" potential energy is $U_f = 0$. Therefore, the work needed to pull the sample out of the field is

$$\begin{aligned} W = \Delta U &= U_f - U_0 \\ &= 0 - (-\mu B_{ext}) \\ &= \mu B_{ext} \\ &= MVB_{ext}, \end{aligned}$$

where M denotes the magnetization and V denotes the volume. In the last step, I used the definition of magnetization as dipole moment per volume: $M = \mu / V$, and hence $\mu = MV$. So, the work needed to remove the substance from the external field is proportional to M. Roughly put, this work comes out positive because the molecules "prefer" to be aligned with the external field, as opposed to randomized and outside the field.

To finish explaining why this work is higher for a colder sample, I must discuss the effect of temperature on the bar-magnet-like molecules. Specifically, I must discuss why heating a substance renders it less susceptible to getting magnetized.

Why does heating a paramagnetic substance render it less susceptible to getting magnetized?

Even in a solid substance, the molecules jiggle. Since the molecules are tiny bar magnets, this jiggling tends to make the bar magnets point in all different directions. That's why a paramagnetic substance, when outside a magnetic field, has no magnetization. Thermal jiggling ensures that the tiny bar magnets become randomly aligned.

So, when immersed in an external field, the tiny bar magnets feel two opposing tendencies. On the one hand, the torque generated by the external field tends to align those tiny bar magnets with the external field. On the other hand, thermal jiggling tends to "randomize" their alignment. Neither of these opposing tendencies wins completely. The tiny bar magnets become partially, but not fully, aligned with the external field.

PARAMAGNETIC MATERIAL IN MAGNETIC FIELD

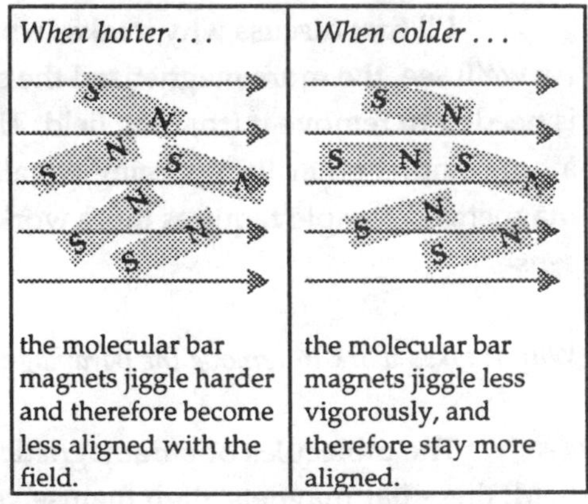

When hotter . . .	*When colder . . .*
the molecular bar magnets jiggle harder and therefore become less aligned with the field.	the molecular bar magnets jiggle less vigorously, and therefore stay more aligned.

Given all this background information, I can finally explain why the hotter sample becomes less magnetized. Hotter molecules jiggle more vigorously. This jiggling tends to randomize the alignment of the molecular bar magnets, as just explained. Therefore, in the hotter sample, the molecular bar magnets become less well aligned with the external field. In other words, the hotter sample becomesless magnetized. As a result, less work is needed to pull the hotter sample out of the external magnetic field.

Your textbook may discuss Curie's law, a precise mathematical relationship between a substance's temperature and its susceptibility to magnetization. This problem tests whether you understand the qualitative ideas behind that quantitative law.

Note To Students

The textbook chapter about magnetic properties of matter contains a smattering of different concepts. I've covered what most professors consider to be the key central ideas. But some professors go into more detail about orbital and "spin" magnetic moments, or about magnetic susceptibility, or about other topics. I focused on more conceptual problems here, because your homework assignment will probably give you practice solving "standard" problems. In any case, find out from your instructor exactly what you need to know about these topics, and at what depth.

43 CHAPTER

Inductance

QUESTION 43-1

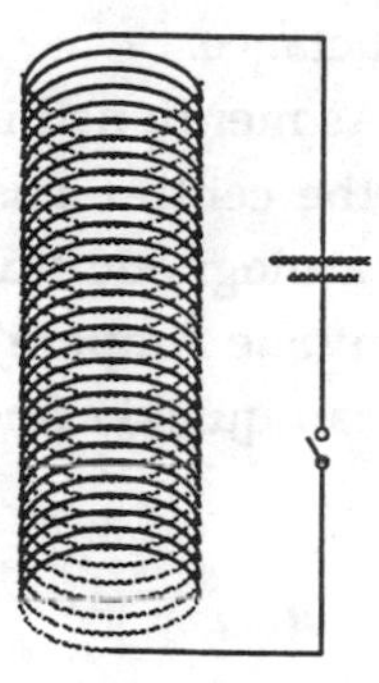

Consider this cylindrical solenoid, consisting of N coils (turns). It has radius r and length D, where $D \gg r$. Wires connect the ends of the solenoid to a battery. But the circuit isn't complete until the switch is closed.

(a) Suppose the battery has voltage V and the solenoid has total resistance R. Ohm's law implies that the current through the solenoid will reach $I_{final} = V/R$ after we close the switch.

But according to actual measurements, when I close the switch, the current does not *immediately* shoot up to I_{final}. The current takes a few tenths of a second to reach I_{final}. By contrast, when I connect a regular resistor to a battery, the current shoots up to I_{final} almost immediately. *Explain in words why the current in the solenoid takes significant time to reach its final value.* Check your answer against mine before proceeding to part (b), because this verbal explanation motivates everything that follows.

(b) As a first step to finding the solenoid's inductance, suppose that arbitrary current I flows through the solenoid. What is the total magnetic flux through *all coils* of the solenoid? Hint: Use Ampere's law to find the magnetic field inside the solenoid.

(c) Suppose the current through the solenoid starts increasing, at known rate dI/dt. What is the corresponding rate of change of the magnetic flux through the solenoid?

(d) What is the induced voltage in the solenoid coils? Does this induced voltage "help" or "hinder" the increasing current?

(e) The inductance L is defined by $\mathcal{E}_{induced} = -L(dI/dt)$, where $\mathcal{E}_{induced}$ denotes the voltage induced by the changing current. Explain in physical terms what inductance means. What is the inductance of this solenoid?

ANSWER 43-1

(a) Many students treat inductance as a mess of formulas. I want to make sure you understand what inductance means, physically. Roughly speaking, we'll see that inductance specifies how "resistant" something is to having the current through it *change*. So, in this case, the inductance prevents the current from immediately shooting up to its final value.

Let's think about what happens when the switch closes, at time $t = 0$. Right before $t = 0$, the current in the solenoid is $I_0 = 0$. After $t = 0$, the battery "wants" the current to reach $I_{final} = V/R$. So, the battery makes the current increase from 0 to I_{final}. But here's the crucial point. As the current increases, the magnetic field it generates inside the solenoid also increases. Consequently, the magnetic flux through the solenoid coils increases. And by Faraday's law, this changing magnetic flux *induces an EMF (voltage)* in the solenoid. The induced EMF always "counteracts" the changing flux. So here, the induced EMF tries to prevent the current in the solenoid from increasing. In other words, the induced EMF is "opposite" to the battery's EMF. Roughly put, the induced EMF "fights" the battery, as the battery increases the current from 0 to I_{final}.

This explains why the current takes several tenths of a seconds—and in some solenoids, over a second—to reach its final value. While increasing from 0 to I_{final}, the current generates an increasing magnetic flux through the solenoid. This increasing flux induces an EMF that "fights" the battery. The backwards induced EMF never "wins"; the battery successfully increases the current from 0 to $I_{final} = V/R$. But the induced EMF delays this process, making it gradual instead of abrupt. The bigger the inductance, the bigger this delay.

(b) For now, I'll just solve the problem. We'll see below how the answer relates to inductance.

When current flows through the solenoid, it generates a magnetic field. As mentioned in Chapter 40, the magnetic field lines inside a solenoid flow along (parallel to) the central axis. Since these field lines pierce through all the coils in the solenoid, they create a magnetic flux through all the coils. To figure out the exact equation relating Φ_B to I, we must invoke Ampere's law to find the magnetic field inside the solenoid. After obtaining the field, we can quickly find the magnetic flux.

Subproblem: Use Ampere's law to find the magnetic field.

According to Ampere's law, if we draw an "Amperian loop," the current enclosed is proportional to the integral of the magnetic field along the boundary of the loop:

$$\oint \mathbf{B} \cdot d\mathbf{s} = \mu_0 I_{encl}. \qquad \textbf{Ampere's law}$$

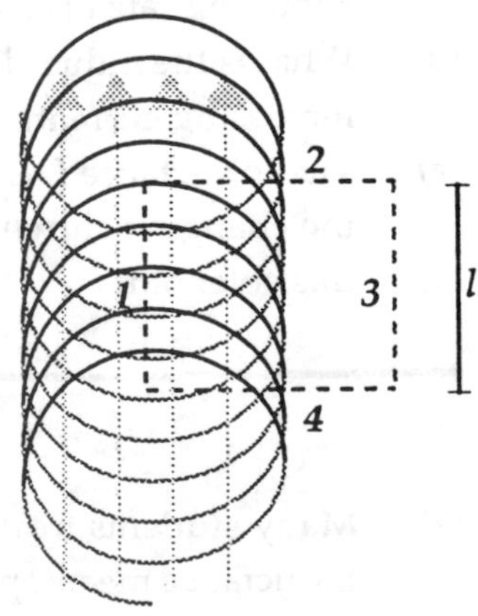

I'll start by simplifying the left-hand side of this equation.

Since the magnetic field is approximately zero outside the solenoid, and flows parallel to the central axis inside the solenoid, the rectangular Amperian loop drawn here works best. Here's why. Along side 3, the magnetic field is approximately zero, and hence $\int_3 \mathbf{B} \cdot d\mathbf{s} = 0$. Sides 2 and 4 run perpendicular to the magnetic field, hence the dot product $\mathbf{B} \cdot d\mathbf{s}$ vanishes along both those sides: $\int_2 \mathbf{B} \cdot d\mathbf{s} = \int_4 \mathbf{B} \cdot d\mathbf{s} = 0$. So, only side 1 "contributes" to the Amperian loop integral, . And since **B** points parallel to side 1, the dot product reduces to a regular product:

$$\oint \mathbf{B}\cdot d\mathbf{s} = \int_1 \mathbf{B}\cdot d\mathbf{s} + \int_2 \mathbf{B}\cdot d\mathbf{s} + \int_3 \mathbf{B}\cdot d\mathbf{s} + \int_4 \mathbf{B}\cdot d\mathbf{s}$$

$$= \int_1 B\,ds + 0 + 0 + 0$$

$$= B\int_1 ds \qquad \text{[since } B \text{ is uniform inside solenoid]}$$

$$= Bl,$$

where l is the arbitrary length of my Amperian loop. You can equate this l to the solenoid's length. Or you can use a smaller l, as I've done. It doesn't matter.

OK, now I'll work on the right-hand side of Ampere's law, by solving for the enclosed current. Let's first figure out how many coils my loop encloses. As drawn, the loop captures five coils. But in general, we can calculate the number of "trapped" coils as in this way: First, define the "linear coil density," n. That's the number of coils per unit length. For instance, n might be 20 coils per centimeter. Here, since the solenoid contains N coils spread out over distance D, the linear coil density is $n = N/D$. Therefore, the number of coils enclosed by my Amperian loop is

$$N_{\text{encl}} = \frac{\text{coils}}{\text{length}} \times (\text{length enclosed}) = nl = \frac{N}{D}l$$

For the rest of this problem, I'll use "n" as an abbreviation for N/D.

Each coil carries the same current, I. Since N_{encl} coils pierce through the Amperian loop, the loop encloses current

$$I_{\text{encl}} = N_{\text{encl}} I = nlI.$$

If this doesn't make sense, please see your instructor, who can help you visualize things three-dimensionally. Unfortunately, my publisher wouldn't buy me three-dimensional paper.

Now that I've simplified both sides of Ampere's law, I can easily solve for the magnetic field:

$$\oint \mathbf{B}\cdot d\mathbf{s} = \mu_0 I_{\text{encl}},$$

$$Bl = \mu_0\, nlI,$$

and hence

$$B = \mu_0\, nI.$$

End of Ampere's law subproblem

So far, I've found the magnetic field inside the solenoid. To finish this problem, we must calculate the magnetic flux through the solenoid coils.

Well, each coil has area $A = \pi r^2$. Since the coils "face" the same direction as **B**, and since **B** is uniform, the dot product $\mathbf{B}\cdot d\mathbf{A}$ simplifies into a simple product. Therefore, the magnetic flux through a *single coil* is

$$\Phi_{B\text{ one coil}} = \int \mathbf{B}\cdot d\mathbf{A} = BA = \mu_0 nI(\pi r^2),$$

where I used my "answer" for B from the Ampere's law subproblem.

A common mistake is to think we've just found the flux through the whole solenoid. But remember, the solenoid consists of N coils. And *each* coil has flux $\Phi_{B\text{ one coil}} = \mu_0 nI\pi r^2$ going through it. Therefore, the *total* magnetic flux through *all* the coils is

$$\begin{aligned}\Phi_B &= N\Phi_{B\text{ one coil}}\\ &= N(\mu_0 nI\pi r^2).\end{aligned}$$

We're done with the physics. But let me rewrite this in prettier form. The linear coil density n is the number of coils per length: $n = N/D$. So, we can rewrite "N" as nD. Also, pull all the constants out front. This gives

$$\Phi_B = (\mu_0 n^2 D\pi r^2)I.$$

Notice that the flux is proportional to I, the current producing the flux. If you double the current, you double the resulting magnetic field (and magnetic flux). This is true for *any* arrangement of wires, not just for solenoids, because it follows from Ampere's law and the Biot-Savart law.

Notice also that all the stuff in front of the I is a *constant* that depends on the size and coil-density of the solenoid. This constant determines the size of the magnetic flux through the solenoid, for a given current. Keep an eye on this constant. It plays a big role for the rest of this problem.

(c) Take our part (b) answer, $\Phi_B = (\mu_0 n^2 D\pi r^2)I$, and differentiate both sides with respect to t. The left hand-side then becomes $d\Phi_B/dt$, the rate of change of flux. That's what we want. On the right-hand side, you only have to differentiate I, because everything else is constant. So,

$$\frac{d\Phi_B}{dt} = (\mu_0 n^2 D\pi r^2)\frac{dI}{dt}.$$

According to this equation, the rate of change of the magnetic flux is directly proportional to the rate of change of the current. The proportionality factor, $\mu_0 n^2 D\pi r^2$, is the same constant we met in part (b). If this constant is small, then changing the current quickly does not make the magnetic flux change too abruptly. But if the proportionality constant is big, then you can increase the flux rapidly by "turning up" the current slowly.

(d) According to Faraday's law, the induced EMF (voltage) around a current loop is given by the rate of change of the magnetic flux: $\mathcal{E}_{\text{induced}} = -d\Phi_B/dt$. So, in part (c), we already found the induced EMF. It's

$$\mathcal{E}_{\text{induced}} = -\frac{d\Phi_B}{dt} = -(\mu_0 n^2 D\pi r^2)\frac{dI}{dt}. \qquad (*)$$

The minus sign reminds us that the induced EMF "counteracts" the change in flux. After the switch gets closed, the battery increases the current from 0 to $I_{\text{final}} = V/R$. But this changing current produces a changing magnetic flux, which induces a backwards EMF. Although it can't actually "cancel" the battery, the induced EMF slows down the process by which the current increases from 0 to I_{final}.

(e) The proportionality constant we met back in part (b), $\mu_0 n^2 D\pi r^2$, is the solenoid's inductance, L. To confirm this, look at Eq. (*) above. (Many textbooks leave out the minus sign, the physical meaning of which I just discussed.)

From Eq. (*), we can infer the meaning of inductance. When the current through a solenoid or other inductor *changes*, an EMF gets induced by the changing magnetic flux. The induced EMF is proportional to how quickly the current changes, dI/dt. The inductance, L, is simply the proportionality constant. If L is small, then changing the current quickly does not result in a very large induced EMF. By contrast, if L is big, then increasing or decreasing the current, even at a slow rate, produces a large induced EMF. Roughly put, inductance is a system's "sensitivity" to changing current. The more sensitive the system (i.e., the higher the inductance), the more strenuously it reacts to a change in current, by inducing an EMF that opposes the change in current.

OK, I want you to get two things out of this problem. One is the physical meaning of inductance, as discussed in parts (a) and (e). Two is the fact that you calculated the inductance way back in part (b), simply by writing an equation for the magnetic flux Φ_B produced by an arbitrary current I. You got $\Phi_B = (L)I$, where L is an abbreviation for that crazy collection of constants in front of the I. By differentiating and using Faraday's law, we saw that the proportionality constant relating Φ_B to I is the *same* proportionality constant relating $\mathcal{E}_{induced}$ to $\frac{dI}{dt}$. This proportionality constant *is* the inductance.

Consequently, you can calculate inductances without ever using Faraday's law or differential calculus. Just do what you did in part (b)! But don't do it thoughtlessly. I want you to understand *why* that proportionality constant is the inductance, and what it means physically.

QUESTION 43-2

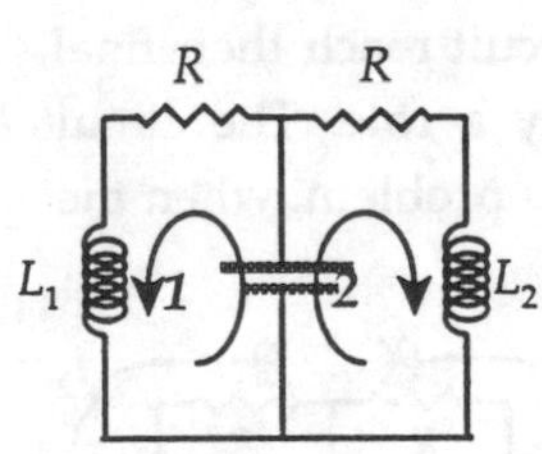

A variable-voltage power supply is connected to two inductors, as drawn here. I've used the standard "battery" icon to indicate the power supply. A small picture of a solenoid always denotes an inductor, whether or not the inductor happens to be a solenoid.

The left inductor has inductance $L_1 = 0.4$ henrys. The right inductor has inductance $L_2 = 0.6$ henrys. The power supply is currently set to $V_0 = 10$ volts. Both inductors are in series with resistors of resistance $R = 2\ \Omega$.

(a) Assuming the power supply has been turned on for a long time, what current flows through the left inductor? The right inductor?

(b) Suddenly, at a moment I'll call $t = 0$, the power supply is abruptly turned down to 0 volts. But current can still flow through its inner workings. Using qualitative reasoning, without doing any calculations, tell me which inductor has more current flowing through it at time $t = 0.1$ s. Explain your reasoning in detail.

(c) A long time after the power supply is set to 0 volts, what current flows through the right inductor? The left inductor? Why?

(d) Now write a differential equation that could be solved for i_1, the current through the left inductor, at arbitrary time t after the power supply is turned to zero. You need not solve the equation, unless you want to. But try to set it up.

(e) Sketch a rough graph of i_1 vs. time and i_2 vs. time. Let $t = 0$ be the moment the power supply was turned to 0 volts. These graphs need not be numerically accurate.

ANSWER 43-2

(a) To answer these questions, you must understand how inductors affect circuits. All the necessary physical insight can be inferred from the solenoid in the previous problem.

In a solenoid or other inductor, a "backwards" EMF gets induced when the current *changes*. That's because a changing current generates a changing magnetic flux, and a changing magnetic flux induces an EMF. Roughly put, when the current through an inductor changes, the inductor tries to "counteract" that change by inducing an EMF. The inductor never succeeds in preventing the current from changing. But it slows down the process of change.

For instance, when the power supply first got turned on, the inductors slowed down the process by which the currents increased from 0 to their final values. Nonetheless, the currents *do* eventually reach their final values, whatever that may be. Therefore, since the power supply has been on for a long time, we can safely assume that the currents in both loops of the circuit have settled to their final values.

What are those final values? You might think the inductors "resist" current, in which case the final currents are lower than they'd otherwise be. But actually, inductors do *not* resist current. An inductor doesn't mind if you pump 1000 amps through it! The inductor "cares" only about *changing* current. That's because only a *changing* current creates a *changing* magnetic flux; and only a *changing* magnetic flux induces an EMF. Therefore, when the current through an inductor stops changing, the inductor becomes "inactive." It just sits there happily, letting current flow through it.

Here's the point: Once the current through an inductor reaches its final value and *stops changing*, no EMF gets induced; the inductor might as well be a straight piece of wire.

That's a tremendously important insight. When the currents in a circuit reach their final, steady-state values (and therefore *stop changing*), inductors no longer play a role. The circuit behaves *as if* the inductors were replaced by regular pieces of wire. So, in this problem, when the power supply has been on for a long time, the inductors have no effect on the current.

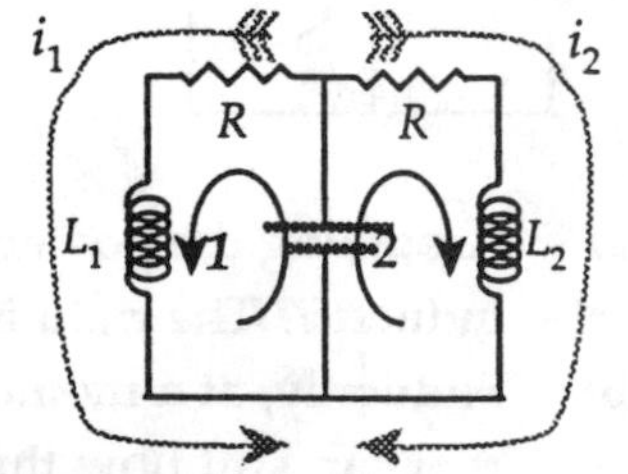

This conclusion makes it easier to find the current through the inductors. Since the inductors no longer "matter," we need to worry only about the power supply and the resistors. Let's solve the circuit using Kirchhoff's laws in the usual way. Specifically, break the circuit into loops, and write separate equations about each loop. Walking around loop 1, you pass through the power supply ($\Delta V = +V_0$), and then you pass through the resistor, walking with the current ($\Delta V = -i_1 R$). Although you also walk through L_1, it contributes nothing to our voltage equation, because the steady current induces no EMF. Similar considerations apply to loop 2. Recall that the total voltage change around any complete loop is 0. So,

Loop 1 (when current is steady) $V_0 - i_1 R = 0.$

Loop 2 (when current is steady) $V_0 - i_2 R = 0.$

Solving for the currents immediately yields

$$i_1 = i_2 = \frac{V_0}{R} = \frac{10\text{ V}}{2\ \Omega} = 5\text{ A}.$$

In summary, since an inductor generates an EMF only when the current through it *changes*, inductors do not enter into your reasoning when you're solving for final (steady-state) currents.

(b) The power supply is what pushes current around the circuit. Therefore, after we turn the power supply down to 0 volts, the current in the circuit eventually drops to zero. If the circuit contained just resistors, with no inductors, then the current would drop to zero *immediately*. But inductors add a new twist. Sure, they can't prevent the current from dropping to 0. But they slow down the process. For instance, the current through L_1 does not immediately drop from $i_1 = 5$ A to $i_{\text{final}} = 0$ A. This drop-off takes a few hundredths or tenths of a second.

For this reason, at time $t = 0.1$ s, current still flows through each inductor. Both currents are less than 5 A, but more than 0. The question is, which inductor carries *more* current at that time?

To answer, think about what inductance means, physically. It's the inductor's "sensitivity" to a change in current. The higher the inductance, the more EMF gets induced when the current changes. Crucially, the induced EMF "counteracts" the changing current.

Let's apply these general conclusions to this particular problem. Each inductor "fights" the process by which the current through it drops from 5 A to 0. But the inductor with more inductance generates a larger EMF, and therefore "fights" the change in current more strenuously. As a result, the inductor with higher inductance does a better job of slowing down the process by which current drops from 5 A to 0. Therefore, since $L_2 > L_1$, i_2 decreases more gradually than i_1 decreases. So, at any given time during the drop-off, i_2 is bigger than i_1.

(c) As just discussed, an inductor cannot prevent the current from changing. All it can do is slow down the process. If this circuit included no inductors, then the current around each loop would drop to zero *immediately* after the power supply gets turned to 0 volts. The inductors make the drop-off occur gradually instead of abruptly. But eventually, the current drops to 0, in both inductors.

So once again, the inductors play no role in a circuit once the current settles down to its final value—i.e., once the current *stops changing*. It's just like part (a), where the power supply was turned on and left on for a long time. In that case, the currents eventually settled at the same value they would have reached if the inductors were replaced by straight wires—namely, 5 amps. Similarly, when the power supply is turned to 0 volts, the current eventually drops to 0, again the same value it would have reached if the inductors were replaced by straight wires. From these examples, you can correctly infer that *inductors have no effect on the long-term (steady state) currents in a circuit*, assuming the circuit also contains a resistor. While keeping this rule of thumb in mind, you should think out each new case individually, too.

(d) When you need to "solve a circuit," use Kirchhoff's laws. We can write an equation about loop 1, and if necessary, about loop 2.

When walking around loop 1, you go through the power supply. But since it's now set to 0 volts, it contributes no voltage. Then you walk through the resistor, in the same direction as the current flows ($\Delta V = -i_1R$). Finally, you walk through the inductor, again with the current. In part (a), the inductor contributed no voltage, because the current was steady ($di_1/dt = 0$). But now, the current in the inductor *is* changing, at some unknown rate di_1/dt. By the definition of inductance, the corresponding induced EMF is $\Delta V = \mathcal{E}_{\text{induced}} = -L_1(di_1/dt)$. *The minus sign indicates that the induced voltage tries to prevent the current from changing.* I'll talk more about the minus sign below.

Sum up these voltage changes and set it equal to zero, to get

Loop 1 (power supply set to 0 volts)
$$0 = \Delta V_{\text{battery}} + \Delta V_{\text{resistor}} + \Delta V_{\text{inductor}}$$
$$= 0 - i_1R - L_1\frac{di_1}{dt}.$$

Without further ado, we can solve this first-order differential equation for $i_1(t)$. We don't need additional information from loop 2. But if we did, equivalent reasoning would give us

Loop 2 (power supply set to 0 volts)
$$0 = -i_2R - L_2\frac{di_2}{dt}.$$

Digression for students who like detailed theoretical stuff

In general, an inductor contributes a $-L(di/dt)$ term to your Kirchhoff loop equations, if you walk in the direction of current flow. Let me explain the minus sign, in this particular case. The induced EMF tries to prevent the counterclockwise current from decreasing. In other words, it tries to push current counterclockwise around loop 1. So, the induced voltage is counterclockwise. Since I walk counterclockwise around the loop, you might think this counterclockwise voltage should count as *positive*. And you'd be right: $\Delta V = -L(di/dt)$ must be positive. If this seems like a contradiction, remember that the current is *decreasing*. In other words, di/dt is negative. Therefore, the minus sign in $\Delta V = -L(di/dt)$ ensures that the overall expression comes out positive, as it must in this case.

(e) At $t = 0$, the current through each inductor is $i_1 = i_2 = 5$ amps. Then, both currents gradually fall to 0. Crucially, i_2 decreases more gradually than i_1, as we saw in part (b). So, you might think the graphs look something like this. That's a good guess.

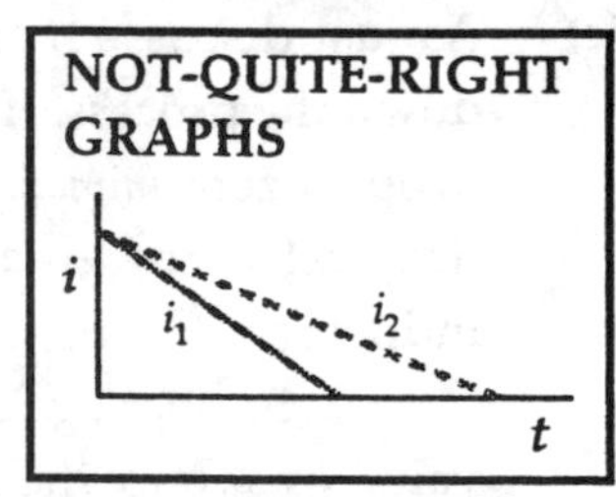

To obtain more accurate graphs, we must solve the differential equations from part (d). The solution will tell us whether the currents decrease at a steady rate or an unsteady rate. Because the mathematics is simple and interesting, I'll work through it in detail.

Interesting calculus starts here. Let's start with our loop 1 equation for i_1, which we can rewrite as

$$L_1\frac{di_1}{dt} = -i_1R.$$

To solve using "separation of variables," we must get all the current terms on one side, and all the time terms on the other. So, multiply through by dt, and divide through by i_1. As long as we're at it, divide through by L_1, too. This gives

$$\frac{di_1}{i_1} = -\frac{R}{L_1} dt.$$

Now integrate both sides, after figuring out the limits of integration. The time starts at $t_0 = 0$ and goes to arbitrary later time t. The current starts at initial value $i_0 = 5$ amps. At later time t, the current has unknown value i_1, which is what we're solving for. So,

$$\int_{i_0=5\text{ A}}^{i_1} \frac{di_1}{i_1} = -\int_0^t \frac{R}{L_1} d$$

Evaluate the integrals to get

$$\ln \frac{i_1}{i_0} = -\frac{R}{L_1} t.$$

Finally, exponentiate both sides, remembering that $e^{\ln x} = x$, and then multiply through by i_0. This gives us

$$i_1 = i_0 e^{-t/\tau_L},$$

where I've defined the time constant τ_L as L_1/R.

End of calculus

We now see that i_1 decays exponentially. At time $t = 0$, the exponential factor equals 1, and hence i_1 quals $i_0 = 5$ A. At very large times, the exponential gets extremely close to zero, indicating that essentially no current flows.

Let's think about the physical meaning of the time constant $\tau_L = L/R$. Mathematically speaking, the bigger this time constant, the more gradually the current drops to zero. Does this make physical sense? Yes! As emphasized above, the bigger the inductance, the more gradually the current changes. Our mathematical expression for τ_L confirms this physical intuition.

CORRECT GRAPHS

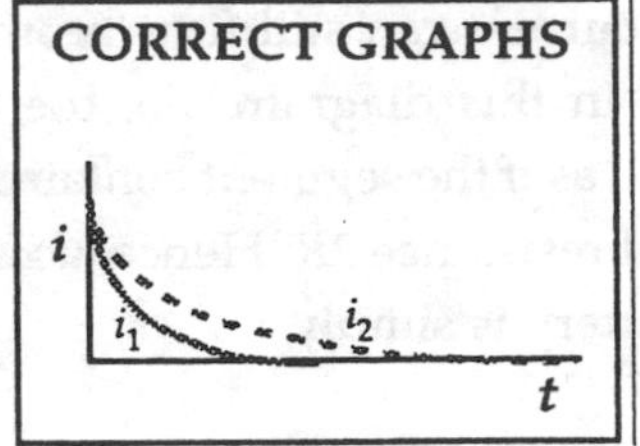

Now we can draw more accurate graphs of i_1 and i_2. Both currents decay exponentially. But i_2 drops off more gradually. Physically, that's because L_2 is bigger than L_1; inductor 2 does a better job of "fighting" the change in current. Mathematically, the time constant for inductor 1, $\tau_L = L_1/R$, is smaller than the time constant for inductor 2, $\tau_L = L_2/R$. So, the current through inductor 1 takes less time to decay.

QUESTION 43-3

Consider the circuit drawn here. Both resistors have the same resistance, R. The switch is closed at time $t = 0$.

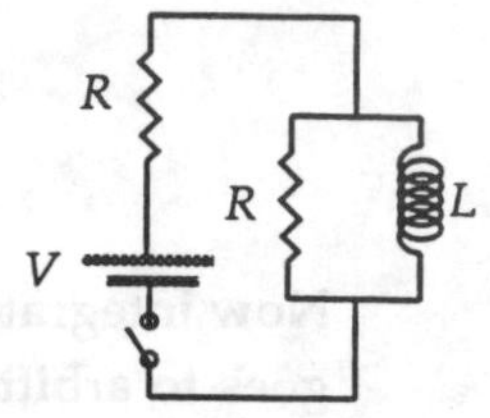

(a) *Immediately* after the switch closes, what current flows through the battery?

(b) *A long time* after the switch closes, what current flows through the battery? Hint: You can answer these questions without carrying out any complicated calculations or derivations. Think physically about how an inductor behaves.

ANSWER 43-3

(a) *Ask your instructor whether you're expected to find the current through this circuit at arbitrary time.* If so, then you need to deal with differential equations like those in question 43-1d and 43-1e. But most instructors tend to ask more conceptual questions about the current *immediately after* the switch is closed, or a *long time* later. You can solve these "limiting case" problems intuitively.

When the current through an inductor *changes*, the changing magnetic flux induces an EMF that opposes the changing current. Although the inductor cannot prevent the current from changing, it delays the process of change, making it more gradual.

For example, before the switch closes, no current flows through the inductor. After the switch closes, the battery tries to increase the current from 0 to some final value, call it i_{final}. If the circuit contained no inductors, the current would shoot up to its final value immediately. But the inductor doesn't let this happen. The inductor counteracts this change in current, by making it gradual instead if abrupt. In other words, the current through the inductor gradually increases from 0 to i_{final}, typically over a time interval of a few centiseconds.

Here's the point. Since the inductor only *gradually* allows the current through it to increase from 0 to i_{final}, the current through the inductor *immediately* after the switch closes is still essentially 0. In general, **when you open or close a switch, the current through the inductor immediately *after* the opening (or closing) equals the current immediately *before* the opening (or closing)**. That's because the current through the inductor changes gradually, not abruptly.

Though I presented this argument intuitively, you can also derive it mathematically.

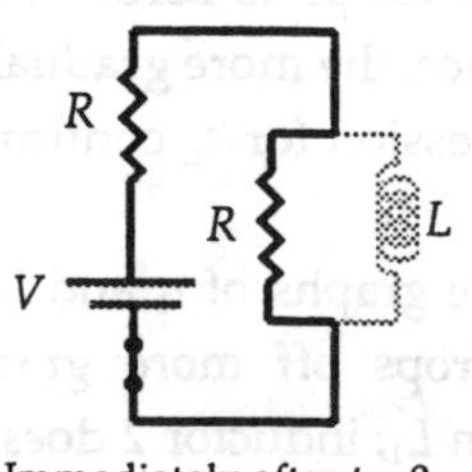

Immediately after $t = 0$, no current flows through inductor (yet).

So far, I've established that, immediately after the switch closes, no current flows through the inductor. But current can still flow around the *other* loop of the circuit, as highlighted in this diagram. So, the circuit behaves as if it contained a single loop, i.e., as if the segment containing the inductor were missing. This part has total resistance $2R$. Hence, according to Ohm's law, the current through the battery is simply

$$i_0 = \frac{V}{2R}.$$

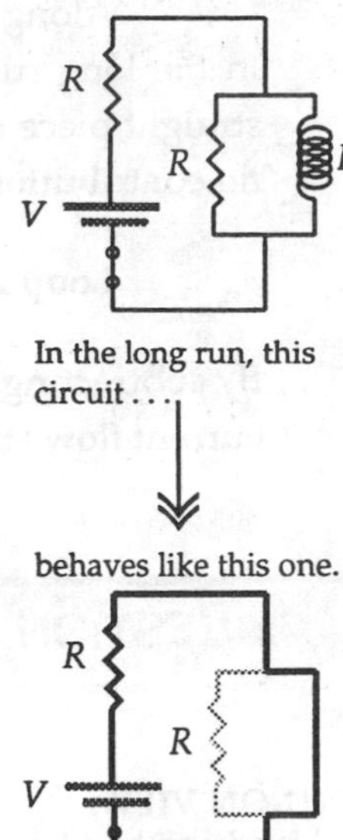

In the long run, this circuit . . .

behaves like this one.

(b) As time passes, the current through the inductor gradually increases to its maximum value, whatever that happens to be. But once the current through the inductor reaches this final value and *stops changing*, no EMF (voltage) gets induced. Remember, an inductor "reacts" only to *changing* current, not to steady current. Therefore, when the current reaches its final steady value, the inductor no longer does anything. It might as well be a straight, resistanceless piece of wire. So, when addressing the circuit at a much later time, think of the inductor as a straight segment of wire. You could even redraw the circuit with that replacement made explicit, as I've done here.

When current reaches the fork in the road (above the second resistor), normally some of it would flow leftward and some would flow rightward. But one of those two paths is resistanceless! Consequently, *all* the current flows down the resistanceless path. To verify this, treat the circuit as a two-loop system, and apply Kirchhoff's laws (with the inductor replaced by straight wire). You'll see that no current flows through the second resistor. At the end of this problem, I'll demonstrate this explicitly. But for now, let me keep going.

So far, I've shown that *all* the current flows along the blackened path in my diagram. This path has resistance R. So, the current in the circuit is

$$i_{\text{final}} = \frac{V}{R},$$

double its value immediately after the switch closed. The current gradually increased from its initial to its final value.

Summary. **In the long run, when the currents reach their final values and stop changing, the inductors play no role in the circuit. When finding these steady-state currents, treat inductors as resistanceless wires.**

Appendix: I'll now prove that, in the long run, all the current flows through the inductor instead of the second resistor.

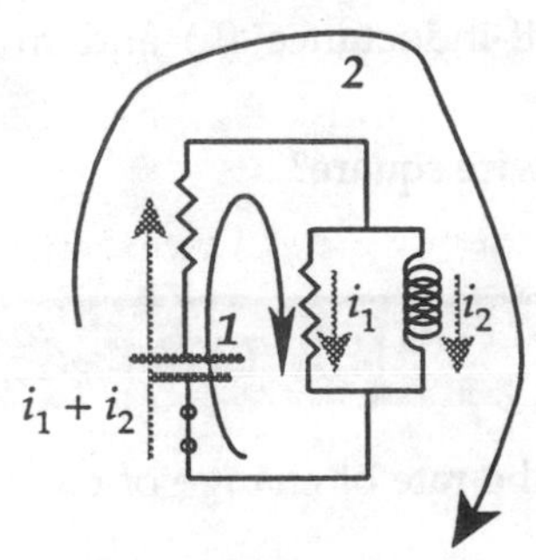

Divide the circuit into two loops. My first loop includes the battery and both resistors. My second loop goes along the "perimeter" of the circuit, including the battery, the first resistor, and the inductor. Let i_1 and i_2 denote the current through the second resistor and through the inductor, respectively. According to the "intersection rule," the current through the battery and first resistor must be $i_1 + i_2$.

Now we can write loop equations in the usual way, by "walking" around each loop, summing up the voltage contributions, and setting them equal to zero. Keep in mind that the current through the first resistor is $i_1 + i_2$. So, for loop 1 we get

Loop 1 $$V - (i_1 + i_2)R - i_1 R = 0.$$

Along loop 2, you walk through the inductor instead of the second resistor. But remember: In the long run, when the currents have settled to their final values, the inductor behaves like a straight piece of wire, because it no longer generates an EMF. So, along loop 2, the inductor makes no contribution. Hence,

Loop 2 (when currents are steady) $V - (i_1 + i_2)R = 0.$

By subtracting the loop 2 equation from the loop 1 equation, you can immediately see that $i_1 = 0$. No current flows through the second resistor. It all flows through the inductor, the path of no resistance.

QUESTION 43-4

END-ON VIEW,
looking down the solenoid tube. It extends distance l into the page.

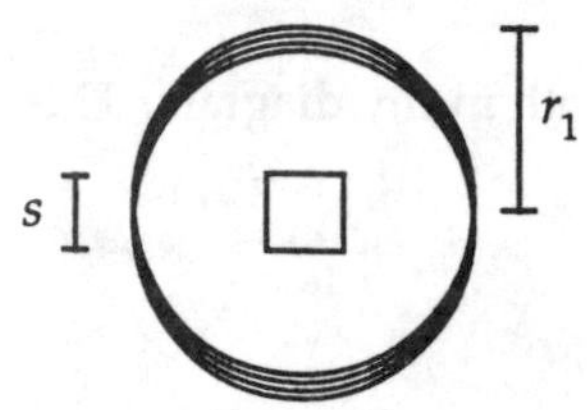

Consider a standard cylindrical solenoid of length l and radius r_1, consisting of N coils. A small metal square of wire, of side length s and resistance R, is placed inside the solenoid. The plane of the square is parallel to the plane of a solenoid coil.

Initially, no current flows through the solenoid. But starting at time $t = 0$, a variable power supply steadily increases the current through the solenoid from 0 to I_f, over time T.

In this problem, you may use answers previously derived in this chapter. You need not rederive them from scratch.

(a) In terms of the given physical quantities, find dI/dt, the rate of change of current *in the solenoid.*

(b) While the current is increasing, what is the induced EMF *in the solenoid?*

(c) Let "M" denote the "mutual inductance" between the solenoid and the square. By definition, it's the proportionality constant relating the rate of change of current *through the solenoid* to the induced EMF (voltage) *in the square:*

$$\mathcal{E}_{\text{induced in square}} = -M\frac{dI_{\text{solenoid}}}{dt}.$$

Find the mutual inductance between the solenoid and the square.

(d) Explain in simple language the physical difference between self-inductance (L) and mutual inductance (M).

(e) Between time $t = 0$ and $t = T$, how much total heat dissipates in the wire square?

ANSWER 43-4

(a) The current *steadily* changes from 0 to I_f in time $\Delta t = T$. Therefore, the rate of change of current is simply

$$\frac{dI}{dt} = \frac{\Delta I}{\Delta t} = \frac{I_f - 0}{T} = \frac{I_f}{T}.$$

If you worried about the induced EMF in the solenoid, keep in mind that the person controlling the power supply "compensates" for the induced EMF, ensuring that I_{solenoid} increases steadily.

(b) We could solve this from scratch, as follows:

1) Starting with Ampere's law, find the magnetic flux through the solenoid in terms of the current: $\Phi_B = (\text{constants})I$.
2) Differentiate both sides with respect to time, to get $\frac{d\Phi_B}{dt} = (\text{constants})\frac{dI}{dt}$.
3) Recall that the rate of change of flux, $d\Phi_B/dt$, gives us the induced EMF in the solenoid: $\mathcal{E}_{\text{induced}} = -\frac{d\Phi_B}{dt} = (\text{constants})\frac{dI}{dt}$.

If this strategy sounds familiar, it's because we already implemented it in question 43-1. From step 3, you can see that "(constants)" is the *inductance*, i.e., the proportionality constant between the rate of change of current and the induced voltage. In question 43-1, we carried out that strategy, and found the inductance of a solenoid to be

$$L_{\text{solenoid}} = \mu_0 n^2 l \pi r^2,$$

where n denotes the linear coil density, the number of coils per length (N/l). Since we already know the inductance, we can immediately find the induced EMF:

$$\mathcal{E}_{\text{induced in solenoid}} = L_{\text{solenoid}}\frac{dI}{dt}$$
$$= (\mu_0 n^2 l \pi r_1^2)\frac{I_f}{T},$$

where I used my part (a) answer.

If you solved this problem from scratch, instead of looking up our earlier answer for L_{solenoid}, it may have been because you didn't realize the physical meaning of inductance. The inductance specifies how "sensitive" the system is to a change in current, i.e., how big an EMF gets induced for a given rate of change of the current.

(c) To solve, we can modify and implement the three-step strategy outlined, but not used, in part (b). The goal of the original strategy was to find the EMF induced *in the solenoid* when the current through the solenoid changes. By contrast, our current goal is to find the EMF induced *in the square* when the current through the solenoid changes. Therefore, we now care about the changing magnetic flux through the square, not the changing magnetic flux through the solenoid. Hence, the three-step strategy gets modified as follows:

1) Starting with Ampere's law, find the magnetic flux through the square in terms of the current in the solenoid: $\Phi_{B\text{ square}} = (\text{stuff})I_{\text{solenoid}}$, where "stuff" is a bunch of constants.
2) Differentiate both sides with respect to time, to get $\frac{d\Phi_{B\text{ square}}}{dt} = (\text{stuff})\frac{dI_{\text{solenoid}}}{dt}$.
3) Recall that the rate of change of flux, $d\Phi_B/dt$, gives us the induced EMF in the solenoid square: $\mathcal{E}_{\text{induced in square}} = -\frac{d\Phi_{B\text{ square}}}{dt} = -(\text{stuff})\frac{dI_{\text{solenoid}}}{dt}$.

Actually, we only need to carry out step 1. Here's why. In step 3, "stuff" is the mutual inductance, by definition:

Definition of mutual inductance $$\mathcal{E}_{\text{induced in square}} = -M\frac{dI_{\text{solenoid}}}{dt}.$$

Amazingly, that same "stuff" appears way back in step 1. For this reason, step 1 alone gives us the mutual inductance. Steps 2 and 3 merely confirm that we've found it. If this seems weird, review question 43-1. We found the solenoid's inductance in part (b), even though we didn't *realize* it until part (e).

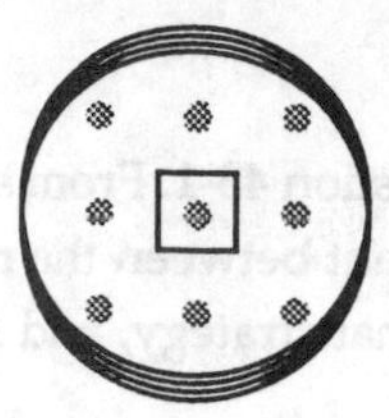

Uniform magnetic field pierces through the square.

OK, let's do step 1. We need to know the magnetic flux through the square when arbitrary current I_{solenoid} flows around the solenoid. Well, using Ampere's law in question 43-1b, we found the magnetic field inside the solenoid. It's $B = \mu_0 n I_{\text{solenoid}}$, where n denotes the number of coils per length. Since that field is uniform, and since it points in the same direction that the square "faces," we can quickly calculate the magnetic flux through the square:

$$\Phi_{B\text{ square}} = \int \mathbf{B}\cdot d\mathbf{A}$$

$$= \int B\,dA \quad \text{[since } \mathbf{B} \text{ and } d\mathbf{A} \text{ face the same direction]}$$

$$= B\int dA \quad \text{[since } B \text{ is uniform over the area of the square]}$$

$$= BA_{\text{square}}$$

$$= \mu_0 n I_{\text{solenoid}} s^2$$

$$= (\mu_0 n s^2) I_{\text{solenoid}},$$

where in the last step, I reordered the constants. Notice that I used the area of the square, not the area of the solenoid.

So, "stuff" equals $\mu_0 n s^2$. As emphasized above, that "stuff" *is* the mutual inductance. To prove this, differentiate both sides of $\Phi_{B\text{ square}} = (\mu_0 n s^2) I_{\text{solenoid}}$ with respect to t, and apply Faraday's law. So, the mutual inductance between the solenoid and the square is

$$M = \mu_0 n s^2,$$

where the linear coil density is $n = N/l$.

(d) Let me write down the relevant equations, for easy reference.

$$\mathcal{E}_{\text{induced in solenoid}} = -L\frac{dI_{\text{solenoid}}}{dt}$$

$$\mathcal{E}_{\text{induced in square}} = -M\frac{dI_{\text{solenoid}}}{dt}$$

In some ways, self-inductance and mutual inductance are similar. They both tell us how "sensitive" a system is to a changing current. Specifically, the inductance specifies how big an EMF gets induced due to a change in current. EMF gets induced because the changing current generates a changing magnetic flux.

Here's the difference between self-inductance and mutual inductance. *Self*-inductance (L) specifies a system's sensitivity to a change in its own current. For instance, the self-inductance of a solenoid tells us how strongly the solenoid "reacts" to a change in *its own* current. By contrast, mutual inductance (M) specifies how sensitive one system is to a change in *another system's* current. Here, M tells us how strongly the *square* reacts to a change in the *solenoid's* current.

(e) Heat is a kind of energy. It gets created when current flows through a resistor. By definition, power is the *rate* at which energy dissipates. As we'll see, in this case the power stays constant. In other words, heat dissipates at a steady rate. So, if we can find the power dissipated in the square, and the time over which it dissipates, then we can easily calculate the heat energy. Mathematically speaking, since $P = \frac{\Delta \text{Energy}}{\Delta t} = \frac{\text{Heat}}{\Delta t}$, we can solve for Heat once we know the power and the time.

Subproblem: Find the power dissipated in the square.

As we've seen before, the power dissipated in a resistor is

$$P = Vi,$$

which can be rewritten as $P = V^2/R$ or $P = i^2R$. These equations "agree," because $V = iR$. I'll use the "$P = V^2/R$" version of the equation.

Crucially, since we're talking about the power in the *square*, I must be careful to use the induced voltage in the square, and the resistance of the square. Don't fall into the common trap of plugging in $\mathcal{E}_{\text{induced in solenoid}}$. We need $\mathcal{E}_{\text{induced in square}}$. But that's straightforward, given part (c). The mutual inductance specifies how big an EMF gets induced in the square due to a changing current in the solenoid:

$$\begin{aligned} \mathcal{E}_{\text{induced in square}} &= -M\frac{dI_{\text{solenoid}}}{dt} \\ &= -(\mu_0 n s^2)\frac{dI_{\text{solenoid}}}{dt} && \text{[using my part (c) answer]} \\ &= -(\mu_0 n s^2)\frac{I_f}{T}. && \text{[using my part (a) answer]} \end{aligned}$$

Now that we know the voltage around the square, we can immediately get the power developed in the square:

$$P = \frac{\mathcal{E}_{\text{square}}^2}{R} = \frac{\mu_0^2 n^2 s^4 I_f^2}{RT^2},$$

where I_f denotes the final current in the *solenoid*.

End of subproblem

At this point, we've found the power dissipated in the square. This dissipation occurs *only* when the changing current in the solenoid generates a changing magnetic flux. In other words, current flows around the square *only* while the solenoid's current *changes* from 0 to I_f. This process takes time T. By contrast, once the solenoid's current settles at its final value and stops changing, the magnetic field in the solenoid stays constant. Therefore, the magnetic flux through the square no longer *changes*. Therefore, no EMF or current gets induced in the square.

Since the power in the square stays "turned on" for time T, and since power is energy per time, the total heat dissipated is

$$Heat = \frac{\text{energy}}{\text{time}} \times \text{time} = PT = \frac{\mu_0^2 n^2 s^4 I_f^2}{RT^2}.$$

QUESTION 43-5

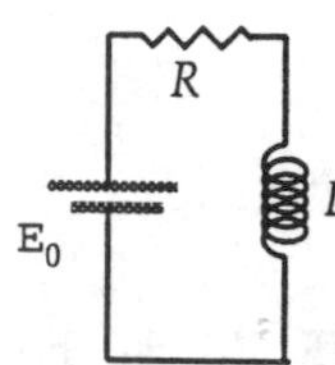

Consider this circuit, consisting of a battery $\mathcal{E}_0 = 9.0$ V, a resistor $R = 0.90\ \Omega$, and an inductor $L = 0.80$ H. (That's a huge inductance.)

(a) The battery has been connected for a long time. What current flows through the circuit?

(b) Suddenly, at time $t = 0$, the battery gets "turned off." How much heat gets dissipated in the resistor *after* $t = 0$. Hint: You can solve quickly, without complicated math.

ANSWER 43-5

(a) When the current through an inductor *changes*, it produces a changing magnetic flux, which induces an EMF (voltage). By contrast, after the battery has been connected for a long time, the current settles to its final value, and stops changing. Therefore, the magnetic flux through the inductor stops changing. As a result, no EMF gets induced; the inductor plays no role in the circuit. It behaves just like a straight piece of wire.

In summary, after the battery has been connected for a long time, the circuit behaves as if it consisted only of a battery and a resistor. The current is therefore

$$i = \frac{\mathcal{E}_0}{R} = \frac{9.0\text{ V}}{0.90\ \Omega} = 10\text{ A}.$$

In general, an inductor can't change the final, steady-state current through a circuit. But the inductor causes the circuit to reach its final current gradually, instead of abruptly.

(b) If the inductor weren't there, the current would *immediately* drop from 10 A to 0 when the battery gets shut off. In the long run, the inductor can't prevent this drop-off from happening. But the inductor "counteracts" the decreasing current by inducing an EMF, an EMF that tries to keep the current flowing. Consequently, the current drops from 10 amps to 0 *gradually* instead of abruptly. This unusually large inductor keeps current flowing for several seconds after the battery shuts off.

To solve for the heat dissipated during those few seconds, you could calculate the power as a function of time (by first solving for the current as a function of time), and then integrate this power over time. What a mess. Fortunately, we can sidestep heavy math by using energy conservation. Here's the crucial physical insight: *An inductor with current flowing through it has potential energy stored in it.* As I'll show below,

$$U_L = \frac{1}{2}Li^2.$$

Let me first use this potential energy formula to solve the problem. Then, I'll explain both intuitively and mathematically where it comes from.

Immediately after the battery gets disconnected at $t = 0$, the total energy stored in the circuit is the inductor's energy, $U_L = (1/2)Li^2$. Remember, the resistor doesn't "store" potential energy; it merely dissipates energy as heat. And the battery, though it stores chemical energy, is no longer part of the system. So, as the current in the circuit drops to zero, the energy initially stored in the inductor goes to zero. By conservation, this energy can't just disappear. *All the energy stored in the inductor converts into heat that dissipates in the resistor* during the few seconds that current continues to flow. In other words, the inductor's potential energy at $t = 0$ equals the heat energy generated after $t = 0$, during the few seconds current continues to flow.

Energy when battery is disconnected = Energy after current stops flowing, a few seconds later

$$\frac{1}{2}Li^2 = \textit{Heat}$$

So,

$$\textit{Heat} = \frac{1}{2}Li^2 = \frac{1}{2}(0.80 \text{ H})(10 \text{ A})^2 = 40 \text{ joules}.$$

Now I'll go back and explain why an inductor has potential energy. "Potential energy" is the energy stored in a system because work was done on it. For instance, if I expend 60 joules of energy lifting a physics book from the floor onto a desk, then that textbook now has 60 joules of gravitational potential energy stored in it. In general, $\Delta U = W_{\text{done on system}}$.

Similarly, in order to increase the current through an inductor, the battery must do *extra* work, above and beyond the work needed to keep current flowing through the resistor. Here's why. When the current through the inductor rises from 0 to i, the inductor generates a "backwards" EMF that counteracts the changing current. In order to fight this backwards EMF, the battery must expend extra energy. In other words, the battery must do extra work. Consequently, the battery "burns out" faster than it otherwise would! Fortunately, the inductor's backwards EMF exists only while the current changes. Once the current settles to it final value, the inductor stops playing a role, and the battery no longer needs to work any harder than usual.

To calculate how much extra work the battery performs while increasing the current in the inductor from 0 to i, I'll reason as follows. First, I'll figure out the extra power the battery must exert to "fight" the inductor's backwards EMF. Then, since power is energy (work) per time, I'll integrate over this power to find the energy (work) expended by the battery.

Well, the inductor generates an EMF $\mathcal{E}_{\text{induced}} = -L(di/dt)$, where the minus sign indicates "backwards." And remember, power is voltage times current: $P = Vi$. So, to "fight" the inductor, the battery must supply extra power

$$P = \mathcal{E}_{\text{induced}} i = L\frac{di}{dt}i.$$

To find the work (energy) expended by the battery during this fight, integrate the power over time. As explained above, this work tells us the potential energy stored in the inductor:

$$\begin{aligned}\Delta U_L = W &= \int P\,dt \\ &= \int Li\frac{di}{dt}dt \\ &= \int Li\,di \qquad \text{[since the } dt\text{'s cancel]} \\ &= \frac{1}{2}Li^2.\end{aligned}$$

Although a test might not require to derive this equation, you should understand *why* an inductor has potential energy stored in it, intuitively.

QUESTION 43-6

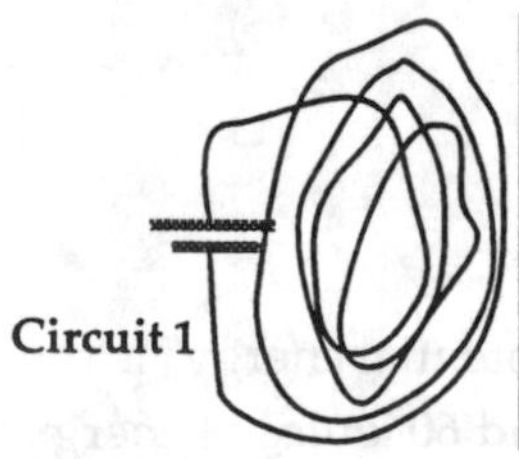

Circuit 1

Circuit 2 $R = 1.5\ \Omega$ $V_0 =$ 3.0 V

Circuit 1 sits next to circuit 2 on a table. Recall from an earlier problem that "mutual inductance" relates the rate of change in current in one system to the EMF induced by that changing current in *another* system:

$$\left|\mathcal{E}_{2\ \text{induced}}\right| = M\frac{dI_1}{dt},$$

where I_1 denotes the current in system 1, and $\mathcal{E}_{2\ \text{induced}}$ denotes the induced voltage in system 2. Here, the mutual inductance between circuit 1 and circuit 2 is $M = 1.0 \times 10^{-4}$ henrys. When I_1 increases, the induced EMF in circuit 2 is clockwise. Circuit 2 has negligible self-inductance

By fiddling with the power supply on circuit 1, I make I_1 fluctuate as graphed here. Notice that the time axis is subdivided into milliseconds.

Current in circuit 1 vs. time

I_1 (amps): 0, 1, 2

t (milliseconds): 2, 4, 6

The battery and resistor in circuit 2 have voltage $V_0 = 3.0$ V and $R = 1.5\ \Omega$.

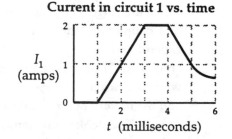

(a) Without using formulas, explain why changing the current in circuit 1 induces an EMF in circuit 2.

(b) (*Very hard*) Sketch a rough graph of the current in circuit 2 as a function of time, between $t = 0$ and $t = 6$ milliseconds. Your graph need not be numerically accurate.

(c) What is the current in circuit 2 at time $t = 2$ milliseconds?

(d) What is the current in circuit 2 at time $t = 3.5$ milliseconds?

(e) **At time $t = 4.5$ milliseconds?**

(f) **If circuit 2 were moved farther away from circuit 1, would the mutual inductance go up, go down, or stay the same? Explain your answer.**

ANSWER 43-6

(a) According to Faraday's law, a changing magnetic flux induces an EMF. For instance, an EMF (voltage) gets induced in circuit 2 when the magnetic flux through that current loop *changes*. Because the current in circuit 1 generates a magnetic field, it contributes to the magnetic flux through circuit 2. As long as I_1 stays constant, the magnetic field it generates stays constant, and therefore the magnetic flux remains constant. But when I_1 *increases*, it creates an increasing magnetic field and flux through circuit 2. This increasing flux induces an EMF around circuit 2.

In summary, a changing current is circuit 1 produces a changing magnetic flux through circuit 2, and hence, an induced EMF in circuit 2. (Similarly, a changing current in circuit 2 would induce an EMF in circuit 1.)

(b) As just discussed, the changing current in circuit 1 induces a voltage in circuit 2. But this induced EMF is not the *only* EMF in circuit 2. Circuit 2 contains a battery of voltage $V_0 = 3$ V. The total voltage around circuit 2 comes from the 3-volt battery *and* from the induced EMF, if any:

$$\mathcal{E}_{2\,\text{total}} = V_0 + \mathcal{E}_{2\,\text{induced}}.$$

When $\mathcal{E}_{2\,\text{induced}} = 0$, the "default" current in circuit 2 is

When there's no induced EMF . . . $\quad i_2 = \dfrac{V_0}{R} = \dfrac{3.0\text{ V}}{1.5\,\Omega} = 2.0\text{ A}.$

The battery makes 2.0 amps of current flow around circuit 2. If we could "turn off" induction, then i_2 would always equal 2 amps. But the induced EMF nudges i_2 above or below this default value.

To draw the i_2 vs. t graph, we must think about how the induced voltage in circuit 2 relates to I_1, the current in circuit 1. Well, as explained in part (a), the induced EMF in circuit 2 is proportional to dI_1/dt. On the I_1 vs. t graph, dI_1/dt is the *slope*. In summary, $\mathcal{E}_{2\,\text{induced}}$ is proportional to the *slope* of the I_1 vs. t graph. So, to figure out how i_2 gets "nudged" away from its default value of 2 amps, we must pay attention to the slope of I_1 vs. t.

Current in circuit 2 vs. time

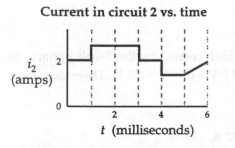

Given this insight, let's address each segment of the graph separately, to avoid getting overwhelmed. During the first millisecond, when no current flows in circuit 1, $\mathcal{E}_{2\,\text{induced}} = 0$. Therefore, i_2 equals its default value, 2.0 A.

Between $t = 1$ and $t = 3$ milliseconds, the current in circuit 1 increases at a *steady rate*. In other words, dI_1/dt is *constant*. Therefore, the magnetic flux through circuit 2 increases at a *steady rate*. So, according to Faraday's law, $\mathcal{E}_{2\,\text{induced}}$ is *constant* during that interval. Because $\mathcal{E}_{2\,\text{induced}}$ is constant, so is i_2. The problem tells us that $\mathcal{E}_{2\,\text{induced}}$ is clockwise when I_1 increases. So, this clockwise induced EMF "knocks" i_2 above its default value of 2 amps.

A common error is to draw i_2 as increasing during this interval. But actually, $\left|\mathcal{E}_{2\,\text{induced}}\right| = M\frac{dI_1}{dt}$ *a steadily increasing current in circuit 1 induces a **constant** EMF in circuit 2.*

From $t = 3$ to $t = 4$ milliseconds, I_1 doesn't change. Therefore, the magnetic flux through circuit 2 stays constant. Therefore, no EMF gets induced in circuit 2. During that interval, the only voltage in circuit 2 comes from the battery. The current returns to its default value, 2.0 amps.

Between $t = 4$ and $t = 5$ μs, the *decreasing* current in circuit 1 induces a counterclockwise EMF around circuit 2. This counterclockwise EMF opposes the 3 volts supplied by the battery. Therefore, during this interval, i_2 is less than its default value. Nonetheless, because I_1 changes at a *steady rate*, $\mathcal{E}_{2\,\text{induced}}$ is *constant*, and hence i_2 is constant.

From $t = 5$ to $t = 6$ milliseconds, I_1 continues to decrease. Hence, a counterclockwise EMF continues to get induced in circuit 2. As a result, i_2 stays below its default value. But notice that I_1 does not decrease at a steady rate. It decreases less rapidly as time goes on. In other words, the slope of the I_1 vs. t graph gets smaller in magnitude. So, the induced EMF gets smaller. For this reason, i_2 gets closer and closer to its default value.

(c) I'll first calculate the induced EMF in circuit 2 caused by the changing current in circuit 1. Then, I'll add that induced voltage to the "permanent" voltage generated by the 3-volt battery. Finally, I'll substitute this total voltage into Ohm's law, to find the current i2.

Find the induced voltage. Start with the mutual inductance equation,

$$\left|\mathcal{E}_{2\,\text{induced}}\right| = M\frac{dI_1}{dt},$$

On a graph, derivatives correspond to slopes. So, dI_1/dt at time $t = 2$ μs is simply the slope of the I_1 vs. t graph at that moment. From the graph, we can easily calculate this slope. It's

$$\frac{dI_1}{dt} = \frac{\text{rise}}{\text{run}} = \frac{2.0\text{ A} - 0}{.003\text{ s} - .001\text{ s}} = 1.0 \times 10^3\text{ A/s}.$$

Substitute this result into the mutual inductance equation, to get

$$\begin{aligned}\left|\mathcal{E}_{2\,\text{induced}}\right| &= M\frac{dI_1}{dt}\\ &= (1.0 \times 10^{-4}\text{ H})(1.0 \times 10^3\text{ A/s})\\ &= 0.10\text{ V}.\end{aligned}$$

This clockwise induced EMF "adds onto" the EMF created by the 3-volt battery. So, the total voltage around circuit 2 is $\mathcal{E}_{\text{total}} = V_0 + \mathcal{E}_{2\,\text{induced}} = 3.0\text{ V} + 0.10\text{ V} = 3.1\text{ V}$. Therefore, according to Ohm's law,

$$i_2 = \frac{\mathcal{E}_{\text{total}}}{R} = \frac{3.1\text{ V}}{1.5\ \Omega} = 2.067\text{ A}.$$

I've "illegally" kept extra significant digits, to make the point that the induced EMF nudges the current above its default value by about 3%.

(d) At $t = 3.5$ μs, the current in circuit 1 is constant. In other words, $dI_1/dt = 0$. Therefore, no EMF gets induced in circuit 2. Physically, the magnetic flux through circuit 2 is stays constant between $t = 3$ μs and $t = 4$ μs. Only a *changing* magnetic flux induces an EMF.

Consequently, at time $t = 3.5$ μs, the total voltage around circuit 2 is simple $\mathcal{E}_{total} = V_0 = 3.0$ V. Therefore, the current equals its default value, 2.0 A.

(e) At $t = 4.5$ μs, the current in circuit 1 is decreasing. The rate of change of I_1 is given by the slope of the I_1 vs. t graph:

$$\frac{dI_1}{dt} = \frac{\text{rise}}{\text{run}} = \frac{1.0\text{ A} - 2.0\text{ A}}{.005\text{ s} - .004\text{ s}} = -1.0 \times 10^3\text{ A/s}.$$

Therefore,

$$\begin{aligned}\mathcal{E}_{2\text{ induced}} &= M\frac{dI_1}{dt}\\ &= (1.0 \times 10^{-4}\text{ H})(1.0 \times 10^3\text{ A/s})\\ &= -0.10\text{ V}.\end{aligned}$$

The minus sign indicates that the induced EMF is counterclockwise, opposing the 3-volt battery instead of helping it. So,

$$i_2 = \frac{\mathcal{E}_{total}}{R} = \frac{3.0\text{ V} - 0.10\text{ V}}{1.5\ \Omega} = 1.933\text{ A}.$$

(f) Mutual inductance tells us how "strongly" circuit 2 reacts to a change in current in circuit 1. To see how M depends on the distance between the two circuits, we must review *why* changing I_1 induces an EMF in circuit 2. I_1 generates a magnetic field. Some of those field lines go through circuit 2. Therefore, when I_1 changes, the magnetic flux through circuit 2 changes. As a result, an EMF gets induced.

OK, now we can think about how the inter-circuit distance affects M. We'll see that moving the circuits farther apart decreases the mutual inductance. To see why, focus your attention on the second segment of the I_1 vs. t graph, when I_1 steadily increases from 0 to 2 amps over time interval 2 microseconds. Let $\Phi_{2\text{ max}}$ denote the magnetic flux through circuit 2 caused by circuit 1, when $I_1 = 2$ amps. So, during those 2 microseconds, the flux through circuit 2 increases from 0 to $\Phi_{2\text{ max}}$. Therefore, during those 2 microseconds, the flux through circuit 2 changes at the following rate:

$$\frac{d\Phi_B}{dt} = \frac{\Delta\Phi_B}{\Delta t} = \frac{\Phi_{2\text{ max}} - 0}{2\ \mu s}.$$

Here's my point. Other things being equal, the rate of change of flux is proportional to the maximum flux. In other words, the bigger the (maximum) magnetic flux through circuit 2, the more rapidly the flux through circuit 2 *changes*, and hence, the bigger an EMF gets induced in circuit 2.

When the circuits are close together, I_1 can generate a large maximum flux through circuit 2. Therefore, as just argued, a large $\mathcal{E}_{2\text{ induced}}$ gets created when I_1 changes. But when we move the circuits far apart, I_1 produces only a small maximum flux through circuit 2. Therefore, a small $\mathcal{E}_{2\text{ induced}}$ gets generated when I_1 changes. Since $|\mathcal{E}_{2\text{ induced}}| = M\frac{dI_1}{dt}$, this all goes to show that moving the circuits farther apart decreases the mutual inductance.

QUESTION 43-7

Consider this circuit, which has $C = 0.50$ F and $L = 0.20$ H. With the switch open (as drawn here), the capacitor is charged up, using a 2-volt battery. The battery is removed. Then, at time $t = 0$, the switch is closed.

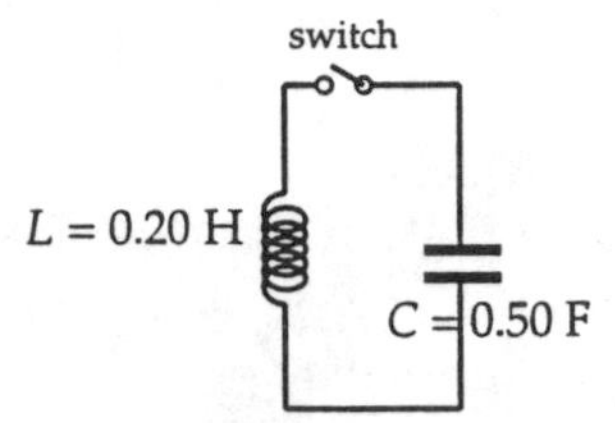

(a) Immediately before the switch is closed, what's the charge on the capacitor? Call your answer "Q_0."

(b) Set up an equation that could be solved for Q, the charge on the capacitor, as a function of time. Do not solve this equation, unless you want to.

(c) Sketch a rough graph of i vs. t (after $t = 0$), where i denotes the current through the inductor. This graph need not be numerically accurate. Just worry about the general shape.

(d) When the charge on the capacitor is $Q_0/2$, what current is flowing through the inductor? Hint: You need not solve the differential equation from part (b). Instead, think about energy.

(e) The period of oscillation of the LC circuit is $T = 2\pi\sqrt{LC}$. (I figured this out by solving the differential equation from part b.) What is the earliest time, after $t = 0$, when the capacitor has no charge on it?

(f) Explain, intuitively, why the period of oscillation increases if we raise the inductance, *or* if we raise the capacitance. It might help you to consider an analogy with a mass on a spring.

ANSWER 43-7

(a) Intuitively, you might guess that a 2-volt battery charges up the capacitor until the voltage across the capacitor is 2 volts. That's exactly right. The battery "enforces" a 2-volt potential difference between the two plates. As a result, the plates keep charging up, until they generate an electric field corresponding o a 2-volt potential difference.

By definition, $C = Q/V$, where Q denotes the charge on the plates, V denotes the potential difference between the plates, and C denotes the capacitance. So, since the initial potential difference across the plates is $V_0 = 2.0$ volts,

$$Q_0 = CV_0 = (0.50\text{ F})(2.0\text{ V}) = 1.0\text{ coulomb}.$$

(b) As usual, to generate an equation about a circuit loop, use Kirchhoff's 2nd law. Let's assume the top plates starts off positively charged. I'll walk around the circuit counterclockwise, the direction in which current initially flows, starting from just above the capacitor. For the rest of the problem, counterclockwise is my positive direction.

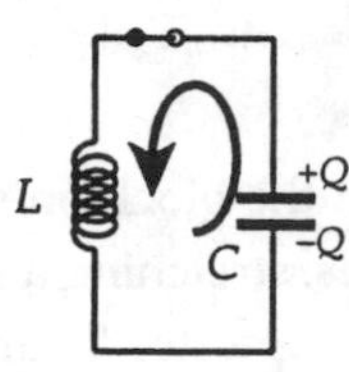

First I walk down through the inductor, with the current. Because the inductor "opposes" changing currents, it induces an EMF that tries to keep the current from increasing. So, the induced EMF is backwards compared to the current. For this reason, the inductor "contributes" a negative voltage. As always, the induced EMF is proportional to the rate of change of current: $\Delta V = \mathcal{E}_{\text{induced}} = -L\frac{di}{dt}$.

I then walk through the capacitor, from the negative to the positive plate. As a result, my potential rises. Here's why. The electric field points from the positive to the negative plate. Therefore, when pushing a positive test charge from the negative to the positive plate, you must do work "fighting" the field. So, the test charge gains potential energy, instead of losing potential energy.

Since $C = Q/V$, my potential rises by $\Delta V = +Q/C$.

Setting the sum of these voltage changes equal to zero gives us

LC loop $$-L\frac{di}{dt} + \frac{Q}{C} = 0.$$

We can't yet solve for Q as a function of time, because the equation contains a second unknown, i. But in this circuit, i relates directly to the charge Q on the capacitor. The current consists of charges flowing from the positive to the negative plate. For instance, if 3 coulombs per second flow off the positive plate and onto the negative plate, then the current is simply 3 coulombs per second. Given this insight, you might think that $i = dQ/dt$. But the current is *positive* (counterclockwise) when the charge on the top plate *decreases*. In other words, i is positive when dQ/dt is negative. So,

$$i = -\frac{dQ}{dt}.$$

Differentiate both sides to get $\frac{di}{dt} = -\frac{d^2Q}{dt^2}$. Then substitute this into the LC loop equation to get

LC loop equation (updated) $$L\frac{d^2Q}{dt^2} + \frac{Q}{C} = 0.$$

This second-order differential equation can be solved for $Q(t)$. See your textbook for the mathematical details.

(c) To answer this without explicitly solving the differential equation, I'll make an analogy between the LC loop equation and the equation describing a block on a spring. To see the analogy, first let me rewrite the loop equation in this way:

LC loop equation (rewritten) $$L\frac{d^2Q}{dt^2} = -\frac{1}{C}Q.$$

Now consider a frictionless block attached to a spring. It feels a net force $F = -kx$, where x denotes the displacement from equilibrium, and k denotes the "stiffness" constant of the spring. By Newton's 2nd law, this force equals mass times acceleration. And acceleration is the derivative of velocity, which is the derivative of position: $a = d^2x/dt^2$. So, according to Newton's 2nd law,

Block on spring $$m\frac{d^2x}{dt^2} = -kx.$$

To understand LC circuits intuitively, focus on the 1:1 correspondence between the block-on-spring equation and the LC loop equation. First of all, x corresponds to Q. In other words, stretching a spring to x is like charging a capacitor to Q. Just as the spring "wants" to unstretch, the capacitor "wants" to discharge. Also, the inductance L "plays the role" of mass. Just as a mass "resists" having its velocity changed, the inductor "resists" having the current changed. Inductance is just like inertia! Finally, the capacitance plays the role of the inverse of the spring constant.

Before exploring these analogies in more depth, let me jump to the punch line. From real-life experience, we know that the solution to the block-on-spring equation is *sinusoidal oscillations*. The block's x oscillates back and forth. Therefore, since the same differential equation describes the LC circuit, the LC circuit also oscillates. Specifically, the charge Q "sloshes" back and forth between the top and bottom plate of the capacitor, just like the block sloshes back and forth. So, if this problem requested a Q vs. t graph, I'd draw a sine or cosine curve.

Unfortunately, the problem asks us to graph i, not Q. We can do so by continuing to focus on the analogy between the LC circuit and the block on a spring. Just like current is the rate of change of charge ($i = -dQ/dt$), velocity is the rate of change of position ($v = dx/dt$). The minus sign doesn't ruin this analogy. So, the block's velocity is analogous to the LC circuit's current. If we can graph v vs. t for a block on a spring, then we've also graphed i vs. t for the LC circuit.

Well, as we saw last semester, the block's velocity oscillates sinusoidally, 90° out of phase with position. So, the i vs. t graph must have the same shape. The only remaining issue is whether to start the curve at 0, at a positive value, or at a negative value.

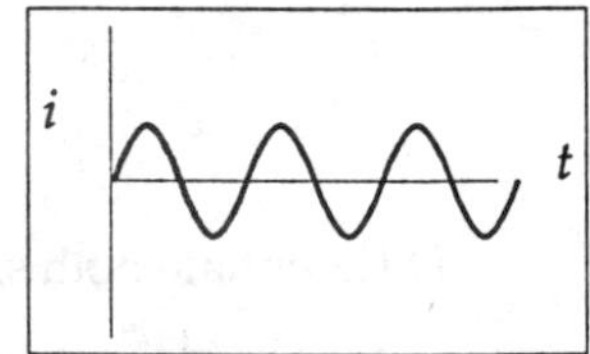

Well, because the inductor "opposes" changing current, it doesn't allow current to start flowing immediately after the switch gets closed. Immediately after $t = 0$, the current is still very close to zero. (Similarly, immediately after you release a block on a spring, its velocity is still very close to zero.) So, the i vs. t curve starts at 0. Negative current here means clockwise.

(d) Again, we can answer by thinking in terms of the block-on-a-spring analogy. This problem asks for the *current* in the circuit when the *charge* on the capacitor is half its initial (maximum) value. That's just like asking for the block's *velocity* when its *displacement* (x) is half its initial (maximum) value. By solving this block problem, we've automatically solved the circuit problem.

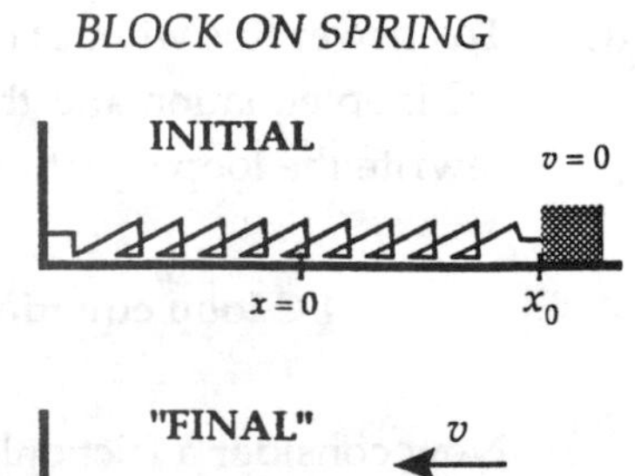

To address the block problem, we could use the differential equation from part (c). But it's easier to use energy conservation. The block has zero initial velocity, and hence, no initial kinetic energy. Recalling that a spring has potential energy $U_s = (1/2)kx^2$, and letting x_0 denote the initial .position, we get

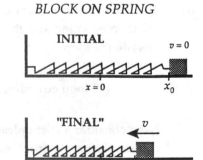

$$Energy_0 = Energy_f$$

$$\frac{1}{2}kx_0^2 = \frac{1}{2}k\left(\frac{x_0}{2}\right)^2 + \frac{1}{2}mv^2.$$

We could immediately solve for v, the block's velocity when its displacement is $x_0/2$.

Energy conservation works equally well for the LC

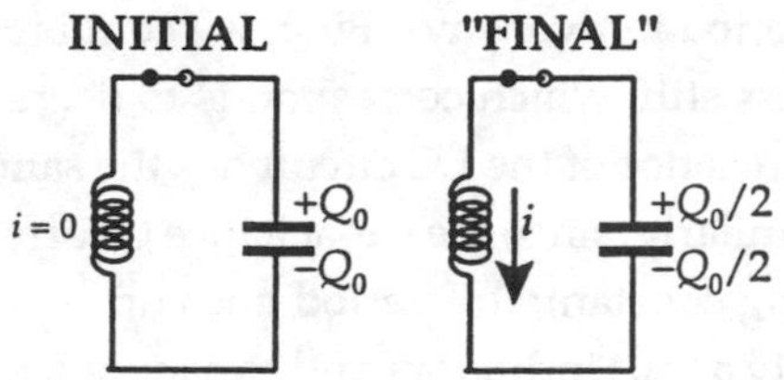

circuit. Recall from previous problems that a capacitor carries potential energy $U_C = (1/2)Q^2/C$, while an inductor carries potential energy $U_L = (1/2)Li^2$. Initially, at $t = 0$, no current flows through the inductor: $i = 0$. At that moment, the capacitor holds charge Q_0, the value we calculated in part (a). When the charge on the top plate has dropped to $Q_0/2$, the current through the inductor is some unknown value, i. By energy conservation,

$$Energy_0 = Energy_f$$

$$\frac{1}{2}Q_0^2 = \frac{1}{2C}\left(\frac{Q_0}{2}\right)^2 + \frac{1}{2}Li^2.$$

Solve for i, to get

$$i = \sqrt{\frac{3Q_0^2}{4LC}} = \sqrt{\frac{3(1 \text{ coulomb})^2}{4(0.20 \text{ H})(0.5 \text{ F})}} = 2.7 \text{ A},$$

where I substituted in our part (a) answer for Q_0.

Before moving on, let me point out something cool. Suppose you forgot the formulas for the potential energy stored in a capacitor and inductor. You could figure them out using our block-on-spring analogy. Recall that inductance corresponds to mass, and current corresponds to velocity. So, the kinetic energy formula, $(1/2)mv^2$, "analogizes" into $(1/2)Li^2$. The inductor's energy plays the role of the block's kinetic energy. Similarly, since charge Q corresponds to displacement x, and since $1/C$ corresponds to the spring constant k, the spring potential energy formula $(1/2)kx^2$ "analogizes" into $(1/2)\frac{1}{C}Q^2$. I can't overemphasize the depth of this analogy between the block on a spring and the LC circuit.

(e) Again, the analogy saves us from mathematical pain. The problem asks for the first time at which $Q = 0$ on the capacitor. That's like asking for the first time at which $x = 0$ for the block. In other words, it's like asking when the block first reaches equilibrium.

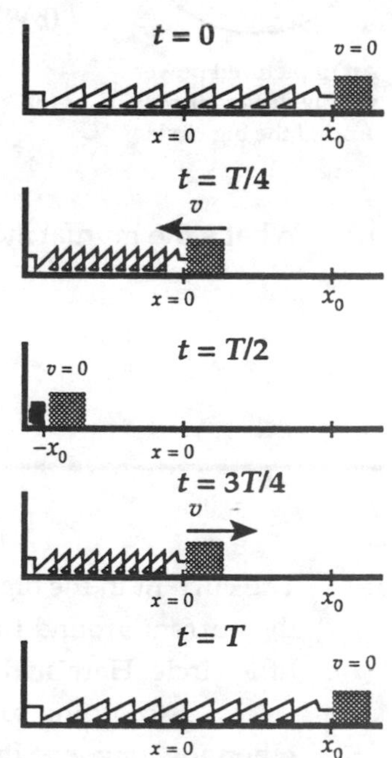

Well, the period, T, is the time the system takes to complete a full oscillation. As this graphic shows, the block takes one fourth of a period to reach $x = 0$. Similarly, the charge on the capacitor reaches $Q = 0$ after a fourth of a full period. See your textbook for the corresponding LC graphic. In any case, the time at which Q first reaches 0 is

$$t = \frac{1}{4}T = \frac{1}{4}[2\pi\sqrt{LC}] = \frac{1}{4}[2\pi\sqrt{(0.20 \text{ H})(0.50 \text{ F})}] = 0.50 \text{ s}.$$

(f) I'll answer this first in terms of the block-on-spring analogy. Big surprise, huh? Then, I'll reason more directly in terms of the LC circuit.

For the block on a spring, we could increase the period in two ways. First, we could make the block more massive. Second, we could make the spring less stiff, which corresponds to decreasing k. Since L corresponds to the block's mass, increasing the inductance of the LC circuit has the same effect as increasing the mass of the block; the period increases. Similarly, since the capacitance C corresponds to $1/k$, *increasing* the capacitance is like *decreasing* the spring constant; the period goes up.

Let me go through this again, talking more directly about inductors and capacitors. Roughly speaking, the inductance specifies how strongly the inductor counteracts a changing current, by inducing an EMF. The inductor doesn't "like" it when the current oscillates. So, the higher the inductance, the more "inertia" the circuit has with respect to the changing current. For this reason, a bigger L results in slower oscillations.

Capacitance is the capacitor's "receptiveness" to having charge piled onto it. The more "willing" the capacitor is to hold charge, the less quickly charge flows off of it. So, by raising C, you slow down the rate at which charge sloshes from plate to plate. In other words, raising the capacitance raises the period.

QUESTION 43-8

Two concentric wire circles sit on a table, as drawn here. The little circle has radius $r = 0.010$ meters and resistance $R = 0.30\ \Omega$. The big circle has radius $10r = 0.10$ meters.

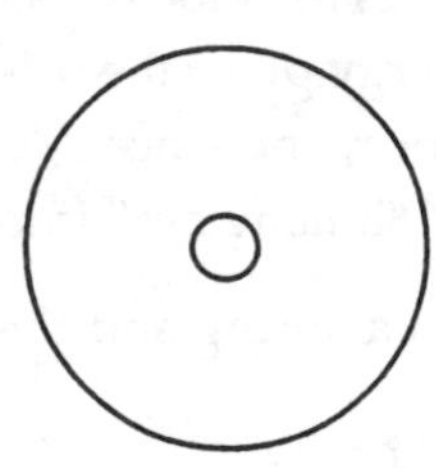

An unpictured power supply drives current around the big circle.

The current in the big circle is steadily increased from 0 to $I_1 = 5$ A. This process happens over a hundredth of second, between $t = 0$ and $t = 0.010$ s. After time $t = 0.010$ s, the current in the big circle stays constant at 5 amps. This current flows clockwise.

Throughout the problem, neglect the self-inductance of the small loop.

(a) At time $t = 1.00$ s, how much current flows around the small circle?

(b) At time $t = 0.005$ s, is the current around the small circle clockwise or counterclockwise? Explain your answer.

(c) Find the current flowing around the small circle at time $t = 0.005$ s. Hint: The magnetic field produced by the big circle is approximately uniform over the whole small circle.

(d) What's the mutual inductance between the big and small circle? Remember, it's defined by

$$\mathcal{E}_{\text{induced in small circle}} = -M\frac{dI_{\text{big circle}}}{dt}.$$

ANSWER 43-8

(a) The current in the big circle changes between $t = 0$ and $t = 0.010$ s. After that, it stays constant. Since the current around the big circle is no longer changing at $t = 1.00$ s, no current gets induced in the little circle. Here's why.

According to Faraday's law, EMF (and therefore current) gets induced in the little circle when the magnetic flux through that circle *changes*. As long as the flux stays constant, no induction happens. Here, the magnetic flux through the little circle gets produced by the current in the big circle. Since that current stays constant after $t = 0.010$ s, the magnetic flux stays constant, too. For this reason, no voltage and current get induced the little circle after that time.

(b) By contrast, at $t = 0.0050$ s, the current in the big circle *is* changing, generating a *changing* magnetic flux through the small circle. As a result, voltage gets induced in the small circle. We must figure out whether this induced voltage, and the corresponding induced current, go clockwise or counterclockwise.

Let's start by thinking about the magnetic field generated by the big circle's clockwise current. It points into the page. To confirm this, use the right-hand rule shortcut introduced earlier. Picking any point on the big circle, point your right thumb in the direction of the current. Then curl your fingertips towards the center of the circle. Your fingertips point into the page.

So, the big circle's current generates an into-the-page magnetic field through the small circle. Since that current increases, the magnetic field increases, too. At time $t = 0.005$ s, the into-the-page magnetic flux through the small circle is getting bigger.

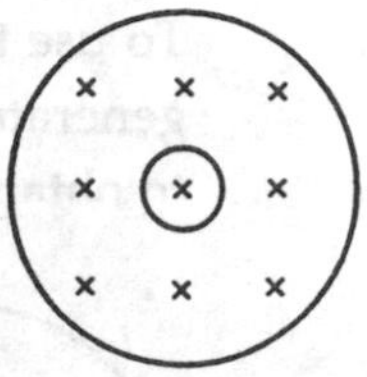

By Lenz' law, the induced current in the small circle tries to "counteract" this changing flux. In order to counteract an increasing into-the-page flux, the small circle creates an out-of-the-page flux. So, this whole problem reduces to the following question: Which way must current flow around the small circle in order to generate an out-of-the-page magnetic field, at points inside the small circle?

By playing around with the right-hand rule shortcut discussed above, you can confirm that the induced current in the small circle must flow *counterclockwise*.

(c) At this stage, we're ready to find the induced EMF and current. Alternatively, you can solve part (d) first, and then use your answer to address part (c). I'll solve the two parts together.

Let $\mathcal{E}_{\text{induced}}$ denote the induced EMF in the *small* circle. And let i denote the current in the *small* circle. According to Ohm's law, $\mathcal{E}_{\text{induced}} = iR$. So, the meat of this problem is finding the induced voltage in the small circle.

To apply Faraday's law, we must find the rate of change of magnetic flux through the *small* circle, $d\Phi_B/dt$. So, I'll first write an expression for the flux, Φ_B. Then I'll differentiate it with respect to time. Let me summarize this strategy, versions of which you've seen before.

Finding an inductance and/or an induced EMF

1) Write the magnetic flux (through the small circle) as a function of the current in the big circle. You should get

$$\Phi_B = MI_{\text{big circle}},$$

where M is a constant. As we'll see below, that constant turns out to be the mutual inductance.

2) Differentiate this equation with respect to time, to get $d\Phi_B/dt$.

3) Set $\mathcal{E}_{\text{induced}} = -d\Phi_B/dt$ to solve for the induced EMF.

Now for the details . . .

Step 1: Find magnetic flux through the small circle, in terms of the current in the big circle.

Since $\Phi_B = \int \mathbf{B} \cdot d\mathbf{A}$, we first need to know the magnetic field through the small circle, i.e., the field at the center of the large circle. Ampere's law can't help us here, because no Amperian loop you draw would allow you to simplify the Amperian loop integral, $\int \mathbf{B} \cdot d\mathbf{s}$. Instead, we must revert to Biot-Savart,

Biot-Savart $$d\mathbf{B} = \frac{\mu_0}{4\pi} I \frac{d\mathbf{l} \times \hat{\mathbf{r}}}{r^2}.$$

To use this law, first color in a small "piece" of the current. Then write $d\mathbf{B}$, the tiny magnetic field generated by that piece of current. Finally, add up (integrate) those infinitesimal field contributions, to obtain the total field.

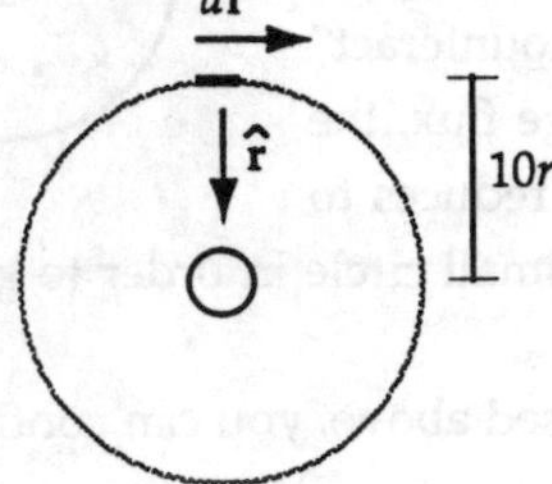

In this case, no matter which piece of current you choose, $d\mathbf{l}$ and $\hat{\mathbf{r}}$ are perpendicular. Therefore, the magnitude of the cross product reduces to a regular product. And remember, $\hat{\mathbf{r}}$ is a unit vector; it has length 1. So, $|d\mathbf{l} \times \hat{\mathbf{r}}| = (dl)(1)\sin 90° = dl$. Finally, recall that "$r$" denotes the distance from the colored-in piece of current to the point at which we're finding the magnetic field. Here, that distance is $10r$, the radius of the big circle. So,

$$dB = \frac{\mu_0}{4\pi} I \frac{|d\mathbf{l} \times \hat{\mathbf{r}}|}{r^2}$$

$$= \frac{\mu_0}{4\pi} I \frac{dl}{(10r)^2},$$

where I denotes the current in the *big* circle. As we saw in part (b), this field points into the page, no matter which piece of the big circle you chose.

To find the total field generated by all the "pieces," integrate around the big circle, which has circumference $2\pi(10r)$. Because all pieces of the big circle are the same distance, $10r$, from the center of that circle, the "$10r$" is a constant we can pull out front.

$$B = \int dB = \int \frac{\mu_0}{4\pi} I \frac{dl}{(10r)^2}$$

$$= \frac{\mu_0}{4\pi(10r)^2} I \int_0^{2\pi(10r)} dl$$

$$= \frac{\mu_0}{4\pi(10r)^2} I[2\pi(10r)]$$

$$= \frac{\mu_0}{4\pi(10r)} I.$$

Let's step back to find our bearings. We just calculated the field produced at the center of the circles when current I flows around the big circle. Using that field, we'll figure out the magnetic flux through the small circle. Finding that flux is the first step towards obtaining the induced EMF in the small circle. See the gray box above.

OK, it's flux time. As the problem tells us, the magnetic field is approximately uniform over the entire area of the small circle. This makes it easier to find the magnetic flux:

$$\text{Flux through small circle} = \Phi_B = \int_{\text{small circle}} \mathbf{B} \bullet d\mathbf{A}$$

$$= \int_{\text{small circle}} B\,dA \quad \text{[since small circle "faces" same direction as } \mathbf{B}]$$

$$= B\int_{\text{small circle}} dA \quad [B \text{ is uniform over whole area of small circle}]$$

$$= BA_{\text{small circle}}$$

$$= \left[\frac{\mu_0}{2(10r)} I\right]\pi r^2$$

$$= MI,$$

where in the last step, I combined aal the constants into one:

$$M = \frac{\mu_0 \pi r^2}{2(10r)} = \frac{\mu_0 \pi r}{20} = \frac{(4\pi \times 10^{-7}\ \text{H/m})\pi(0.010\ \text{m})}{20} = 2.0 \times 10^{-9}\ \text{H}.$$

I didn't pull this M out of a hat. It's just the name I gave to all the constants on the right-hand side of the flux equation. This combined constant turns out to be the mutual inductance, as we'll see below. For the rest of this problem, I'll write "M" instead of "$\mu_0\pi r/20$."

Step 2: Differentiate the flux with respect to time, to get $d\Phi_B/dt$.

Since M is "made of" constants, it's oblivious to differentiation:

$$\frac{d\Phi_B}{dt} = M\frac{dI}{dt}.$$

Keep in mind that I denotes the current in the big circle, while Φ_B denotes the flux through the small circle. According to this equation, a changing current in the big circle produces a changing magnetic flux through the small circle.

From the given information, we can calculate dI/dt, the rate of change of current. The big circle's current *steadily* increases from 0 to $I_1 = 5.0$ amps in time $t_1 = 0.01$ seconds. So,

$$\frac{dI}{dt} = \frac{\Delta I}{\Delta t} = \frac{I_1 - 0}{t_1} = \frac{5.0\ \text{A}}{0.010\ \text{s}} = 500\ \text{A/s}.$$

Therefore, the above equation for $d\Phi_B/dt$ becomes

$$\frac{d\Phi_B}{dt} = M\frac{dI}{dt}$$

$$= \frac{\mu_0 \pi r}{20}\frac{I_1}{t_1}$$

$$= (2.0\times10^{-9}\text{ H})(500\text{ A/s})$$

$$= 1.0\times10^{-6}\text{ volts.}$$

Step 3: Find the induced current by setting $E_{\text{induced}} = -d\Phi_B/dt$, *and then using Ohm's law.*

This isn't really a separate step. It's just a reminder that, according to Faraday's law, the rate of change of flux gives us the induced EMF (voltage): $E_{\text{induced}} = -d\Phi_B/dt$. The minus sign means that the induced EMF "opposes" the changing flux.

In step 2, we found this induced EMF to be 1.0×10^{-6} volts. By Ohm's law, the induced current in the small circle is

$$i = \frac{\mathcal{E}_{\text{induced}}}{R} = \frac{1.0\times10^{-6}\text{ V}}{0.020\ \Omega} = 3.3\times10^{-6}\text{ A.}$$

(d) We found the mutual inductance way back in step 1, when we obtained

$$\Phi_B = MI,$$

with

$$M = \frac{\mu_0 \pi r}{20} = 2.0\times10^{-9}\text{ H.}$$

You can confirm that this "M" really is the mutual inductance, by looking over steps 2 and 3. In step 2, we differentiated to get $\frac{d\Phi_B}{dt} = M\frac{dI}{dt}$. In step 3, we used Faraday's law, $\mathcal{E}_{\text{induced}} = -\frac{d\Phi_B}{dt}$. Put those steps together to get

$$\mathcal{E}_{\text{induced}} = -M\frac{dI}{dt}.$$

By definition, the M in that equation is the mutual inductance. So, I've just proven that the "M" from step 1 is indeed the mutual inductance.

44 CHAPTER

AC Circuits

Note to students

Physics instructors differ radically in their views towards AC circuits. Some want you to understand the mathematical details I'll bleep over. Others want you to focus on the most important qualitative concepts. Others emphasize the major formulas. In my experiences, most instructors gravitate towards more qualitative questions, covering the concepts and simple equations I've emphasized in this chapter. A few of my problems may be harder than your instructor's exam questions. I'm including them just to be "safe." Ask your instructor exactly what you need to know, and at what level.

QUESTION 44-1

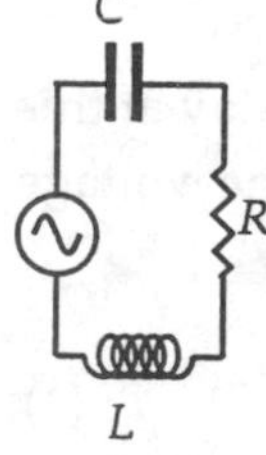

This alternating circuit contains an alternating power supply of frequency $f = 60$ Hz and maximum voltage $\mathcal{E}_0 = 5.0$ V; a resistor of resistance $R = 0.50\ \Omega$; a capacitor $C = 20\ \mu$F (microfarads); and an inductor $L = 0.40$ H.

(a) What is the period of oscillation of this circuit? In other words, how much time does the current take to slosh back and forth, completing one cycle?

(b) During these oscillations, what is the biggest current that ever flows through the inductor?

(c) What is the largest power (rate of heat dissipation) reached in the resistor?

(d) *(Extra hard. Ask your instructor if you need to understand this.)* Suppose the power supply reaches its maximum voltage at time t_1. How long *after* time t_1 does the current through the resistor first reach its maximum value?

(e) The power supply is suddenly turned to zero. (But current can still flow through its inner workings.) Describe in words what happens to the current in the circuit over the next few tenths of a second. Sketch a rough, non-numerical graph of i vs. t. Is the period still the value you calculated in part (a)?

ANSWER 44-1

Because the mathematics gets so complicated, many students have trouble following the textbook's discussion of "forced alternating" circuits, which contain an alternating power supply, along with the usual gang of circuit elements (resistors, capacitors, inductors). As you know, I usually emphasize concepts, as opposed to rote problem-solving. Unfortunately, since introductory physics courses typically spend only a week on AC circuits, you might not have time to develop a deep understanding. You may have to rely on "canned" problem-solving procedures. I'll present those procedures, and the accompanying intuitions (to the extent possible).

Let's begin by writing out the most general "solution" for this kind of circuit. Then, in parts (a) through (e), we'll see what these formulas mean physically, and how to apply them.

In alternating circuits, the power supply produces a voltage that oscillates sinusoidally. If we let $\mathcal{E}_m$ denote the power supply's maximum voltage, then

$$\mathcal{E} = \mathcal{E}_m \sin \omega t, \qquad (1)$$

where ω is the angular frequency, which is 2π times the oscillation frequency.

The circuit elements "impede" the current. The higher the "impedance," Z, the less current flows around the circuit. Intuitively, we expect that a higher resistance leads to higher impedance. We also expect that raising the inductance increases Z, because the inductor opposes changes in current. The formula for Z confirms these intuitions:

$$Z = \sqrt{R^2 + (\omega L - 1/\omega C)^2} \qquad \textbf{Impedance} \qquad (2)$$

As you might expect, the maximum current in the circuit is proportional to the maximum voltage, and inversely proportional to the impedance:

$$i_m = \frac{\mathcal{E}_m}{Z} \qquad \textbf{Maximum current} \qquad (3)$$

This is the AC analog of Ohm's law, $i = V/R$. But in a forced AC circuit, the current doesn't stay at this maximum value. The power supply makes the current oscillate, at the same frequency the voltage oscillates:

$$i = i_m \sin(\omega t - \phi), \qquad \textbf{Current} \qquad (4)$$

where

$$\phi = \tan^{-1}\frac{\omega L - 1/\omega C}{R} \qquad \textbf{"Lag constant"} \qquad (5)$$

I'll discuss the physical meaning of this lag factor in part (d) below. For now, let me just point out that Eqs. (2) through (5) can be combined into one monster equation. I'll write it out, just for the record:

$$i = \frac{\mathcal{E}_m}{\sqrt{R^2 + (\omega L - 1/\omega C)^2}} \sin\left[\omega t - \left(\tan^{-1}\frac{\omega L - 1/\omega C}{R}\right)\right] \qquad (2) - (5) \text{ combined}$$

When solving problems, you should keep the equations separate, instead of combining them, to avoid getting bogged down.

With these five equations, you can find almost anything you need to know about *any* forced AC circuit loop. We'll practice using these equations in parts (b) through (d) below.

(a) The oscillation frequency is 60 s^{-1}. (Hertz are inverse seconds.) So, the voltage and current oscillate back and forth 60 times per second. You're asked for the period, i.e., the time needed to complete one oscillation. Well, intuitively, if the circuit oscillates 60 times per second, then each oscillation takes one sixtieth of a second. Period is the inverse of frequency:

$$T = \frac{1}{f} = \frac{1}{60\ \text{Hz}} = 0.017\ \text{s}.$$

Part (a) isn't really about AC circuits. It's about the relationship between period and frequency in any oscillating system.

Below, we'll use the system's angular frequency,

$$\omega = 2\pi f = 2\pi(60\ \text{s}^{-1}) = 377\ \text{s}^{-1}.$$

(b) Because the circuit consists of a single loop, the current through the inductor equals the current through any other circuit element. Unless the current "branches" at an intersection, it has the same value at all points on the circuit loop. Otherwise, charge would "build up" somewhere in the circuit (other than the capacitor). This can't happen, as I explained in the DC circuits chapter. The general rules governing DC circuits apply also to AC circuits.

We're looking for the maximum current. It's given by Eq. (3). But to use Eq. (3), we must first calculate the impedance, with Eq. (2).

Well, in part (a), I found the angular frequency. It's $\omega = 2\pi f = 377\ \text{s}^{-1}$. So,

$$\begin{aligned} Z = \sqrt{R^2 + (\omega L - 1/\omega C)^2} &= \sqrt{(0.50\ \Omega)^2 + \left[(377\ \text{s}^{-1})(0.40\ \text{H}) - \frac{1}{(377\ \text{s}^{-1})(20\times10^{-6}\ \text{F})}\right]^2} \\ &= \sqrt{(0.50\ \Omega)^2 + [(151\ \text{H/s}) - (133\ \text{H/s})]^2} \\ &= \sqrt{(0.50\ \Omega)^2 + (18.2\ \Omega)^2} \\ &= 18.2\ \Omega. \end{aligned}$$

From this calculation, notice that almost all of the impedance comes from the capacitor and inductor, not from the resistor. That's not a general rule. It just happens to be the case here.

Substitute this impedance into Eq. (3), to find the maximum current:

$$i_m = \frac{\mathcal{E}_m}{Z} = \frac{5.0\ \text{V}}{18.2\ \Omega} = 0.27\ \text{A}.$$

By contrast, if the inductor and capacitor weren't there, the maximum current would be $i_m = \mathcal{E}_m/R = 10$ amps. Apparently, the capacitor and inductor can play a big role in keeping the current down. To understand why intuitively, think in terms of the block-on-a-spring analogy discussed in your textbook. If you grab the block and push it back and forth, its mass can "impede" your motion. So can the spring.

(c) As we've seen before, the power dissipated in any circuit element is $P = Vi$. Since the voltage across a resistor is $V = iR$, the power dissipated is

$$P = i^2R.$$

So, as expected, the power reaches it highest value when the current hits its largest value. And in part (b), we already found the maximum current. Just substitute it into the power equation, to get

$$P_{\max} = i_m^2 R = (0.275 \text{ A})^2(0.50\ \Omega) = 0.038 \text{ watts.}$$

In a later problem, I'll carefully distinguish the *maximum* power from the *average* power.

A common mistake is to think $P_{\max}$ equals $\mathcal{E}_0 i_m$. But the "V" in $P = Vi$ refers to the voltage (potential change) across the resistor, not the voltage across the power supply. These two voltages differ, because according to Kirchhoff's 2nd law, the power supply's voltage equals the voltage drop across all the other circuit elements *combined*. So, the power supply's $\mathcal{E}$ equals the potential drop across the resistor *plus* the potential drop across the capacitor *plus* the potential drop across the inductor. Consequently, the voltage across the resistor is *less* than $\mathcal{E}$.

(d) OK, this is where things get messy. We have to take a closer look at Eqs. (4) and (5) above. According to Eq. (4), the current oscillates at angular frequency ω, the *same* angular frequency with which the voltage of the power supply oscillates. This makes sense, since the voltage "drives" the current. But for nonzero, ϕ, *the current oscillates "out of sync" with the voltage.* In other words, the current does not reach its maximum value at the same moment the power supply's voltage reaches its maximum value. Instead, the current "lags behind" the voltage, or "leads" the voltage. The lag constant ϕ tells us how much.

For instance, if $\phi = 90°$, then the current is a quarter of a cycle out of phase with the voltage. In other words, the current reaches its maximum value a quarter of a period after the voltage reached its maximum value.

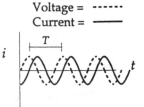

IF $\phi = 90°$ . . .

the current reaches its peak a quarter of a period after the voltage reaches its peak.

We've implicitly figured out the relationship between ϕ and the "time lag." Since 360° corresponds to a full cycle (period), 90° corresponds to a quarter cycle. Actually, we should work in radians. So, 2π corresponds to a full period, and $\pi/2$ corresponds to a quarter period. In general, to find by what fraction of a full period the current lags or leads the voltage, divide ϕ by the angle of a full period. For instance, in this case,

$$\phi = \tan^{-1}\frac{\omega L - 1/\omega C}{R} = \tan^{-1}\left[\frac{(377 \text{ s}^{-1})(0.40 \text{ H}) - \dfrac{1}{(377 \text{ s}^{-1})(20 \times 10^{-6} \text{ F})}}{0.50\ \Omega}\right] = \tan^{-1} 36.35 = 88°.$$

As a fraction of a full period, this lag factor is simply

$$\frac{\phi}{360°} = \frac{88°}{360°} = 0.245.$$

From part (a), $T = 0.017$ s. Therefore, the time by which the current lags the voltage is

$$t_{\text{lag}} = \frac{\phi}{360°}T = (0.246)(0.017 \text{ s}) = 0.0041 \text{ s}.$$

(e) In this part, we turn off the alternating voltage. Nevertheless, current continues to oscillate back and forth. As we saw in question 43-7 above, and the current in an LC circuit oscillates, just like a block on a spring oscillates. You can *force* the block to oscillate at a certain frequency, by grabbing it and pushing it back and forth. But when you let go of the block, it continues to oscillate, not at your "push" frequency, but at its own "natural" frequency. Similarly, you can force the current in an LC circuit to oscillate at whatever frequency you want, using a power supply. But turning off the alternating voltage is like letting go of the block; the current continues to oscillate, but at its own natural frequency.

Current in circuit after the power supply's voltage is turned off.

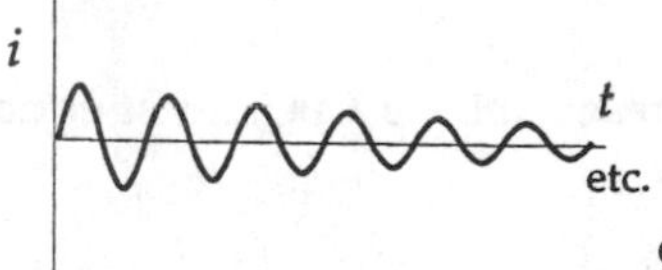

This circuit, unlike the one in question 43-7, contains a resistor. It's an RLC circuit, not just an LC circuit. In terms of the block-on-a-spring analogy, the resistor plays the role of friction. For a real-life block on a spring, the oscillations get smaller and smaller, as friction "steals away" energy. Similarly, in this circuit, current does not oscillate back and forth forever. The resistor steals away energy as current sloshes through it. Consequently, the oscillations get smaller and smaller, until they die out completely.

Crucially, these unforced oscillations occur at the "natural" frequency of the system, not at the frequency supplied by the power supply before it was turned off. I'll discuss unforced oscillations in more detail later.

QUESTION 44-2

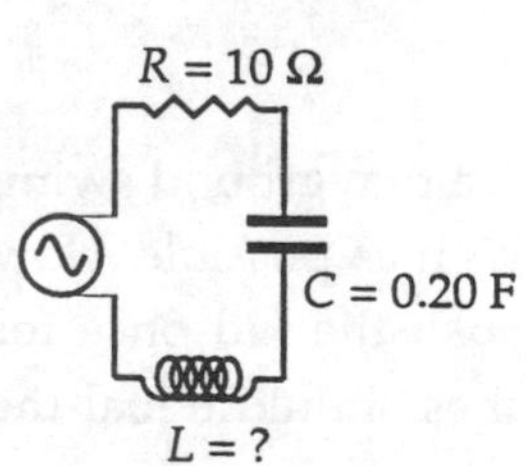

An inventor is building an electric heater. It will plug into standard American household electrical outlets, which supply an oscillating EMF of frequency 60 Hz and root-mean-square voltage $\mathcal{E}_{\text{rms}} = 120$ V. The inventor wants the heater to contain one resistor, one capacitor, and one inductor, as drawn here. From an army surplus store, she bought a capacitor with $C = 0.20$ F. Her resistor has resistance $R = 10\ \Omega$.

(a) In order for her device to generate as much heat as possible, what inductance should the inductor have?

(b) If she uses the inductor you suggested in part (a), how much heat does this device produce in 1 minute?

(c) Explain why commercially available heaters use only resistors, with no capacitors or inductors.

(d) If you plugged the inventor's device, for one minute, into an electrical outlet that supplies $\mathcal{E}_{\text{rms}} = 120$ V at $f = 50$ Hz instead of 60 Hz, would the device generate more heat, less heat, or the same heat as you found in part (b)?

ANSWER 44-2

(a) In general, power is the rate of energy transfer. Here, it's the rate at which electrical energy converts into heat energy. So, the inventor wants to maximize the power. Since $P = i^2R$, it follows that she wants to maximize the current through the resistor. In other words, she wants the peak current through the circuit, i_m, to be as large as possible.

I see two ways of addressing this problem. We can use the forced RLC circuit formulas summarized in the previous problem. Or, we can think more physically about "resonance" in oscillating circuits. These two methods yield the same answer, because they're really two different ways of saying the same thing. I'll demonstrate both methods. Make sure you understand the connections between the two.

Method 1: Forced RLC equations

In analogy with Ohm's law ($i = V/R$) for DC circuits, the maximum current that flows through an AC circuit is $i_m = \mathcal{E}_m/Z$, where Z denotes the impedance. So, for a given voltage, we can maximize the current by minimizing the impedance. The impedance is

$$Z = \sqrt{R^2 + (\omega L - 1/\omega C)^2}$$

To make Z as small as possible for a given R, the stuff inside the parentheses, $\omega L - 1/\omega C$, must equal zero. Therefore, we can set

$$\omega L - 1/\omega C = 0,$$

and solve for the inductance. Since $\omega = 2\pi f = 2\pi(60\text{ s}^{-1}) = 377\text{ s}^{-1}$, we get

$$L = \frac{1}{\omega^2 C} = \frac{1}{(377\text{ s}^{-1})^2(0.20\text{ F})} = 3.52 \times 10^{-5}\text{ H}.$$

Method 2: Forced oscillations and resonance

Now for some intuitive physical reasoning. Imagine a kid swinging on a playground swing set. By periodically giving him a brief push, you plan to make him swing as high as possible. How often should you push? As experienced swingers will testify, you should push the kid once for every complete swing he makes. In other words, the frequency of your pushes should equal the "natural" frequency of oscillation of the swing.

This is a general property of forced (pushed) oscillatory motion. **The motion achieves the highest amplitude when the push frequency matches the natural ("resonance") frequency of the system.** When the push frequency "agrees" with the system's natural frequency, the pushes reinforce the natural motion, making the oscillations bigger than they'd otherwise be. The most dramatic example of this is the Tacoma-Narrows bridge. It collapsed partly because gusts of wind rocked it at its natural frequency. As a result, the bridge's oscillations got bigger and bigger. Your textbook probably has a picture.

As a more mundane example, consider a block on a spring. We could attach the block to a machine that pushes it back and forth. The machine makes the block oscillate most vigorously if the machine's "push frequency" matches the natural frequency of the block on the spring. As we saw last semester, the natural angular frequency of a block on a spring is

Block on spring $$\omega = \sqrt{\frac{k}{m}}.$$

The same reasoning applies to the forced AC circuit. As discussed in question 43-7, current oscillates back and forth in an LC (or RLC) circuit, even when there's no power supply. The circuit naturally oscillates at a certain frequency. When the same circuit gets plugged into an electrical outlet, and therefore undergoes "forced" oscillations, the oscillations are biggest when the frequency supplied by the power source matches the natural frequency of the system. Here, we can't control the frequency of the power supply. But by choosing the "right" inductor, we can tweak the natural frequency of the circuit, so that it matches the frequency supplied by the electrical outlet. When this matching occurs, the current undergoes the largest oscillations possible. In other words, the peak current, i_m, becomes as large as possible.

To find the natural frequency of the circuit, make an analogy with the block on a spring. Remember, m is analogous to L, because mass and inductance both supply "inertia" against oscillations. The spring constant k is analogous to $1/C$, because a stiffer spring that hates getting stretched (high k) corresponds to a capacitor that hates getting charged up (low C). Since the angular frequency of a block on a spring is $\omega = \sqrt{k/m}$, the corresponding angular frequency of an LC circuit must be

Unforced LC circuit $$\omega = \sqrt{\frac{1/C}{L}} = \frac{1}{\sqrt{LC}}.$$

We want the inductance to be such that the circuit's natural angular frequency matches the ω supplied by the electrical outlet. So, solve this equation for L, and set $\omega = 377\ \text{s}^{-1}$, to get

$$L = \frac{1}{\omega^2 C} = \frac{1}{(377\ \text{s}^{-1})^2(0.20\ \text{F})} = 3.52 \times 10^{-5}\ \text{H}.$$

the same answer obtained using method 1 above.

Let me tie all this together. By matching the circuit's natural oscillation frequency to the power supply's oscillation frequency, you minimize the impedance, and therefore achieve the largest oscillations (i.e., the biggest possible current) in the circuit.

(b) As mentioned above, power is the rate of energy transfer—in this case, the rate of heat dissipation. So, to find the total heat, multiply the average power by the time: *Heat* = $\bar{P}\,\Delta t$. Since $\Delta t = 60$ seconds, we just need to find the average power dissipated in the resistor. (No power dissipates in the other circuit elements; the capacitor and inductor "trade" potential energy back and forth.)

As mentioned in part (a), a resistor dissipates power $P = i^2R$. In an AC circuit, the current isn't constant. It oscillates back and forth, getting bigger and smaller. Since the current is usually less than its peak value, i_m, the average power is less than i_m^2R. *The average power turns out to be exactly one half of this peak value,* as your textbook shows.

Let me relate this to "root mean square." Root mean square voltages and currents are defined so that, when you use them to calculate power, they yield the correct *average* power. For instance, the root-mean-square current relates to the root-mean-square voltage exactly as you'd expect:

$$i_{rms} = \frac{\mathcal{E}_{rms}}{Z}.$$

Given this root-mean-square current, the average power is simply

$$\overline{P} = i_{rms}^2 R.$$

To apply these equations, we first need the impedance, $Z = \sqrt{R^2 + (\omega L - 1/\omega C)^2}$. But in part (a), we deliberately chose L to ensure that the stuff inside the parentheses equals 0, thereby making Z as small as possible (for given R). See my "method 1" in part (a). So,

$$Z = \sqrt{R^2 + (0)^2} = R = 10\ \Omega.$$

Here's the point. When the circuit's natural oscillation frequency matches the "driving" frequency, the inductor and capacitor don't contribute to the impedance. In that case, the impedance reduces to the resistance. But otherwise, the impedance is bigger.

Substitute this impedance into the above formula for root-mean-square current to get

$$i_{rms} = \frac{\mathcal{E}_{rms}}{Z} = \frac{120\text{ V}}{10\ \Omega} = 12\text{ A}.$$

Therefore, the average power is

$$\overline{P} = i_{rms}^2 R = (12\text{ A})^2(10\ \Omega) = 1440\text{ watts (joules per second)}.$$

Now that we know the average rate at which heat gets dissipated, just multiply this rate by the time, in order to obtain the total heat:

$$Heat = \overline{P}\Delta t = (1440\text{ J/s})(60\text{ s}) = 86000\text{ J}.$$

(c) In this answer, I'll show the following: When trying to generate a certain power, you're better off using a resistor alone, instead of a resistor along with a capacitor and inductor.

Let me frame the question more precisely. Suppose you have an AC circuit consisting of nothing except a power supply and a resistor. As a result, a certain oscillating current flows through the circuit. By adding capacitors and inductors to the circuit (in series), could we increase this current? No. Here's why.

Adding a capacitor and inductor to the circuit would increase the current only if they cause the impedance to decrease. That's because current is inversely proportional to impedance: $i_m = \mathcal{E}_m/Z$. But from the formula

$$Z = \sqrt{R^2 + (\omega L - 1/\omega C)^2},$$

we see that capacitors and inductors never lower the impedance. They can only raise Z, or leave Z unchanged, because $(\omega L - 1/\omega C)^2$ can only be positive or zero. Physically, that's because inductors tend to "oppose" changes in current; and capacitors don't allow current to flow through them, when they get fully charged.

So, if you're trying to keep the impedance low, using a resistor alone always works at least as well as using that same resistor along with a capacitor and inductor. For this reason, real heaters use only resistors, not capacitors or inductors.

(d) In part (a), we chose the inductor so that the natural frequency of the LC circuit would match the 60 Hz frequency supplied by the electrical outlet. In other words, we "tuned" the circuit to work best when driven at 60 Hz. So, when we plug the circuit into a 50 Hz outlet, its natural frequency (60 Hz) no longer matches the driving frequency (50 Hz). Therefore, we get a smaller current and power.

To confirm this result, calculate the new impedance. In part (b), the impedance was $Z = \sqrt{R^2 + (0)^2} = R = 10\Omega$. But now, the electrical outlet drives the system at $f = 50\ \text{s}^{-1}$, which corresponds to angular frequency $\omega = 2\pi f = 2\pi(50\ \text{s}^{-1}) = 314\ \text{s}^{-1}$. So, the new impedance is

$$\begin{aligned} Z = \sqrt{R^2 + (\omega L - 1/\omega C)^2} &= \sqrt{(10\ \Omega)^2 + \left[(314\ \text{s}^{-1})(3.52\times10^{-5}\ \text{H}) - \frac{1}{(314\ \text{s}^{-1})(0.20\ \text{F})}\right]^2} \\ &= \sqrt{(10\ \Omega)^2 + (0.011\ \text{H/s} - 0.016\ \text{H/s})^2} \\ &= \sqrt{(10\ \Omega)^2 + (-0.0049\ \Omega)^2}, \end{aligned}$$

which is a little above 10 ohms. Hence, the rms current is a little below 12 A.

In this case, the impedance barely changed. But don't draw any general conclusions from this. Some circuits are more "sensitive" to the driving frequency.

QUESTION 44-3

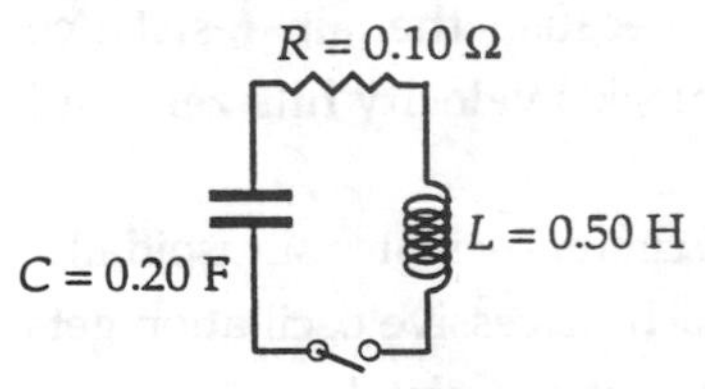

An undriven (battery-less) RLC circuit has $L = 0.50$ H, $C = 0.20$ F, and $R = 0.10\ \Omega$. The capacitor is charged up to $Q_0 = 1.0$ coulomb, with the positive charge on the top plate. The switch is closed at time $t = 0$.

(a) *Immediately* after $t = 0$, what current flows through the resistor?

(b) At time $t = 100$ s, what current flows through the resistor?

(c) Sketch a rough, non-numerical graph of the current through the resistor as a function of time.

(d) What is the earliest time at which there's no charge on the capacitor plates?

(e) How much total heat dissipates in the resistor before the current dies away entirely? Hint: You need not get mired in math.

ANSWER 44-3

To visualize what happens, think of the analogy between an LC (or RLC) circuit and a block on a spring. As we've seen before, the charge sloshes back and forth between the top and bottom capacitor plate, just like the block sloshes back and forth.

(a) The circuit consists of just one loop. Therefore, at any moment, the same current flows through all three circuit elements.

Before the switch closes, the current is 0. If the circuit contained no inductor, then closing the switch would *abruptly* change the current from 0 to some big value. In other words, the current would immediately shoot up from 0 to some positive number. But the inductor prevents this from happening. Specifically, the inductor generates a "backwards" EMF in response to a *changing* current. As a result, when the switch closes, the current does not immediately increase from 0 to some positive value. The current increases *gradually* instead of abruptly. So, *immediately* after $t = 0$, the current is still essentially 0.

If this sounds weird, think of a block on a spring. Recall that the charge Q on the top capacitor plate corresponds to x, the displacement of the block from equilibrium. Therefore, the current $i = -dQ/dt$ corresponds to the block's velocity, $v = dx/dt$. So, this problem is analogous to the following question: If a block on a spring is stretched and then released from rest, what is the block's velocity *immediately* after getting released? As you know, the block's velocity increases gradually, not abruptly. *Immediately* after getting released, the block's velocity is still essentially 0. The same conclusion applies to the current in an RLC circuit.

(b) In disguised form, this question asks for the current a *long time* after the switch closes. If the circuit were resistanceless, then current would slosh back and forth forever, just like a *frictionless* block on a spring. But a real-life block experiences friction or air resistance. As a result, the block's oscillations gradually become smaller and smaller, until dying out entirely. Similarly, in an RLC circuit, the resistor plays the role of air resistance. Although charge sloshes back and forth between the two plates for a while, the oscillations get smaller and smaller. Eventually, the capacitor plates end up with no charge, and the current dies away completely.

In summary, after a long time, i hits zero and stays there. The circuit is "dead."

(c) In part (a), we saw that the current in an RLC circuit corresponds to the velocity of a block on a spring. So, this question is analogous sketching the velocity vs. time graph of a block on a spring (with air resistance). If you've studied damped harmonic motion, then you know that the velocity oscillates sinusoidally, just like the displacement (though with a different phase). But the peak velocity reached during each successive oscillation gets smaller and smaller, because the air resistance "steals away" energy. Eventually, the block's velocity hits zero and stays there.

i or v — t — etc.

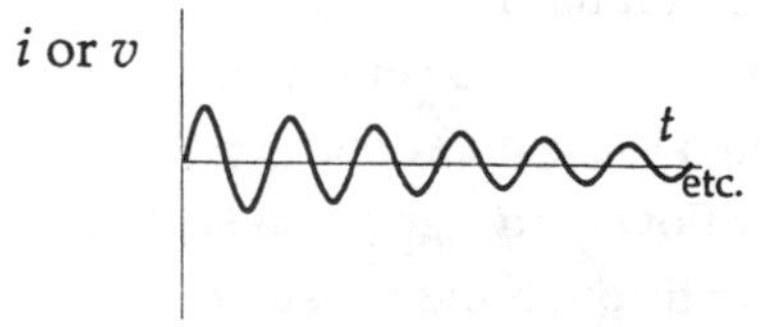

This graph shows the velocity of a block on a spring (with friction), or the current in an undriven LRC circuit.

Similarly, the current in the RLC circuit oscillates sinusoidally. But the "peak" current reached during each successive oscillation gets smaller and smaller, until the current dies out completely.

As we found in part (a), the current right at $t = 0$ is $i = 0$. So, the graph starts at the origin.

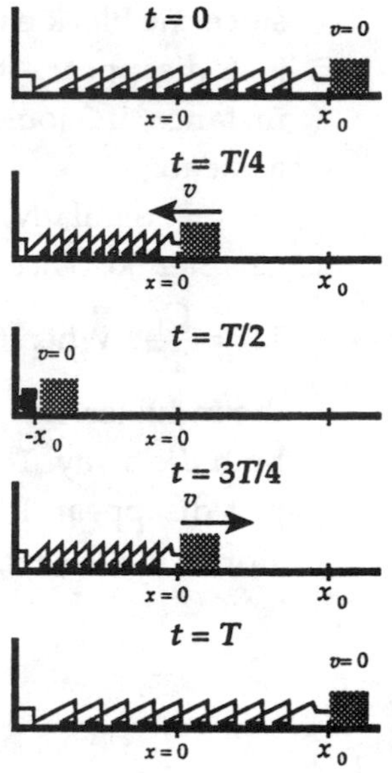

(d) The charge Q on the capacitor corresponds to the displacement x of the block on a spring. Therefore, this circuit question corresponds to the following: When a block on a stretched spring is released, how much time does it take to reach $x = 0$?

Well, as this shamelessly recycled graphic shows, the block takes one quarter of a whole period (cycle) to reach $x = 0$. Similarly, the charge on the capacitor first reaches $Q = 0$ after a quarter period. Your textbook displays the appropriate graphic. So, we just need to find the period of our RLC circuit, and divide it by 4.

Does the presence of the resistor change the period of an LC circuit? Strictly speaking, yes. But if the "damping" is small, then the resistor affects the period only negligibly. By slogging through the math, you can show that the "small-damping" approximation holds when $1/LC \gg (R/L)^2$, which is true here.

In summary, we need to find the period of the LC circuit, without worrying about the resistor. To figure out the formula, let's again use the block-on-spring analogy.

Since $T = 2\pi/\omega$, and the angular frequency of a block on a spring is $\omega = \sqrt{k/m}$, we get

Block on spring $$T = 2\pi\sqrt{\frac{m}{k}}.$$

As explained in question 44-2, the block's mass m plays the same role as the inductance L. And the spring constant k plays the same role as $1/C$. Therefore, the angular frequency of the LC circuit must be $\omega = 1/\sqrt{LC}$, with corresponding period

LC circuit $$\begin{aligned} T &= 2\pi\sqrt{LC} \\ &= 2\pi\sqrt{(0.50\text{ H})(0.20\text{ F})} \end{aligned}$$

Therefore, as the above graphic shows, the charge on the capacitor first reaches zero at time

$$t_{Q=0} = \frac{1}{4}T = \frac{1}{4}(2.0\text{ s}) = 0.50\text{ s}.$$

(e) You could solve the long way by first deriving the equation for the current as a function of time; using that current to obtain the power dissipated in the resistor as a function of time; and then integrating that power over time to obtain the total heat energy. But it's easier to use conservation of energy.

To see why, consider the analogous question about a block on a spring: How much total energy gets dissipated by air resistance during the block's oscillations? Well, the system initially has potential energy stored in the spring. During the oscillations, this energy transfers back and forth between potential and kinetic. But eventually, all the potential and kinetic energy vanishes, since the block ends up at rest at $x = 0$. According to energy conservation, the initial energy didn't just disappear. It converted into heat, as the block "frictionally" rubbed against the air. For instance, if 3 joules of energy were initially stored in the spring, then 3 joules of energy dissipate as heat.

Similarly, in the RLC circuit at $t = 0$, the capacitor "stores" energy $\mathcal{E} = \mathcal{E}_m \sin \omega t$, as shown in Chapter 36. Since $i = 0$ at $t = 0$, the inductor initially has no energy. So, the system has initial energy $U_0 = \frac{Q_0^2}{2C}$. When the switch closes, this energy transfers back and forth between the capacitor and the inductor. But in the long run, the charge on the capacitor and the current through the inductor both die away. Those circuit elements end up with no energy. But the initial potential energy didn't just disappear. It dissipated as heat in the resistor. Therefore, however much energy was initially stored in the capacitor, that's how much heat energy dissipates in the resistor:

$$Energy_0 = Energy_f$$

$$\frac{Q_0^2}{2C} = Heat.$$

So,

$$Heat = \frac{Q_0^2}{2C} = \frac{(1 \text{ coulomb})^2}{2(0.20 \text{ F})} = 2.5 \text{ joules.}$$

Here's my meta-point. When dealing with an RLC circuit, think of the analogy between that circuit and a damped harmonic oscillator, such as a block on a spring (with air resistance). Most students find the oscillating block easier to visualize.

QUESTION 44-4

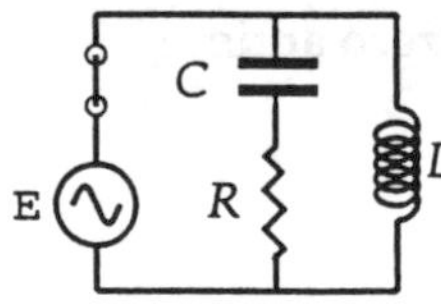

In this circuit, $C = 0.50$ F, $R = 0.10\ \Omega$, and $L = 0.80$ H. The switch next to the battery is initially closed, as drawn here. So, current flows through the circuit. The power supply provides an EMF

$$\mathcal{E} = 3\sin(10\pi t) \text{ volts,}$$

with t in seconds.

(a) What is the peak current through the capacitor? (The current oscillates back and forth. I'm looking for the largest value it reaches during those oscillations.)

(b) What is the peak current through the inductor?

(c) What is the peak current through the power supply? Do not complete the math. Just set things up.

(d) Suddenly, I open the switch. This "disconnects" the power supply, leaving us with an undriven RLC circuit. Is the new period of oscillation longer than, shorter than, or the same as it was when the power supply was connected?

ANSWER 44-4

You *cannot* treat this circuit like a single-loop RLC circuit, because the capacitor and inductor are in parallel, not in series. A given bit of current flows through the capacitor *or* the inductor, but not both. Intuitively, the current "splits" at the intersection above the capacitor, and re-merges at the intersection below the resistor. Therefore, as with DC circuits, you must think *separately* about the different loops. Remember Kirchhoff's laws? (No groaning, please!)

As always, you can choose your loops however you want. Many students choose the left loop and the right loop. Although this works, it's more efficient to include the power supply in both your loops. So, loop 1 contains the power supply, capacitor, and resistor. Loop 2 goes around the perimeter of the circuit, and contains just the power supply and the inductor.

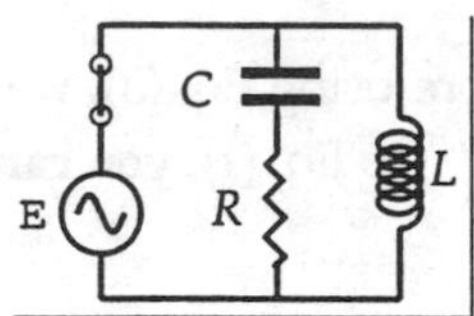

Let i_1 denote the current through the capacitor, and i_2 denote the current through the inductor. If you use Kirchhoff's 2nd law to write equations for loops 1 and 2, the loop 1 equation contains i_1, but not i_2. And the loop 2 equation contains i_2, but not i_1. Therefore, i_1 and i_2 are independent; they don't affect each other. In other words, i_1 behaves the same as if loop 2 were missing. Similarly, i_2 behaves the same as it would if loop 1 were missing. For this reason, we can answer questions about i_1 and i_2 by treating the two loops as separate circuits.

(a) As just noted, loop 1 behaves independently from loop 2. To find the peak current though the capacitor, treat loop 1 as a separate circuit. This circuit contains a power supply, capacitor, and resistor. So, in our AC formulas, we'll set $L = 0$.

For easy reference, let me summarize the five crucial formulas for forced (driven) AC circuits. I introduced these equations in question 44-1.

$$\mathcal{E} = \mathcal{E}_m \sin \omega t, \tag{1}$$

$$Z = \sqrt{R^2 + (\omega L - 1/\omega C)^2} \quad \textbf{Impedance} \tag{2}$$

$$i_m = \frac{\mathcal{E}_m}{Z} \quad \textbf{Maximum current} \tag{3}$$

$$i = i_m \sin(\omega t - \phi), \quad \textbf{Current} \tag{4}$$

$$\phi = \tan^{-1}\frac{\omega L - 1/\omega C}{R} \quad \textbf{"lag constant"} \tag{5}$$

Here, we want the maximum (peak) current. But to use Eq. (3), we first need to find the impedance, using Eq. (2). The impedance depends on R (which we know), L (which equals 0 for loop 1), C (which we know), and the angular frequency, ω. We don't know ω. But we can figure it out, because the problem tells us that

$$\mathcal{E} = 3\sin(10\pi t) \text{ volts,} \tag{*}$$

with t in seconds. Notice the "10π" in front of the t. From your general knowledge of oscillatory motion, and by comparing this expression to Eq. (1), you can see that the angular frequency is $\omega = 10\pi\ \text{s}^{-1}$. For loop 1, we also know that $R = 0.10\ \Omega$, $C = 0.50$ F, and $L = 0$. Substitute all this information into Eq. (2), to get

$$Z_1 = \sqrt{(0.10\ \Omega)^2 + \left[0 - \frac{1}{(10\pi\ \text{s}^{-1})(0.50\ \text{F})}\right]^2} = 0.119\ \Omega.$$

Before using Eq. (3), we still need $\mathcal{E}_m$, the peak voltage supplied by the power source. By comparing Eq. (*) to Eq. (1), you can see that the amplitude of the voltage oscillations is $\mathcal{E}_m = 3$ V. So, from Eq. (3),

$$i_{1m} = \frac{\mathcal{E}_m}{Z_1} = \frac{3.0\ V}{0.119\ \Omega} = 25\ \text{A}.$$

(b) We can recycle the reasoning of part (a), expect now we're talking about loop 2. A common mistake is to set $C = 0$, since there's no capacitor in loop 2. But if you substitute $C = 0$ into Eq. (2), it "blows up." You get an infinite impedance, which means no current flows around loop 2. This makes no physical sense.

When a circuit contains no capacitor, C does not equal 0. Instead, C equals ∞. To see why, think about the role played by a capacitor in a circuit. When fully charged up, a capacitor no longer allows current to flow through it, unless the current switches direction. And C is the "willingness" of the capacitor to hold charge. So, the higher the capacitance, the more charge can flow around the circuit before the capacitor gets fully charged. Therefore, a higher C implies that current can flow around the circuit in one direction for a longer time.

If the circuit contains no capacitor, then charge can flow around the circuit in one direction *indefinitely*. As just shown, the longer current can flow in one direction, the higher C is. So, the *absence* of a capacitor corresponds to $C = \infty$, not $C = 0$. As a result, the $1/\omega C$ term in the impedance formula goes to zero.

For loop 2, $R = 0$ (since the loop contains no resistor), $C = \infty$, and $L = 0.80$ H. Therefore,

$$Z_2 = \sqrt{(0)^2 + [(10\pi\text{s}^{-1})(0.80\ \text{H}) - 0]^2} = 25.1\ \Omega,$$

and hence

$$i_{2m} = \frac{\mathcal{E}_m}{Z_2} = \frac{3.0\ V}{25.1\ \Omega} = 0.12\ \text{A}.$$

In summary, the absence of a capacitor corresponds to the absence of the $1/\omega C$ term in the impedance.

(c) According to the "intersection rule," the current through the power supply—call it i_0—equals the sum of the currents through the capacitor and the inductor:

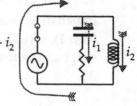

$$i_0 = i_1 + i_2.$$

From this fact, you may want to conclude that the peak current through the power supply is simply $i_{0m} = i_{1m} + i_{2m}$. That would be true if i_1 and i_2 reached their peaks in sync. But they don't. When i_1 reaches its peak value of i_{1m}, i_2 does *not* equal its peak value of i_{2m}. Similarly, when i_2 reaches its peak value (i_{2m}), i_1 does not equal its peak value. **Since the two currents never hit their peaks at the same time, the sum of those currents never reaches $i_{1m} + i_{2m}$.**

If this seems counterintuitive, consider the money in my piggy bank account, which reaches \$10 at its peak. My wife's piggy bank balance reaches \$20 at its peak. Does this mean we sometimes have \$30 total? Not necessarily. My piggy bank balance reaches its peak every Sunday. My wife's piggy bank balance reaches its peak every Tuesday. So, whenever I have \$10, she has less than \$20. Because our piggy bank balances reach their peaks "out of sync," we never have \$30 combined. Similarly, since i_1 and i_2 reach their peaks "out of sync," the total current never reaches $i_{1m} + i_{2m}$.

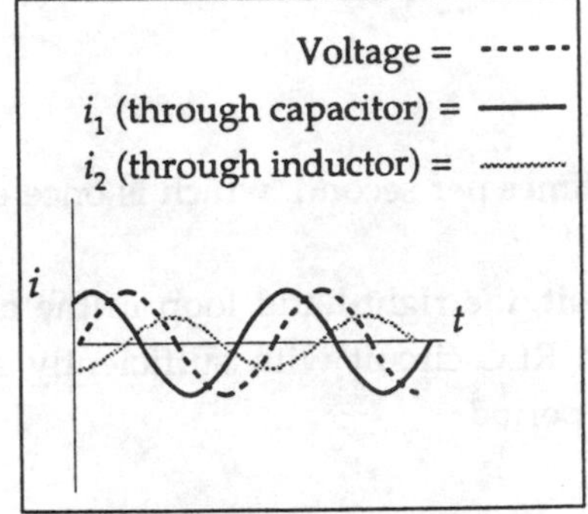

Notice that i_1 reaches higher peaks than i_2 does. More important, those two currents reach their peaks out of sync.

I'll now briefly explain why i_1 and i_2 don't reach their peak values together. The current through an AC circuit loop does not usually oscillate in sync with the power supply. Recall the "lag constant" ϕ in Eqs. (4) and (5) above. As your book explains, the current through the inductor lags behind the power supply, because the inductor opposes changes in current. By contrast, the current through the capacitor "gets ahead" of the power supply, because the capacitor naturally discharges. The accompanying graph attempts to clarify this pattern. Current i_1 reaches its peak *before* the power supply voltage reaches it peak. In Eqs. (4) and (5), this corresponds to a negative ϕ. By contrast, i_2 reaches its peak *after* the voltages hits its peak. This corresponds to positive ϕ.

By contrast, when a single circuit loop contains both an inductor and a capacitor, the current through both elements must be the same. That current can either "lag" or "lead" the voltage, depending on whether the inductor or the capacitor " dominates."

The main lesson I want you to take home from this problem is why $i_{0m} \neq i_{1m} + i_{2m}$. To find the actual answer, we must use the maximization/minimization strategy from calculus. For a review of this strategy and the intuitions underlying it, see question 33-6b. Here, we must write $i_1 + i_2$ as a function of time, differentiate with respect to time, and set the derivative equal to zero.

Well, from Eq. (4), the current through the power supply is

$$i = i_1 + i_2 = i_{1m}\sin(\omega t - \phi_1) + i_{2m}\sin(\omega t - \phi_2).$$

We know i_{1m} and i_{2m} from parts (a) and (b). We also know that $\omega = 10\pi\ \text{s}^{-1}$. And we can easily use Eq. (5) to obtain the lag factors, ϕ_1 and ϕ_2. So, I'll now set

$$\begin{aligned} 0 &= \frac{di}{dt} \\ &= \omega\, i_{1m}\cos(\omega t - \phi_1) + \omega\, i_{2m}\cos(\omega t - \phi_2). \end{aligned}$$

By solving this equation for t, and then substituting that t back into the equation for $i_1 + i_2$, we could find the maximum current through the power supply. To save trees, I'll skip the messy math.

(d) We can't guess the answer intuitively. The power supply might be set to a higher period or a lower period than the circuit's "natural" (undriven) period. To find out, I'll first calculate the period of the driven circuit, with the power supply connected. Then I'll find the period of the undriven RLC circuit. By the way, we could equally well compare the angular frequencies of the driven and undriven circuits. Higher frequency means lower period, and vice versa.

When driven by the power supply, the circuit's angular frequency is $\omega = 10\pi\ s^{-1}$. Angular frequency is 2π times the regular frequency, f. And frequency is one over the period, T. Putting this together gives

$$T_{driven} = \frac{2\pi}{\omega} = \frac{2\pi}{10\pi s^{-1}} = 0.20\ s.$$

The power supply makes the current slosh back and forth 5 times per second, which is once every 0.20 seconds.

Now let's find the period of the undriven RLC circuit, the right-hand loop in the above drawings. As shown in the previous problem, an undriven RLC circuit with sufficiently small resistance has angular frequency $\omega = 1/\sqrt{LC}$, and therefore period

$$T_{undriven} = \frac{2\pi}{\omega} = 2\pi\sqrt{LC} = 2\pi\sqrt{(0.80\ H)(0.50\ F)} = 4.0\ s.$$

So, the period becomes longer when we disconnect the power supply. That's not a general result. With different numbers, we'd get the reverse.

QUESTION 44-5

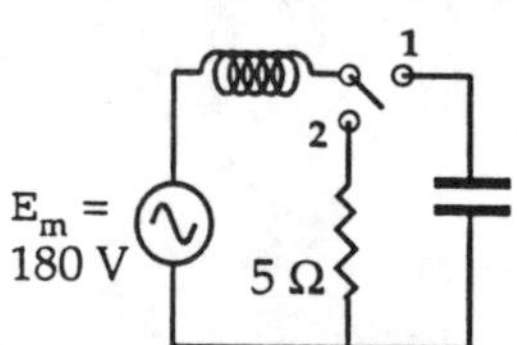

(*Very hard*) This alternating circuit contains a power supply whose maximum voltage is permanently set to 180 V, but whose frequency can be adjusted by the user. The resistor's resistance is known to be $R = 5.0\ \Omega$. The capacitance and inductance are unknown. In this problem, you're going to figure them out, from the following observations.

When the switch is set to position 1, the peak current through the circuit depends on the frequency of the power supply. But the peak current, i_m, is *highest* when the power supply frequency is set to 10 Hz.

With the power supply still set to 10 Hz, the switch is flicked to position 2. In 20 seconds, 2.60×10^4 joules of heat dissipate in the resistor.

What is the capacitance of the capacitor and the inductance of the inductor?

ANSWER 44-5

To solve for the two unknowns (L and C), we'll need two equations. We can extract those equations from the two observations.

With the switch set to position 1, the resistor is "cut off," leaving us with a driven LC circuit. The current is biggest when the driving frequency is 10 Hz. What does this tell us about the capacitance and inductance?

Well, a kid on a swing goes highest if you push her with a frequency that matches the natural frequency of her swinging. The Tacoma-Narrows bridge collapsed because the frequency of the wind gusts matched the natural oscillation frequency of the bridge. Similarly, the peak current in the driven LC circuit goes highest if the driving frequency matches the natural ("resonance") frequency of the undriven circuit. Since the peak current is biggest with the power supply set to 10 Hz, the natural frequency of the undriven LC circuit must be 10 Hz.

In an earlier problem, by making an analogy between a block on a spring and an LC circuit, we found the angular frequency to be $\omega = 1/\sqrt{LC}$. The frequency of oscillation is therefore

$$f_{\text{undriven LC}} = \frac{\omega_{\text{undriven LC}}}{2\pi} = \frac{1}{2\pi\sqrt{LC}}.$$

As just discussed, for this circuit, the resonance frequency is 10 Hz. So,

$$\frac{1}{2\pi\sqrt{LC}} = 10 \text{ Hz}. \tag{1}$$

So far, we have one equation in two unknowns (L and C). We need a second equation.

To get it, consider what happens when the switch gets flicked to position 2. This cuts off the capacitor, leaving us with a forced LR circuit. A total of $\Delta E = 2.60 \times 10^4$ joules of heat energy dissipate in time $\Delta t = 20$ seconds. Therefore, the average power is $\overline{P} = \frac{\Delta E}{\Delta t} = \frac{2.6 \times 10^4 \text{ J}}{20 \text{ s}} = 1300 \text{ watts}$.

From this power, we can calculate the current. And from the current, we can deduce the inductance. Actually, I'll combine these two steps of reasoning into one, as follows.

Since the current oscillates sinusoidally, the power fluctuates as well. As your textbook shows, the *average* power during these fluctuations is one half the peak power, $P_{\text{peak}} = i_m^2 R$. So,

$$\begin{aligned}
\overline{P} &= \frac{1}{2} i_m^2 R \\
&= \frac{1}{2}\left(\frac{\mathcal{E}_m}{Z}\right)^2 R \qquad \left[\text{since } i_m = \frac{\mathcal{E}_m}{Z}, \text{ according to "AC Ohm's law"}\right] \\
&= \frac{1}{2}\left[\frac{\mathcal{E}_m}{\sqrt{R^2 + (\omega L)^2}}\right]^2 R
\end{aligned} \tag{2}$$

where in the last step, I used the general expression for impedance, $Z=\sqrt{R^2+(\omega L-1/\omega C)^2}$, with the capacitance term omitted. Remember, when the switch occupies position 2, the capacitor gets "cut off" from the circuit. The absence of a capacitor corresponds to $C=\infty$, not $C=0$. See question 44-4.

In this equation, L is the only unknown! Remember, we found $\bar{P}=300$ W above. And we know that $R=5.0\ \Omega$, $\mathcal{E}_m=180$ V, and $\omega=2\pi f=2\pi(10\ \text{s}^{-1})=62.8\ \text{s}^{-1}$. So, we can solve Eq. (2) for L, and then substitute it into Eq. (1) to solve for C. We're done, except for algebra and number-crunching.

Algebra starts here. Starting with Eq. (2), multiply through by $[R^2+(\omega L)^2]$ to get

$$\bar{P}[R^2+(\omega L)^2]=\frac{1}{2}\mathcal{E}_m^2R.$$

Now subtract $\bar{P}R^2$ from both sides, divide through by $\bar{P}\omega^2$, and then take the square root to get

$$L=\sqrt{\frac{\frac{1}{2}\mathcal{E}_m^2R-\bar{P}R^2}{\bar{P}\omega^2}}=\sqrt{\frac{\frac{1}{2}(180\ \text{V})^2(5.0\ \Omega)-(1300\ \text{W})(5.0\ \Omega)^2}{(1300\ \text{W})(62.8\ \text{s}^{-1})^2}}=0.097\ \text{H}.$$

Now solve Eq. (1) for C, and substitute in this value for the inductance, to get

$$C=\frac{1}{(2\pi)^2(10\ \text{s}^{-1})^2L}=\frac{1}{(2\pi)^2(10\ \text{s}^{-1})^2(0.097\ \text{H})}=0.0026\ \text{F}.$$

QUESTION 44-6

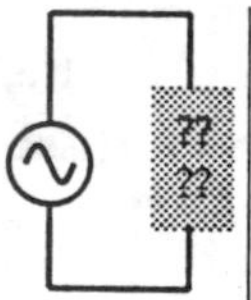

(*Hard but cool*) An AC power supply is connected to a "mystery box," as drawn here. You can't look to see what's inside. But you can take measurements on the wire "leads" coming out of the box. Your voltmeter and ammeter tell you that the voltage and current across the box are

$$\mathcal{E}=100\sin(3\pi t)\ \text{volts}$$

and

$$i=20\sin(3\pi t-30^\circ)\ \text{amps},$$

where t is in seconds. From this information, you can figure out something—but not everything—about what's inside the mystery box. (Assume the box contains no power supply.)

(a) Can you tell whether there's an inductor inside? Explain.

(b) (*Extremely hard*) Can you tell whether there's a resistor inside? Explain.

(c) Can you tell whether there's a capacitor inside? Explain.

(d) (*Challenge problem*) *Without looking inside the mystery box,* what experiment could you do to gather more information about whether the box contains a resistor, inductor, and capacitor?

ANSWER 44-6

Let's begin by listing all the information we're given. The maximum (peak) voltage is $\mathcal{E}_m = 100$ V, and the peak current is $i_m = 20$ A. Also, the angular frequency of oscillation must be $\omega = 3\pi$, and the "lag constant" is $\phi = 30°$.

(a) The current "lags" behind the voltage by 30°. In other words, the oscillating current doesn't reach its peak value at the same moment the power supply's voltage reaches its peak. The current reaches its peak a little later.

For this reason, the mystery box must contain an inductor. Because an inductor opposes a change in current, it makes the current lag behind whatever the power supply tries to make it do. By contrast, a capacitor tends to make the current reach its peak *before* the externally-applied voltage reaches its peak. See your textbook for the details about. Here, since the current lags the voltage (instead of "leading" it), the mystery box must contain an inductor. If the box also contains a capacitor, the inductor "dominates."

This physical argument corresponds directly to the mathematical expression

$$\phi = \tan^{-1}\frac{\omega L - 1/\omega C}{R}.$$

Here, we know that ϕ is positive, namely +30°. Therefore, L must be positive.

(b) I don't see a direct, intuitive physical argument as to why the box must contain a resistor. We must rely largely on mathematical reasoning. I could prove mathematically that power dissipates in the box, and then note that power dissipates only in a resistor. But instead, I'll base my reasoning on the formula for the lag constant,

$$\phi = \tan^{-1}\frac{\omega L - 1/\omega C}{R}.$$

Using this expression, I'll prove that the box *must* contain a resistor, or else the lag constant couldn't equal 30°. Here's the argument . . .

Suppose the box contained no resistor, and hence, $R \approx 0$. Then $\frac{\omega L - 1/\omega C}{R}$ approaches either positive infinity or negative infinity. Therefore, $\phi = \tan^{-1}\frac{\omega L - 1/\omega C}{R}$ approaches either +90° or −90°. (It's highly unlikely that ωL exactly equals $1/\omega C$. And if it did, then the current would "blow up," unless the box contained a resistor.)

In summary, if $R \approx 0$, then the current either lags or leads the voltage by 90°, depending on whether the inductance or the capacitance "dominates." Since $\phi = 30°$, the resistance must not be zero. The mystery box contains a resistor.

(c) The box definitely contains an inductor and a resistor. Can we figure out whether it also contains a capacitor? No. L and R might have values that fully account for the observed lag constant and peak current. Or, L and R might have different values, in which case the box must also contain a capacitor. We can't figure it out, from the given information.

(d) We know that the box contains an inductor and resistor. The only open question is whether it also contains a capacitor.

To "test for" a capacitor, we could turn off the power supply in such a way that current can still flow around the circuit. For instance, we could throw this switch from position 1 to position 2. Doing so cuts off the power supply, but leaves a complete circuit around which current could flow. By hooking up an ammeter, we could measure the current in that new loop. I'll now explain why this measurement tells us whether the box houses a capacitor.

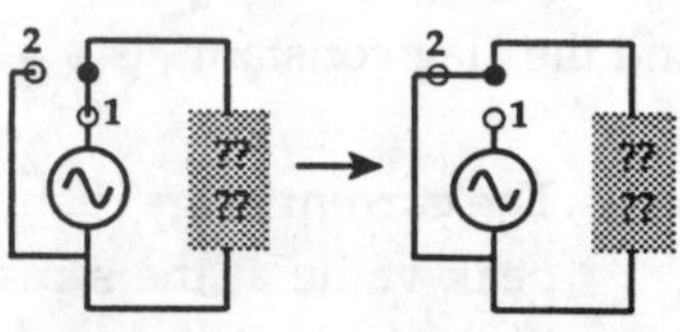

Throwing the switch "cuts off" the power supply, but leaves a complete circuit intact.

If the mystery box contains only an inductor and resistor, then current keeps flowing after the power supply gets cut. But that current flows in *one direction*. Remember, when the power supply gets cut off, the current "wants" to drop to zero. The inductor opposes this change in the current. Although the inductor can't stop the current from dropping to zero, the inductor makes this process happen gradually instead of instantaneously.

By contrast, if the mystery box contains a capacitor along with the inductor and resistor, then it becomes an undriven RLC circuit when the power supply gets cut. As we've seen before, the current in an RLC circuit oscillates back and forth while gradually dying out.

In summary: After the power supply gets cut off, the current in the circuit gradually dies out. But if the circuit contains a capacitor, the current *oscillates* while dying. By contrast, if the box contains no capacitor, the current keeps flowing in one direction while dying. Cute, huh? (I've over-simplified slightly, to avoid the case of "overdamping.")

Displacement Current and Maxwell's Equations

45 CHAPTER

NOTE TO STUDENTS

Many introductory courses cover this material superficially, if it all. Find out at what level you need to understand it. In all likelihood, these two problems cover everything you need to know about induced magnetic fields—or maybe too much. Check with your instructor. Given my page limitations, I'd rather include more problems about core material such as Gauss' law, circuits, Ampere's law, etc. That's why this chapter is short.

QUESTION 45-1

TOP-DOWN VIEW

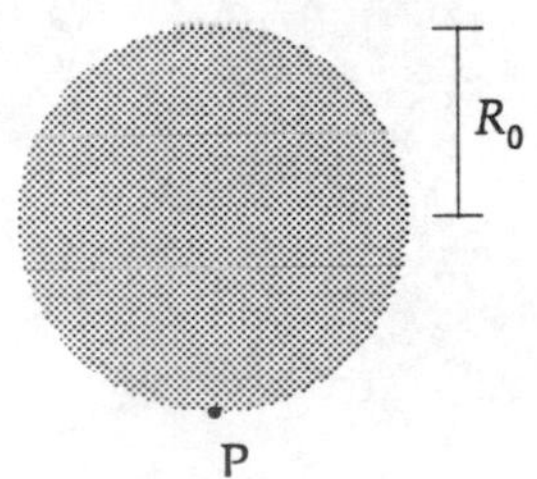

This is a top-down view of a circular table top, of radius $R_0 = 1.0$ m. An electric field points upward through the table top, perpendicular to its surface (out of the page in this drawing). The field is strongest near the center of the table and weakest near the edge:

$$E(r) = (2.0\ \text{V/m}^2)(R_0 - r),$$

where r denotes the radial distance from the center.

Suppose this field is steadily reduced to zero over a 3-second time interval.

(a) Show that the initial electric flux through the table top is $\Phi_E = \frac{2}{3}\pi R_0^3$, where the "2/3" has units of volts/meters^2.

(b) During the 3-second interval, what is the magnitude of the *magnetic* field at point P, on the edge of the table top?

(c) During the 3-second interval, what is the displacement current through the tabletop?

(d) Suppose we want to generate a magnetic field at point P using an infinite wire of current, instead of a changing electric field. The wire runs through the center of the table top, perpendicular to its surface. What current must flow through the wire to generate the *same* magnetic field (at point P) that we found in part (b)?

ANSWER 45-1

This problem illustrates the central new concept of the chapter. Just as a *changing magnetic flux* generates an electric field (and corresponding EMF), a *changing electric flux* generates a magnetic field. This problem also clarifies the *physical* meaning of displacement current.

(a) Roughly speaking, the flux specifies how many electric field lines pierce through the table top. If E were uniform, we could just multiply it by the table's area. But E is non-uniform. It's stronger near the center and weaker towards the edge. Our flux calculation must take this variation into account.

To evaluate the flux integral, $\Phi_E = \int \mathbf{E}\cdot d\mathbf{A}$, divide the table into infinitesimal rings. I've "blackened" a ring of arbitrary radius r and width dr. Since E depends only on r, the electric field has the same value at all points *within* a given ring. Also, since the table "faces" the same direction as $\mathbf{E}$, the dot product reduces to a regular product.

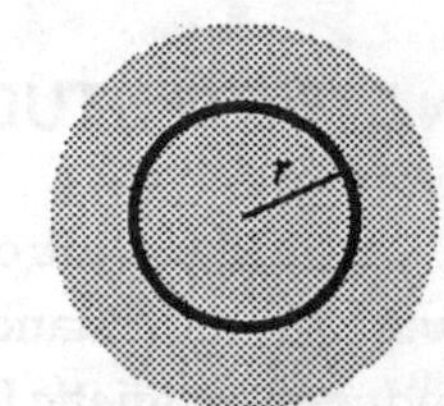

The blackened ring has length (circumference) $2\pi r$, and width dr. Therefore, it has area dA = length × width = $2\pi r dr$. So, the infinitesimal flux through the ring is

$$d\Phi_E = E dA = E[2\pi r dr]$$
$$= (2.0\ \text{V/m}^2)(R_0 - r)\,[2\pi r dr].$$

Now we just have to "add up" the tiny flux contributions from all the rings. The smallest ring has radius $r = 0$, while the largest one has radius $r = R_0$. Those are the limits of integration.

$$\Phi_E = \int E dA = \int_0^{R_0} (2.0\ \text{V/m}^2)(R_0 - r)2\pi r dr$$
$$= (2.0\ \text{V/m}^2)2\pi \int_0^{R_0} (R_0 r - r^2)dr$$
$$= (2.0\ \text{V/m}^2)2\pi \left(R_0 \frac{r^2}{2} - \frac{r^3}{3} \right)\Bigg|_0^{R_0}$$
$$= (2.0\ \text{V/m}^2)2\pi \left(R_0 \frac{R_0^2}{2} - \frac{R_0^3}{3} \right)$$
$$= \frac{(2\ \text{V/m}^2)}{3}\pi R_0^3$$

(b) Let's guess the relevant formula by making an analogy with Faraday's law. The voltage around a real or imaginary loop, caused by a changing magnetic field, is

Induced electric field
$$-\frac{d\Phi_B}{dt} = \mathcal{E}_{\text{induced}}$$
$$= \oint \mathbf{E}\cdot d\mathbf{s},$$

where in the last step, I remembered the relationship between potential changes and fields. In Faraday's law, that integral must cover a complete loop.

Since a changing magnetic flux induces an electric field, we expect a changing electric flux to induce a magnetic field. And indeed, that's the case:

$$\textbf{Induced magnetic field} \quad \varepsilon_0\mu_0 \frac{d\Phi_E}{dt} = \oint \mathbf{B}\cdot d\mathbf{s}.$$

Notice the symmetry between these two equations. The only difference is a minus sign and some constants.

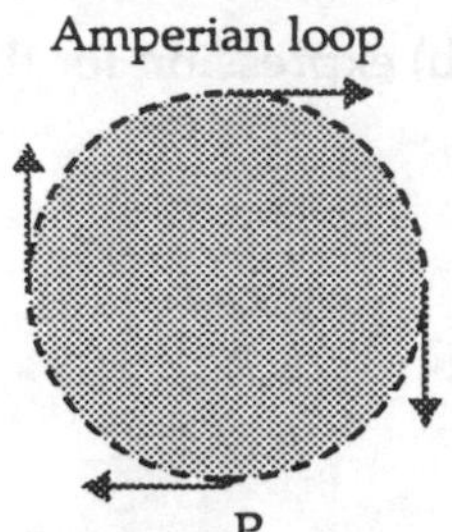

To calculate a magnetic field using this new induction formula, we must begin by choosing an Amperian loop. This is the loop through which we'll find the rate of change of electric flux, and along which we'll integrate the magnetic field ($\oint \mathbf{B}\cdot d\mathbf{s}$).

Since we're looking for the magnetic field at point P, let's make sure the loop includes point P. As always, $\oint \mathbf{B}\cdot d\mathbf{s}$ simplifies most readily if the magnetic field has the same strength over the whole loop. In that case, we can pull *B outside the integral*. Well, by radial symmetry, the magnetic field is equally strong at all points around the edge of the table. So, I'll choose the dashed Amperian loop drawn here. Because the field lines turn out to be circular, **B** always points along my loop, and hence

$$\oint \mathbf{B}\cdot d\mathbf{s} = \int_0^{2\pi R_0} \mathbf{B}\cdot d\mathbf{s} = B\int_0^{2\pi R_0} ds = B(2\pi R_0).$$

Now I'll simplify the other side of the induced magnetic field equation, by finding the rate of change of electric flux through my Amperian loop. The loop captures the whole table top. And in part (a), we found the initial electric flux through the table top. Over a 3-second interval, this flux steadily drops to zero. So,

$$\frac{d\Phi_E}{dt} = \frac{\Delta\Phi_E}{\Delta t} = \frac{0 - \dfrac{(2\text{V/m}^2)}{3}\pi R_0^3}{t_1} = -\frac{(\frac{2}{3}\text{V/m}^2)\pi R_0^3}{(3\ s)}$$

Now that we've simplified both sides of the induced magnetic field equation, we can tie it all together, and solve for B.

$$\varepsilon_0\mu_0 \frac{d\Phi_E}{dt} = \oint \mathbf{B}\cdot d\mathbf{s}$$

$$-\varepsilon_0\mu_0 \frac{(\frac{2}{3}\text{V/m}^2)\pi R_0^3}{(3\ s)} = B(2\pi R_0),$$

and hence

$$B = -\frac{\varepsilon_0\mu_0(\frac{2}{3}\text{V/m}^2)\pi R_0^3}{(3\ s)2\pi R_0} = \frac{\varepsilon_0\mu_0(\frac{2}{3}\text{V/m}^2)R_0^2}{2(3\ s)}$$

$$= \frac{(8.85\times10^{-12}\ \text{F/m})(4\pi\times10^{-7}\text{H/m})(\frac{2}{3}\text{V/m}^2)(1.0\ \text{m})^2}{2(3.0\ s)}$$

$$= -1.2\times10^{-18}\ \text{tesla}.$$

Notice how weak this field is. Induced magnetic fields are appreciable only when the electric field changes very rapidly.

So far, we haven't figured out whether the minus sign indicates a "clockwise" or "counter-clockwise" magnetic field around the table. Well, way back in part (a), I chose out-of-the-page as the positive direction. So, point your right thumb out of the page. Your fingertips curl counterclockwise. So, counterclockwise is the positive direction along the boundary of the table. (This "right-hand rule" comes from Stokes' law in math.) The minus sign in front of the magnetic field means that it points clockwise.

(c) By definition, the displacement current is $i_d = \varepsilon_0 \frac{d\Phi_E}{dt}$. So, using our part (b) expression for the rate of change of electric flux, I get

$$\begin{aligned} i_d &= \varepsilon_0 \frac{d\Phi_E}{dt} \\ &= (8.85\times10^{-12}\,\text{F/m})\frac{(\frac{2}{3}\,\text{V/m}^2)\pi R_0^3}{(3\ s)} \\ &= 6.2\times10^{-12}\,\text{A}. \end{aligned}$$

(d) For reasons explained below, the current here *must* equal the displacement current from part (c)! Before discussing why, I'll solve the long way, using Ampere's law.

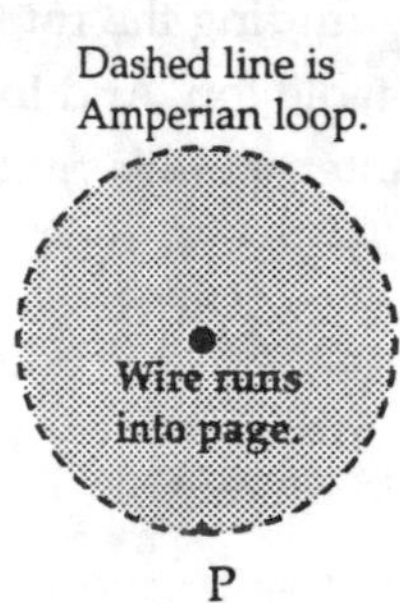
Dashed line is
Amperian loop.
Wire runs
into page.
P

Imagine an infinite wire running through the center of the table top, perpendicular to its surface. We want to know how much current is needed to create a certain magnetic field, at distance $R_0 = 1.0$ m away. To find the relevant relationship between the current and the magnetic field, use Ampere's law. Since I already applied Ampere's law to an infinite wire in Chapter 40, let me skip most of the details. Using the Amperian loop from part (b), I get

Ampere's law $\oint \mathbf{B}\cdot d\mathbf{s} = \mu_0 i_{\text{encl}}$

$$B(2\pi r) = \mu_0 i.$$

Here, we want the current to create a field of strength $B = 1.2\times10^{-18}$ teslas at a distance $r = R_0 = 1.0$ m from the wire. So, solve for i, and substitute in those numbers, to get

$$i = \frac{B(2\pi r)}{\mu_0} = \frac{(1.2\times10^{-18}\,\text{T})2\pi(1.0\text{ m})}{(4\pi\times10^{-7}\,\text{T}\cdot\text{m/A})} = 6.2\times10^{-12}\,\text{A},$$

which equals the displacement current from part (c).

Indeed, that's the whole point of this problem. A changing electric flux, like an infinite wire of current, generates a magnetic field! If a changing electric flux corresponds to a displacement current of 3 amps, then 3 amps of actual current would generate the *same* magnetic field that the changing electric flux induces. In general, the current needed to reproduce the magnetic field caused by $d\Phi_E/dt$ equals the displacement current corresponding to $d\Phi_E/dt$. Displacement current is *defined* to make this true.

Late, I'll give a deeper mathematical explanation of why $i_d = \varepsilon_0 d\Phi_E/dt$ plays the role of a current in Maxwell's equations. Here, I'm just trying to give you the rough intuitive idea.

QUESTION 45-2

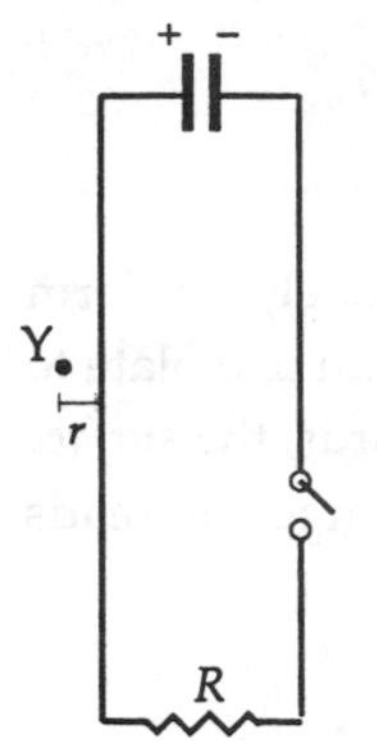

This problem reviews concepts from the entire semester.

In the RC circuit pictured here, the capacitor is initially charged to Q_0. The capacitor consists of two circular plates of radius L and plate-separation s, where $s \ll L$. The resistor has resistance R. At time $t = 0$, the switch is closed, allowing the capacitor to discharge.

(a) What is the capacitance of this capacitor? Figure it out from scratch, without looking up any parallel-plate capacitor formulas.

(b) Write an equation that could be solved for $Q(t)$, the charge on the capacitor as a function of time. In your equation, let "C" denote the answer to part (a). Don't actually write out your expression for C.

(c) Prove that the solution to your equation from part (b) is $Q(t) = Q_0 e^{-t/RC}$.

(d) Write an expression for $E(t)$, the electric field between the capacitor plates as a function of time.

(e) Point P is between the plates, a distance r from the center, as drawn here. At arbitrary time t, what is the strength of the magnetic field at point P?

(f) Point Y sits a distance r from a straight segment of wire, as drawn above. Because r is tiny compared to that length of wire, and because the other parts of the circuit are far away, the magnetic field at point Y approximates the field generated by an infinite wire. At arbitrary time t, what is the magnetic field strength at point Y?

(g) (*Very hard*) Explain why your part (e) answer is less than your part (f) answer. For what r are those two answers the same? Why is this so?

CLOSE-UP VIEW OF CAPACITOR

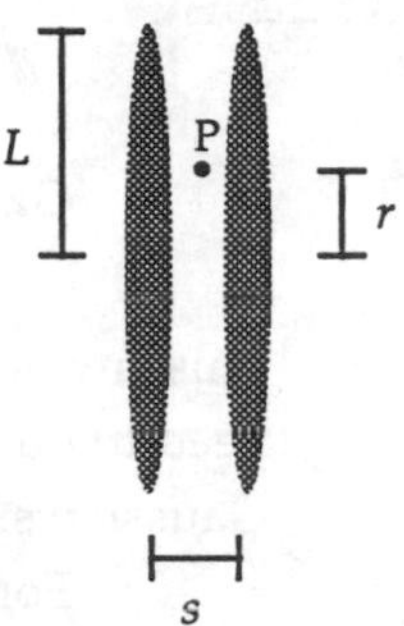

ANSWER 45-2

(a) "Capacitance" is the capacitor's willingness to hold charge. More specifically, it's the charge carried by the capacitor *per volt of applied potential difference between the plates*: $C = Q/V$. Therefore, to find C, we can calculate the potential difference between the plates generated by charge Q, and then divide Q by V. Let me summarize this strategy, first introduced in Chapter 36.

Capacitance-finding strategy

1) Assign an arbitrary charge $+q$ ands $-q$ to the two plates.
2) Using Gauss' law or other techniques, find the electric field between the plates.
3) From that field, calculate the potential difference between the plates: $V = -\int \mathbf{E} \cdot d\mathbf{s}$.
4) Calculate the capacitance using $C = q/V$. The arbitrary charge q you chose in step 1 should cancel out.

OK, let's do it.

Step 1: Assign the plates arbitrary charge +q and −q.

You can use the actual charge, Q_0, or an arbitrary charge. It doesn't matter, because this charge cancels out when you calculate the capacitance.

Step 2: Find the electric field created by those charges, using Gauss' law.

When two capacitor plates are very close, the electric field is approximately uniform between them, and approximately zero outside the plates. The field points directly from one plate to the other, partly because the charge on each plate spreads out uniformly. In other words, the *surface charge density*, σ = charge/area, is approximately constant. Here, since charge q spreads out over plate area πL^2, the positive plate carries surface charge density

$$\sigma = \frac{q}{\pi L^2},$$

while the negative plate carries an equal and opposite surface charge density.

Let's invoke Gauss' law, according to which the electric flux flowing out of a closed surface is proportional to the enclosed charge:

Gauss' Law $$\oint \mathbf{E}\cdot d\mathbf{A} = \frac{q_{\text{encl}}}{\varepsilon_0}.$$

This law holds true, no matter what Gaussian surface you draw. But to solve for the electric field, we need to pull E outside the flux integral. Therefore, we want the field to be uniform over (part of) the Gaussian surface.

For this reason, I'll use a "Gaussian cylinder," as drawn here. Like a soda can, it consists of a tube (side) and two "ends." The ends are parallel to the capacitor plates. Notice that field lines only pierce through the right end of the tube, not through the left end or the tube. (The field lines "skim along" the tube of the cylinder, without actually piercing *through* the tube.)

GAUSSIAN CYLINDER

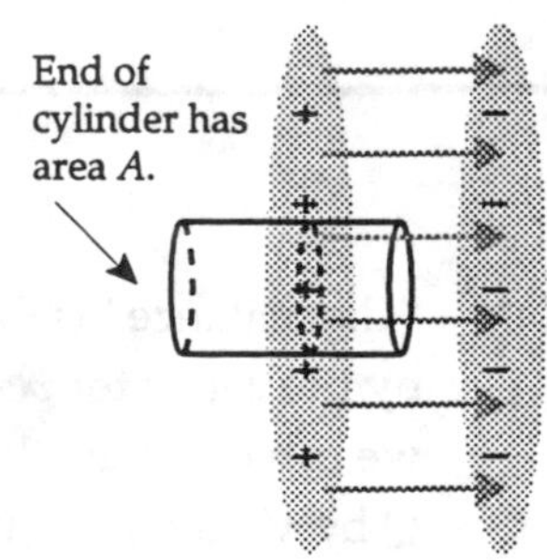
End of cylinder has area A.

Given this Gaussian surface, I'm ready to simplify both sides of Gauss' law. Each end of the cylinder has arbitrary area A. So, the cylinder encloses area A of the positive capacitor plate. Using the surface charge density calculated above, we can easily find the charge "trapped" inside the Gaussian surface. It's

$$q_{\text{encl}} = \frac{\text{charge}}{\text{area}} \times (\text{area enclosed}) = \sigma A = \frac{q}{\pi L^2} A.$$

Now I'll address the flux integral. As mentioned above, field lines flow only through the right end of the Gaussian surface. In other words, only that end "contributes" to the flux. And since the field in constant over that surface, we can pull E outside the integral:

$$\oint_{\text{Gaussian cylinder}} \mathbf{E}\cdot d\mathbf{A} = \int_{\text{right-hand end}} \mathbf{E}\cdot d\mathbf{A} = E \int_{\text{right-hand end}} dA = EA.$$

Therefore, by Gauss' law,

$$\oint_{\text{Gaussian surface}} \mathbf{E} \cdot d\mathbf{A} = \frac{q_{\text{encl}}}{\varepsilon_0}$$

$$EA = \frac{q}{\pi L^2 \varepsilon_0} A.$$

Cancel the A's to get

$$E = \frac{q}{\pi L^2 \varepsilon_0},$$

which is simply a special case of the general result for parallel-plate capacitors, $E = \sigma / \varepsilon_0$.

OK, let's step back to get our bearings. I just used Gauss' law to find the electric field between the plates when they carry charge $\pm q$. That was step 2 of our capacitance-finding strategy, outlined above.

Step 3: Find the potential difference between the plates by integrating over the electric field.

Here, the integral is easy, because **E** is uniform. I'll integrate along a straight line connecting the plates, the dashed segment on this drawing. To ensure that V comes out positive, integrate from the negative to the positive plate. Since **E** points antiparallel to this path, the dot product $\mathbf{E} \cdot d\mathbf{s}$ reduces to a *negative* regular product. So,

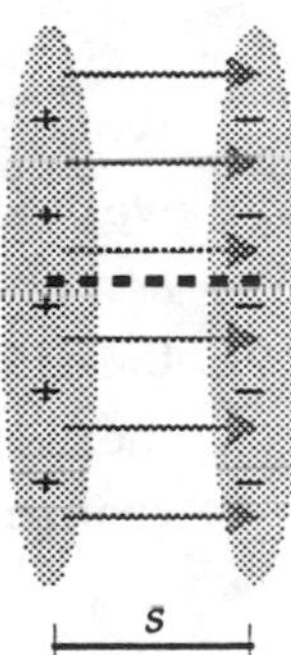

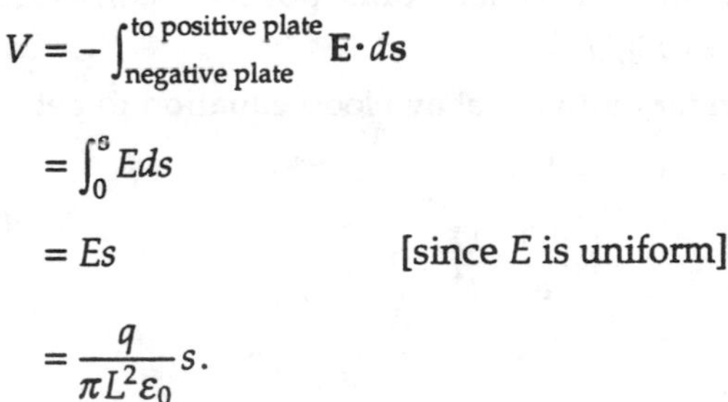

$$V = -\int_{\text{negative plate}}^{\text{to positive plate}} \mathbf{E} \cdot d\mathbf{s}$$

$$= \int_0^s E ds$$

$$= Es \qquad \text{[since } E \text{ is uniform]}$$

$$= \frac{q}{\pi L^2 \varepsilon_0} s.$$

Step 4: Calculate the capacitance using $C = q/V$.

$$C = \frac{q}{V} = \frac{q}{\left(\frac{qs}{\varepsilon_0 \pi L^2}\right)} = \varepsilon_0 \frac{\pi L^2}{s}.$$

As always, the "q's" cancel. So, the capacitance is a constant. If you double the charge on the plates, you double the potential difference. The *ratio* of charge to voltage—i.e., the capacitance—stays the same. By the way, we just derived the textbook formula, $C = \varepsilon_0 A/s$, for parallel-plate capacitors.

(b) As usual, to generate useful equations about a circuit loop, use Kirchhoff's 2nd law. I'll "walk" around the circuit, adding up the potential differences across each circuit element, and setting that sum equal to zero. (The sum must equal zero, because your final potential equals your initial potential, when you return to your starting point.) Starting from the upper left hand corner, I'll walk in the direction of current flow, counterclockwise.

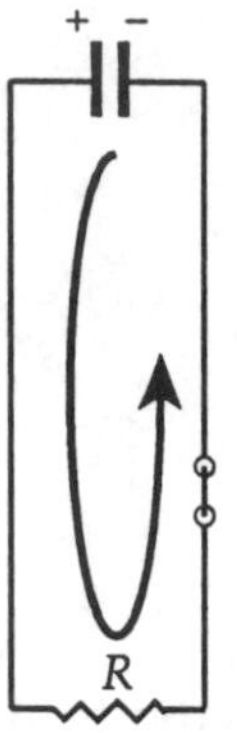

First, I walk through the resistor, with the current. In the "cars-on-a-roadway" analogy from see chapters 37–38, this corresponds to walking downhill. So, the voltage drops: $\Delta V = -iR$.

Next, I walk through the capacitor, from the negative to the positive plate. This corresponds to a positive potential difference, since work is needed to push a positive test charge from the negative to the positive plate. By definition, $C = q/V$. So, the potential difference across the capacitor is $\Delta V = +Q/C$, where Q denotes the charge on the capacitor at arbitrary time. We must use a general Q, instead of Q_0, because the charge on the capacitor changes with time. Hence, the voltage changes, too.

Now add up these potential differences, and set the sum to zero. This gives

$$-iR + \frac{Q}{C} = 0.$$

We can't yet solve for $Q(t)$, because the equation contains another unknown, namely i. Fortunately, we can write i in terms of Q. Here, the current is "made of" charges flowing off the positive plate and onto the negative plate. The faster those charges transfer from one plate to the other, the bigger the current. Given this reasoning, you might write $i = dQ/dt$. But i is *positive* (counterclockwise) when the charge on the positive plate *decreases*. In other words, positive i corresponds to a negative rate of change of charge on the plates: $i = -dQ/dt$.

Substitute this expression for i into the above loop equation to get

$$R\frac{dQ}{dt} + \frac{Q}{C} = 0.$$

This first-order differential equation can be solved for $Q(t)$, using "separation of variables."

(c) To prove that $Q_0e^{-t/RC}$ solves this differential equation, just substitute $Q_0e^{-t/RC}$ into the equation, to see if it "works":

$$\begin{aligned} R\frac{dQ}{dt} + \frac{Q}{C} &= R\frac{d(Q_0e^{-t/RC})}{dt} + \frac{Q_0e^{-t/RC}}{C} \\ &= R\left(-\frac{1}{RC}\right)Q_0e^{-t/RC} + \frac{Q_0e^{-t/RC}}{C} \\ &= -\frac{1}{C}Q_0e^{-t/RC} + \frac{Q_0e^{-t/RC}}{C} \\ &= 0. \end{aligned}$$

So, when Q equals $Q_0e^{-t/RC}$, it's true that $R\frac{dQ}{dt}+\frac{Q}{C}=0$. So, $Q_0e^{-t/RC}$ is *a* solution to the differential equation. But do multiple solutions exist? And if so, is $Q_0e^{-t/RC}$ the "right" one?

I don't want to get bogged down in mathematical details. Let me just summarize the main results. The most general solution to that differential equation is $Q(t) = Ke^{-t/RC}$, where K is any constant. But in this particular case, the constant must be Q_0. Here's why. At time $t = 0$, the capacitor carries charge $Q(0) = Q_0$. Substitute $t = 0$ into $Ke^{-t/RC}$, and you get $Q(0) = K$. So, K must equal Q_0.

(d) In part (a), while finding the capacitance, we used Gauss' law to figure out the electric field between the plates. We got

$$E=\frac{q}{\pi L^2\varepsilon_0},$$

where q denotes the charge on the plates. So, E is proportional to q. As the capacitor discharges, the charge on the plates gets smaller and smaller, and therefore E gets weaker and weaker. To find the exact expression for $E(t)$, substitute $q = Q_0e^{-t/RC}$ into our equation for the electric field. This gives

$$E(t)=\frac{q(t)}{\pi L^2\varepsilon_0}=\frac{Q_0e^{-t/RC}}{\pi L^2\varepsilon_0}.$$

E decreases exponentially.

(e) Recall the main point of this chapter: A magnetic field gets generated not only by currents, but also by *changing* electric fields. So, a magnetic field gets induced at point P, because the electric field at that point changes. According to Maxwell's equations,

Induced magnetic field $$\oint \mathbf{B}\cdot d\mathbf{s}=\mu_0 i_d, \qquad \text{where } i_d=\varepsilon_0\frac{d\Phi_E}{dt}.$$

This looks exactly like Ampere's law, except that a "displacement current" (corresponding to a changing electric flux) replaces an actual current. When the capacitor discharges, the electric field between the plates changes, inducing a magnetic field. By contrast, before the capacitor begins discharging, and after it finishes discharging, the electric field stays constant, and therefore no magnetic field gets induced.

OK, let's solve for the magnetic field. When using either Ampere's law *or* the induced magnetic field equation, begin by choosing the Amperian loop around which $\mathbf{B}\cdot d\mathbf{s}$ gets integrated. To pull B outside that integral, we must use an Amperian loop along which B is constant.

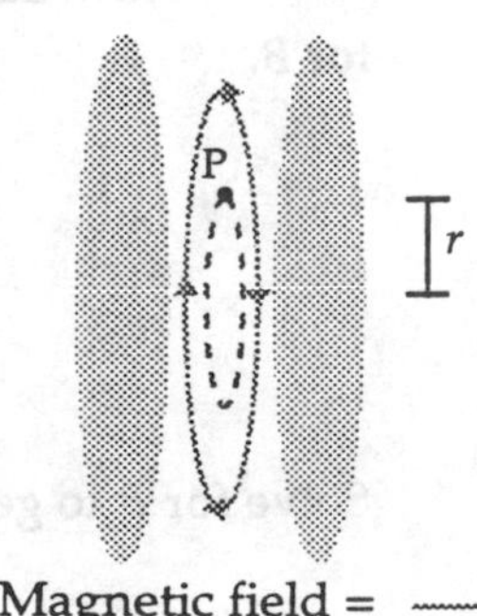

Here, point P sits a distance r from the center. By radial symmetry, B is uniform along the circular Amperian loop I've drawn here, since that loop consists of points equidistant from the center. Of course, B varies in time. But at any given moment, the magnetic field has the same strength over the entire Amperian loop.

As your textbook shows, the magnetic field lines trace out circles. They're just like the magnetic field lines around a wire. Therefore, the magnetic field always points along the Amperian loop, reducing $\mathbf{B}\cdot d\mathbf{s}$ to a regular product:

$$\oint \mathbf{B}\cdot d\mathbf{s} = \oint B ds = B(2\pi r)$$

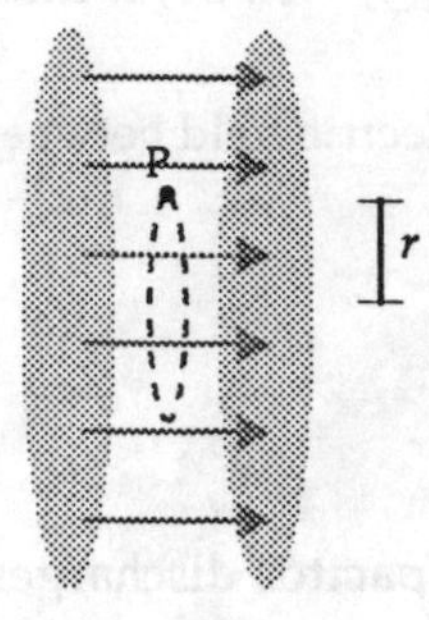

This Amperian loop "encloses" some electric flux.

So far, I've simplified the left-hand side of the induced magnetic field equation. Now let's work on the right-hand side, by finding the displacement current enclosed by the Amperian loop.

The displacement current depends on $d\Phi_E/dt$, the rate of change of electric flux *through the Amperian loop*. So, let's first write down the electric flux through the Amperian loop, and then differentiate it with respect to t. As we saw in part (d), the electric field is uniform between the plates, with strength

$$E(t) = \frac{Q_0 e^{-t/RC}}{\pi L^2 \varepsilon_0}.$$

Since the Amperian loop defines a disk of area $A_{\text{Amperian loop}} = \pi r^2$, the electric flux *through the Amperian loop* is

$$\Phi_E = \int_{\text{Amperian loop}} \mathbf{E}\cdot d\mathbf{A} = EA_{\text{Amperian loop}} = \frac{Q_0 e^{-t/RC}}{\pi L^2 \varepsilon_0}\pi r^2.$$

Therefore, the displacement current is

$$i_d = \varepsilon_0 \frac{d\Phi_E}{dt} = \varepsilon_0 \frac{d}{dt}\left(\frac{Q_0 e^{-t/RC}}{\pi L^2 \varepsilon_0}\pi r^2\right)$$

$$= \varepsilon_0\left(-\frac{1}{RC}\right)\frac{Q_0 e^{-t/RC}}{\pi L^2 \varepsilon_0}\pi r^2$$

$$= \frac{Q_0 e^{-t/RC}}{RCL^2} r^2.$$

Now that we've simplified both sides of the induced magnetic field equation, we can solve for B.

$$\oint \mathbf{B}\cdot d\mathbf{s} = \mu_0 i_d$$

$$B(2\pi r) = -\mu_0 = \frac{Q_0 e^{-t/RC}}{RCL^2} r^2.$$

Solve for B to get

$$B = \frac{\mu_0 Q_0 e^{-t/RC} r}{2\pi RCL^2}.$$

Since the problem asks only for the strength of B, I won't figure out whether the minus sign indicates clockwise or counterclockwise around the Amperian loop.

(f) I'll use Ampere's law, treating the wire segment as infinite. An infinite wire generates circular magnetic field lines. So, if we draw a circular Amperian loop, the field always points along the loop. Furthermore, since the loop consists of points equidistant from the wire, the field has the same strength over the entire loop. Therefore,

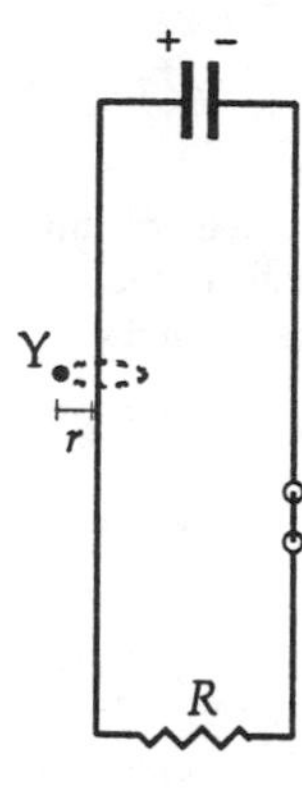

Ampere's law $\oint \mathbf{B} \cdot d\mathbf{s} = \mu_0 i_{\text{encl}}$

$$B(2\pi r) = \mu_0 i.$$

To solve for B, we need i, the current in the circuit.

Well, we know the charge on the capacitor as a function of time: $Q(t) = Q_0 e^{-t/RC}$. Since the current is the rate at which charge flows *off* the positive plate, we get

$$i = -\frac{dQ}{dt} = -\frac{\mu_0 Q_0 e^{-t/RC}}{dt} = \frac{1}{RC} Q_0 e^{-t/RC}.$$

Solve the above Ampere's law equation for B, and substitute in this expression for i, to get

$$B = \frac{\mu_0 i}{2\pi r} = \frac{\mu_0 Q_0 e^{-t/RC} r}{2\pi RCr}.$$

(g) Let B_P denote the magnetic field at point P, which is between the capacitor plates. Let B_Y denote the field at point Y, near the wire. To see that B_Y is bigger, take our part (f) answer, and multiply the numerator and denominator by r. This gives

$$B_Y = \frac{\mu_0 Q_0 e^{-t/RC} r}{2\pi RCr^2},$$

which we can directly compare to our part (e) answer,

$$B_P = \frac{\mu_0 Q_0 e^{-t/RC} r}{2\pi RCL^2}.$$

These answers "disagree" only about the r^2 vs. L^2 in the denominator. Since $r < L$, B_Y has a smaller denominator. So, B_Y is bigger.

With question 45-1 fresh in your mind, you might have expected these fields to be the same. After all, Ampere's law and the induced magnetic field equation have the same form:

Ampere's law $\oint \mathbf{B} \cdot d\mathbf{s} = \mu_0 i_{\text{encl}}$

Induced magnetic field Eq. $\oint \mathbf{B} \cdot d\mathbf{s} = \mu_0 i_d.$

So, if the displacement current between the capacitor plates equals the actual current flowing around the circuit, the fields at point P and point Y should come out the same, right?

To see the subtle flaw in this reasoning, recall that my displacement current from part (e), $i_d = \varepsilon_0 d\Phi_E/dt$, refers to the electric flux *through the Amperian loop*. This Amperian loop has radius r, which is smaller than the plate radius L. Therefore, the Amperian loop does not "capture" all the electric flux between the plates. In other words, the displacement current used in part (e) is less than the *total* displacement current between the plates.

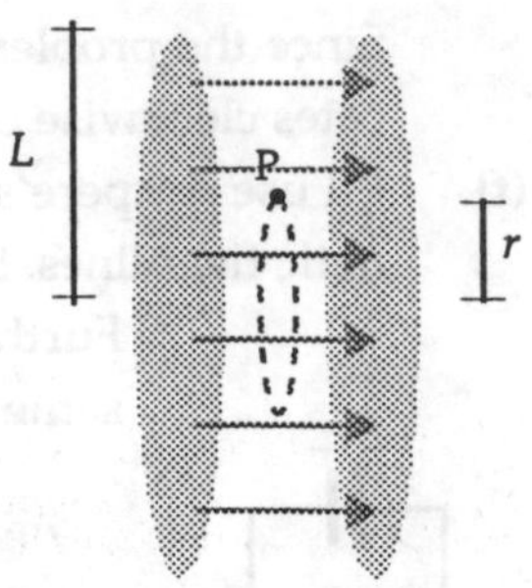

This Amperian loop captures some, but not all, of the electric flux between the plates.

In order to capture *all* of the electric flux and therefore *all* of the displacement current between the plates, the Amperian loop must have radius L. And indeed, if we set r equal to L in the above equations for B_Y and B_P, those two fields become equal! So, the *total* displacement current between the plates must equal the actual current flowing through the wire, since those currents generate the same magnetic field.

Electromagnetic Waves

46 CHAPTER

Note to Students

Some or all of this material may get left out of your current course, and included instead in a later course covering waves. This chapter emphasizes the concepts most likely to show up on *electromagnetism* tests at this level. But it's likely that I'll cover some material you don't need to know. Find out from your instructor exactly what you need to study.

QUESTION 46-1

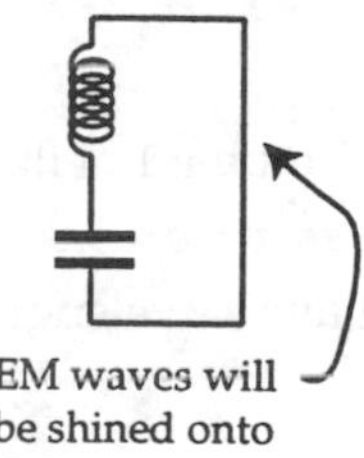

EM waves will be shined onto this wire.

An LC circuit, with $L = 2.0 \times 10^{-6}$ H and $C = 8.0 \times 10^{-6}$ F, initially has no charge on the capacitor, and no current. In order to get current flowing, I plan to shine an electromagnetic plane wave on one wire of the circuit. I need to know whether it's best to use visible light, x-rays, radio waves, or some other part of the spectrum.

Assuming I want a big current to oscillate back and forth in the circuit, what's the best wavelength for me to use? Hint: How should the frequency of the electromagnetic waves relate to the "natural" oscillation frequency of the circuit?

ANSWER 46-1

Electromagnetic waves consist of oscillating electric and magnetic fields. The oscillating electric fields cause electrons in the wire to slosh back and forth. In other words, the electromagnetic waves play the role of an AC power supply, by "driving" the current back and forth. For this reason, the frequency of the electromagnetic waves is the driving frequency of the LC circuit.

Which driving frequency gives us the biggest alternating current? As we've seen before, an oscillating system undergoes the biggest swings when the driving frequency matches the natural frequency of the (undriven) system. For instance, if a block on a spring naturally oscillates 3 times per second, you can produce the biggest oscillates by pushing it 3 times per second.

An LC circuit is analogous to a block on a spring. Just as a block on a spring oscillates, the charge in an LC circuit oscillates back and forth between the two capacitor plates. To create the largest oscillations, and hence the biggest peak current, the driving frequency should equal the natural oscillation frequency of the LC circuit. And here, the electromagnetic waves supply the driving frequency. So,

frequency of electromagnetic waves = natural oscillation frequency of LC circuit.

Using this insight, I'll solve for the frequency of the electromagnetic waves. Then, I'll calculate the corresponding wavelength.

First, we need the natural frequency of the circuit. Since I can't remember the formula, I'll figure it out by analogy with a block on a spring. For a block, the angular frequency is $\omega = \sqrt{k/m}$, and hence, the frequency is

Block on spring $$\nu = \frac{\omega}{2\pi} = \frac{1}{2\pi}\sqrt{\frac{k}{m}}.$$

Recall from Chapter 44 that inductance is analogous to mass, since the inductor puts up "inertia" to a changing current. And capacitance is analogous to the inverse spring constant, because a higher capacitance corresponds to a looser spring. Since $L \leftrightarrow m$, and also $k \leftrightarrow 1/C$, the block-on-spring equation corresponds to

LC circuit $$\nu = \frac{1}{2\pi}\sqrt{\frac{1}{LC}} = \frac{1}{2\pi}\sqrt{\frac{1}{(2.0\times 10^{-6}\text{ H})(8.0\times 10^{-6}\text{ F})}} = 4.0\times 10^{4}\text{ s}^{-1}.$$

To produce the biggest possible current, the electromagnetic wave frequency must match this natural oscillation frequency of the circuit. But how does frequency relate to wavelength? As some of you saw last semester, *all* waves obey the relation

$$v = \lambda\nu,$$

where v denotes velocity, λ denotes wavelength, and ν denotes frequency. Question 21-1b in Volume 1 of this study guide gives an intuitive derivation. For light waves and all other electromagnetic waves, $v = c = 3.00 \times 10^8$ m/s, the "speed of light." Therefore, electromagnetic waves of frequency $\nu = 4.0 \times 10^4\text{ s}^{-1}$ have wavelength

$$\lambda = \frac{c}{\nu} = \frac{3.00\times 10^{8}\text{ m/s}}{4.0\times 10^{4}\text{ s}^{-1}} = 7500\text{ m},$$

a typical radio wavelength.

By the way, a simple radio works using the principle illustrated in this problem. When turning the "tuning" knob, you're tweaking the inductance or capacitance of an LC circuit. The signal best received by the radio corresponds to the natural oscillation frequency of that LC circuit.

QUESTION 46-2

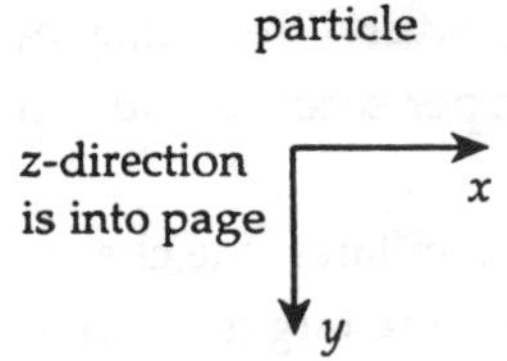

A tiny charged dust particle floats in mid-air. A high-intensity plane polarized electromagnetic wave propagates in the z-direction, into the page. The wave is polarized so that the electric field always points rightward or leftward, in the $\pm x$-direction.

(a) At a certain moment, the electric field at the particle points in the *negative x*-direction. What is the direction of the magnetic field at the particle, at that same moment?

(b) Will there be any time or times at which the particle feels *no* electric field? If so, what is the direction of the magnetic field at that same time and place?

(c) As the electromagnetic waves pass by the particle, the biggest electric field it feels is 2000 V/m. What is the biggest magnetic field felt by the particle?

(d) When the electric field strength is 1000 V/m, what's the magnetic field strength?

(e) Assuming the dust particle carries a net charge, what motion does it undergo?

ANSWER 46-2

In this problem, we'll review several key characteristics of electromagnetic waves.

(a) EM waves consist of oscillating electric and magnetic fields. The waves are *transverse*, which means they point *perpendicular* to the direction of motion. You can create transverse waves by stretching a slinky and shaking one end sideways. The waves travel down the slinky. But the individual coils oscillate sideways, perpendicular to the wave's direction of motion. Similarly, since this EM wave travels in the z-direction, the electric and magnetic fields point in the x- and y-directions.

So far, I've reviewed the sense in which electromagnetic waves resemble other transverse waves. By contrast, this next characteristic follows specifically from Maxwell's equations, and therefore applies specifically to EM waves: *The electric and magnetic fields are perpendicular to each other.* Therefore, if the waves travel in the z-direction, and the electric field points in the negative x-direction, then the magnetic field must point in the y-direction. Or maybe the negative y-direction. We haven't figured out which. Here's how to do so.

Maxwell's equations imply that the cross product $\mathbf{E} \times \mathbf{B}$ points in the direction of motion of the waves. (More on this later, when we meet the Poynting vector). Here, the waves travel in the z-direction. So, when **E** points in the negative x-direction, **B** must be such that the cross product points in the $+z$-direction. Mathematically speaking,

$$-\hat{\mathbf{x}} \times \hat{\mathbf{B}} = \hat{\mathbf{z}}.$$

To solve this equation for the direction of **B**, use the right hand rule. First, point your right-hand fingers in the direction of $-\hat{\mathbf{x}}$. When you curl your fingertips in the direction of **B**, your thumb must point in the z-direction, into the page. By playing around, you'll see that you must curl your fingertips up the page, in the negative y-direction, to make your thumb point into the page. So, **B** points in the negative y-direction.

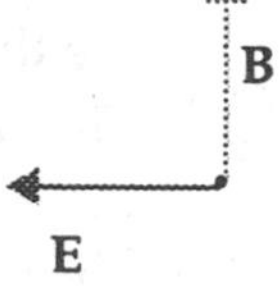

To double check this result, take the cross product of **E** with **B**, as drawn here. The cross product indeed points into the page, the direction of motion of the waves.

B

E

x

y

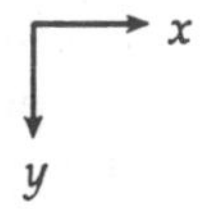

(b) This "snapshot" of the electric and magnetic fields hides the fact that the fields oscillate in time and space. At any given point, such as the dust particle, the fields oscillate in time.

The following graphic shows eight snapshots of the dust particle, taken an eighth of a period apart. Notice a couple of things. First, the electric and magnetic fields oscillate "in phase"; they get bigger and smaller in sync with each other. Second, $\mathbf{E} \times \mathbf{B}$ always points in the direction of motion, which is into the page. Whenever **E** switches direction, so does **B**, ensuring that the cross product does *not* switch direction.

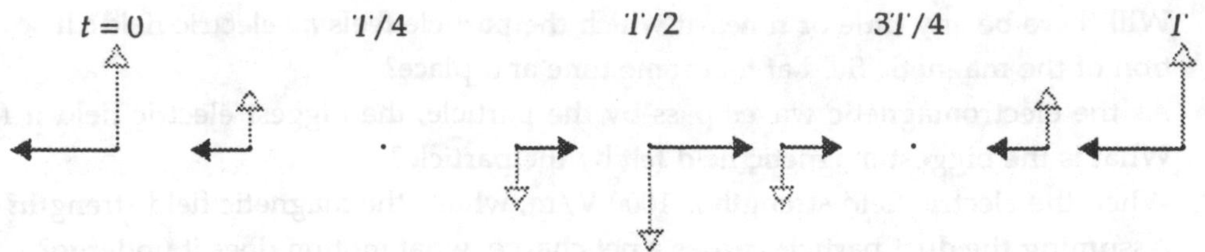

From this graphic, we see that the electric field vanishes twice per period, as represented by the dots at $T/4$ and $3T/4$. And whenever the electric field hits zero, so does the magnetic field. Whenever the dust particle feels no electric field, it also feels no magnetic field.

(c) The precise equations corresponding to the above graphic are

$$E = E_m \sin(kz - \omega t),$$

$$B = B_m \sin(kz - \omega t),$$

where E_m and B_m denote the maximum field strengths, corresponding to $t = 0$, $t = T/2$, and $t = T$ in the above graphic. From Maxwell's equations, your textbook derives the fact that

$$E = cB,$$

where c is the speed of light. This relationship between the electric and magnetic fields holds at all times, including when the fields reach their maximum values. So,

$$B_m = \frac{E_m}{c} = \frac{2000 \text{ V/m}}{3.00 \times 10^8 \text{ m/s}} = 6.67 \times 10^{-6} \text{ T}.$$

(d) As just noted, the electric and magnetic fields oscillate in phase, increasing and decreasing in sync. The magnetic field strength is always proportional to the electric field strength: $E = cB$. In this problem, we're considering a moment such as $t = T/8$, when the electric field is half its maximum value. At that instant, B must be half its maximum value, too: $B = B_m/2 = 3.33 \times 10^{-6}$ T.

To double check this answer, use the formula $E = cB$ directly. Solve for B to get

$$B = \frac{E}{c} = \frac{1000 \text{ V/m}}{3.00 \times 10^8 \text{ m/s}} = 3.33 \times 10^{-6} \text{ T}.$$

Notice once again that the above "snapshots" can help you solve problems. You can immediately see that when E reaches half its maximum value, so does B.

(e) The charged dust particle gets "pushed" by an electric field. Here, the electric field always points in the $\pm x$-direction, oscillating back and forth. So, we expect the particle to oscillate back and forth in the x-direction. Furthermore, the particle's frequency of oscillation should equal the EM wave's frequency, whatever that happens to be. That's because a "driven" oscillating system always ends up oscillating at the driving frequency.

But wait. Does the magnetic field also play a role? Recall that magnetic fields don't act on stationary charges. They affect only moving charges. So, when the dust particle first starts to move, no magnetic force acts on it. But once the particle is already moving in the x-direction, the magnetic field *does* act on it, in accordance with the magnetic force formula

$$\mathbf{F}_{mag} = q\mathbf{v} \times \mathbf{B}.$$

According to this equation, the magnetic force pushes the particle perpendicular to its direction of motion (**v**), and also perpendicular to the magnetic field. Therefore, since the particle's velocity is in the $\pm x$-direction, and the magnetic field points in the $\pm y$-direction, the magnetic force pushes the particle in the $\pm z$-direction. Of course, once the particle acquires some z-directed motion, the magnetic field now pushes it partially in the x-direction. So, the motion becomes complicated. But here's the main point. Because of the magnetic field, the particle oscillates not only in the x-direction, but also in the z-direction.

QUESTION 46-3

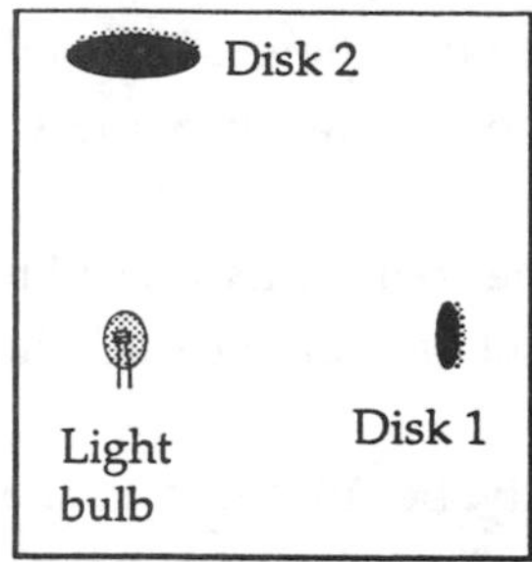

A small light bulb radiates light uniformly in all directions. Consider two black disks, both facing the light source, and both the same distance away from the light source. The disks are similar, except that disk 2 has twice the radius of disk 1. The disks absorb essentially all the light that reaches them. After the disks are put in place, the light bulb is turned on for 30 seconds, and then turned off.

The radius of each disk is much less than its distance from the light bulb.

(a) Which disk absorbs more electromagnetic energy: disk 1, disk 2, or both equally? Justify your answer intuitively, without formulas.

(b) Which disk, if either, ends up at a higher temperature? Justify your answer intuitively, without formulas.

(c) Which disk, if either, experiences a higher electromagnetic *power*? Explain your answer, relating it to part (a) or part (b).

(d) At which disk, if either, is the *intensity* greater? Explain your answer, relating it to part (a) or part (b).

(e) How far away from the light bulb would we have to move disk 2 to ensure that it receives the same power as disk 1? Three times as far away as its current position? Eight times as far away? Justify your answer.

ANSWER 46-3

This problem aims to illustrate the intuitive difference between power and intensity.

(a) Intuitively, the light is equally bright at both disks, because they're equidistant from the bulb. But disk 2 is bigger. As this picture shows, more light rays hit disk 2, *not* because the rays are more concentrated at disk 2, but simply because disk 2 is bigger. Consequently, disk 2 absorbs more energy. In fact, since disk 2 has four times the area of disk 1, it absorbs four times as much energy.

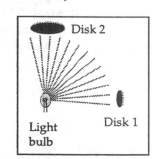

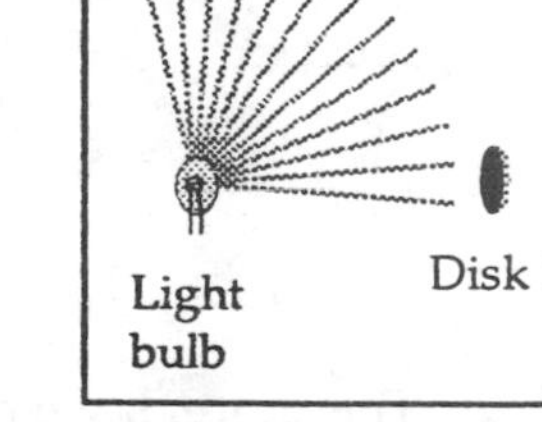

Here's an equivalent way of saying the same thing. The light bulb spews out energy equally in all directions. Because disk 2 is bigger, it grabs a bigger share of the total energy.

To visualize this more easily, imagine that the bulb is a sprinkler spewing out water instead of electromagnetic waves. In this analogy, the disks are sponges that absorb water instead of energy. Sponge (disk) 2 would absorb more water, not because the water is more concentrated there, but simply because sponge 2 is bigger.

(b) As just noted, disk 2 absorbs four times as much energy as disk 1, because it has four times the area. So, you might think that disk 2 ends up hotter. But exactly because disk 2 is four times bigger, it *needs* four times as much energy to heat up by the same amount. For this reason, both disks end up with the same temperature.

Let me explain this result in another way. Although disk 2 absorbs more total energy, both disks absorb the same energy *per area*. In other words, the light is equally bright (concentrated) at both disks. A square centimeter of disk 2 absorbs the same energy as a square centimeter of disk 1. Therefore, each square centimeter heats up by the same amount, no matter which disk it's part of. That's why both disks heat up equally.

(c) *Power* is energy per time: $P = \Delta Energy/\Delta t$. As we saw in part (a), when both disks spend equal time in the light, disk 2 absorbs four times as much energy as disk 1 absorbs. So, disk 2 receives more energy per time. It experiences a higher power. This goes to show that the "power" carried by electromagnetic waves depends not just on the brightness, but also on the area under consideration.

For instance, suppose I cover the roof of my apartment, and also the roof of my car, with solar panels. The panels on the roof of my apartment generate more power, not because the sunlight is brighter up there, but simply because those panels cover more area.

(d) *Intensity* corresponds to brightness. Both disks experience the same intensity, because they're the same distance from the light bulb. From part (b), we see that "brightness" corresponds not to power, but to power *per area*. That's why it has units of watts per square meter. Since the intensity is the same at both disks, a square centimeter of disk 1 absorbs the same power as a square centimeter of disk 2. Disk 2 absorbs more total power, but only because it contains more square centimeters.

(e) We want to move disk 2 far enough away so that it absorbs the same power as disk 1. Disk 2 has four times the area. Therefore, we must place disk 2 someplace where the power *per area* is 1/4 the power *per area* absorbed by disk 1.

Power per area is intensity, as explained in part (d). So the intensity at disk 2 must be one fourth the intensity at disk 1, in order for both disks to absorb the same power. For instance, if the light intensity at disk 1 is 10 watts/m^2, then the light at disk 2 had better be only 2.5 watts/m^2 bright.

Given all this, we can reduce the problem to the following question: Compared to the current position of disk 1, where is the light one fourth as bright (intense)? Do we need to go twice as far away, three times as far away, or what? In other words, how does light intensity depend on the distance from the light source?

As I'll explain below, intensity decreases with the *square* of your distance from the light source:

$$I \sim \frac{1}{r^2}.$$

Therefore, if we place disk 2 *twice* as far away from the bulb as disk 1, disk 2 will absorb light that's *four* times less bright (intense), and hence, both disks will absorb the same power.

I'll now explain, in two different ways, why $I \sim 1/r^2$.

Explanation #1 of why intensity goes down with the square of distance

The light bulb radiates light equally in all directions. Imagine that it's surrounded by two glass spheres, as drawn here. Sphere 2 has twice the radius of sphere 1. Intuitively, *all* the light emitted by the bulb passes through sphere 1, because it has nowhere else to go. And all the light that passes through sphere 1 also passes through sphere 2, for the same reason. So, *the same power goes through both spheres.*

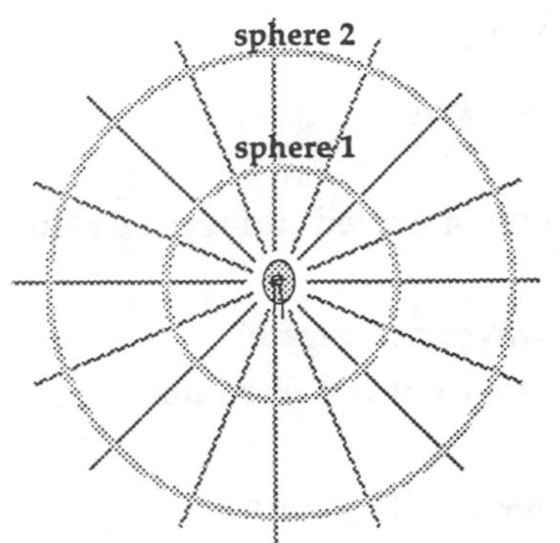

Using this insight, we can compare the intensity (power per area) at the two spheres. Because a sphere has surface area $A = 4\pi r^2$, sphere 2 has *four times* the area of sphere 1. Therefore, the power passing through sphere 2 is spread out over four times the area of the power passing through sphere 1.

In summary, the same power passes through both spheres. But at sphere 2, that power is spread out over four times as much area. So, the power *per area* at sphere 2 is one fourth the power *per area* at sphere 1. When you get twice as far from the light bulb, the light appears one fourth as intense (bright).

Using this reasoning, we can even derive the exact formula for the intensity created by a uniform light source. Suppose the bulb emits power P_0. If you draw a sphere of radius r around that light source, then power P_0 passes through a sphere of area $A = 4\pi r^2$. So, the intensity at distance r from the light source is

$$I = \frac{\text{power}}{\text{area}} = \frac{P_0}{4\pi r^2}.$$

Explanation #2 of why intensity goes down with the square of distance

Imagine placing disk 2 directly behind disk 1, as drawn here, so that disk 1 blocks all the light heading toward disk 2. In this picture, disk 1 absorbs a certain amount of power. Suppose we suddenly remove disk 1. Then disk 2 receives all the power formerly received by disk 1, because disk 2 gets illuminated by the same "wedge" of light.

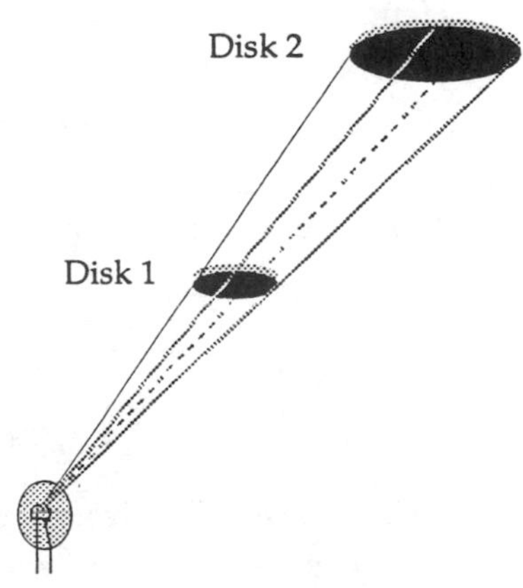

In order for disk 2 to receive that same wedge of light, it must be twice as far from the bulb. That's because disk 2 has twice the radius of disk 1. (Your textbook proves this, using trigonometry.) In summary, disk 2 must be twice as far from the bulb, in order to receive the same power as disk 1. Since disk 2 has four times the area, the intensity (power per area) at disk 2 must be four times weaker.

QUESTION 46-4

COORDINATE AXES

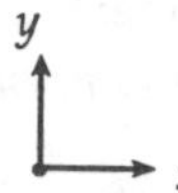

z-direction is out of the page.

A polarized plane wave is described by

$$\mathbf{E} = E_m \sin[(10^{14}\ \text{s}^{-1})t - kz]\hat{\mathbf{x}}$$

$$\mathbf{B} = (10^{-6}\ \text{T})\sin[(10^{14}\ \text{s}^{-1})t - kz]\hat{\mathbf{y}},$$

where $\hat{\mathbf{x}}$ and $\hat{\mathbf{y}}$ are unit vectors. Notice that I haven't specified numerical values for two of the constants, E_m and k.

(a) In which direction do these electromagnetic waves propagate?

(b) What is the numerical value of E_m, in volts per meter? You can figure it out from the information provided.

(c) What is the wavelength of this radiation, in meters? Could you see it with your eyes?

(d) What is the intensity of this radiation?

ANSWER 46-4

(a) Electromagnetic waves travel in the direction given by $\mathbf{E} \times \mathbf{B}$. Below, we'll see that $\mathbf{E} \times \mathbf{B}$ is proportional to the "Poynting vector," which specifies the electromagnetic wave intensity. The *direction* of the Poynting vector (i.e., the direction of $\mathbf{E} \times \mathbf{B}$) tells you the direction of motion of the waves. Remember, electromagnetic waves are transverse; the fields point perpendicular to the waves' direction of motion. Therefore, since $\mathbf{E} \times \mathbf{B}$ is perpendicular to both $\mathbf{E}$ and $\mathbf{B}$, it makes sense that the direction of motion corresponds to that cross product.

Here, whenever and wherever $\mathbf{E}$ points in the x-direction (for instance, at $z = 0$, $t = 10^{-14}$ s), $\mathbf{B}$ points in the y-direction. So, as you can confirm using the right-hand rule, their cross product points in the z-direction, out of the page. That's the direction in which the waves propagate.

(b) From the expression $B = (10^{-6}\ \text{T})\sin[(10^{14}\ \text{s}^{-1})t - kz]$, we know that the maximum magnetic field strength is $B_m = 10^{-6}$ T. Maxwell's equations imply that $E_m = cB_m$, for all electromagnetic waves. Hence,

$$E_m = cB_m = (3.00 \times 10^8\ \text{m/s})(10^{-6}\ \text{T}) = 300\ \text{V/m}.$$

(c) The general expression for electromagnetic plane waves is

$$E_m = E_m \sin(\omega t - kz)$$

$$B_m = B_m \sin(\omega t - kz),$$

where the *angular frequency* ω is related to the oscillation frequency ν by $\nu = \omega/2\pi$; and the *wave constant* k is related to the wavelength by $\lambda = 2\pi/k$. See question 21-3b of this study guide series, for a detailed explanation of these relationships.

Since we don't know the numerical value of k, we can't use it to calculate the wavelength. But we *can* calculate the frequency, because the problem tells us the value of ω. It's 10^{14} s^{-1}. So,

$$\nu = \frac{\omega}{2\pi} = \frac{10^{14}\ \text{s}^{-1}}{2\pi} = 1.59 \times 10^{13}\ \text{s}^{-1}.$$

For *all* waves, electromagnetic or otherwise, $v = \lambda\nu$. Since all EM waves travel at the speed of light, c, we can solve for λ to get

$$\lambda = \frac{c}{\nu} = \frac{3.00 \times 10^{8}\ \text{m/s}}{1.59 \times 10^{13}\ \text{s}^{-1}} = 1.89 \times 10^{-5}\ \text{m},$$

which is 18900 nanometers. (A nanometer is 10^{-9} meters.)

Visible light wavelengths range from 400 nanometers to 700 nanometers, with 400 nm corresponding to violet and 700 nm corresponding to red. This electromagnetic wave is much longer then visible light. From your textbook's diagram about the electromagnetic spectrum, you can see that electromagnetic radiation with wavelength 10^{-5} m is infrared.

(d) Intensity is given by the Poynting vector,

$$\mathbf{S} = \frac{1}{\mu_0}\mathbf{E} \times \mathbf{B}.$$

The magnitude of this vector tells you the instantaneous intensity, i.e., the power per area carried by the wave. And the direction of **S** tells you the direction in which this power gets carried, i.e., the direction of motion of the waves.

For the electromagnetic plane waves in this problem, we know that

$$\mathbf{E} = E_m \sin(\omega t - kz)\hat{\mathbf{x}}$$
$$\mathbf{B} = B_m \sin(\omega t - kz)\hat{\mathbf{y}},$$

where $B_m = 10^{-6}$ T and $E_m = 300$ V/m, from part (b). Taking the cross gives

$$\mathbf{S} = \frac{1}{\mu_0}\mathbf{E} \times \mathbf{B} = \frac{1}{\mu_0}[E_m \sin(\omega t - kz)\hat{\mathbf{x}}] \times [B_m \sin(\omega t - kz)\hat{\mathbf{y}}]$$
$$= \frac{1}{\mu_0}E_m B_m \sin^2(\omega t - kz)\hat{\mathbf{x}} \times \hat{\mathbf{y}}$$
$$= \frac{1}{\mu_0}E_m B_m \sin^2(\omega t - kz)\hat{\mathbf{z}}.$$

Notice that the magnitude of this vector oscillates in time. Since $\sin^2(\cdots)$ oscillates between 0 and 1, the instantaneous intensity dips as low as 0 and as high as $\frac{1}{\mu_0}E_m B_m$. These intensity oscillations happen incredibly quickly, because the frequency is so high. For that reason, when a problem asks for "intensity," it wants the *average* intensity, i.e., the *average* magnitude of the Poynting vector. So, we need to find the time-averaged value of S.

Well, as $\sin^2(\cdots)$ oscillates between 0 and 1, its average value turns out to be 1/2. So, the average value of $S = \frac{1}{\mu_0} E_m B_m \sin^2(\omega t - kz)$ is

$$\bar{S} = \frac{1}{2}\left(\frac{1}{\mu_0} E_m B_m\right)$$

$$= \frac{1}{2}\left[\frac{1}{4\pi \times 10^{-7}\ \text{H/m}}(300\ \text{V/m})(10^{-6}\ \text{T})\right]$$

$$= 120\ \text{watts/m}^2.$$

That's the intensity.

QUESTION 46-5

A thin cylindrical rod of radius r_0 and length $l \gg r_0$ is coated with uniform linear charge density $+\lambda$. (Here, "λ" denotes linear charge density, *not* wavelength.) The rod moves upward at constant speed v, as drawn here.

(a) What is the direction of the Poynting vector at point P, at the moment when the edge of the cylinder "brushes by" that point, as drawn here?

(b) What is the direction of the Poynting vector at point W on the other side of the cylinder, when the cylinder "brushes by" it?

(c) Explain, intuitively, why the Poynting vector points in this direction.

(d) (*Extra credit*) What is the (approximate) magnitude of the Poynting vector at point P? Answer in terms of λ, v, r_0, and any universal constants you need.

ANSWER 46-5

(a) The Poynting vector,

$$\mathbf{S} = \frac{1}{\mu_0}\mathbf{E} \times \mathbf{B},$$

is perpendicular to the electric and magnetic fields. To find its direction, we can figure out the direction of **E** and **B** separately, and then cross 'em.

Well, a uniformly charged rod spews out radial electric field lines, except near the ends of the rod. In this picture, dark lines represent the electric field. So, at P, the electric field points rightward.

What about the magnetic field? Well, current is nothing more than moving charges. The moving charged rod constitutes a current, much like the current you'd find in a wire. We know that a current-carrying wire generates circular magnetic field lines. So here, the magnetic field lines encircle the rod. In the side view drawing, these circles go into and out of the page.

To determine whether these circular field lines point clockwise or counterclockwise, use the right-hand rule shortcut, as follows. Point your thumb in the direction of the current, and curl your

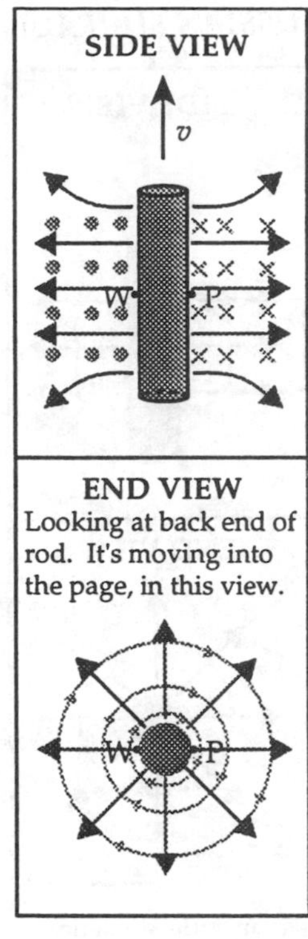

fingertips. They curl in the direction of the magnetic field lines. So here, the magnetic field points into the page on the right side of the rod, and out of the page on the left side.

To find the direction $\mathbf{E} \times \mathbf{B}$ at point P, use the "full-blown" right-hand rule. Point your right fingers along the electric field lines. Then curl your fingertips to point along the magnetic field line. Your thumb now points in the direction of $\mathbf{S} = \frac{1}{\mu_0}\mathbf{E} \times \mathbf{B}$. To double check yourself, apply this reasoning to the side view *and* the end view. Your answers should agree that **S** points in the rod's direction of motion, up the page in the side view, and into the page in the end view.

(b) We've already figured out the directions of **E** and **B** at point W. So, to find the direction of $\mathbf{S} = (1/\mu_0)\,\mathbf{E} \times \mathbf{B}$, we must apply the right-hand rule to that cross-product. I'll use the side view. Pointing my fingers leftward (along **E**), and then curling my fingertips out of the page (along **B**), I find that my thumb points up the page, which is the rod's direction of motion. You can confirm this result using the end view. So, at P *and* W, the Poynting vector points along the rod's direction of motion.

(c) This makes sense, for the following reason. The direction of **S** corresponds to the direction in which power propagates. Remember, energy is stored in the electric and magnetic fields around the rod, whether or not those fields constitute electromagnetic waves. As the rod moves, the fields, which are "centered" on the rod, move with it. So, the energy contained in those fields gets dragged along with the rod. Since power is the rate at which energy gets transferred, the power flow follows the rod. In other words, **S** points in the rod's direction of motion.

(d) To obtain the magnitude of $\mathbf{S} = (1/\mu_0)\mathbf{E} \times \mathbf{B}$, we need the strength of the electric and magnetic field at point P. Since those fields point perpendicular to each other, the magnitude of the cross product reduces to a regular product,

$$S_P = \frac{1}{\mu_0} E_P B_P,$$

where subscript "P" means "at point P."

Let's divide this problem into two subproblems. First, find the electric field at point P. Then, find the magnetic field at point P.

Subproblem 1: Find electric field at point P.

Since $r_0 \ll l$, the rod "looks" almost infinitely long, from P's perspective. For this reason, the electric field is sufficiently symmetric that Gauss' law can help us.

In order to find the electric field using Gauss' law,

$$\oint_{\text{Gaussian surface}} \mathbf{E} \cdot d\mathbf{A} = \frac{q_{\text{encl}}}{\varepsilon_0},$$

the electric field must be uniform over (at least part of) the Gaussian surface. That way, we can pull **E** outside the Gaussian flux integral. In this case, a Gaussian cylinder does the trick. Since field lines pierce only through the "tube" (side) of this Gaussian surface, and since the electric field has uniform strength over that tube (by radial symmetry), the left-hand side of Gauss' law reduces to

$$\oint_{\text{Gaussian surface}} \mathbf{E}\cdot d\mathbf{A} = E\oint_{\text{Tube of Gaussian surface}} dA = E(2\pi rL),$$

where in the last step I used the surface area of the tube (side) of a cylinder of radius r and length L.

Now I'll work on the other side of Gauss' law, by finding the charge enclosed by my Gaussian cylinder. Since linear charge density is charge per length,

$$q_{\text{encl}} = \frac{\text{charge}}{\text{length}} \times (\text{length of rod enclosed}) = \lambda L.$$

Substituting all this information into Gauss' law gives us

$$\oint_{\text{Gaussian surface}} \mathbf{E}\cdot d\mathbf{A} = \frac{q_{\text{encl}}}{\varepsilon_0}$$

$$E(2\pi rL) = \frac{\lambda L}{\varepsilon_0}.$$

GAUSSIAN SURFACE

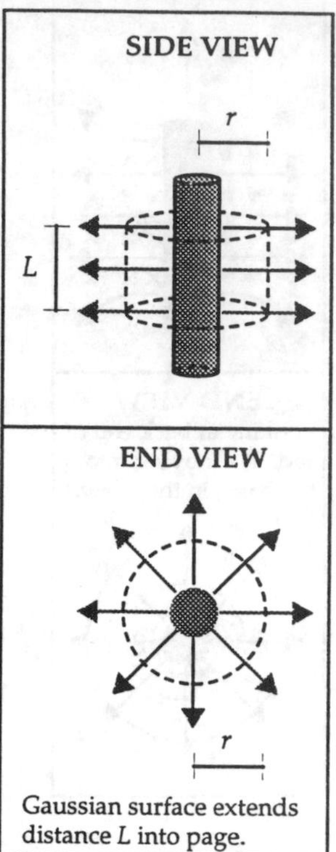

Gaussian surface extends distance L into page.

Cancel the L's and solve for E to get

$$E = \frac{\lambda}{2\pi r_0 \varepsilon_0}.$$

That's the electric field outside the rod, a distance r from the center. Since point P sits a distance r_0 from the rod's center,

$$E_P = \frac{\lambda}{2\pi r_0 \varepsilon_0}.$$

So far, I've found the electric field at point P. Now let's figure out the magnetic field, so that we can calculate the Poynting vector.

Subproblem 2: Find the magnetic field at point P.

Again, approximate the rod of charge as infinitely long, from P's perspective. Since the rod moves at constant speed, it constitutes a steady current, similar to the current in an infinite wire. So, we can use Ampere's law,

$$\oint_{\text{Amperian loop}} \mathbf{B}\cdot d\mathbf{s} = \mu_0 i_{\text{encl}}.$$

To solve Ampere's law for B, we must pick an Amperian loop along which the magnetic field is constant. That way, we can pull B outside the Amperian loop integral. Well, by symmetry,

AMPERIAN LOOP

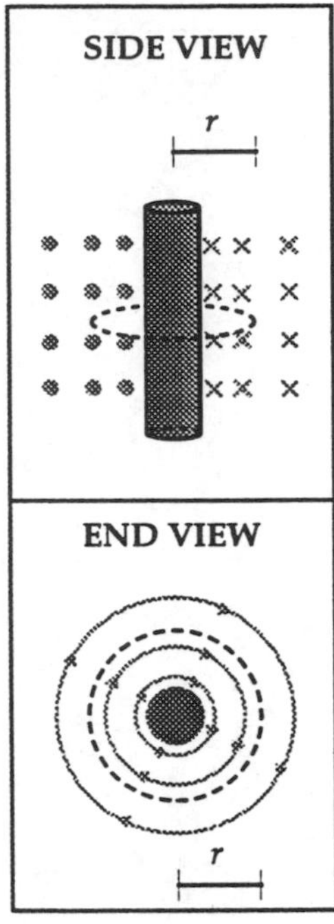

B has the same strength at all points equidistant from the rod. So, I'll draw a circular Amperian loop.

Since **B** always points along the Amperian loop, the dot product $\mathbf{B} \cdot d\mathbf{s}$ reduces to a regular product. And we can pull B outside the integral, since it's constant over the whole loop:

$$\oint_{\text{Amperian loop}} \mathbf{B} \cdot d\mathbf{s} = \oint_{\text{Amperian loop}} B ds = B\int_0^{2\pi r} ds = B(2\pi r).$$

Now I'll simplify the other side of Ampere's law, by finding the enclosed current.

Current is the charge per time: $i = dq/dt$. As the rod travels at speed v, how much charge, dq, passes an imaginary line in time dt? Well, in time dt, the *length* of rod passing the line is $dx = vdt$. Since linear charge density is charge per length ($\lambda = dq/dx$), the charge contained in that length of rod is

$$dq = \lambda dx = \lambda(vdt).$$

Divide this equation through by dt to get

$$\frac{dq}{dt} = \lambda v.$$

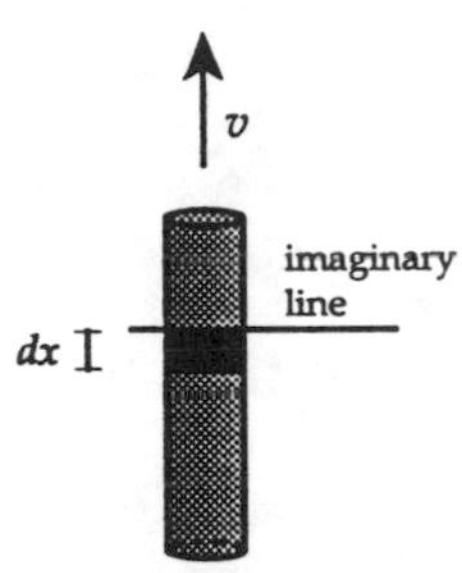

The blackened segment passes the imaginary line in time dt.

That's the current.

Now that we've simplified both sides of Ampere's law, we have

$$\oint_{\text{Amperian loop}} \mathbf{B} \cdot d\mathbf{s} = \mu_0 i_{\text{encl}}$$

$$B(2\pi r) = \mu_0 \lambda v.$$

Solve for B, and substitute in $r = r_0$, to get

$$B_P = \frac{\mu_0 \lambda v}{2\pi r_0}.$$

End of subproblem 2

At this stage, we've found the electric and magnetic fields at point P. Since those fields are perpendicular, the magnitude of the Poynting vector is simply

$$\begin{aligned} S_P &= \frac{1}{\mu_0}|\mathbf{E}_P \times \mathbf{B}_P| \\ &= \frac{1}{\mu_0} E_P B_P \\ &= \frac{1}{\mu_0}\left(\frac{\lambda}{2\pi r_0 \varepsilon_0}\right)\left(\frac{\mu_0 \lambda v}{2\pi r_0}\right) \\ &= \frac{\lambda^2 v}{4\pi^2 r_0^2 \varepsilon_0}. \end{aligned}$$

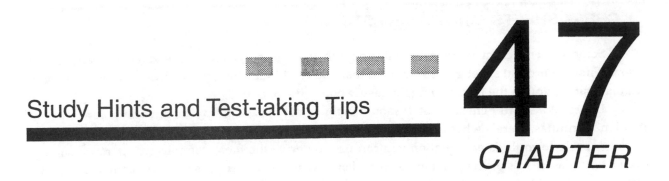

Study Hints and Test-taking Tips

In this chapter, I'll offer some thoughts about how to make the best use of your study time. I'll also discuss how to write an effective "cheat sheet" (for those of you allowed to bring notes into the exam), and how to take tests well.

STUDY HINTS

Start studying several days in advance. You can't "cram" physics. Believe me, I've tried. Studying 3 hours per day for 4 days is better than studying 12 hours for one day, because your brain needs "down time" to process the knowledge.

For each topic, try to do the following:

(1) Review your notes from class, focusing on the main intuitive ideas and the sample problems. Don't spend lots of time rereading the textbook; you'll learn more by solving problems.

(2) Rework the relevant chapters in this book, skipping algebra to save time. Just set things up. Don't read the answer until you've tried to solve the problem on your own.

(3) Rework the relevant homework assignments. Check your answers against your instructor's solutions. Some of the homework problems are not worth reviewing, because they cover a topic that won't appear on the exam, or because they're too hard. Ask your instructor or teaching assistant for a list of homework problems you need not study.

(4) Do the appropriate review problems in this book; see the table of contents. Alternatively, some students do review problems in step 2 or 3 above, to find out what they need to study the most.

(5) Take practice tests. This book contains several sample midterms and final exams, described in the table of contents. Also, ask your instructor for old exams.

- Whatever you do, *don't* stay up all night studying. In physics, that extra bit of knowledge won't compensate for a lack of clear-headedness. You can score well on many problems by thinking incisively, even if you don't know every last detail.
- Find out what you're allowed to bring into the test (e.g., cheat sheet, calculator, aardvark, etc.).
- If you're allowed a cheat sheet, spend the time needed to make a good one. The process of preparing a cheat sheet helps you learn the material, even if you never refer to the sheet during the test itself. If cheat sheets aren't permitted, skip the next section, and go right to "test-taking tips."

CHEAT SHEETS: ANDY'S SECRETS

Some students cram every last formula onto their cheat sheets. I recommend against this, for two reasons. First, your cheat sheet gets so cluttered that you can't find what you need. Second, with so many formulas staring you in the face, you might misuse one of them.

I suggest a shorter cheat sheet. It should include all the basic formulas, but only a few "special-situation" formulas. Here's what I mean.

A basic formula is an equation you can use to derive the more complicated or more context-specific formulas. It's a starting point in your problem solving. Basic formulas from this course include the electric force ($\mathbf{F} = q\mathbf{E}$), the definition of electric flux ($\int \mathbf{E} \cdot d\mathbf{A}$), Gauss' law, the definition of linear and surface charge density, the definition of potential ($U = qV$ and hence, $\Delta V = -\int \mathbf{E} \cdot d\mathbf{s}$) the definition of capacitance ($C = q/V$), formulas for capacitors and inductors in series and in parallel, the potential energy stored in a capacitor $\left(U_C = \frac{1}{2}QV = \frac{1}{2}CV^2 = \frac{1}{2}\frac{Q^2}{C}\right)$, the energy stored in an electric field $\left(U = \int \frac{1}{2}\varepsilon_0 E^2 d^3\mathbf{x}\right)$, the energy stored in an inductor $\left(U_L = \frac{1}{2}Li^2\right)$, dielectric modifications to Gauss' law and capacitance, definitions of current, current density, and resistance, Ohm's law, Kirchhoff's laws, the magnetic force acting on a point charge ($\mathbf{F}_{\mathbf{mag}} = q\mathbf{v} \times \mathbf{B}$) and on a piece of current-carrying wire ($d\mathbf{F}_{\mathbf{mag}} = id\mathbf{l} \times \mathbf{B}$), Ampere's law and the Biot-Savart law, definition of permeability, magnetic flux and Faraday's law, defining equations for inductance, the five basic formulas describing AC circuits (see Chapter 44), the "induced magnetic field" formula and the definition of displacement current from Chapter 45, and the basic electromagnetic wave formulas from Chapter 46.

You may notice that lots of important formulas are conspicuously absent from my list, such as the capacitance of a parallel-plate capacitor, and the electric field generated by a point charge. These are "special-situation" formulas. They apply to one particular physical context. Of course, some contexts show up so frequently that you should include the corresponding formulas on your cheat sheet. *But label them carefully*, so that you won't use them in situations where they don't apply.

On your cheat sheet, you might also include some problem-solving "cues," little hints to yourself about how to approach problems. These cues won't help if you haven't worked lots of practice problems. But if you have, the cues can help you overcome a temporary mental block.

On the next page, I showcase these hints by presenting two sample cheat sheets. They're designed for a midterm exam about electrostatics.

NOT-SO-GOOD CHEAT SHEET

$\mathbf{F} = q\mathbf{E} = \dfrac{Qq}{4\pi\varepsilon_0 r^2}$

$\mathbf{F}_{\text{total}} = \Sigma_n \mathbf{F}_n$

$\underset{\text{closed surface}}{\oint} \mathbf{E}\cdot d\mathbf{A} = \dfrac{q_{\text{encl}}}{\varepsilon_0}$

Linear charge density = λ.

$E = \dfrac{\lambda}{2\pi\varepsilon_0 r}$

Surface charge density = σ.

$E = \dfrac{\sigma}{2\varepsilon_0}$

$V = -\int \mathbf{E}\cdot d\mathbf{s} = \dfrac{q}{4\pi\varepsilon_0 r}$

$U = qV$

$\frac{1}{2}mv^2 = q\Delta V$

Capacitors

$C = \frac{q}{V},\ U_C = \frac{1}{2}QV = \frac{1}{2}CV^2 = \frac{1}{2}\frac{Q^2}{C}$

$C = \dfrac{\varepsilon_0 A}{d}$

$E = \dfrac{\sigma}{\varepsilon_0}$

$C_{\text{in parallel}} = C_1 + C_2 + \cdots$

$1/C_{\text{in series}} = 1/C_1 + 1/C_2 + \cdots$

Dielectrics

$E = \dfrac{E_0}{\kappa}$

$C = \kappa C_0$

BETTER CHEAT SHEET

$\mathbf{F} = q\mathbf{E}$

Flux = $\underset{\text{closed surface}}{\oint} \mathbf{E}\cdot d\mathbf{A} = \dfrac{q_{\text{encl}}}{\varepsilon_0}.$ (water analogy)

GAUSSIAN SURFACES

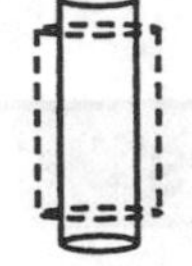

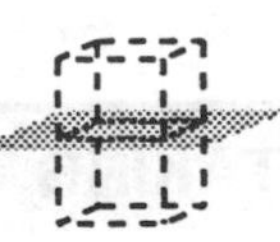

sphere cylinder plane

Linear charge density = charge/length = λ.
Surface charge density = charge/area = σ.
To find **E** . . .
1) Gauss' law if symmetric, OR
2) Break into pieces, use $dE = \dfrac{dq}{4\pi\varepsilon_0 r^2}$. Vectors.

Potential = U of 1-coulomb charge:
$U = qV$, therefore $\Delta V = -\int \mathbf{E}\cdot d\mathbf{s}$
Contour map analogy. Path independence.

Point charges only:

$E = \dfrac{q}{4\pi\varepsilon_0 r^2},\ F = \dfrac{Qq}{4\pi\varepsilon_0 r^2},\ V = \dfrac{q}{4\pi\varepsilon_0 r}$

Capacitors

$C = \frac{q}{V},\ U_C = \frac{1}{2}QV = \frac{1}{2}CV^2 = \frac{1}{2}\frac{Q^2}{C}$

(1) "Pretend" $\pm q$ on plates, (2) Find E,
(3) Find $V = -\int \mathbf{E}\cdot d\mathbf{s}$, (4) $C = q/V$

$C_{\text{in parallel}} = C_1 + C_2 + \cdots$
$1/C_{\text{in series}} = 1/C_1 + 1/C_2 + \cdots$

Parallel plate : $C = \varepsilon_0 \dfrac{A_{\text{plate}}}{d_{\text{between plates}}},\ E = \dfrac{\sigma}{\varepsilon_0}$

Dielectrics dampen E by κ
Example: $C_{\text{with dielectric}} = \kappa C_{\text{without}}$

Your cheat sheet should differ from mine, because no two courses emphasize exactly the same material. I'm just trying to show what a good vs. a not-so-good cheat sheet looks like. On the not-so-good one, general and situation-specific formulas get mixed together. This could cause you to misapply a formula on the test. Also, the not-so-good cheat sheet lists the electric field due to a line charge and a charged plane, while the better cheat sheet gives cues about what Gaussian surfaces can help you *derive* those formulas. As noted above, it's fine to put situation-specific formulas on your cheat sheet. But when solving difficult problems, you'll sometimes have to go back to basics, instead of just plugging in pre-derived results.

TEST-TAKING TIPS

- Keep in mind the problem-solving skills you've practiced all semester, such as making sketches, formulating a strategy, etc. Sometimes students panic and revert to formula-plugging. To reduce your anxiety level, take practice exams ahead of time.
- Don't waste precious time completing algebra at the expense of getting to every problem. When running out of time, you can earn partial credit by setting up the relevant equations, and describing in words your problem-solving strategy. Check with your instructor to determine how best to weight your time between setting things up and completing the math.
- If you get stuck on a problem, go on to the next one. Your unconscious mind keeps processing that old problem, even while you work on the new one. As a result, when you return to the old problem, fresh insights often pop into your mind. This sounds crazy, I know. But several students confirmed that it works. And it's often worked for me.
- Sometimes, test problems suffer from ambiguous phrasing. If you're the slightest bit confused, ask the instructor to clarify. Students routinely misinterpret problems because they were unwilling to bother the instructor. This is no time to be shy.
- Bringing an aardvark to the exam can calm your nerves.

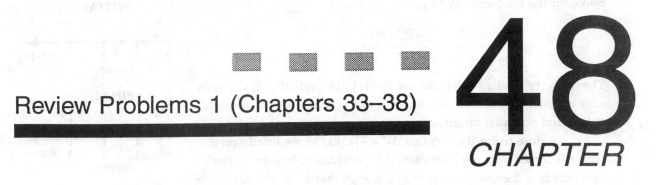

Review Problems 1 (Chapters 33–38)

CHAPTER 48

- These problems cover electrostatics and circuits. I've labeled "electrostatics only" the questions that cover chapters 33–36 only.

REVIEW PROBLEM 1-1

(*Electrostatics only*) A funky charge distribution generates an electric field $\mathbf{E} = \frac{b}{x^3}\hat{\mathbf{x}}$, where $\hat{\mathbf{x}}$ denotes a unit vector in the x-direction, and $b = 5.0$ volts·meters^2. This field fills my room. I place a frictionless particle of charge $q = 4.0 \times 10^{-6}$ C and mass $m = 3.0 \times 10^{-8}$ kg on the floor, and release it from rest, from $x = 2.0$ m.

(a) Initially, what is the net force on the particle?

(b) How fast is the particle moving when it reaches $x = 3.0$ m?

ANSWER

(a) In general, a charge q in an electric field $\mathbf{E}$ feels a force $\mathbf{F}_{\text{elec}} = q\mathbf{E}$. So here, the particle feels an x-directed force of magnitude

$$F_{\text{elec}} = qE = q\frac{b}{x^3} = (4.0 \times 10^{-6}\ \text{C})\frac{5.0\ \text{V}\cdot\text{m}^2}{(2.0\ \text{m})^3} = 2.5 \times 10^{-6}\ \text{newtons.}$$

Since no friction acts on the particle, and the floor's normal force cancels the downward gravitational force, F_{elec} is the net force.

(b) Given part (a), many students get sucked into the following reasoning: Calculate the particle's acceleration using $F_{\text{net}} = ma$, and then invoke a kinematic formula such as $v^2 = v_0^2 + 2a\Delta x$, to obtain the velocity. This strategy doesn't work, because the particle feels a non-constant force. As x gets bigger, the field and force get weaker. Since F_{net} changes, so does the acceleration. Therefore, we can't use those old constant-acceleration kinematic formulas.

Conservation of energy comes to the rescue. As the particle "falls" from $x = 2$ to $x = 3$ m, it loses potential energy, and gains an equal amount of kinetic energy. Recalling that $U_{\text{elec}} = qV$, we get

$$K_0 + U_0 = K_f + U_f$$

$$0 + qV_0 = \frac{1}{2}mv^2 + qV_f.$$

Solve for the final velocity to get

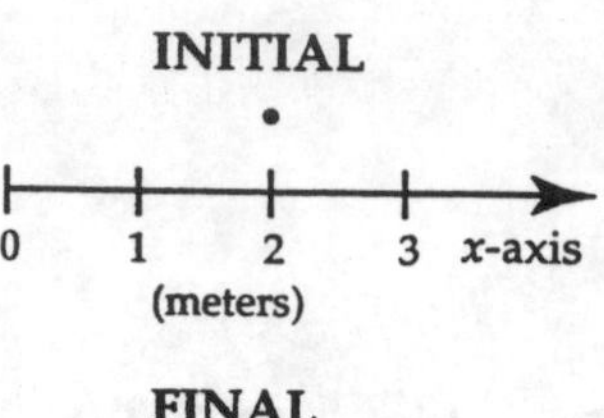

$$v = \sqrt{\frac{2q(V_0 - V_f)}{m}}.$$

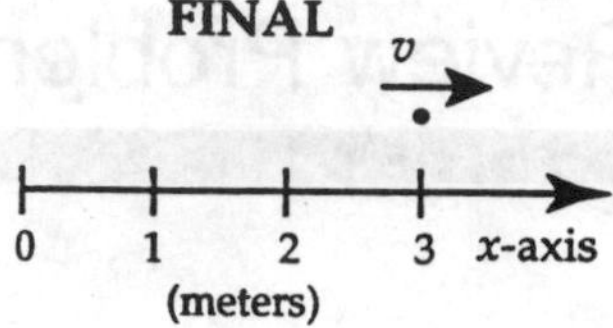

The velocity does not depend on V_0 or V_f individually. It depends only on their *difference*. So, to complete this problem, we just need calculate that potential difference.

Before using the formula $\Delta V = -\int \mathbf{E}\cdot d\mathbf{s}$, let me briefly review where it comes from. By definition, the "potential difference" between two points is the change in potential energy that a 1-coulomb charge would experience, if moved from one point to the other. And the change in potential energy is simply the work needed to accomplish the move. Well, since work is force integrated over distance, we get

$$\Delta V = W_{\text{to move a 1-coulomb charge}} = \int \mathbf{F}_{\text{on a 1-coulomb charge}}\cdot d\mathbf{s} = \int q\mathbf{E}\cdot d\mathbf{s} = \int \mathbf{E}\cdot d\mathbf{s},$$

since $q = 1$ for a 1-coulomb charge. I've played fast and loose with units, to keep this "derivation" simple.

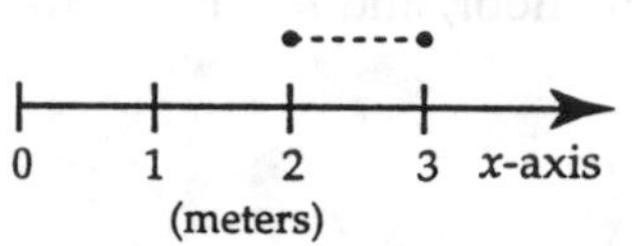

Now I'll use this formula to calculate ΔV for the particle. When evaluating $\int \mathbf{E}\cdot d\mathbf{s}$, you must choose a path, though your answer won't depend on your choice. I'll pick the path actually followed by the particle: the straight line connecting its initial and final position, as drawn here. So, $d\mathbf{s}$ turns into $d\mathbf{x}$. Since $\mathbf{E}$ also points in the x-direction, the dot product $\mathbf{E}\cdot d\mathbf{x}$ reduces to a regular product. The path starts at $x = 2$ m and ends at $x = 3$ m. So,

$$\Delta V = \int_{\text{initial}}^{\text{final}} \mathbf{E}\cdot d\mathbf{s} = \int_{2\text{ m}}^{3\text{ m}} E\,dx$$

$$= -\int_{2\text{ m}}^{3\text{ m}} \frac{b}{x^3}\,dx$$

$$= \left.\frac{b}{2x^2}\right|_{2\text{ m}}^{3\text{ m}}$$

$$= \frac{5.0\text{ V}\cdot\text{m}^2}{2(3.0\text{ m})^2} - \frac{5.0\text{ V}\cdot\text{m}^2}{2(2.0\text{ m})^2} \qquad [\text{since } b = 5.0\text{ V}\cdot\text{m}^2]$$

$$= 0.278\text{ V} - 0.625\text{ V}$$

$$= -0.347\text{ V}.$$

The minus sign indicates that the positive particle *loses* potential energy while "falling" from $x = 2.0$ m to $x = 3.0$ m. If you're worried about plugging a negative number into our above expression for v, remember that ΔV is the *final* minus the *initial* potential: $\Delta V = V_f - V_0$. Fortunately, v depends on $V_0 - V_f = -\Delta V$:

$$v = \sqrt{\frac{2q(V_0 - V_f)}{m}} = \sqrt{\frac{2q(-\Delta V)}{m}} = \sqrt{\frac{2(4.0\times10^{-6}\text{ C})(0.347\text{ V})}{3.0\times10^{-8}\text{ kg}}} = 9.6\text{ m/s}.$$

REVIEW PROBLEM 1-2

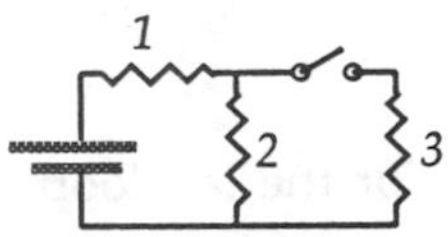

In this circuit, the three resistances are all different, and all unknown. So is the battery's voltage. Nonetheless, you can answer the following questions. At first, the switch is open, as drawn here. When the switch gets closed,

(a) Does the current through the resistor 1 go up, go down, or stay the same?

(b) Does the current through resistor 2 go up, go down, or stay the same?

(c) Now let's assume all three resistors have the *same* resistance. By what percentage does the current through resistor 2 change, when the switch gets closed?

ANSWER

(a) The current through resistor 1 equals the current through the battery. So, we're talking about the total current through the circuit. When the switch closes, what happens to the total current?

It goes up, because the total equivalent resistance of the circuit goes down. To see why, look at the intersection above resistor 2. With the switch open, all current reaching the intersection must flow down through resistor 2. But with the switch closed, current can split itself between resistors 2 and 3. It's just like opening an extra lane on a highway. With the extra lane available, more total "car current" can flow. Similarly, with resistors 2 and 3 both available, more total electrical current can flow.

A common mistake is to reason as follows: "The battery supplies a fixed amount of current. Closing the switch allows that current to branch, but doesn't create *more* current." This reasoning doesn't work, because it views the battery as the "source" of current. But actually, the charges constituting the current are valence electrons from the metal wires. The battery merely pushes these charges around the circuit. It doesn't "provide" the charges in the first place. When the switch closes, trillions of valence electrons in resistor 3 (and in the wire around resistor 3) become part of the current. It's as if you opened the extra lane of the highway, and a bunch of cars were already parked in that lane, just waiting to go.

Mathematical reasoning confirms my conclusion that closing the switch lowers the total equivalent resistance. With the switch open, all the current flows through resistor 2, which has resistance R_2. But with the switch closed, resistors 2 and 3 are in parallel. They have an equivalent resistance given by $\frac{1}{R_{eq}} = \frac{1}{R_2} + \frac{1}{R_3}$. For all possible values of R_2 and R_3, you can confirm that R_{eq} is less than R_2.The equivalent resistance of two in-parallel resistors is always *less* than the individual resistances, because opening another "path" (lane) allows more current to flow.

(b) Given part (a), you could sensibly guess that the current through R_2 goes up, because more total current flows around the circuit. Or, you could sensibly guess that the current through R_2 goes down, since resistor 3 now "steals" some of the current that used to flow through resistor 2. This second tendency "wins"; closing the switch *decreases* the current through resistor 2. Here's why.

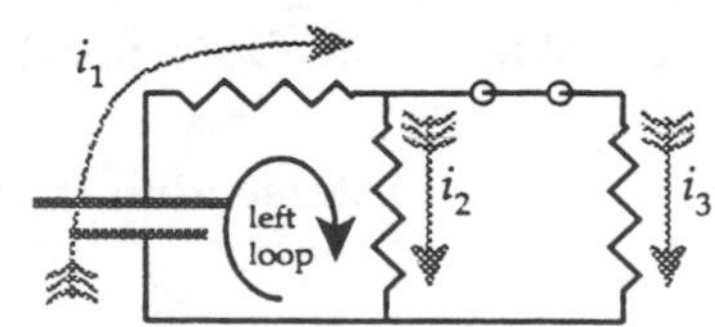

According to Kirchhoff's 2nd law, the total voltage change around any loop is zero. Applying this law to the left loop gives me

Left loop equation $\mathcal{E} - i_1 R_1 - i_2 R_2 = 0.$

When the switch closes, i_1 increases, as explained in part (a). Therefore, in order for the left loop equation to remain true, i_2 must decrease. In other words, since the voltage drop across resistor 1 increases, the voltage drop across resistor 2 must decrease.

(c) I'll solve for i_2 both before and after the switch closes. Comparing those two answers, we'll see by what percentage i_2 decreases.

Subproblem 1: Find the current through resistor 2 *before* the switch closes.

Since all three resistances are equal, the above left loop equation reduces to

Left loop equation $\mathcal{E} - i_1 R - i_2 R = 0,$

where $\mathcal{E}$ denotes the battery's voltage. Before the switch closes, the left loop is the *only* loop, and hence, $i_1 = i_2$. Solving for that i gives us

$$i_{2\text{ before switch closes}} = \frac{\mathcal{E}}{2R}.$$

Subproblem 2: Find the current through resistor 2 *after* the switch closes.

After the switch closes, things get more complicated. I see two ways of solving. On the one hand, we could calculate the total equivalent resistance of the circuit, in order to solve for i_1. Given that i_1, we could use the left loop equation to find i_2. On the other hand, we could write a second loop equation. I'll demonstrate both methods.

Method 1: Write another loop equation. If I walk counterclockwise around the right loop, then I pass through resistor 3 with the current. Therefore, my potential changes by $-i_3 R$. Remember, walking through a resistor with the current corresponds, in the car analogy, to rolling downhill. By contrast, when I walk through resistor 2, I'll moving against the current (i.e., "uphill"). So, $\Delta V = +i_2 R$. Since these voltage changes add up to zero,

Right loop equation $-i_3 R - i_2 R = 0,$

and hence, $i_3 = i_2$. This information seems useless, until you remember the intersection rule: The current entering an intersection equals the current leaving that intersection. Applying this rule to the intersection above resistor 2 gives

Intersection rule $i_1 = i_2 + i_3.$

Using the left loop equation, right loop equation, and intersection rule, we can solve for i_2.

Algebra starts here. Since $i_2 = i_3$, the intersection rule reduces to $i_1 = i_2 + i_2 = 2i_2$. Intuitively, i_2 is half of i_1 because i_1 splits into two equal portions at the intersection. In any case, substitute this result into the left loop equation, to get

$$\mathcal{E} - 2i_2R - i_2R = 0,$$

and hence,

$$i_{2\text{ before switch closes}} = \frac{\mathcal{E}}{2R}.$$

Before finishing off the problem, let me show another way to obtain $i_{2\text{ after switch closes}}$. Then, I'll finally compare $i_{2\text{ before switch closes}}$ to $i_{2\text{ after switch closes}}$.

Method 2 Equivalent resistances. The equivalent resistance of the circled resistors is given by

$$\frac{1}{R_{eq}} = \frac{1}{R} + \frac{1}{R} = \frac{2}{R},$$

and hence $R_{eq} = R/2$. This "equivalent resistor" is in series with a resistor R. So, the circuit has total resistance

$$R_{tot} = R_{eq} + R = \frac{R}{2} + R = \frac{3}{2}R.$$

Therefore, the current through the battery, which equals the current through resistor 1, is

$$i_1 = \frac{\mathcal{E}}{R_{tot}} = \frac{\mathcal{E}}{(3R/2)} = \frac{2\mathcal{E}}{3R}.$$

Substitute this result into the above left loop equation, to get

$$\mathcal{E} - \frac{2\mathcal{E}}{3R}R - i_2R = 0,$$

which we can solve for i_2 to get $i_{2\text{ before switch closes}} = \dfrac{\mathcal{E}}{3R}$.

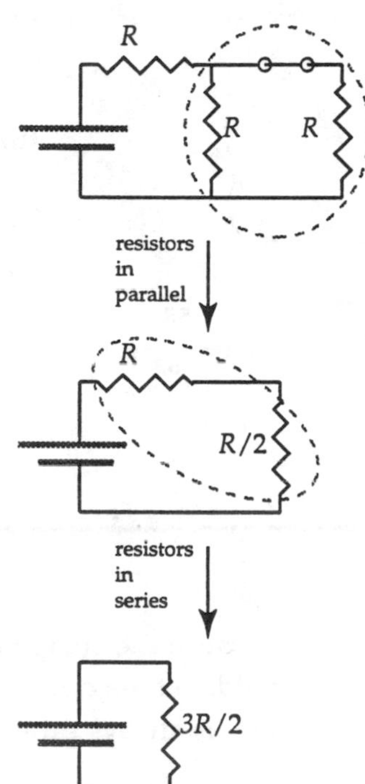

End of subproblem 2

OK, now that we know i_2 both before and after the switch closes, we can calculate by what percentage it drops. In absolute terms, i_2 drops by

$$|\Delta i_2| = i_{2\text{ before switch closes}} - i_{2\text{ after switch closes}} = \frac{\mathcal{E}}{2R} - \frac{\mathcal{E}}{3R} = \frac{\mathcal{E}}{6R}.$$

Therefore, in fractional terms, i_2 drops by

$$\text{fractional drop} = \frac{|\Delta i_2|}{i_{2\text{ before switch closes}}} = \frac{\frac{\mathcal{E}}{6R}}{\frac{\mathcal{E}}{2R}} = \frac{1}{3},$$

about 33%.

REVIEW PROBLEM 1-3

END VIEW
The cylindrical rod and shell extend distance L into the page.

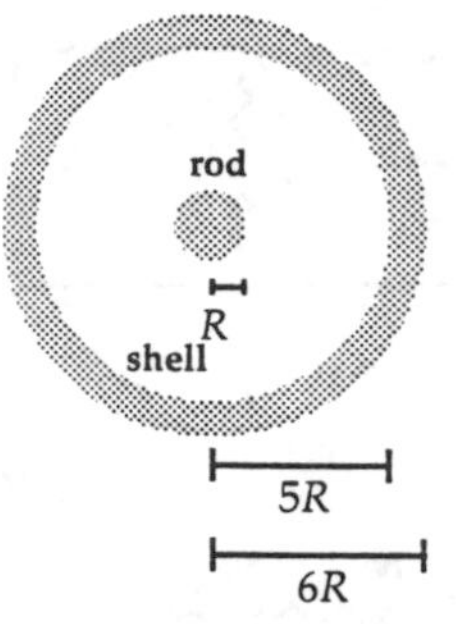

(*Electrostatics only*) Consider a cylindrical shell of inner radius $5R$ and outer radius $6R$. Inside the shell, concentric with its central axis, is a solid cylindrical rod of radius R. The rod and shell both have length L, where $L \gg R$.

The rod is given charge $+Q$, and the shell is given charge $-3Q$.

(a) What is the electric field a distance $2R$ from the central axis?

(b) How much work is required to push a particle of positive charge q_1 and mass m from the inner surface of the shell (at $r = 5R$) to the edge of the rod (at $r = R$)? Neglect gravity.

(c) That same particle is held a distance $10R$ from the central axis, and thrown directly towards the cylindrical shell with initial speed v_0. What is its speed immediately before crashing into the cylindrical shell?

(d) As mentioned above, charge $-3Q$ "lives" on the cylindrical shell. What percentage of that charge lives on the *inner* surface of the shell (at $r = 5R$)?

ANSWER

(a) Because the charge distribution is highly symmetric, Gauss' law allows us to solve for the electric field. Let me clarify a crucial point. Gauss' law is always *true*; the electric flux through a closed surface is always proportional to the charge enclosed by that surface:

$$\oint \mathbf{E} \cdot d\mathbf{A} = \frac{q_{\text{encl}}}{\varepsilon_0}.$$

Unfortunately, we can solve this equation for the electric field only if we can pull E outside the flux integral. Which happens only when E is constant over (part of) the Gaussian surface.

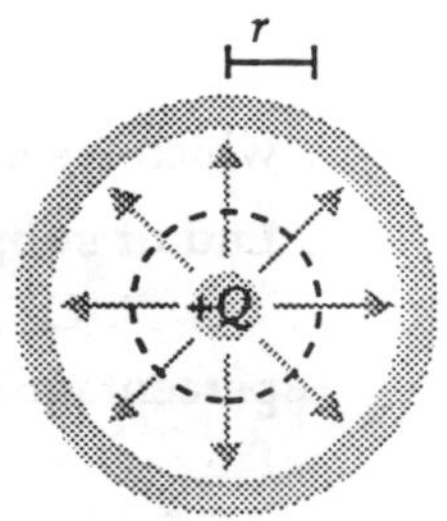

The dashed Gaussian surface extends distance L into the page.

Fortunately, in this case, we can draw a Gaussian surface over which the field stays constant. Since we want the field between the rod and shell, I've drawn a Gaussian cylinder in that region, of length L and arbitrary radius r. (Later, I'll set $r = 2R$.) By symmetry, the field has the same strength at any two points equidistant from the central axis. So, E is constant over the entire side (tube) of this Gaussian surface. Furthermore, no field lines pierce through the "ends" of my Gaussian cylinder. The field lines merely skim along those surfaces, as explained in question 34-3. Therefore, all the flux flows through the tube:

$$\oint \mathbf{E}\cdot d\mathbf{A} = \int_{\text{tube}} E dA = E\int_{\text{tube}} dA = E(2\pi r L$$

where in the last step, I used the surface area of a cylindrical tube.

So far, we've simplified the left-hand side of Gauss' law. Now I'll work on the right-hand side. This Gaussian surface encloses the entire rod, but none of the shell. So, $q_{\text{encl}} = Q$. Therefore, according to Gauss' law,

$$\oint \mathbf{E}\cdot d\mathbf{A} = \frac{q_{\text{encl}}}{\varepsilon_0}$$

$$E(2\pi r L) = \frac{Q_0}{\varepsilon_0},$$

and hence

$$E = \frac{Q}{2\pi\varepsilon_0 L r}.$$

So, at distance $r = 2R$ from the central axis,

$$E(2R) = \frac{Q}{2\pi\varepsilon_0 L(2R)} = \frac{Q}{4\pi\varepsilon_0 L R}.$$

By the way, since the rod carries linear charge density (charge per length) $\lambda = Q/L$, we can rewrite the above equation as $E = \frac{\lambda}{2\pi\varepsilon_0 r}$, the textbook formula for the field generated by a long charged rod.

(b) Starting with the definition of work, $W = -\int \mathbf{F}\cdot d\mathbf{s}$, you could integrate over the electric force on the particle, $\mathbf{F}_{\text{elec}} = q_1\mathbf{E}$. My technique, though essentially equivalent, gives me an excuse to review the meaning of "potential." Here goes . . .

Recall the relationship between potential energy and work. If I do 80 joules of work lifting a block from the floor onto a shelf, then I've given the block 80 joules of energy. In other words, it now has 80 joules of potential energy stored in it, energy you could "release" by nudging the block off the shelf. This goes to show that $\Delta U = W_{\text{by you}}$. In words, the work you do on an object equals its change in potential energy, assuming you don't give it kinetic energy or some other kind of energy instead. So, to solve this problem, we just need to find ΔU, the change in the particle's potential energy.

As emphasized in Chapter 35, the potential denotes the potential energy of a 1-coulomb test charge. For instance, suppose a 1-coulomb charge, when moved to the middle of my room, has electrostatic potential energy 5 joules. In other words, the center of my room has potential $V = 5$ volts. Then a 2-coulomb charge placed in the same location would acquire potential energy 10 joules. A 3-coulomb charge in the same spot would have $U = 15$ joules. And so on. In general, potential relates to potential energy by $U = qV$.

Putting all this together, we get $W_{\text{by you}} = \Delta U = q\Delta V$. Therefore, we can solve this problem as follows. First, calculate ΔV, the potential difference between the inner surface of the shell (at $r = 5R$)

and the edge of the rod (at $r = R$). Multiply this potential difference by q_1 to get $\Delta U = q_1 \Delta V$, the particle's change in potential energy. As just discussed, that's the work needed to push the particle from one point to the other.

So, the meat of this problem is finding the potential difference, ΔV. To figure out the relevant formula, use the fact that $W_{\text{by you}} = \Delta U = q\Delta V$:

$$q\Delta V = W_{\text{by you}} = -\int \mathbf{F}_{\text{elec}} \cdot d\mathbf{s} = -\int q\mathbf{E} \cdot d\mathbf{s}.$$

Divide this equation through by q to get $\Delta V = -\int \mathbf{E} \cdot d\mathbf{s}$, the formula we can use to solve this problem.

Since the particle gets pushed from the shell to the rod, we must find the potential difference betweenthose two objects. In part (a), we found the field between the rod and shell: $E = \dfrac{\lambda}{2\pi\varepsilon_0 r}$, This field points radially outward. Therefore, as my path of integration, I'll choose a straight radial line from the shell to the rod. Along that path, $d\mathbf{s}$ reduces to dr, and hence, the dot product $\mathbf{E} \cdot d\mathbf{s}$ reduces to a regular product. This path starts at $r = 5R$ and ends at $r = R$. So,

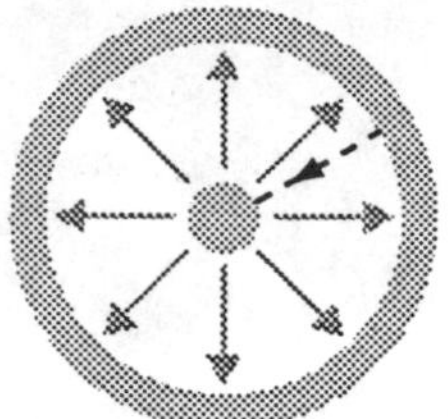
Dashed line is the path of integration.

$$\begin{aligned}
\Delta V &= -\int_{\text{outer shell}}^{\text{rod}} \mathbf{E} \cdot d\mathbf{s} \\
&= -\int_{5R}^{R} \frac{Q}{2\pi\varepsilon_0 L r} dr \\
&= -\frac{Q}{2\pi\varepsilon_0 L} \ln r \Big|_{5R}^{R} \\
&= -\frac{Q}{2\pi\varepsilon_0 L} \ln \frac{R}{5R} \\
&= +\frac{Q}{2\pi\varepsilon_0 L} \ln \frac{5R}{R} \\
&= \frac{Q}{2\pi\varepsilon_0 L} \ln 5.
\end{aligned}$$

Therefore, when pushed from the shell to the rod, the particle acquires potential energy

$$\Delta U = q_1 \Delta V = q_1 \frac{Q}{2\pi\varepsilon_0 L} \ln 5.$$

Since $W_{\text{by you}} = \Delta U$, that's the work.

When evaluating the above integral for ΔV, a common mistake is to plug in our part (a) answer for the electric field at $r = 2R$, or to plug in $r = R$ or $r = 5R$. Let me explain why this doesn't work. As the particle moves from the shell to the rod, it passes through all the points on the dashed line on my picture. The electric field has a different strength at all those different points. Mathematically, E depends on r, the particle's distance from the center. As r changes, so does the field. Your integral must reflect this. If you used $\Delta V = -\int_{5R}^{R} \frac{Q}{2\pi\varepsilon_0 L} dr$, or something like that, then you accidentally calculated how much the potential would change *if the particle "felt" the same electric field,* $\frac{Q}{2\pi\varepsilon_0 L(2R)}$, for its entire journey. But the electric field varies.

(c) At first glance, you might try to use force and acceleration. But as the particle "falls" toward the shell, the electric force on it increases. Therefore, it experiences a non-constant acceleration. Those old constant-acceleration kinematic formulas don't apply.

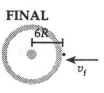

As emphasized last semester, when forces get too messy, we can rely on conservation laws instead. Here, since no heat or other dissipative energy gets produced (until the particle crashes), we can use energy conservation. Let V_0 and V_f denote the initial potential (at $r = 10R$) and the final potential (at $r = 6R$). We want v_f, the particle's final speed just before crashing. So, by energy conservation,

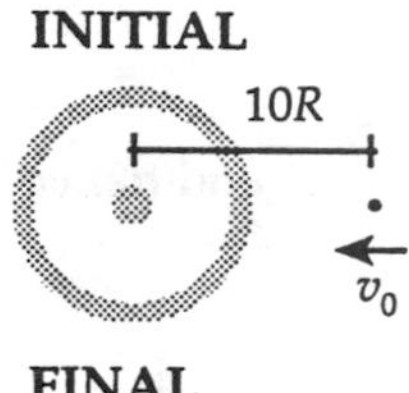

$$K_0 + U_0 = K_f + U_f$$

$$\frac{1}{2}mv_0^2 + q_1 V_0 = \frac{1}{2}mv_f^2 + q_1 V_f,$$

and hence

$$v_f = \sqrt{v_0^2 + \frac{2q_1(V_0 - V_f)}{m}}.$$

We're given v_0, q_1, and m. So, to finish solving for v_f, we need to find the initial and final potential. Actually, that's wrong. All we need is the potential *difference*, since v_f depends only on $V_0 - V_f$.

Well, as shown in part (b), the potential difference is given by $\Delta V = -\int \mathbf{E}\cdot d\mathbf{s}$. So, to find ΔV, we need to know the electric field outside the shell. That's a classic Gauss' law problem.

Before continuing, let me summarize the problem-solving strategy that just emerged.

1. Using Gauss' law, find the electric field outside the shell.
2. From that field, calculate the potential difference between the particle's "initial" and "final" position, using $\Delta V = -\int_{\text{initial}}^{\text{final}} \mathbf{E}\cdot$
3. Substitute that potential difference into the above energy-conservation equation, to find v_f.

Subproblem 1: Use Gauss' law to find the electric field outside the shell.

We can't just recycle our E from part (a), because there we found the field between the shell and the rod. Here, we need the field outside the shell.

Once again, I'll draw a Gaussian cylinder of arbitrary radius r, and length L. By radial symmetry, the field has the same strength over the entire side (tube) of this surface. So, as in part (a), we can pull E outside the flux integral:

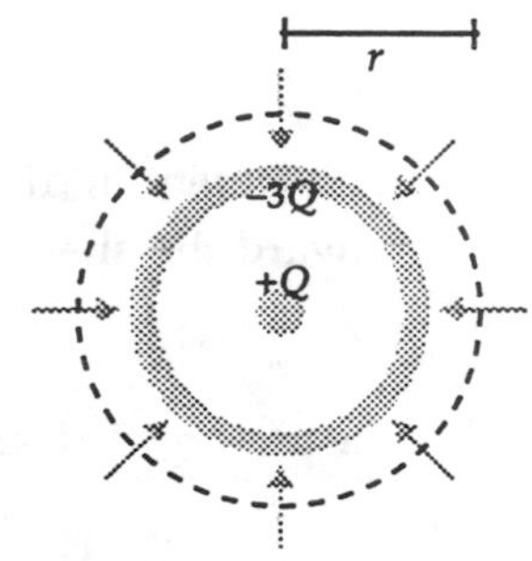

The dashed Gaussian surface extends distance L into the page.

$$\oint \mathbf{E}\cdot d\mathbf{A} = \int_{\text{tube}} E\,dA = E\int_{\text{tube}} dA = E(2\pi rL),$$

But unlike in part (a), the Gaussian surface now encloses the rod *and* the shell, a net charge of $-2Q$. So, Gauss' law gives us

$$\oint \mathbf{E}\cdot d\mathbf{A} = \frac{q_{\text{encl}}}{\varepsilon_0}$$

$$E(2\pi rL) = \frac{-2Q}{\varepsilon_0},$$

and hence

$$E = -\frac{Q}{2\pi\varepsilon_0 Lr}.$$

The minus sign indicates that the field lines flow into the Gaussian surface, not out of the Gaussian surface.

Subproblem 2: Using this electric field, find the potential difference between the particle's "initial" and "final" positions.

Notice that I found a general expression for the field at arbitrary distance r from the central axis. I didn't plug in $10R$ or $6R$. When we evaluate the potential difference using $\Delta V = -\int \mathbf{E}\cdot d\mathbf{r}$, we must allow to E to vary as a function of r.

As my path of integration, I'll choose a straight radial line from $r = 10R$ to $r = 6R$, i.e., the path along which the particle actually travels:

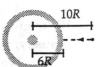

Dashed line is path of integration.

$$\begin{aligned}\Delta V &= -\int_{\text{initial position}}^{\text{outer edge of shell}} \mathbf{E}\cdot d\mathbf{s} \\ &= -\int_{10R}^{6R} \frac{-Q}{\pi\varepsilon_0 Lr}\,dr \\ &= \frac{Q}{\pi\varepsilon_0 L}\ln r\Big|_{10R}^{6R} \\ &= \frac{Q}{\pi\varepsilon_0 L}\ln\frac{6}{10} \\ &= -\frac{Q}{2\pi\varepsilon_0 L}\ln\frac{10}{6}.\end{aligned}$$

This potential difference comes out negative, because the particle *loses* potential energy as it "falls" toward the shell. The positive particle feels attracted to the negative shell.

Subproblem 3: Use that potential difference in our energy conservation equation.

Way back at the beginning of this answer, we used energy conservation to find an expression for the particle's "final" velocity when it reaches the outer shell:

$$v_f = \sqrt{v_0^2 + \frac{2q_1(V_0 - V_f)}{m}}.$$

Now that we know the potential difference, let's substitute it into this equation. But wait. We can't plug a negative expression into a square root. Fortunately, we don't have to. The change in potential, $\Delta V = -\frac{Q}{\pi\varepsilon_0 L}\ln\frac{10}{6}$, is the *final* minus the *initial* potential: $\Delta V = V_f - V_0$. So, the square root contains $-\Delta V$, not ΔV itself:

$$v_f = \sqrt{v_0^2 + \frac{2q_1(V_0 - V_f)}{m}} = \sqrt{v_0^2 + \frac{2q_1(-\Delta V)}{m}} = \sqrt{v_0^2 + \frac{2q_1\left(\frac{Q}{\pi\varepsilon_0 L}\ln\frac{10}{6}\right)}{m}}.$$

(d) In this problem, we must figure out how the charge on the cylindrical shell divides itself up between the inner and outer surfaces of the shell. How can we even get started? By remembering that *the electric field inside a conducting material is zero.* A nonzero electric field inside the conducting material would push valence electrons around. Therefore, when the electrons settle down and stop moving, the electric field inside the conducting material must vanish. See your textbook for more details.

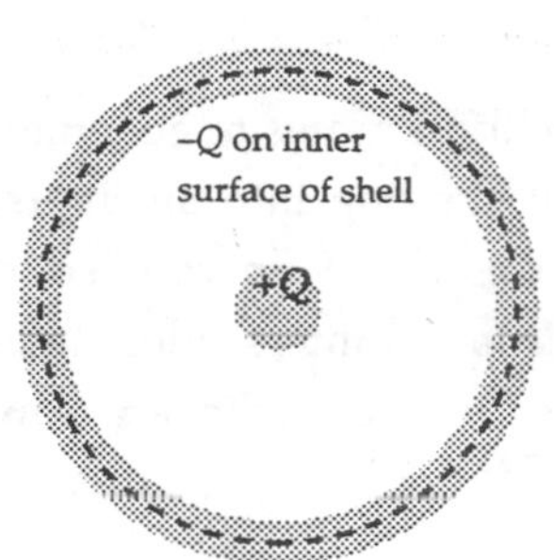

In any case, since E inside a conductor is zero, the field must vanish along the dashed Gaussian surface drawn here. Therefore, no electric flux can flow through this Gaussian surface: $\int \mathbf{E}\cdot d\mathbf{A} = 0$. Therefore, according to Gauss' law, my Gaussian surface encloses zero net charge. So, $-Q$ must "live" on the inner surface of the shell, in order to cancel the $+Q$ on the rod.

Let me summarize this argument. Because the electric field—and hence, the electric flux—inside a conducting material is zero, Gauss' law implies that the Gaussian surface drawn here contains zero net charge. Therefore, charge $-Q$ must live on the inner surface of the shell. The remaining $-2Q$ of charge on the shell must inhabit the outer surface.

REVIEW PROBLEM 1-4

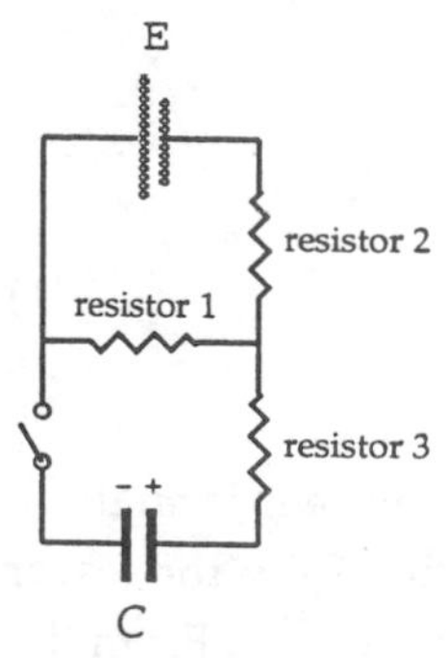

In this circuit, the battery has voltage $\mathcal{E}_0 = 2.0$ V, and each resistor has resistance $R = 10\ \Omega$. The capacitor, which has capacitance $C = 1.0 \times 10^{-12}$ F, carries initial charge 3.0×10^{-12} C, with the positive charge on the right plate. The switch is closed at time $t = 0$.

(a) Immediately after $t = 0$, what current flows through resistor 1?

(b) A long time later, what current flows through resistor 1?

(c) (*Hard but interesting*) Sketch a rough graph of the charge on the right capacitor plate, as a function of time.

ANSWER

CHOOSE YOUR LOOPS

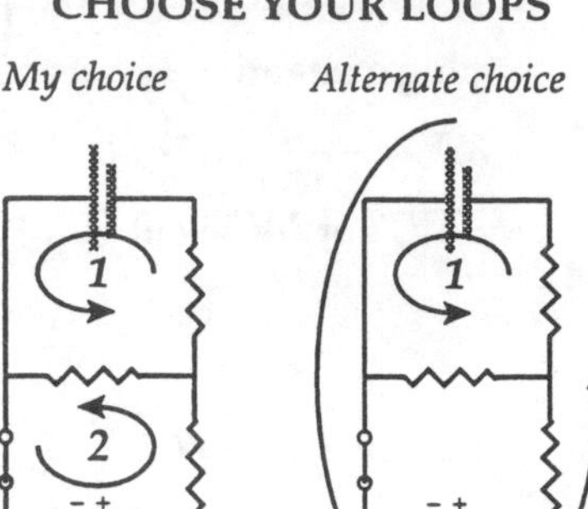

When the switch closes, the system becomes a multi-loop circuit. This should trigger you to use Kirchhoff's laws. First, I'll invoke those laws to generate equations that apply to the circuit at any time. Then, I'll use those equations to address parts (a) and (b).

When using Kirchhoff's laws, begin by choosing your loops. For instance, I'll choose the top and bottom loop, as drawn here. But you could equally well choose the top loop and the "whole-circuit" loop, or the bottom loop and the whole-circuit loop.

Next, label the currents in the different segments of the circuit, keeping in mind that the current changes only at an intersection.

LABEL THE CURRENTS

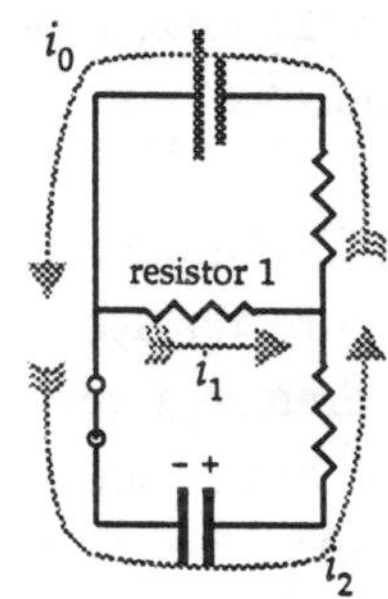

Here, we're solving for i_1 at $t \approx 0$ and at $t = \infty$. As always, I've guessed the direction of each current. If I'm wrong, then the current comes out negative. So, it doesn't matter if I mis-guess.

Given these loops and currents, we can use Kirchhoff's laws to generate equations. Let's start with Kirchhoff's 1st law: The current flowing into an intersection equals the current flowing out of that intersection. (If this law were violated, charge would either build up in an intersection or disappear from an intersection, neither of which can happen.) Look at the intersection to the left of resistor 1. Since i_0 flows into that intersection, while $i_1$and i_2 flow out, we get

Intersection rule $$i_0 = i_1 + i_2 \qquad (0)$$

You obtain the same equation by considering the intersection to the right of resistor 1.

Now we can apply Kirchhoff's 2nd law to each loop: The voltage change around a complete loop is zero. Intuitively, if you "walk" around a loop and end up back where you started, then your potential energy didn't change, overall.

Let's begin with loop 1, the top loop. I'll walk around it counterclockwise, starting from the upper right corner. So first, I walk through the battery, from negative to positive terminal. Therefore, my potential rises by the battery's voltage: $\Delta V = +\mathcal{E}_0$. Next, I walk through resistor 1, in the direction of current flow. Recall from the car-and-roadway analogy that charges flowing through a resistor correspond to cars "flowing" down a rocky hill. So, since I walk with the current, instead of against it, I'm walking "downhill." Therefore, my potential drops: $\Delta V = -i_1R$.

Finally, I walk through resistor 2, again with the current. So, $\Delta V = -i_0R$.

Since the net voltage change equals zero,

Loop 1 $$\mathcal{E}_0 - i_1R - i_0R = 0. \qquad (1)$$

A clockwise walk would yield the same final equation.

Since we have three unknowns (i_0, i_1, and i_2) but only two equations, we need more information. I'll generate an equation about loop 2 by "walking" around it counterclockwise, starting from the lower left corner. So first, I pass through the capacitor, moving from the negative to the positive plate. From the definition of capacitance, $C = Q/V$, we know that the potential changes across a capacitor by $V = Q/C$. But does the potential rise or fall? To find out, imagine carrying a positive test charge as you cross the capacitor. It "wants" to fall toward the negative plate. So, you must do work to carry it from the negative to the positive plate. In other words, the potential rises: $\Delta V = +Q/C$, where Q denotes the charge on the plates.

Next, I walk through resistor 3, in the same direction that i_2 flows. So, $\Delta V = -i_2R$. Finally, I walk through resistor 1, in the *opposite* direction from the current, according to my guess about i_1. This corresponds to walking uphill, against the flow of traffic. So, my potential *rises*: $\Delta V = +i_1R$.

Set the sum of these potential changes equal to zero, to get

Loop 2 $$\frac{Q}{C} - i_2R + i_1R = 0. \tag{2}$$

Given these three equations, we can address parts (a) through (c).

(a) *Immediately* after the switch closes, very little charge has flowed onto or off of the capacitor. So, the capacitor still carries its initial charge, $Q_0 = 3.0 \times 10^{-12}$ C. This crucial physical insight, which commonly shows up on exams, allows you to solve the problem.

Since we know the three resistances, as well as the battery's EMF, equations (0) through (2) contain only three unknowns, the three currents. Therefore, we can solve for i_1. It's just a matter of algebra.

Algebra starts here. Solve Eq. (1) for i_0 to get $i_0 = \frac{\mathcal{E}_0}{R} - i_1$. Solve Eq. (2) for i_2 to get $i_2 = \frac{Q}{CR} + i_1$. Now substitute those expressions for i_0 and i_2 into Eq. (0), and set $Q = Q_0$, to get

$$\frac{\mathcal{E}_0}{R} - i_1 = i_1 + \left(\frac{Q_0}{CR} + i_1\right).$$

Solve for i_1:

$$\begin{aligned} i_i &= \frac{1}{3}\left[\frac{\mathcal{E}_0}{R} - \frac{Q_0}{CR}\right] \\ &= \frac{1}{3}\left[\frac{2.0\text{ V}}{10\ \Omega} - \frac{3.0 \times 10^{-12}\text{ C}}{(1.0 \times 10^{-12}\text{ F})(10\ \Omega)}\right] \\ &= \frac{1}{3}[0.20\text{ A} - 0.30\text{ A}] \\ &= -0.033\text{ A}. \end{aligned}$$

End of algebra.

The negative answer indicates that the current through resistor 1 initially flows leftward, not rightward as I guessed. From the algebraic steps, we can see why. The battery tries to make current flow rightward through resistor 1. But the discharging capacitor tries to make current flow leftward through that resistor. Initially, the capacitor "wins." But later on, when the capacitor has lost some of its initial charge, the battery "wins," as we'll see in part (b).

(b) Let's begin by thinking physically. As time passes, the capacitor discharges. In other words, positive charges flow off the positive plate, counterclockwise around the circuit, and onto the negative plate. As a result, the charge on the capacitor decreases.

If the circuit contained no battery, then the capacitor would discharge completely, until Q reaches zero. But here, the battery supplies a voltage that "keeps" some charge on the capacitor. The capacitor doesn't discharge completely. Instead, it reaches some final charge, Q_f, and stays there.

When the capacitor "settles" to its final charge, no more current flows through it. Remember, charge flows "through" a capacitor when it's charging or discharging. Indeed, i_2 is the rate at which current flows onto or off of the capacitor: $i_2 = -dQ/dt$. When Q stops changing, i_2 disappears.

That's the physical insight needed to solve the problem. Just set $i_2 = 0$, and use our Kirchhoff's law equations from above. We don't even need to find Q_f, because Eqs. (0) and (1) allow us to solve for i_1, without further ado. Since $i_2 = 0$, Eq. (0) tells us that $i_1 = i_0$. Substitute this conclusion into Eq. (1), to get

$$\mathcal{E}_0 - i_1R - i_1R = 0,$$

and hence

$$i_1 = \frac{\mathcal{E}_0}{2R} = \frac{2.0\ \text{V}}{2(10\ \Omega)} = 0.10\ \text{A}.$$

"Circuit reduction"

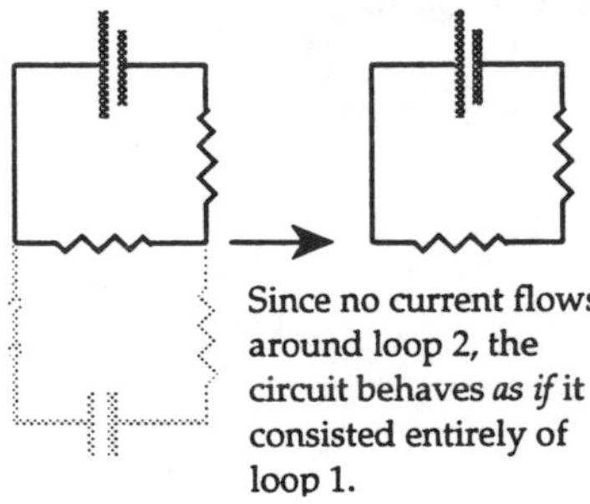

Since no current flows around loop 2, the circuit behaves *as if* it consisted entirely of loop 1.

Notice that i_1 is now positive; the battery makes current flow rightward through resistor 1.

Let me show you what these mathematical steps mean, physically. As explained above, after the capacitor reaches its final charge, i_2 drops to zero. In other words, current no longer flows around loop 2. Since loop 2 becomes a "dead end," the circuit behaves *as if* loop 1 were the whole system. As this diagram shows, if loop 1 is the whole circuit, then $\mathcal{E}_0 - 2iR = 0$, and hence, $i = \mathcal{E}_0/2R$.

(c) Most students don't realize that Q_f, the final charge on the right capacitor plate, is *negative*. The left plate ends up with the positive charge. Let me explain how I know, in both physical and mathematical terms. Mathematically, by starting with the loop 2 equation and setting $i_2 = 0$, you get

$$\frac{Q}{C} + i_1R = 0.$$

Since i_1 ends up positive, as we saw in part (b), Q must end up negative.

OK, here's a more intuitive argument along the same lines. The battery tries to make current flow counterclockwise around the whole circuit, and hence, upward through resistor 3. But when the capacitor's charge settles to its final value, no current flows through resistor 3. Therefore, the capacitor must generate a clockwise voltage around loop 2, to cancel the battery's counterclockwise voltage. In order for the capacitor to generate a clockwise voltage, the left plate must carry positive charge, and the right plate must carry negative charge.

At this stage, we know that the charge on the right capacitor plate starts out positive and ends up negative. But how does it get from one point to the other? As you can confirm by playing around with the differential equations describing RC circuits, when a capacitor charges or discharges, the process happens exponentially. For instance, in a regular RC circuit, Q decays exponentially from its initial value to zero. Similarly, in this case, Q "decays" exponentially from its initial to its final value.

Charge on right capacitor plate

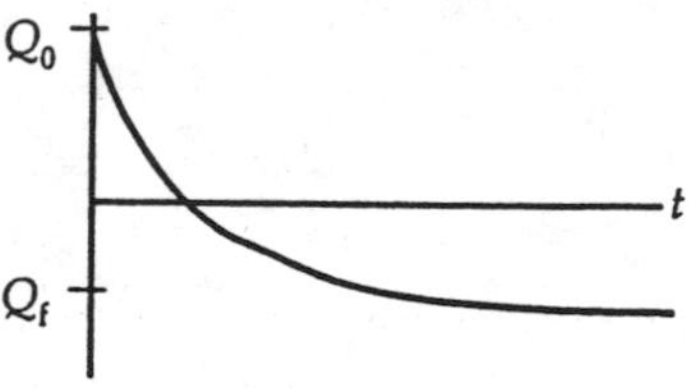

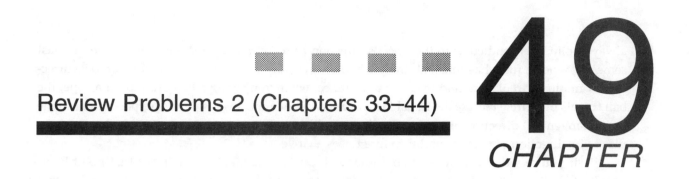

Review Problems 2 (Chapters 33–44)

These problems cover everything except displacement current and electromagnetic waves, which many courses leave off the final exam. If your exam covers those topics, then please work through chapters 45 and 46.

REVIEW PROBLEM 2-1

Attached to the floor is a particle of charge 3.0×10^{-14} C. Next to it on the floor, exactly 0.010 eters away, is a small frictionless marble of mass 0.0040 kg, and charge 8.0×10^{-16} C. The marble is released from rest, and slides away. Eventually, when the marble gets a *very large distance* away from the other particle, it enters a uniform magnetic field of strengthtrace the trajectory of the marble after it enters the magnetic field of strength 7.0 T, pointing upward from the floor to the ceiling. (That's a huge field, obtained only from superconducting electromagnets.) The particle enters the field at point P, as drawn here.

TOP-DOWN VIEW (from ceiling)

particle nailed to floor

marble

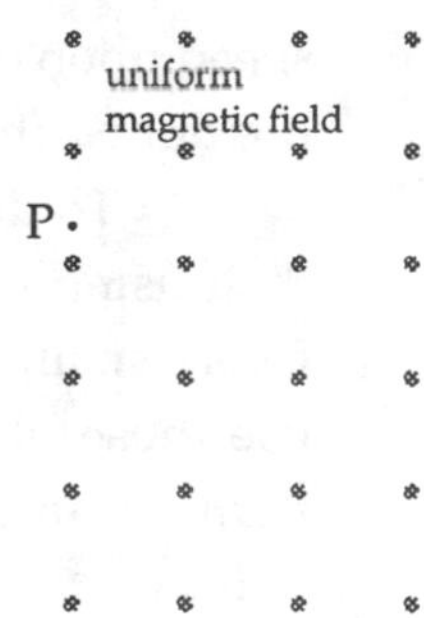

(a) On this diagram, trace the trajectory of the marble after it enters the magnetic field. Please check your answer before doing part (b).

(b) How far from point P does the marble leave the magnetic field?

(c) How much work does the magnetic field do on the marble?

ANSWER

(a) The marble enters the field at point P, moving rightward. The magnetic field makes it travel along a circular path. Here's why. The cross product of two vectors points perpendicular to both vectors. So, the magnetic force,

$$\mathbf{F}_{\text{mag}} = q\mathbf{v} \times \mathbf{B},$$

always points perpendicular to the object's direction of motion, the direction of **v**. As we saw last semester, when a force has constant strength and always pushes the object perpendicular to its direction of motion, circular motion results. For instance, when you swing a ball around on a rope, the ball travels in a circle *because* the radial force exerted by the rope always points perpendicular to the ball's (tangential) direction of motion. Similarly, inside the field, the marble gets pushed perpendicular to its direction of motion, and therefore traces a circle.

If you're worried that an electric force also pushes the marble, remember that point P sits *very* far from the nailed-down particle. Therefore, the marble feels a negligible electric force at that point.

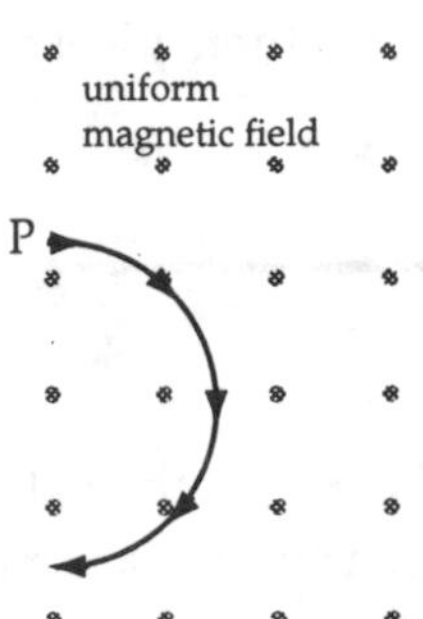

To draw the marble's exact trajectory, we must decide whether it traces a clockwise or counterclockwise circle. In other words, we must decide whether the marble gets deflected "up the page" or "down the page." To figure out the direction of the magnetic force at point P, apply the right-hand rule to the cross-product in $\mathbf{F}_{\text{mag}} = q\mathbf{v} \times \mathbf{B}$. When the marble enters the field, **v** points rightward and **B** points out of the page. So, point your fingers rightward, and then curl your finger tips out of the page. Your thumb now points down the page. That's the direction of $\mathbf{F}_{\text{mag}}$. The marble deflects the down the page, and traces a clockwise semicircle before emerging from the field.

(b) First, using conservation of energy, I'll find the marble's speed when it enters the magnetic field. Then, by considering the magnetic force on the marble, I'll use circular-motion reasoning to figure out the radius of the semicircle.

OK, let me explain this more fully. The radius of the semicircle depends on the marble's speed upon entering the field. Intuitively, a speedier marble traces a bigger circle. So, we need to figure out v_P, the marble's speed at point P.

To do so, you might try calculating the electrostatic force, obtaining the acceleration, and then using kinematics to find v_P. But as the marble gets farther from the nailed-down particle, it feels a smaller and smaller force. Since the force varies, so does the acceleration. Therefore, we can't use those old constant-acceleration kinematic formulas. As always, when forces get too messy, think about conservation laws. Energy is conserved as the marble moves from its initial position to point P.

Given all this, we can divide the problem into two subproblems, as mentioned above. First, invoke energy conservation to find the marble's speed at point P. Then, use circular-motion force reasoning to figure out the radius of the semicircle. If we call that radius R, then the marble emerges from the field a distance $2R$ from point P, as drawn here.

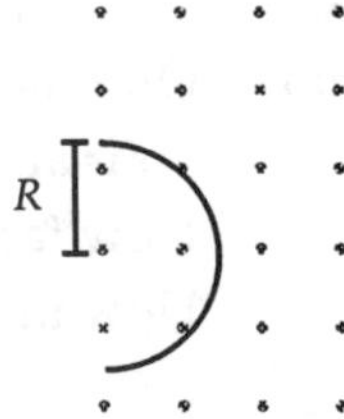

Subproblem 1: Energy conservation to find the marble's speed at point P

By "initial," I mean the moment the marble gets released; and by "final," I mean when the marble reaches point P. According to energy conservation,

$$K_0 + U_0 = K_f + U_f$$
$$O + qV_0 = \frac{1}{2}mv_P^2 + qV_f,$$

where q denotes the *marble's* charge, and V_0 and V_f refer to its initial and final potential. Notice that I used $U = qV$, the defining equation for potential.

We know q and m. To solve for v_P, we just need to find the initial and final potential. Actually, the final answer for v_P depends only on the potential *difference*. But here, we can find ΔV efficiently by *separately* finding V_0 and V_f, and subtracting.

In general, to calculate a potential, you integrate the electric field over distance: $V = \int \mathbf{E} \cdot d\mathbf{s}$. In this particular case, the marble feels the field generated by the nailed-down particle, whose charge I'll call Q. Since' it's a point charge, the particle generates a Coulomb field, $E = \frac{Q}{4\pi\varepsilon_0 r^2}$. Therefore, the nailed-down particle generates a potential

$$V_{\text{point charge}} = -\int E dr = -\int \frac{Q}{4\pi\varepsilon_0 r^2} dr = \frac{Q}{4\pi\varepsilon_0 r},$$

where I obeyed the usual textbook convention of setting the integration constant equal to zero. Remember, this is the potential generated *by* the nailed-down particle, which has charge $Q = 3.0 \times 10^{-14}$ C.

Since the marble starts a distance $r = 0.010$ m from the particle,

$$V_0 = \frac{Q}{4\pi\varepsilon_0 r_0} = \frac{3.0 \times 10^{-14}\text{ C}}{4\pi(8.85 \times 10^{-12}\text{ F/m})(0.010\text{ m})} = 0.027\text{V}.$$

What about the final potential? It's zero! Point P is *very far* from the nailed-down particle. Therefore, $r \approx \infty$. When you plug $r \approx \infty$ into our expression for V, you get 0. Physically, this means that two charged objects have no potential energy when located *very* far apart.

Now that we know V_0 and V_f, we can solve the above energy-conservation equation for v_P. In this equation, q denotes the charge of the *marble.*

$$K_0 + U_0 = K_f + U_f$$
$$O + qV_0 = \frac{1}{2}mv_P^2 + qV_f,$$

and hence

$$v_P = \sqrt{\frac{2q(V_0 - V_f)}{m}} = \sqrt{\frac{2(8.0 \times 10^{-6}\text{C})(0.027\text{ V} - 0)}{0.0040\text{kg}}} = 0.0104\text{m/s},$$

about 1.0 centimeter per second. That's the marble's speed upon entering the magnetic field.

Subproblem 2: Circular motion reasoning to find the radius of the semicircle

As we saw last semester, when an object travels in a circle, it has radial (centripetal) acceleration $a_c = v^2/R$, where R denotes the radius of the circle. Don't confuse this "R" with the r's in subproblem 1 above. According to Newton's 2nd law, the radial (centripetal) force needed to create this radial acceleration is $F_c = mv^2/R$. Remember, the centripetal force isn't a *separate* force on your free-body diagram. It's just the name given to whatever force makes the marble move in a circle. Here, the magnetic force causes the circular motion. So,

$$\begin{aligned} m\frac{v^2}{R} &= F_{mag} \\ &= q|\mathbf{v}\times\mathbf{B}| \\ &= qvB\sin\theta && [\text{where } \theta \text{ is the angle between } \mathbf{v} \text{ and } \mathbf{B}] \\ &= qvB. && [\text{since } \theta = 90^\circ, \sin\theta = 1] \end{aligned}$$

Solve for R to get $R = \dfrac{mv}{qB}$. Here, the marble enters the field with speed v_p, as found in subproblem 1. So,

$$R = \frac{mv_p}{qB} = \frac{(0.0040 \text{ kg})(0.0104 \text{ m/s})}{(8.0\times 10^{-6}\text{C})(7.0\text{T})} = 0.74 \text{ meters.}$$

Therefore, the marble emerges from the field a distance $2R = 1.5$ meters from point P.

(c) A magnetic field *never* does work on a charged particle, even when it exerts a huge force. Here's why. By definition, $W = \int \mathbf{F}\cdot d\mathbf{s} = \int F_{\parallel} ds$, where $F_{\parallel}$ denotes the component of force *along* the direction of motion. Remember, a dot product "picks off" the parallel components of vectors. But the magnetic force acts entirely *perpendicular* to the direction of motion. *No* component of the magnetic force ever points along the direction of motion. For this reason, a magnetic force can do no work on a charged particle.

Because the magnetic field does no work, it can't make the marble speed up or slow down. In other words, it can't give energy to the marble, or take energy away. A magnetic field can only change the direction of motion.

REVIEW PROBLEM 2-2

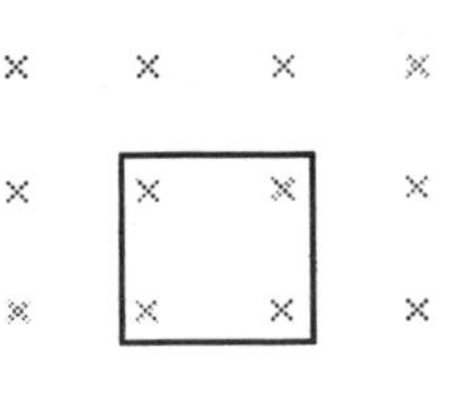

A metal square consists of four straight wires of length s and resistance R. The square sits inside a uniform external magnetic field pointing into the page. Initially, the field has strength B_0. But starting at time $t = 0$, the field changes at a steady rate. Specifically, in time T, the field "flips," so that it ends up pointing out of the page with strength B_0.

(a) Between $t = 0$ and $t = T$, what current flows around the square?

(b) At $t = T/2$, the external magnetic field is momentarily zero. What current flows around the square at that moment?

(c) Between $t = 0$ and $t = T/2$, the external field points into the page, and decreases from B_0 to 0. During this time interval, does the current in the square flow clockwise or counterclockwise?

(d) Between $t = T/2$ and $t = T$, the external field increases from 0 to B_0, and points out of the page. During this interval, does the current in the square flow clockwise or counterclockwise?

(e) At time $t = T/2$, what is the magnetic field at the center of the square? If you end up with a nasty integral, *don't evaluate it;* just leave the answer in the form of an integral.

ANSWER

(a) According to Faraday's law, an EMF—and therefore, a current—gets induced in the square when the magnetic flux through the square *changes*:

Faraday's law $$\mathcal{E}_{\text{induced}} = -\frac{d\Phi_B}{dt},$$

where $\Phi_B = \int \mathbf{B}\cdot d\mathbf{A}$. Roughly speaking, the flux tells us how many field lines flow through the square. For more intuitions about flux, please review the first few problems in Chapter 34.

To find the induced voltage and current using Faraday's law, I must figure out how quickly the flux changes. So first, after finding the initial flux and the final flux, I'll use them to calculate the change in flux: $\Delta\Phi_B = \Phi_{B\text{ final}} - \Phi_{B\text{ initial}}$. Since the flux changes over time $\Delta t = T$, its average rate of change is

$$\frac{\Delta\Phi_B}{\Delta t} = \frac{\Phi_{B\text{ final}} - \Phi_{B\text{ initial}}}{\text{T}}.$$

And because the flux changes steadily, this average rate of change equals the instantaneous rate of change at all moments between $t = 0$ and $t = T$. So, the meat of this problem is calculating the initial and final flux.

Subproblem: Find the initial and final flux.

Since **B** is uniform, and since it points in the same direction that the square "faces," we can pull it outside the flux integral to get

$$\Phi_B = \int_{\text{square}} \mathbf{B}\cdot d\mathbf{A} = \text{B} \int_{\text{square}} dA = BA_{\text{square}} = Bs^2.$$

You may wonder why I pulled B outside the integral, even though it isn't constant. Although B changes *in time,* it's uniform *in space.* Let me explain. As time passes, B changes. So, we could not pull B outside a time integral such as $\int B dt$. But at any given instant in time, B has the same strength over the entire area of the square. In other words, B is spatially uniform. Therefore, since we're integrating over the area of the square, we can treat B as a constant.

OK, now we're ready to write down expressions for the initial and final flux. I'll choose "out of the page" as my positive direction. So, the initial field is negative. Hence, $\Phi_{B\text{ initial}} = -B_0 s^2$. By contrast, since the field ends up pointing out of the page with strength B_0, the final flux through the square is $\Phi_{B\text{ final}} = +B_0 s^2$.

End of subproblem

Now that we know the initial and final flux, we can calculate the rate of change of flux:

$$\frac{\Delta\Phi_B}{\Delta t} = \frac{\Phi_{B\,\text{final}} - \Phi_{B\,\text{initial}}}{T} = \frac{B_0 s^2 - (-B_0 s^2)}{T} = \frac{2B_0 s^2}{T}.$$

So, according to Faraday's law, the induced EMF is simply

$$\mathcal{E}_{\text{induced}} = \frac{d\Phi_B}{dt} = -\frac{2B_0 s^2}{T}.$$

The minus sign reminds us that the current flows so as to generate a magnetic field that "opposes" the changing flux. More on this below.

Given the induced voltage, we can calculate the current using Ohm's law. Since the square consists of four resistors in series, it has total resistance $R_{\text{tot}} = 4R$. So, from $V = iR$, we get

$$i = \frac{\mathcal{E}_{\text{induced}}}{4R} = \frac{B_0 s^2}{2RT}.$$

I'm omitting the minus sign for now, because I'll figure out whether the current flows clockwise or counterclockwise in parts (c) and (d).

(b) At $t = T/2$, the external magnetic flux through the square momentarily hits zero. So, you might be tempted to say that no voltage or current gets induced. But remember, the induced voltage does not depend on the *size* of the magnetic flux. It depends on the rate at which the flux *changes*. Even at $t = T/2$, the magnetic flux is in the process of *changing*. Here's why. Immediately before $t = T/2$, the field points into the page. Immediately after $t = T/2$, the field points out of the page. So, right at $t = T/2$, the flux is changing from "into the page" to "out of the page." And its rate of change is the same as always. That's why the voltage and current we found in part (a) apply at *all* times between $t = 0$ and $t = T$, including $t = T/2$.

(c) According to Lenz' law, the induced current creates its own magnetic field that counteracts the changing flux. Between $t = 0$ and $t = T/2$, the flux "faces" into the page, and decreases. The induced current "fights" this decrease, by generating its own field *pointing into the page*.

To generate an into-the-page field, does the induced current flow clockwise or counterclockwise? The right-hand rule shortcut can tell us. According to that rule, if you point your right thumb in the direction that current flows through a wire, then your curled fingertips represent the magnetic field lines. So here, pick any side of the square, point your thumb along that side, and curl your fingertips towards the center of the square. By trial and error, you can confirm that your fingertips point into the page only if your thumb "assumes" a clockwise current. So, the induced current flows clockwise.

(d) Because the rate of change of flux stays constant, the induced voltage and current should also stay constant, according Faraday's law. Therefore, the induced current should continue flowing clockwise. That's right! But it might seem counterintuitive, because the external field changes direction. Let me reconcile this answer with intuition, by applying Lenz' law again.

Between $t = T/2$ and $t = T$, the external field points *out* of the page, and increases. The induced current opposes this increase. To oppose an increasing field that points out of the page, you must generate a field pointing *into* the page. So, as in part (c), the induced current produces an into-the-page magnetic field. Therefore, the induced current flows clockwise, just as it did in part (c).

Digression: Changing flux vs. acceleration. If this seems weird, think about the rate of change of velocity—i.e., the acceleration Ñ of a ball thrown straight upward. As the ball rises, its upward velocity decreases. This corresponds to a downward acceleration. Later, as the ball falls back down, its downward velocity increases. Again, this corresponds to a downward acceleration. Even at its peak, the ball has a downward acceleration, because the velocity is *changing*. So, even though the ball switches directions, its *rate of change* of velocity stays the same throughout the process. Similarly, even though the magnetic field switches directions, its rate of change stays the same throughout the process. To oppose this changing field, the induced current must flow in the same direction the whole time.

(e) As emphasized above, even though the external field vanishes at $t = T/2$, voltage still gets induced around the square, because the external flux is still *changing*. The induced EMF creates a current, which generates its own magnetic field. Because the external field momentarily disappears at $t = T/2$, the field generated by the induced current is the *total* field. So, this problem asks us to find the field produced by a square current loop.

We know two ways to find the field produced by a current. When the current distribution is sufficiently symmetric, you can use Ampere's law, the magnetic analog of Gauss' law. But here, you can't pull B outside the line integral in Ampere's law, no matter how cleverly you draw the Amperian loop. So, we're stuck using Biot-Savart, the magnetic analog of Coulomb's law.

Before diving into the details of Biot-Savart, let's simplify things by thinking physically. Each side of the square generates a magnetic field. The center of the square is equidistant from the mid-point of each side. Therefore, each side "contributes" the same field at the center. Do those four fields add together or cancel? To find out, apply the right hand rule to the cross product in Biot-Savart,

$$d\mathbf{B} = \frac{\mu_0}{4\pi}\frac{i d\mathbf{l} \times \hat{\mathbf{r}}}{r^2}.$$

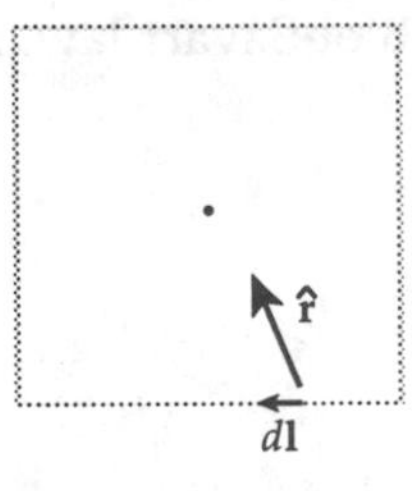

$d\mathbf{l}$ is a tiny "piece" of the clockwise current.

In this formula, $d\mathbf{l}$ refers to a tiny "piece" of the current, and $\hat{\mathbf{r}}$ denotes a unit vector pointing from that piece to the location at which we're finding the field, in this case the center of the square. So, point your fingers along $d\mathbf{l}$, and then curl your fingerstips until they point toward the center of the square. Your thumb now points into the page. Furthermore, you get this same result no matter which side of the square you choose. So, all four sides generate a field *in the same direction*.

Let me summarize what we've figured out so far. By symmetry, each side of the square generates the same magnetic field at the center. And those four fields point in the same direction. Therefore, at the center of the square, the total field is four times the field generated by any one side:

$$B_{total} = 4B_{one\ side}.$$

Given this simplification, we just need to find the field produced by *one* side of the square. But how can we calculate it? Well, when figuring out the electric field produced by a continuous charge distribution, we "broke up" the object into pieces, figured out the infinitesimal electric field generated by each piece, and added up those tiny fields. Similarly, in this case, we can break the wire into pieces, calculate the infinitesimal magnetic field generated by each piece, and then add up those tiny field contributions. Let me summarize the strategy.

Biot-Savart strategy

1) Blacken a tiny piece of the wire (of length $d\mathbf{l}$), and draw in the vector $\mathbf{r}$ pointing from that blackened piece to the location at which you're finding the magnetic field.

2) Using Biot-Savart, find the magnetic field, $d\mathbf{B}$, generated by that piece of wire.

3) Add up those infinitesimal magnetic fields by integrating over the wire, using vector components if necessary. (Fortunately, components won't be needed here.)

OK, now I'll apply this strategy to the bottom side of the square.

Step 1: On a carefully-labeled diagram, blacken an infinitesimal piece of the wire.

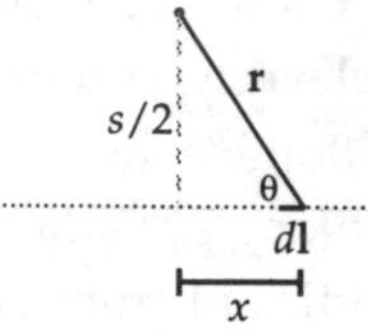

Don't choose a "special" piece of the wire, such as the midpoint or the end. Notice that I've made up a name, x, for the distance from my blackened piece to the midpoint. Notice also that the center of the square is a distance $s/2$ from the midpoint of the side. By θ, I mean the angle between $d\mathbf{l}$ and $\mathbf{r}$.

Step 2: Write an expression for $d\mathbf{B}$, the infinitesimal field generated by that piece of wire.

As shown above, the field points into the page. According to the Biot-Savart law, it has magnitude

$$\begin{aligned} d\mathbf{B} &= \frac{\mu_0}{4\pi}\frac{i\,|\,d\mathbf{l}\times\hat{\mathbf{r}}\,|}{r^2} \\ &= \frac{\mu_0}{4\pi}\frac{i\,dl\sin\theta}{r^2} \\ &= \frac{\mu_0}{4\pi}\frac{i\,dx\sin\theta}{x^2+(s/2)^2}, \end{aligned}$$

where in the last step I applied Pythagorean theorem to the triangle in the diagram, and also noticed that "dl" is the same as dx. In the second step, I used the general formula for the magnitude of a cross product, $|\mathbf{a}\times\mathbf{b}| = ab\sin\theta$. Remember, $\hat{\mathbf{r}}$ is a unit vector. It points in the direction of $\mathbf{r}$, but has length 1. That's why $|d\mathbf{l}\times\hat{\mathbf{r}}| = dl(1)\sin\theta = dl\sin\theta$.

Step 3: Add up (integrate) these tiny fields to find the total field generated by the wire.

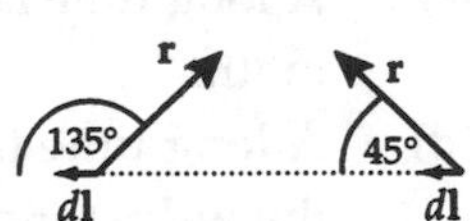

We can integrate over dx or $d\theta$. It's your choice. If you work in terms of x, the limits of integration are $-s/2$ to $s/2$, since $x = 0$ corresponds to the middle of the side. As this new diagram shows, if you work in terms of θ, the integral ranges from $\theta = 45°$ to $\theta = 135°$. But I'll integrate over dx:

$$B_{\text{one side}} = \int dB = \frac{\mu_0}{4\pi}\int_{-s/2}^{s/2} \frac{i\,dx\sin\theta}{x^2+(s/2)^2}.$$

You *cannot* evaluate this integral, because it contains two variables that depend on each other. The angle θ depends on x. In other words, as x changes, so does θ. Therefore, before integrating, you must rewrite θ in terms of x; or vice versa. But that's not too hard, given the triangle in the step 1 diagram. From that triangle,

$$\sin\theta = \frac{s/2}{r} = \frac{s/2}{\sqrt{x^2+(s/2)^2}}.$$

Substitute this expression into the integral, and pull the i out front, to get

$$B_{\text{one side}} = \frac{\mu_0}{4\pi} i \int_{-s/2}^{s/2} \frac{dx\sin\theta}{x^2+(s/2)^2} = \frac{\mu_0}{4\pi} i \int_{-s/2}^{s/2} \frac{(s/2)dx}{\left[x^2+(s/2)^2\right]^{3/2}}.$$

Find out whether your instructor expects you to evaluate this kind of integral on a test. Many instructors would give you an integral table.

From part (a), we know the current to be $i = \frac{B_0 s^2}{2RT}$. And the total magnetic field at the center of the square is four times the field generated by one side. So,

$$B_{\text{total}} = 4B_{\text{one side}} = 4\frac{\mu_0}{4\pi}\frac{B_0 s^2}{2RT}\int_{-s/2}^{s/2} \frac{(s/2)dx}{\left[x^2+(s/2)^2\right]^{3/2}}.$$

REVIEW PROBLEM 2-3

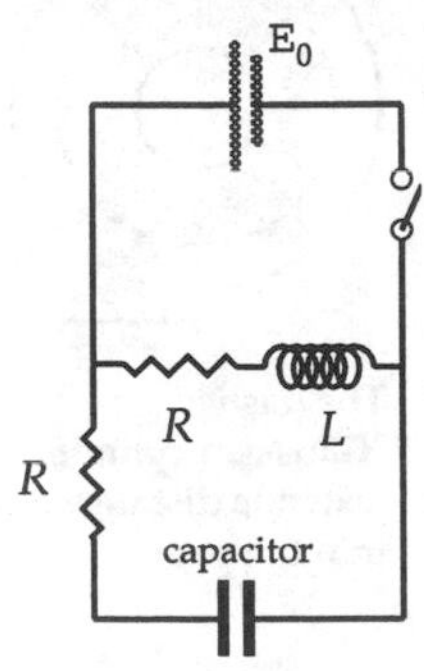

In this circuit, the capacitor consists of two concentric cylindrical shells, of length l. The inner and outer shell have radius s and $3s$, respectively, where $s \ll l$. The battery generates EMF ε_0, the inductor has inductance L, and the resistors have resistance R. The capacitor starts off uncharged. The switch is closed at time $t = 0$.

Magnified, end view of capacitor

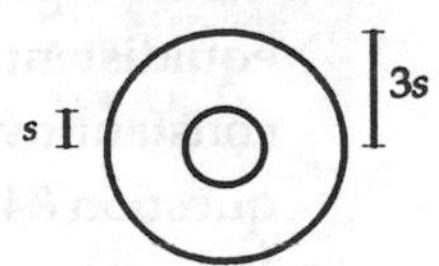

These cylinders extend distance *l* into the page.

(a) What is the capacitance of this capacitor? Derive your answer; don't just plug in a textbook formula about concentric cylinders.

(b) *Immediately* after $t = 0$, what current flows through the battery? You can answer this without solving differential equations.

(c) A long time later, what current flows through the battery? Again, you can answer without too much math.

(d) A long time later, at $t = t_1$, the switch is again opened. Sketch a rough graph of the current through the inductor as a function of time, starting at t_1.

(e) (*Very hard, but conceptually interesting*) After t_1, how much total heat dissipates in the circuit before it dies out completely?

ANSWER

(a) When a capacitor charges up, its two "plates" (in this case, cylinders) hold equal and opposite charge. The capacitance is the system's willingness to carry charge. Specifically, it's the charge that accumulates on the plates *per volt* of applied potential difference: $C = q/V$. So, if you double the potential difference between the plates, then you double how much charge gathers on the plates. For a given capacitor, the *ratio* of charge to voltage stays the same. This suggests a strategy:

> **Capacitance-finding strategy**
>
> 1) Assign arbitrary charge q to your plates.
> 2) Using Gauss' law or other techniques, find the electric field between the plates.
> 3) From that field, calculate the potential difference between the plates: $V = \int \mathbf{E}\cdot d\mathbf{s}$.
> 4) Figure out the capacitance using $C = q/V$. The arbitrary charge q you chose in step 1 should cancel out.

Because you've seen this before, I'll rush through the steps. See Chapter 36 for examples worked out in detail.

Steps 1 and 2: Assign charge to plates, and calculate the electric field between them.

GAUSSIAN SURFACE

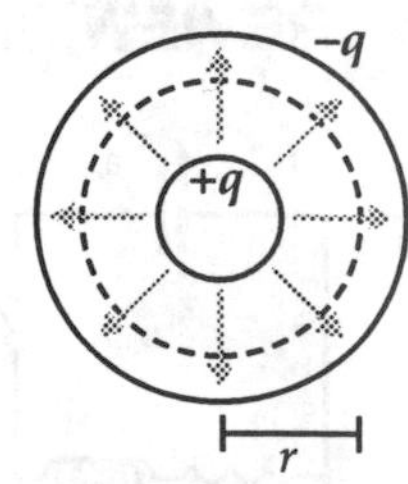

The dashed "Gaussian cylinder" extends distance l into the page.

When using Gauss' law, $\oint \mathbf{E}\cdot d\mathbf{A} = \frac{q_{encl}}{\varepsilon_0}$, to find an electric field, we must be able to pull E outside the flux integral. In other words, E must be constant over (at least some faces of) the Gaussian surface. Here, we can treat the cylindrical shells as infinitely long, since $l \gg s$. Therefore, by symmetry, the field points radially outward, and has the same strength at any two points equidistant from the center. So, if I draw a "Gaussian cylinder," the field stays constant over the entire side (tube) of the Gaussian cylinder, as explained in question 34-3b. So, the flux integral simplifies as follows:

$$\oint \mathbf{E}\cdot d\mathbf{A} = \int_{\text{tube of Gaussian cylinder}} \mathbf{E}\cdot d\mathbf{A} = E \int_{\text{tube}} dA = E(2\pi r l),$$

where in the last step I used the surface area of a cylindrical tube, i.e., the circumference times the length.

This Gaussian surface encloses the entire inner shell, but none of the outer shell. So, the surface encloses charge q, the "pretend" charge I placed on the plates. Therefore, from Gauss' law,

$$\oint \mathbf{E} \cdot d\mathbf{A} = \frac{q_{\text{encl}}}{\varepsilon_0}$$

$$E(2\pi r l) = \frac{q}{\varepsilon_0},$$

and hence

$$E = \frac{q/l}{2\pi\varepsilon_0 r}.$$

Notice that the numerator contains the linear charge density, $\lambda = q/l$.

Step 3: Integrate over the field between the plates to find the potential difference.

Remember, we're finding the capacitance, i.e., the charge per volt carried by the plates. So, we must find the voltage between the plates when they hold charge q.

To ensure that V comes out positive, I'll integrate from the negative to the positive shell. As my path of integration, I'll choose the radial segment drawn here. That way, the dot product $\mathbf{E} \cdot d\mathbf{s}$ reduces to a regular product:

$$\begin{aligned} V &= -\int_{\text{outer shell}}^{\text{inner shell}} \mathbf{E} \cdot d\mathbf{s} \\ &= -\int_{3s}^{s} \frac{q/l}{2\pi\varepsilon_0 r} dr \\ &= -\frac{q/l}{2\pi\varepsilon_0 r} \ln r \Big|_{3s}^{s} \\ &= -\frac{q/l}{2\pi\varepsilon_0 r} \ln \frac{s}{3s} \\ &= +\frac{q/l}{2\pi\varepsilon_0 r} \ln 3. \end{aligned}$$

Step 4: Calculate the capacitance.

Now that we know the potential difference between the plates when they carry charge q, we can find the ratio of charge to voltage:

$$C = \frac{q}{V} = \frac{q}{\left(\frac{q/l}{2\pi\varepsilon_0 r} \ln 3\right)} = \frac{2\pi\varepsilon_0 l}{\ln 3}.$$

The q canceled, reflecting the fact that the ratio of charge to voltage (i.e., the capacitance) stays the same, no matter what charge you dump onto the plates.

(b) The battery pushes current through both the capacitor and the inductor. To formalize this statement, let i_0, i_1, and i_2 denote the current through the battery, inductor, and capacitor, respectively. See my drawing. According to Kirchhoff's 1st law, the current flowing into an intersection equals the current flowing out:

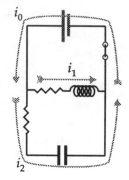

$$i_0 = i_1 + i_2 .$$

So, to calculate the current through the battery, we can find i_1 and i_2, and then add them up.

At this stage, you could use Kirchhoff's 2nd law to write two loop equations. But we can solve parts (b) and (c) without "solving the circuit" completely, by thinking *physically* about how inductors and capacitors behave.

When the current through an inductor *changes*, an EMF gets induced that counteracts the change. Here, when the switch closes, the battery increases the current through the inductor from 0 to some final value. The inductor opposes this change; it wants the current to stay at 0. In the long run, the inductor can't prevent the current from attaining its final value. But the inductor slows down the process; the current increases *gradually* instead of *abruptly*. Therefore, *immediately* after $t = 0$, the current through the inductor is still essentially zero. In symbols, $i_1 \approx 0$ at $t \approx 0$.

So, immediately after $t = 0$, all current flows through the capacitor. To complete this problem, we just need to find that current.

Well, in general, the voltage across a capacitor is $V = Q/C$, where Q is the charge on the capacitor plates. But immediately after $t = 0$, essentially no charge has had time to flow onto the capacitor. So, $Q \approx 0$, and therefore the capacitor generates negligible voltage. At $t \approx 0$, since the capacitor generates no voltage and has no resistance, it plays no role in the circuit. It behaves just like a resistanceless piece of wire. This is no longer true *after* charge flows onto the capacitor. But *immediately* after $t = 0$, the capacitor is essentially uncharged.

This diagram summarizes the argument I just used to "reduce" the circuit.

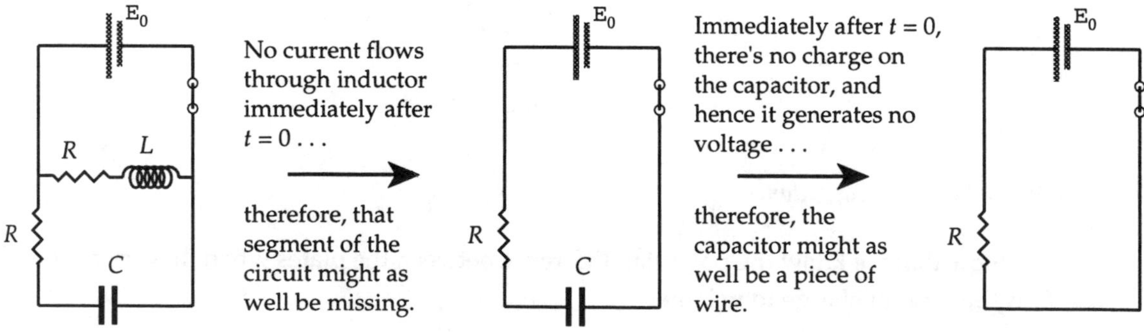

These steps of reasoning apply *only* during the moments immediately after the switch closes. But during those moments, the circuit behaves as if it contained just a battery and a resistor. Therefore, according to Ohm's law, the current through the battery is simply

$$i_0 = \frac{\mathcal{E}_0}{R}.$$

When solving for a current immediately after a switch opens or closes, or a long time afterward, always look for simplifications. You need not always find $i(t)$, the current at arbitrary time.

(c) Taking my own advice, I'll look for ways to "simplify" the circuit.

After current has flowed for a long time, the capacitor is fully charged. In other words, charge no longer flows onto or off of the plates. Therefore, no current flows through the capacitor. Here's why. As explained in Chapter 38, when current flows "through" a capacitor, charge is actually flowing onto one plate and off the other plate. Individual charges don't "jump the gap" between the plates. Therefore, when charge stops flowing onto the positive plate (and off the negative plate), the current stops: $i_2 = 0$ at $t \approx \infty$.

Since $i_2 = 0$, the system "reduces" to a one-loop circuit containing a battery, inductor, and resistor. We can simplify matters even further, by thinking physically about the inductor. As emphasized in part (a), the inductor delays the process by which current increases from 0 to its final value. But the inductor can't change what that "final value" is. To see why, remember that an inductor "reacts" only to changing current. Once the current settles at its final value, and therefore *stops changing*, the inductor generates no EMF. (Mathematically, when the current through the inductor stops changing, $\mathcal{E}_{\text{induced}} = -L di/dt = 0$.) In other words, when the current reaches its final value, the inductor behaves as if it were a resistanceless wire.

Again, I'll summarize all this reasoning in a diagram.

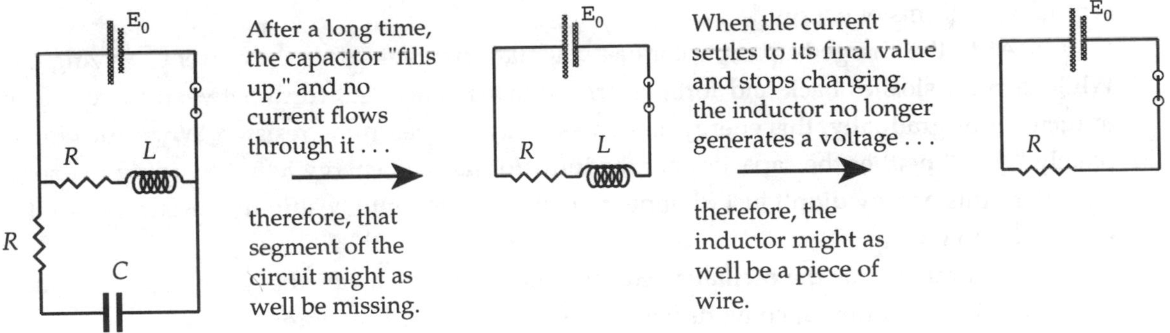
E0
R
L
R
C
After a long time, the capacitor "fills up," and no current flows through it . . .
therefore, that segment of the circuit might as well be missing.
E0
R
L
When the current settles to its final value and stops changing, the inductor no longer generates a voltage . . .
therefore, the inductor might as well be a piece of wire.
E0
R

A long time after the switch closes, the circuit behaves as if it consisted entirely of a battery and one resistor. By Ohm's law,

$$i_0 = \frac{\mathcal{E}_0}{R},$$

the same current through the battery immediately after the switch closes. Don't draw any general conclusions from this coincidence. Between $t = 0$ and $t = \infty$, the current fluctuates. It just happens to start and end at the same value.

(d) Opening the switch cuts off the battery from the rest of the circuit. In other words, the top loop becomes a dead end around which current can't flow. Current can flow only around the bottom loop drawn here.

The bottom loop is an unforced (battery-less) RLC circuit. As explained in Chapter 44, an RLC circuit is analogous to a block on a spring with air resistance. Just as a block resists having its velocity changed (due to inertia), the inductor "resists" having its current changed (due to inductance). Just as air resistance "saps" energy from the spring-and-block system and dissipates that energy as heat, the resistor saps electrical energy from the circuit and dissipates that energy as heat. Just as the spring provides the force that makes the system oscillate back and forth, the capacitor plays the same role in an RLC circuit. This analogy isn't just a rough intuition. It's mathematically exact, as you could confirm by setting up the equation of motion for both systems.

Here's my point. The current in an RLC circuit behaves just like a block's velocity when it's attached to a spring (with air resistance). By graphing that block's velocity, we've automatically graphed the RLC circuit's current.

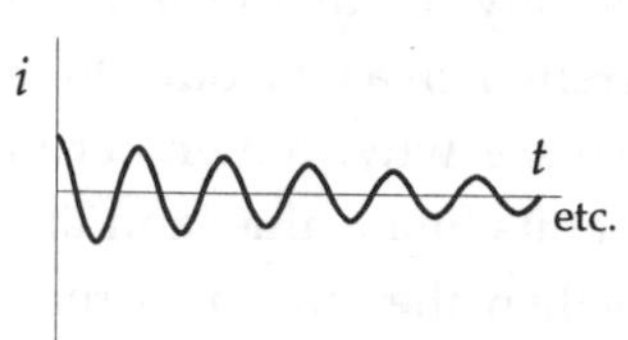

Well, the block's velocity oscillates back and forth sinusoidally. But because of air resistance, the oscillations gradually get smaller and smaller until dying out completely. Similarly, the current in the RLC circuit oscillates sinusoidally. But because of the energy dissipation in the resistor, these oscillations gradually get smaller and smaller until dying out.

(e) By solving the differential equation describing the RLC circuit, we could write down the power dissipated in the resistor as a function of time. Integrating this power over time would then tell us the energy dissipated in the resistor as heat. Fortunately, we can avoid this messy math by thinking conceptually, in terms of energy.

At t_1, the charged-up capacitor has potential energy, as does the current-carrying inductor. While current sloshes back and forth, energy transfers back and forth between the capacitor and inductor. But gradually, this energy dissipates away as heat in the resistor. When the circuit completely "dies," neither the capacitor nor the inductor has any energy left. According to energy conservation, this energy didn't just disappear. It all converted into heat in the resistor, since it had no where else to go.

In summary, as the oscillating current gradually dies away, the energy stored in the capacitor and inductor at time t_1 converts into heat in the resistor. Therefore,

$$\text{Heat dissipated in resistor after } t_1 = U_{\text{capacitor}}(t_1) + U_{\text{inductor}}(t_1)$$

So, to solve this problem, we just need to find the energy stored in the capacitor and inductor at t_1.

Subproblem 1: Find energy stored in capacitor at t_1.

As explained in Chapter 36, a capacitor carries carries potential energy $U_{\text{capacitor}} = \frac{1}{2}QV$. Since $C = Q/V$, we can rewrite this expression as $\frac{1}{2}\frac{Q^2}{C}$ or $\frac{1}{2}CV^2$. In part (a), we already found the capacitance. So, to figure out $U_{\text{capacitor}}$, we just need to know the charge on the capacitor, or the voltage across it, at time t_1.

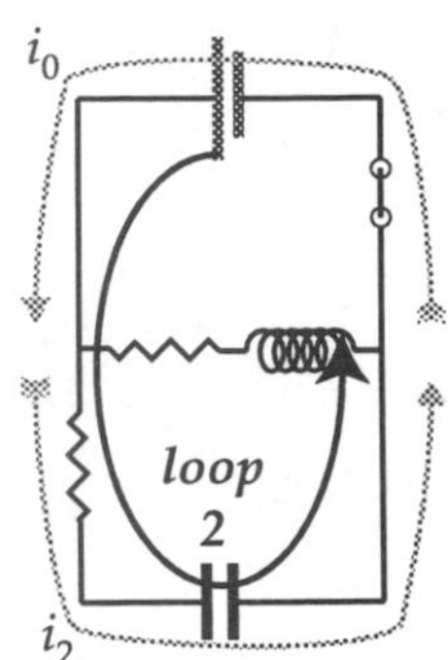

Well, using Kirchhoff's 2nd law, we can write an equation about "loop 2" in this diagram. Loop 2 goes around the perimeter of the circuit. (We could equally well use the "bottom" loop.) With luck, we can solve this loop equation for Q at time t_1.

According to Kirchhoff's 2nd law, the total voltage change around the whole loop is zero. I'll walk counterclockwise, starting from the upper right corner. So first, I walk through the battery, from negative to positive terminal. This changes my potential by $\Delta V = \mathcal{E}_0$. Next, I walk through a resistor, in the same direction that i_2 flows. So, $\Delta V = -i_2R$. Finally, I walk through the capacitor, from the positive to the negative plate. Because a positive test charge would naturally "fall" in this direction, my potential decreases: $\Delta V = -Q/C$.

Since these potential changes must sum to zero,

Loop 2 $$\mathcal{E}_0 - i_2R - \frac{Q}{C} = 0.$$

Fortunately, we can simplify this equation. At t_1, the capacitor is "fully charged," and hence, no current flows across it: $i_2(t_1) = 0$. Therefore, at time t_1, the loop 2 equation simplifies to

$$\mathcal{E}_0 - \frac{Q}{C} = 0,$$

and hence $Q = C\mathcal{E}_0$. Therefore, the potential energy stored in the capacitor at t_1 is

$$U_{\text{capacitor}}(t_1) = \frac{1}{2}\frac{Q^2}{C} = \frac{1}{2}\frac{(C\mathcal{E}_0)^2}{C} = \frac{1}{2}C\mathcal{E}_0^2.$$

If you obtained this expression simply by plugging the battery's voltage into $\frac{1}{2}CV^2$, then you got a little lucky. Only by writing the loop 2 equation, and realizing that $i_2(t_1) = 0$, could you confirm that the voltage across the capacitor equals the battery's voltage. When $i_2 \neq 0$—i.e., when the capacitor is charging or discharging—the capacitor's voltage does *not* equal the battery's voltage.

So far, we've figured out the energy stored in the capacitor at t_1. Now we must calculate the energy stored in the inductor at that same moment. The sum of these potential energies equals the total heat dissipated in the resistor, after time t_1.

Subproblem 2: Find the energy stored in inductor at t_1.

A current-carrying inductor "stores up" $U_{\text{inductor}} = \frac{1}{2}Li^2$. In part (c), we found the current through the inductor (and battery) a long time after $t = 0$. It's $i = \frac{\mathcal{E}_0}{R}$. So, at time t_1,

$$U_{\text{inductor}}(t_1) = \frac{1}{2}Li^2 = \frac{1}{2}L\left(\frac{\mathcal{E}_0}{R}\right)^2.$$

Therefore, at t_1, the total energy stored in the circuit is

$$\begin{aligned} U_{\text{total}}(t_1) &= U_{\text{capacitor}}(t_1) + U_{\text{inductor}}(t_1) \\ &= \frac{1}{2}C\mathcal{E}_0^2 + \frac{1}{2}L\left(\frac{\mathcal{E}_0}{R}\right)^2 \\ &= \frac{1}{2}\left(C + \frac{L}{R^2}\right)\mathcal{E}_0^2, \end{aligned}$$

where "C" denotes the capacitance found in part (a). As discussed above, all this energy gradually dissipates away as heat in the resistor. So, $U_{\text{total}}(t_1)$ is the total heat generated after t_1.

REVIEW PROBLEM 2-4

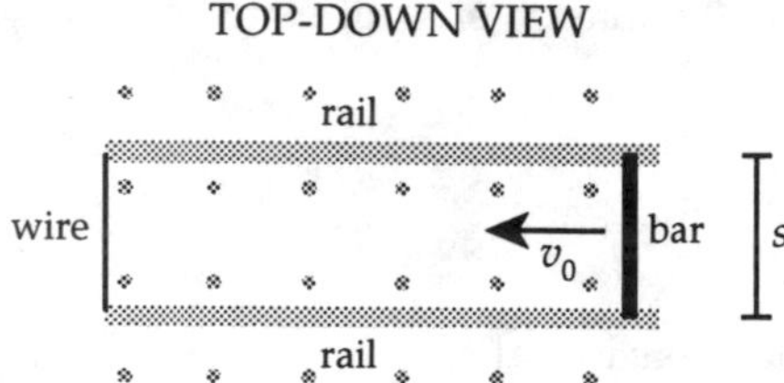

A resistanceless wire connects two metal rails of negligible resistance, as drawn here. The rails are $s = 0.10$ meters apart. A bar of length s, mass $M = 0.020$ kg, and resistance $R = 5.0 \times 10^{-4}\ \Omega$ is pushed leftward at constant speed $v_0 = 9.0$ m/s. The bar slides frictionlessly over the rails. The room is filled with a uniform magnetic field of strength $B = 3.0 \times 10^{-4}$ T, pointing upward.

(a) While the bar gets pushed at constant speed v_0, does the current through the bar increase, decrease, or stay the same? Explain

(b) Find the current through the bar immediately after it starts getting pushed at speed v_0. In which direction does this current flow?

(c) What force must you exert to keep the bar moving at this speed?

(d) Now, instead of pushing the bar continuously, someone gives it a quick leftward shove and then lets it slide freely. Describe the subsequent motion of the bar in words. Does it speed up or slow down?

(e) (*Very hard*) By writing and solving a differential equation, find the bar's speed as a function of time.

ANSWER

(a) To answer this, we must review *why* current flows through the bar. The bar, rails and wire form a complete loop around which current can flow. According to Faraday's law, the induced EMF around that loop is given by the *rate of change of magnetic flux* through the loop:

Faraday's law $$\mathcal{E}_{\text{induced}} = -\frac{d\Phi_B}{dt}$$

If the bar were sitting still, the flux through the loop would stay constant, and hence no voltage or current would get induced. But here, the bar slides leftward, reducing the area of the current loop. Therefore, the flux through the loop decreases. This changing flux induces an EMF, which drives current around the loop.

Let's get more exact. Since the bar slides at constant speed, the area of the current loop decreases at a constant rate. Therefore, the magnetic flux decreases at a constant rate. In other words, $d\Phi_B/dt$ is constant. So, $\mathcal{E}_{\text{induced}}$ is constant, too. As a result, the induced current is constant.

(b) The reasoning of part (a) suggests a strategy for finding the current. First, figure out the rate at which the flux decreases, due to the bar's leftward motion. That rate of change of flux tells us the induced EMF. Finally, calculate the corresponding current, using Ohm's law, $\mathcal{E}_{\text{induced}} = i_{\text{induced}}R$.

To find the rate of change of flux, $d\Phi_B/dt$, I'll write an expression for the flux at arbitrary time. Then I'll differentiate it with respect to t.

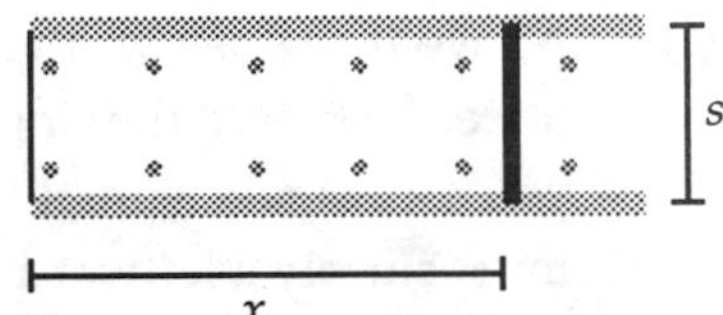

In general, $\Phi_B = \int \mathbf{B}\cdot d\mathbf{A}$. Let x denote the bar's distance from the wire. The current loop has area $A = sx$. Since the magnetic field points in the same direction that the current loop "faces," the dot product $\mathbf{B}\cdot d\mathbf{A}$ reduces to a regular product. And since the field is uniform over the entire loop, we can pull it outside the integral:

$$\Phi_B = \int \mathbf{B}\cdot d\mathbf{A} = B\int dA = BA = Bsx.$$

When differentiating this flux with respect to time, B and s are constants, since they don't change. But x changes as the bar slides leftward:

$$\frac{d\Phi_B}{dt} = \frac{d(Bsx)}{dt} = Bs\frac{dx}{dt} = -Bsv_0.$$

In the last step of reasoning, I remembered that velocity is the rate of change of position. The minus sign indicates that a positive velocity corresponds to a *decreasing* x, in this particular case: $v = -dx/dt$.

Now that we know the rate of change of flux, Faraday's law gives us the induced voltage. It's

$$\mathcal{E}_{\text{induced}} = -\frac{d\Phi_B}{dt} = Bsv_0$$

Therefore, by Ohm's law,

$$i_{\text{induced}} = \frac{Bsv_0}{R} = \frac{(3.0\times10^{-4}T)(0.10\text{m})(9.0\,\text{m/s})}{5.0\times10^{-4}\Omega} = 0.54\text{ A}.$$

To figure out whether this current flows clockwise or counterclockwise around the loop, use Lenz' law. According to Lenz' law, the induced current generates its own magnetic field that "counteracts" the changing flux. Here, the flux through the loop "points" upward, but decreases. The induced current tries to oppose this decrease, by generating its own upward magnetic field.

So far, I've found that the induced current produces an upward (out-of-the-page) magnetic field. Now I'll figure out whether the induced current must flow clockwise or counterclockwise, to create such a field. According to the right-hand rule shortcut, if you point your right thumb in the direction of current flow, then your curled fingertips represent he magnetic field lines produced by that wire. So, point your thumb along the bar, and curl your fingertips toward the middle of the current loop. By trial and error, you can confirm that your thumb must point up the page, to ensure that your fingertips curl out of the page. Therefore, current must flow counterclockwise, as drawn here, in order to generate an out-of-the-page field.

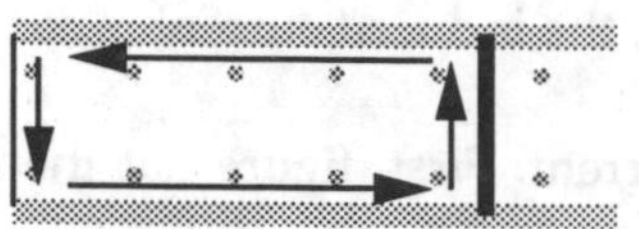

Arrows show direction of current.

(c) As we'll see below, the bar experiences a rightward magnetic force. To keep it moving at constant velocity (zero acceleration), you must make sure it feels *zero* net force. In other words, your must supply a leftward push force exactly big enough to cancel the rightward magnetic force. In symbols, $F_{push} = F_{mag}$. So, by figuring out the magnetic force, we've automatically figured out the push force.

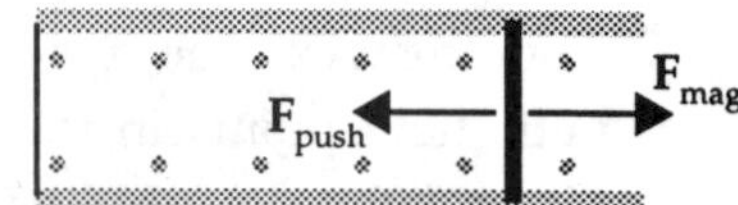

As shown in Chapter 39, a tiny "piece" of a current-carrying wire or bar feels magnetic force

$$d\mathbf{F}_{mag} = i d\mathbf{l} \times \mathbf{B},$$

where $d\mathbf{l}$ points in the direction of current flow. Let's begin by confirming that the magnetic force points rightward, as claimed above. Apply the right-hand rule to the cross product $d\mathbf{l} \times \mathbf{B}$. First, point your fingers in the direction that current flows through the rod. Then, curl your fingertips in the direction of **B**, out of the page. Your thumb now points rightward. That's the direction of $d\mathbf{F}_{mag}$.

To calculate the magnitude of this force, notice that $d\mathbf{l}$ points perpendicular to **B**, and hence, the magnitude of this cross product reduces to a regular product. Also, since **B** is uniform over the whole bar, we can pull it outside the integral:

$$F_{mag} = \int_{bar} i(dl)B = iB\int_0^s dl = iBs,$$

since the bar has length s. Now just substitute in the numbers, including our part (b) answer, to get

$$F_{push} = F_{mag} = iBs = (0.54\,\text{A})(3.0\times10^{-4}\,\text{T})(0.10\,\text{m}) = 1.62\times10^{-5}\,\text{N},$$

which you could easily supply by blowing lightly on the bar. By the way, in symbolic form, this answer is $F_{mag} = iBs = \frac{Bsv_0}{R}Bs = \frac{B^2s^2v_0}{R}$.

(d) The bar slows down. As just discussed, the leftward-moving bar feels a rightward magnetic force. Unless you cancel this force by pushing the bar, the bar gradually loses velocity.

Does it slow down at a steady rate (constant acceleration)? No! As just found, the magnetic force is proportional to the bar's speed: $F_{\text{mag}} = \frac{B^2s^2v}{R}$. Therefore, as the bar slows down, it experiences a smaller and smaller force. The bar keeps slowing down, but more and more gradually.

(e) To solve for the bar's motion, use Newton's 2nd law, $F_{\text{net}} = Ma$. Here, since you stop pushing the bar after the initial nudge, the magnetic force is the net force. If we count leftward as the positive direction, then this force is negative: $F_{\text{net}} = -\frac{B^2s^2v}{R}$. So,

$$F_{\text{net}} = Ma$$
$$-\frac{B^2s^2v}{R} = \frac{dv}{dt},$$

since acceleration is the rate of change of velocity.

We can solve this differential equation by separation of variables. Divide through by v, and multiply through by dt. Then integrate both sides, remembering that the velocity starts out at v_0, and the time starts out at $t = 0$:

$$-\int_0^t \frac{B^2s^2}{R}dt = \int_{v_0}^{v} M\frac{dv}{v}$$
$$-\frac{B^2s^2}{R}\int_0^t dt = M\int_{v_0}^{v}\frac{dv}{v}$$
$$-\frac{B^2s^2}{R}t = M\ln\frac{dv}{v_0}.$$

Now divide through by M, and then exponentiate both sides, recalling that $e^{\ln z} = z$. This gives

$$e^{-\frac{B^2s^2t}{MR}} = \frac{v}{v_0},$$

and hence

$$v = v_0e^{-\frac{B^2s^2t}{MR}}.$$

The velocity "decays" exponentially in time. Here, the velocity decreases *very* slowly. To see why, rewrite v in the form

$$v = v_0e^{-t/\tau},$$

where in this case the "time constant" t is given by

$$\tau = \frac{MR}{B^2 s^2} = \frac{(0.020\text{ kg})(5.0\times10^{-4}\Omega)}{(3.0\times10^{-4}\text{ T})^2(0.10\text{ m})^2} = 11000\text{ seconds,}$$

about 3 hours. That's how long it would take the bar to slow down to $1/e = 37\%$ of its initial speed. Realistically, the bar would reach the end of the rails before slowing down noticeably.

Practice Midterm 1: Electrostatics

50

CHAPTER

◆ *Please read the "study hints" in Chapter 47 before taking this practice exam.*

Three problems. You have 60 minutes.

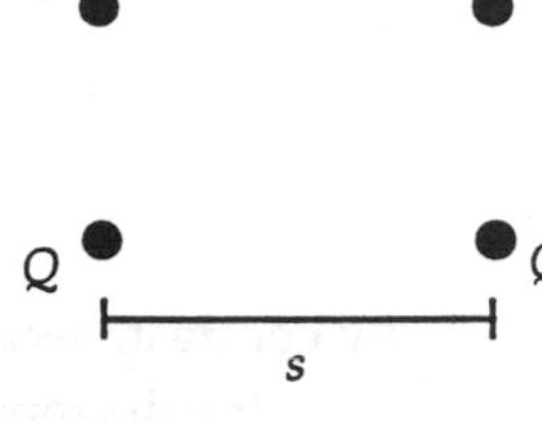

1) Consider a square of side-length s, with a point charge Q at each of the four corners.
   (a) What is the magnitude and direction of the force on the charge in the upper right corner?
   (b) How much work would it take to push that charge to the center of the square, assuming the other three charges are fixed in place?

2) A point charge Q_0 sits on top of my desk. I also have a drinking glass, currently filled with root beer. The glass is a cylinder of radius R and height h, with a closed bottom but an open top.

   glass

   desk

   point charge

   Suddenly, I finish drinking the root beer, turn the glass upside down, and place it over the point charge. The point charge is at the center of the circle defined by the rim of the glass.

   What is the electric flux through the glass?

   Hint: If you find yourself doing lots of math, you're missing something.

3) Two concentric conducting spherical shells are hooked up to a 9.0-volt battery (not drawn), so that the inner shell acquires positive charge and the outer shell acquires negative charge. The capacitance of the system is $C = 1.0 \times 10^{-12}$ F. The outer shell has twice the radius of the inner shell.

   Spherical shells

   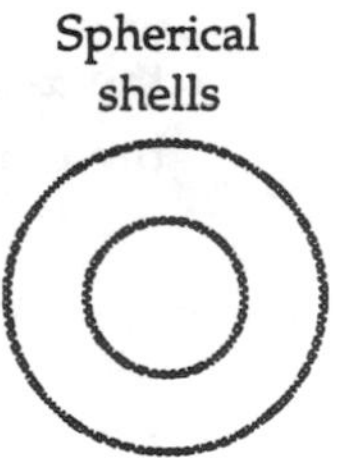

   (a) How much charge builds up on the inner shell?
   (b) A particle of charge $q_0 = 2.0\ \mu\text{C}$ (microcoulombs) and mass $m = 4.0 \times 10^{-7}$ kg gets fired from the outer shell towards the inner shell. What is the smallest initial velocity the particle must have, to ensure that it reaches the inner shell? Neglect gravity.
   (c) What's the radius of the inner sphere? Derive any formulas you use from basic laws.

ANSWER 1

(a) I'll number the charges, for easy reference. We want the net force on charge 1. Well, the other three charges all generate an electric field at charge 1. Charge 1 "feels" the sum of these fields. In other words, the net force on charge 1 is the sum of the forces exerted by charges 2 through 4. Notice that we can reason in terms of fields or forces, since they relate in a simple way: F = QE. I'll work with the fields.

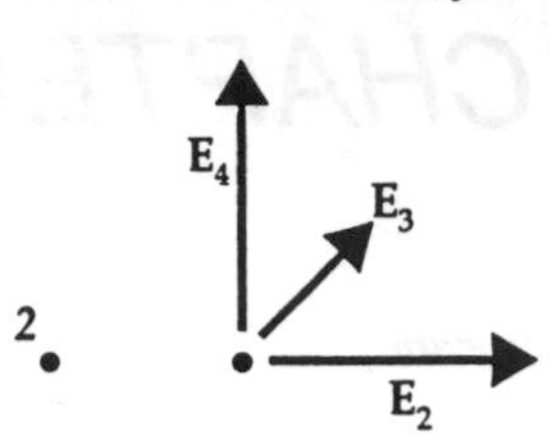

The charges repel each other. As this diagram shows, charge 2 pushes charge 1 rightward; charge 4 pushes charge 1 upward; and charge 3 pushes it at a 45° angle. To obtain the net field (or force), we can solve for these three fields (or forces) *separately*, and then add up, vectorially. *You can't add these fields or forces like numbers, because they point in different directions.*

According to Coulomb's law, since charges 2 and 4 sit the same distance from charge 1, they generate equally strong fields at charge 1:

$$E_2 = E_4 = \frac{q}{4\pi\varepsilon_0 r^2} = \frac{Q}{4\pi\varepsilon_0 s^2}.$$

By contrast, because charge 3 sits farther away, it generates a weaker field at charge 1. Specifically, since the diagonal of this square has length $\sqrt{2}s$, charge 3 produces a field

$$E_3 = \frac{Q}{4\pi\varepsilon_0(\sqrt{2}s)^2} = \frac{1}{2}\frac{Q}{4\pi\varepsilon_0 s^2},$$

half the field strength generated by charge 2 or charge 4.

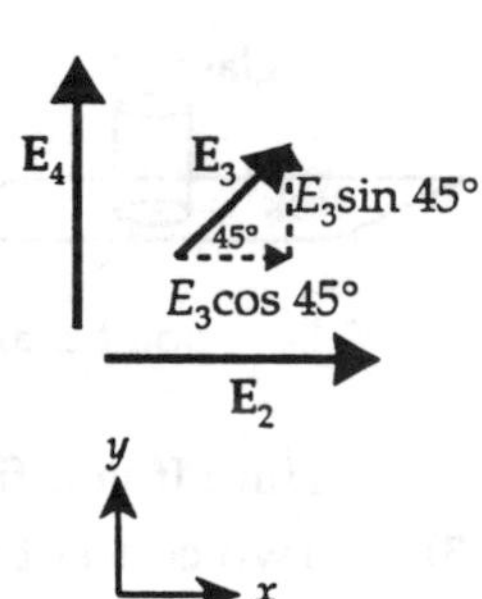

Now that we know the three fields, let's add them vectorially. Intuitively, the net field points in the same direction as $\mathbf{E}_3$.

I'll work in terms of x- and y-coordinates. Since $\mathbf{E}_2$ points purely in the x-direction, and $\mathbf{E}_4$ points purely in the y-direction, we only need to break up $\mathbf{E}_3$. I've done so here. From the diagram,

$$E_x = E_2 + E_3 \cos 45° = \frac{Q}{4\pi\varepsilon_0 s^2} + \frac{1}{2}\frac{Q}{4\pi\varepsilon_0 s^2}\cos 45°$$

$$E_y = E_4 + E_3 \cos 45° = \frac{Q}{4\pi\varepsilon_0 s^2} + \frac{1}{2}\frac{Q}{4\pi\varepsilon_0 s^2}\sin 45° = E_x,$$

since cos 45° = sin 45°.

By Pythagorean theorem, the overall field has strength

$$E = \sqrt{E_x^2 + E_y^2}$$

$$= \sqrt{E_x^2 + E_y^2} \qquad \text{[since } E_y = E_x\text{]}$$

$$= \sqrt{2}\,E_x$$

$$= \sqrt{2}\left(\frac{Q}{4\pi\varepsilon_0 s^2} + \frac{1}{2}\frac{Q}{4\pi\varepsilon_0 s^2}\cos 45^\circ\right)$$

$$= \frac{Q}{4\pi\varepsilon_0 s^2}\left(\sqrt{2} + \frac{1}{2}\right).$$

Therefore, charge 1 feels a net force

$$F = QE = \frac{Q^2}{4\pi\varepsilon_0 s^2}\left(\sqrt{2} + \frac{1}{2}\right).$$

This force points at a 45° angle.

(b) Given the force from part (a), you might try using $W = \int \mathbf{F} \cdot d\mathbf{s}$. But the electric force on charge 1 *changes* as it gets pushed toward the center of the square. We would need to calculate the force as a function of position, and integrate over it. Another technique might get us to the answer more quickly.

Recall the relationship between work and potential energy: $W_{\text{by you}} = \Delta U$. In words, if you raise an object's energy by 5 joules, then you've performed 5 joules of work on it. Furthermore, "potential" relates to potential energy by $U = qV$. Therefore, by calculating the potential difference between the upper right hand corner and the center of the square, we can find the work, using

$$W = \Delta U = Q\Delta V.$$

As emphasized in part (a), the total electric field at charge 1 equals the sum of the fields generated by charges 2 through 4. This *superposition principle* applies to potentials, too. The potential at charge 1 equals the sum of the potentials generated by charges 2 through 4. And since potentials are scalars, not vectors, they're easier to add.

To make further progress, we need to know the potential generated by a point charge. Starting with the basic formula for potential, and substituting in the electric field generated by a point charge, I get

$$V_{\text{point charge}} = -\int \mathbf{E}_{\text{point charge}} \cdot d\mathbf{s} = -\int \frac{Q}{4\pi\varepsilon_0 r^2} dr = \frac{Q}{4\pi\varepsilon_0 r},$$

where I followed the textbook convention of setting the integration constant equal to 0.

OK, now we can calculate the initial and final potential of charge 1. Initially, it sits a distance s from charges 2 and 4, and a distance $\sqrt{2}s$ from charge 3. So,

$$V_i = V_{2i} + V_{3i} + V_{4i}$$

$$= \frac{Q}{4\pi\varepsilon_0 s} + \frac{Q}{4\pi\varepsilon_0 \sqrt{2}s} + \frac{Q}{4\pi\varepsilon_0 s}$$

$$= \frac{Q}{4\pi\varepsilon_0 s}\left(2 + \frac{1}{\sqrt{2}}\right).$$

INITIAL

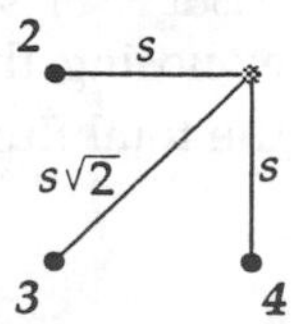

2 s
$s\sqrt{2}$ s
3 4

At the center of the square, charge 1 sits half a diagonal away from each of the other three charges. That distance is $\frac{1}{2}\sqrt{2}s = \frac{1}{\sqrt{2}}s$. So, charge 1 experiences a final potential

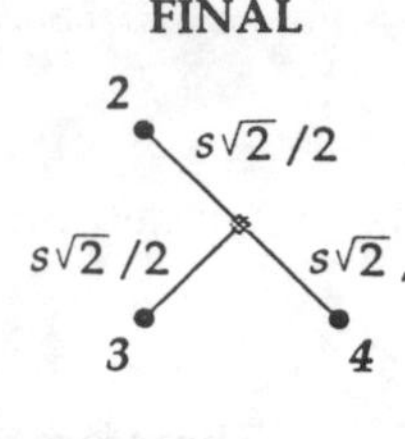

$$\begin{aligned} V_f &= V_{2f} + V_{3f} + V_{4f} \\ &= \frac{Q}{4\pi\varepsilon_0(s/\sqrt{2})} + \frac{Q}{4\pi\varepsilon_0(s/\sqrt{2})} + \frac{Q}{4\pi\varepsilon_0(s/\sqrt{2})} \\ &= \frac{Q}{4\pi\varepsilon_0 s} 3\sqrt{2}. \end{aligned}$$

Therefore, the work needed to move charge 1 from the upper right corner to the center is

$$\begin{aligned} W = \Delta U &= Q\Delta V \\ &= Q(V_f - V_i) \\ &= Q\frac{Q}{4\pi\varepsilon_0 s}\left(3\sqrt{2} - 2 - \frac{1}{\sqrt{2}}\right) \\ &= \frac{Q^2}{4\pi\varepsilon_0 s}\left(\frac{5\sqrt{2}}{2} - 2\right). \end{aligned}$$

ANSWER 2

By seeing the "trick," you can solve this problem in 30 seconds. Otherwise, a messy flux integral beckons.

Remember the Gaussian water analogy? A charge "spewing out" electric field lines is like a sprinkler spewing out water streamlines. More specifically, a point charge is like a small spherical sprinkler shooting water equally in all directions. In this analogy, gravity "turns off"; the water flows in straight lines, without arcing down. Of course, electric field lines pass through desks and glasses, while water streamlines do not. But for the sake of this analogy, pretend that water streamlines flow right through solid objects. The issue is this: How much water per second—i.e., how much water flux—passes through the glass? For instance, if the "point sprinkler" emits 10 liters of water per minute, how much of it flows through the glass?

Picturing the situation three dimensionally, you can see that exactly half the water emitted by the sprinkler flows through the glass. Here's one way of visualizing this. By symmetry, half of the sprinkler's water goes above the plane of the desk, while half the water goes below the plane of the desk. And *all* the streamlines flowing above the desk pass through the glass, as this picture shows. So, exactly half of the total water flows through the glass.

By analogy, exactly half the field lines generated by the point charge—and hence, half the electric flux—passes through the glass.

Half the field lines pass through the glass.

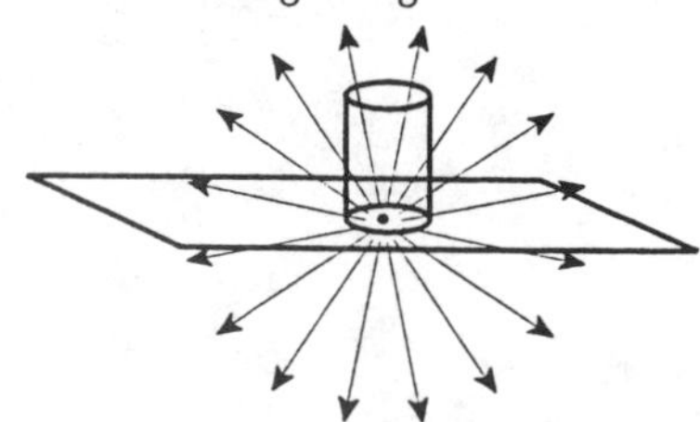

According to Gauss' law, the point charge emits total flux $\Phi_{total} = Q_0/\varepsilon_0$. (That's the flux passing through *any* Gaussian surface enclosing the point charge.) Therefore, since the glass captures half the total flux,

$$\Phi_{\text{through glass}} = \frac{1}{2}\Phi_{\text{total}} = \frac{Q_0}{2\varepsilon_0}.$$

ANSWER 3

(a) By definition, capacitance specifies the "willingness" of a system to carry charge, *per volt of applied potential difference*. In symbols,

$$C = \frac{Q}{V}.$$

Here, we know the potential difference between the two shells. We also know the capacitance. Just solve for the charge, to get

$$Q = CV = (1.0 \times 10^{-12}\ \text{F})(9.0\ \text{V}) = 9.0 \times 10^{-12}\ \text{coulombs}.$$

(b) The inner shell carries positive charge, while the outer shell holds negative charge. Therefore, between the two shells, the electric field points radially outward. As the positive particle travels inward, this field exerts a "backwards" force, slowing the particle down. The initial velocity must be large enough to ensure that the particle reaches the inner shell before stopping completely.

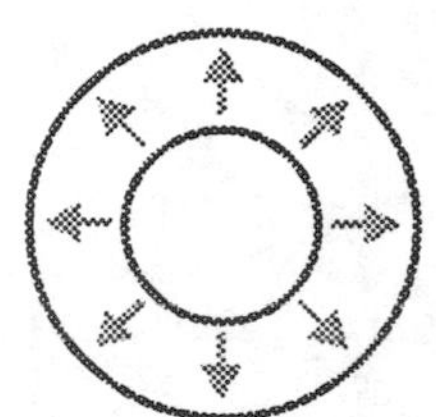

Given this intuitive description, you might try solving for the particle's deceleration, and then using kinematics. But the electric field between the shells isn't uniform. For this reason, the particle feels a non-constant force, and hence, a non-constant deceleration. We can't use those constant-acceleration kinematic formulas.

Last semester, when force-and-acceleration reasoning got too messy, we resorted to conservation laws. In this case, conservation of energy makes life easier. As the particle travels inward, it loses kinetic energy, but gains potential energy. Since we're looking for the smallest initial speed needed to ensure that the particle reaches the inner shell, we may assume that the particle *barely* reaches the inner shell. In other words, the particle loses all its velocity by the time it reaches the inner shell: $v_f = 0$. So, the final kinetic energy goes to 0. But the initial kinetic energy is $K_0 = \frac{1}{2}mv_0^2$. We're solving for v_0.

INITIAL

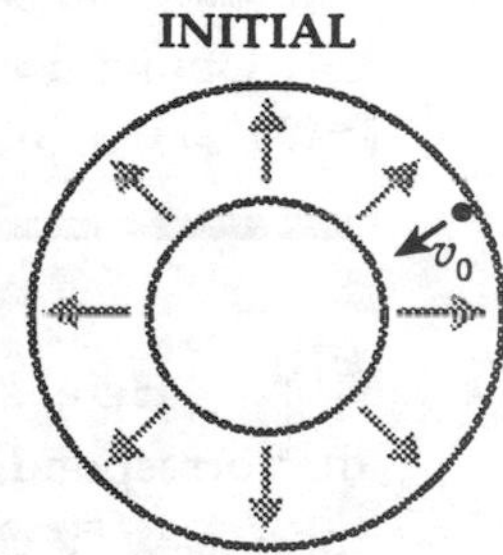

FINAL

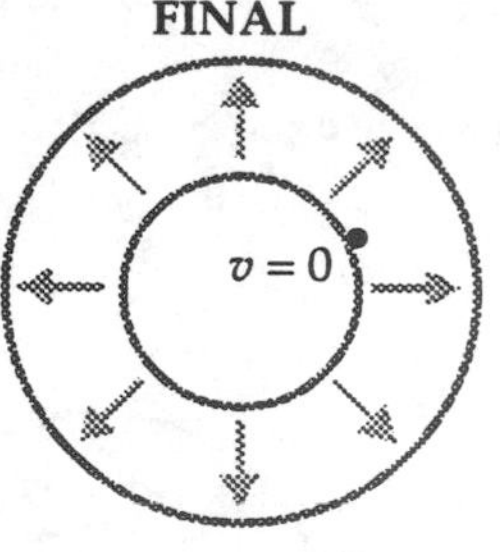

What about potential energy? Well, "potential" is the potential energy of a 1-coulomb test charge. In this case, the potential difference between the shells is $\Delta V = 9.0$ volts. Here's what that means, physically. If a 1-coulomb test charge travels from the outer to the inner shell, its potential energy rises by 9.0 J. Since this particle has charge q_0, its potential energy rises by

$$\Delta U = q_0 \Delta V = (2.0 \times 10^{-6}\ \text{C})(9.0\ \text{V}) = 1.8 \times 10^{-5}\ \text{J}.$$

OK, now apply energy conservation.

$$K_0 + U_0 = K_f + U_f$$

$$\frac{1}{2}mv_0^2 + U_0 = 0 + U_f.$$

Notice that v_0 depends only on the *change* in potential energy, $U_f - U_0 = \Delta U$. We just calculated that ΔU. So, we can solve for v_0 to get

$$v_0 = \sqrt{\frac{2(U_f + U_0)}{m}} = \sqrt{\frac{2\Delta U}{m}} = \sqrt{\frac{2(1.8 \times 10^{-5}\ \text{J})}{4.0 \times 10^{-7}\ \text{kg}}} = 9.5\ \text{m/s}.$$

(c) Normally, a problem would ask for the capacitance, *given* the radii of the two spherical shells. Here, the question asks for the radii of the shells, *given* the capacitance. So, this is a backwards version of the standard problem. By invoking the usual capacitance-finding strategy, we can find the capacitance in terms of the spheres' radii. Then, we can "invert" this equation to solve for the radii in terms of the capacitance. So, the meat of this problem is deriving a symbolic expression for the capacitance, in terms of the spheres' radii.

Before summarizing the capacitance-finding strategy, let me review the underlying idea. Capacitance is charge per voltage: $C = q/V$. This ratio stays the same, no matter how you fiddle with the charge or voltage. Therefore, when calculating the capacitance, you may assign any arbitrary charge to the plates. You need not use the "correct" charge, because the ratio q/V comes out the same no matter what.

Capacitance-finding strategy

1) Assign arbitrary charge q to your plates, giving them equal and opposite charges.
2) Use Gauss' law or other techniques to find the electric field between the plates.
3) Obtain the potential dzifference between the plates, using $V = -\int \mathbf{E} \cdot d\mathbf{s}$.
4) Get the capacitance using $C = q/V$. The arbitrary charge q you chose in step 1 should cancel out.

Here, since you know the voltage, you might be tempted to take V as given and solve for the corresponding q. But it's easier to "pretend" you know q and solve for the corresponding V. The ratio $C = q/V$ comes out the same either way.

Since I applied this strategy to spherical shells in question 36-3, I'll condense my presentation.

Steps 1 and 2: Assign charge $\pm q$ to the plates, and figure out the electric field between them.

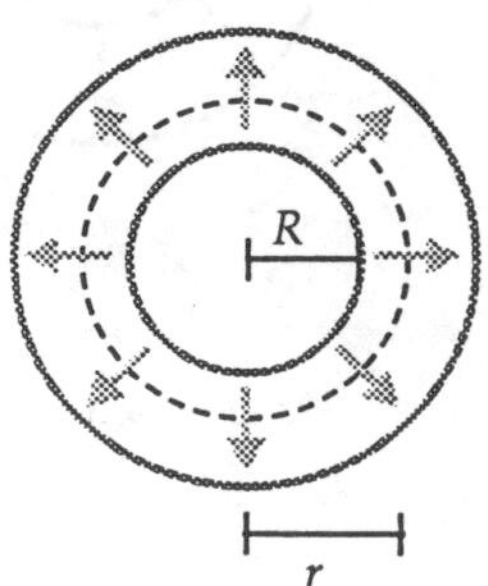

From now on, assume the inner shell carries charge $+q$, while the outer shell carries charge $-q$. The inner and outer shell have radius R and $2R$. Ultimately, we're solving for R.

We want the electric field *between* the spherical shells, because this field determines the potential difference. Fortunately, the spherical symmetry invites us to try Gauss' law.

By choosing a spherical Gaussian surface, drawn here with a dashed line, I ensure (by symmetry) that the electric field has the same strength over the entire surface. This Gaussian sphere has arbitrary radius r, where r is between R and $2R$. Remember, when integrating over E to find the potential difference, we'll need to know $E(r)$, the field as a general function of r. Therefore, you should reason in terms of a general "r," instead of plugging in the radius of one of the spheres.

Now we can apply Gauss' law,

$$\oint \mathbf{E}\cdot d\mathbf{A} = \frac{q_{\text{encl}}}{\varepsilon_0}.$$

The Gaussian surface encloses only the inner shell. So, it encloses charge $+q$.

Let's simplify the left-hand side of Gauss' law. Since **E** points radially outward, the same direction in which the Gaussian surface "faces," $\mathbf{E}\cdot d\mathbf{A}$ reduces to a regular product. Furthermore, the field strength stays constant over the whole surface, allowing us to pull E outside the integral:

$$\int_{\text{Gaussian sphere}} \mathbf{E}\cdot d\mathbf{A} = E\int_{\text{Gaussian sphere}} dA = EA_{\text{Gaussian sphere}} = E(4\pi r^2).$$

So, we have

$$\oint \mathbf{E}\cdot d\mathbf{A} = \frac{q_{\text{encl}}}{\varepsilon_0}$$

$$E(4\pi r^2) = \frac{q}{\varepsilon_0},$$

and hence, $E = \dfrac{q}{4\pi\varepsilon_0 r^2}$.

Step 3: Find the potential difference between the plates, using the electric field just calculated.

You can ensure that $V = -\int \mathbf{E}\cdot d\mathbf{s}$ comes out positive by integrating from the *negative* to the *positive* capacitor plate. I'll choose a straight radial path of integration, from the outer to the inner shell. So, "$d\mathbf{s}$" becomes dr, and the dot product reduces to a regular product:

$$\begin{aligned} V &= -\int_{\text{outer shell}}^{\text{inner shell}} \mathbf{E}\cdot d\mathbf{s} \\ &= -\int_{2R}^{R} \frac{q}{4\pi\varepsilon_0 r^2}\,dr \\ &= \frac{q}{4\pi\varepsilon_0 r}\Big|_{2R}^{R} \\ &= \frac{q}{4\pi\varepsilon_0 R} - \frac{q}{4\pi\varepsilon_0 (2R)} \\ &= \frac{q}{8\pi\varepsilon_0 R}. \end{aligned}$$

A common mistake is to integrate from 0 to R, or from R to 0. But the limits of integration *have* to be R and $2R$. Here's why. In the electric field formula $E = \dfrac{q}{4\pi\varepsilon_0 r^2}$, the r refers to the distance *from the center of the spheres*. Since we're integrating over dr, our limits of integration must be consistent with this meaning of r.

Step 4: Calculate the capacitance, with $C = q/V$.

Using our result from step 3, we get

$$C = \frac{q}{V} = \frac{q}{\left(\frac{q}{8\pi\varepsilon_0 R}\right)} = 8\pi\varepsilon_0 R$$

The q's cancel, proving that it doesn't matter how much charge we assign to the plates, for the purposes of finding the capacitance.

OK, now for the final twist. Normally, we'd solve for C in terms of a given R. But here, we're solving for R in terms of the given C:

$$R = \frac{C}{8\pi\varepsilon_0} = \frac{1.0 \times 10^{-12}\ \text{F}}{8\pi(8.85 \times 10^{-12}\ \text{F/m})} = 4.5 \times 10^{-3}\ \text{m},$$

about half a centimeter.

51 CHAPTER

Practice Midterm 2: More Electrostatics

Three questions, 70 minutes.

1) Consider an equilateral triangle made of three thin rods, each of charge Q and length L. The charge on each rod is uniformly distributed.

(a) What is the electric field at the center of the triangle? Show your reasoning.

(b) What is the electric potential at the center of the triangle? (Take a point infinitely far away as the "reference point" where $V = 0$.) You need not complete the math, if it gets too messy. But set everything up.

2) These equipotential lines are "created" by an unpictured but funky charge distribution. A, B and C are imaginary points, not charges.

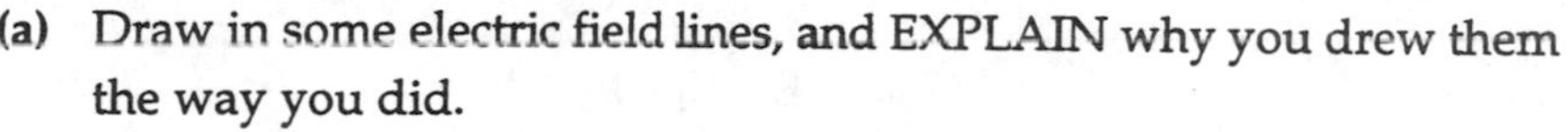

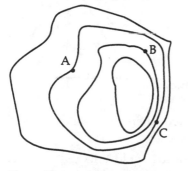

(a) Draw in some electric field lines, and EXPLAIN why you drew them the way you did.

(b) At which of the three labeled points is the electric field largest? Explain your answer.

(c) The potential of point B is higher than the potential of point A. Suppose you push a positive test charge along a straight line from A to C. Would the total work exerted by you on the test charge be positive, negative, or zero? Explain.

3) Two square metal plates, of side-length L and plate separation D, are hooked up to opposite terminals of a battery. As a result, the top and bottom plate acquire charge $+Q_0$ and $-Q_0$, respectively. *The battery is then disconnected, and left disconnected for the remainder of this problem.* By the way, $D \ll L$.

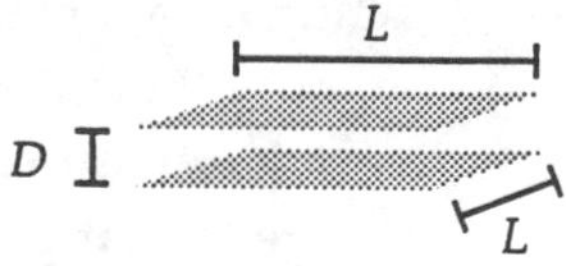

(a) What was the voltage of the battery? Please derive your answer from basic laws, instead of plugging in a pre-derived formula. Express your answer in terms of L, D, Q_0, and any universal constants you need.

(b) A particle of charge $-q_1$ is placed between the plates, and released. Instead of rising or falling, it just floats there. What is its mass? Express your answer in terms of L, D, Q_0, q_1, and other constants you need.

(c) The particle from part (b) is removed. A dielectric slab of length L, width L, and thickness $D/2$ is placed between the plates, resting on the bottom plate. So, the dielectric fills half the space between the plates, as drawn

Part (c)

SIDE VIEW, looking between the two plates "edge-on"

dielectric

here. When the dielectric gets inserted, does the potential difference between the plates go up, go down, or stay the same? Explain.

(d) (*Very hard*) How much work would it take to remove the dielectric slab from between the plates?

ANSWER 1

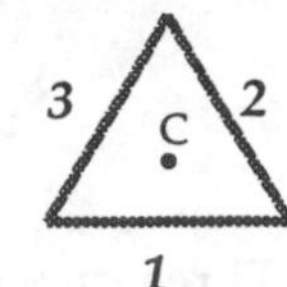

(a) The electric field vanishes at the center, because the three rods create fields that cancel out. To explain why, I'll label the three rods, and I'll let C denote the center.

Consider rod 1, the bottom of the triangle. The field it creates at C points straight up, with no leftward or rightward component. That's because the left half of the rod creates a field with a rightward component; and the right half of the rod creates a field with a leftward component. By symmetry, those rightward and leftward components cancel. In summary, $\mathbf{E}_1$ points directly away from the midpoint of rod 1.

By equivalent reasoning, rods 2 and 3 also create fields pointing directly away from their midpoints. These three field contributions ($\mathbf{E}_1$, $\mathbf{E}_2$, and $\mathbf{E}_3$) all have the same magnitude, because the three rods are identical, and point C is the same distance from each one.

Point C feels a total electric field

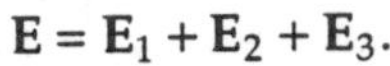

$$\mathbf{E} = \mathbf{E}_1 + \mathbf{E}_2 + \mathbf{E}_3.$$

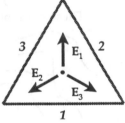

As this diagram shows, the three field vectors add up to zero; they cancel out.

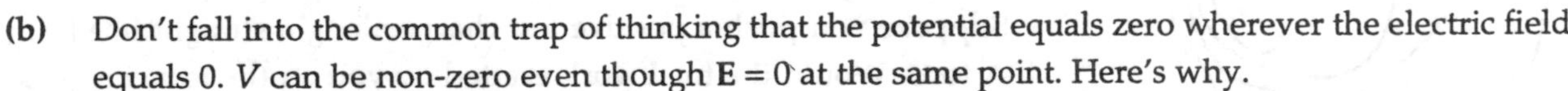

(b) Don't fall into the common trap of thinking that the potential equals zero wherever the electric field equals 0. V can be non-zero even though $\mathbf{E} = 0$ at the same point. Here's why.

From the fundamental electrostatic force law, $\mathbf{F} = q\mathbf{E}$, the electric field tells us the force acting on a 1-coulomb test charge. For instance, a test charge at point C would feel no force.

"Potential" means something different. By definition, V_C is the potential energy of a 1-coulomb test charge at C. And remember, potential energy is the work needed to move the test charge to that point. Therefore, V_C is the *work required to push a 1-coulomb test charge to the center of the triangle, from very far away.*

At first glance, you might think that pushing a test charge to point C takes no work, because $\mathbf{E} = 0$ at the center. But the electric field is nonzero everywhere *except* point C. When pushing the test charge towards point C, you fight that field. Therefore, you do work while moving the test charge to point C. In other words, point C has non-zero potential.

Given this insight, you could solve the problem by first calculating the electric field around the triangle, and then evaluating $V = -\int_{\infty}^{C} \mathbf{E} \cdot d\mathbf{s}$. But we can work (excuse the pun) reasoning as follows. The superposition principle applies not just to fields, but also to potentials. By adding up the potentials created by each rod, you can find the potential generated by the whole triangle:

$$V_C = V_1 + V_2 + V_3,$$

where V_1 denotes the potential produced by rod 1.

To simplify things further, remember that each rod is identical, and point C sits equidistant from the midpoint of each rod, along an imaginary perpendicular line through the rod's midpoint. Therefore, by symmetry, all three rods generate the same potential at C: $V_1 = V_2 = V_3$. Combine this insight with the above superposition equation, to get

$$\begin{aligned} V_C &= V_1 + V_2 + V_3 \\ &= 3V_1. \end{aligned}$$

Fortunately, potentials are scalars, not vectors. Unlike electric fields, potentials can't point in different directions. That's why you can add them up without worrying about vector components.

In summary, the potential at point C is simply three times the potential generated by any one of the rods. Therefore, we must solve the following subproblem.

Subproblem: Calculate the potential due to a single rod.

We could find the field generated by a single rod, and then evaluate $V = -\int_{\infty}^{C} \mathbf{E}_{\text{due to one rod}} \cdot d\mathbf{s}$. But it's easier to calculate the potential "directly". Think of the rod as a bunch of point charges. A single point charge q creates a potential $V = \frac{q}{4\pi\varepsilon_0 r}$. By breaking the rod into tiny "pieces" of charge, and writing down the tiny potential generated by each piece, we can then add up those potentials to obtain the total potential. You used a similar strategy in Chapter 33, to calculate electric fields.

Calculating the potential due to a continuous charge distribution

1) Blacken an arbitrary infinitesimal "piece" of the charge distribution. Don't choose a special piece, such as the end or the midpoint. Label the distance from that piece to the point at which you're finding the potential.
2) Write an expression for dq, the infinitesimal charge of that piece.
3) Using $dV = \frac{dq}{4\pi\varepsilon_0 r}$, calculate the infinitesimal potential at the point of interest, generated by the "piece" you just colored in.
4) Integrate dV over the whole charge distribution, to obtain the total potential.

I'll now apply this strategy to rod 1, the bottom of the triangle.

Step 1: Color in dq.

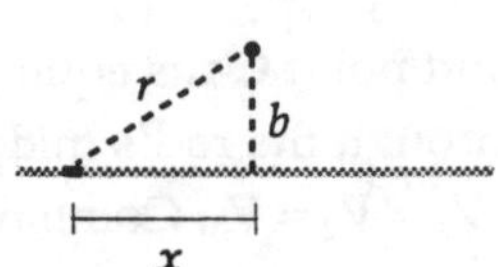

This blackened piece of the rod sits an arbitrary distance x from the midpoint, and a distance $r=\sqrt{x^2+b^2}$ from point C. the piecehas infinitesimal length dx. Later, I'll figure out b in terms of the rod's length. But for now, I'll keep working in terms of "b".

Step 2: Write dq in terms of dx, dy, or whatever.

Since the rod carries charge Q uniformly spread over length L, the charge per length (linear charge density) is $\lambda = Q/L$. Therefore, a piece of length dx contains charge

$$
\begin{aligned}
dq &= \frac{\text{charge}}{\text{length}} \times (\text{length of the blackened piece}) \\
&= \lambda dx \\
&= \frac{Q}{L} dx.
\end{aligned}
$$

Step 3: Write expression for dV, the tiny potential generated by the blackened piece.

Given our results from steps 1 and 2, this is straightforward:

$$
\begin{aligned}
dV &= \frac{dq}{4\pi\varepsilon_0 r} \\
&= \frac{(Q/L)dx}{4\pi\varepsilon_0\sqrt{x^2+b^2}}.
\end{aligned}
$$

If this were a field, we'd break it into vector components. But a potential has no direction. It's just a number.

Step 4: Add up these tiny potentials to obtain the whole potential, by integrating over the rod.

In the above diagram, "x" is measured from the midpoint of the rod. Therefore, when we integrate over the rod, x ranges from $-L/2$ to $+L/2$:

$$
\begin{aligned}
V = \int dV &= \int_{-L/2}^{L/2} \frac{(Q/L)dx}{4\pi\varepsilon_0\sqrt{x^2+b^2}} \\
&= \frac{Q}{4\pi\varepsilon_0 L}\int_{-L/2}^{L/2}\frac{dx}{\sqrt{x^2+b^2}} \\
&= \ldots \text{ difficult integral} \ldots.
\end{aligned}
$$

That's the potential generated at point C by one rod. Many instructors would supply a table of integrals, or would allow you to leave the integral unevaluated. Below, I'll evaluate the integral. But first, let me complete the problem.

As shown above, the total potential at point C is three times the potential generated by one rod:

$$V_C = 3V_{\text{one rod}} = 3\frac{Q}{4\pi\varepsilon_0 L}\int_{-L/2}^{L/2}\frac{dx}{\sqrt{x^2+b^2}}.$$

We know Q and L. but what about b, the distance from the midpoint of a rod to the center of the triangle? Well, as this diagram shows, tan $30° = \frac{b}{L/2}$, and hence

$$b = \frac{L\tan 30°}{2} = \frac{L}{2\sqrt{3}}.$$

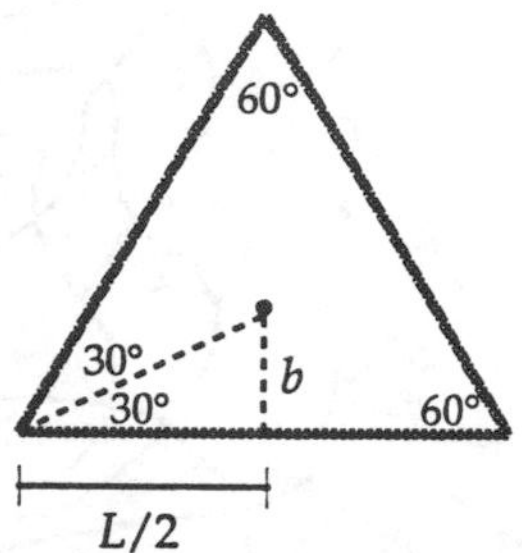

We're done. On a real midterm, you probably wouldn't need to evaluate the integral. But just for the record . . .

Mathematical appendix (for calculus fiends). Use hyperbolic trig substitution, by setting $x = b\sinh\theta$. Then $dx = b\cosh\theta\, d\theta$. We must also change the limits of integration. Since $x = b\sinh\theta$, we can find the "angle" corresponding to $x = L/2$ by setting $x = L/2$ and solving for θ. Well, from $\frac{L}{2} = b\sinh\theta = \frac{L}{2\sqrt{3}}\sinh\theta$, we immediately get $\sinh\theta = \sqrt{3}$, and hence, $\theta = \sinh^{-1}\sqrt{3}$. So, the limits of integration are $\pm\sinh^{-1}\sqrt{3}$. Putting all this together, I get

$$\int_{-L/2}^{L/2}\frac{dx}{\sqrt{x^2+b^2}} = \int_{-\sinh^{-1}\sqrt{3}}^{\sinh^{-1}\sqrt{3}}\frac{b\cosh\theta\, d\theta}{\sqrt{b^2\sinh^2\theta+b^2}}$$

$$= \int_{-\sinh^{-1}\sqrt{3}}^{\sinh^{-1}\sqrt{3}}\frac{\cosh\theta\, d\theta}{\sqrt{\sinh^2\theta+1}}$$

$$= \int_{-\sinh^{-1}\sqrt{3}}^{\sinh^{-1}\sqrt{3}}\frac{\cosh\theta\, d\theta}{\sqrt{\cosh\theta}} \quad [\sinh^2\theta+1=\cosh^2\theta]$$

$$= \int_{-\sinh^{-1}\sqrt{3}}^{\sinh^{-1}\sqrt{3}} d\theta$$

$$= 2\sinh^{-1}\sqrt{3}$$

$$\approx 2.6.$$

ANSWER 2

(a) We must figure out the relationship between equipotential lines and field lines. To gain insight, imagine pushing a test charge along an equipotential line. Does this require any work? No, because the potential energy ($U = qV$) doesn't change. Since $W_{\text{by you}} = \Delta U$; if an object's potential energy stays constant, then no net work is getting done on the object.

Let me explain why this matters. As we saw last semester, $W = \int \mathbf{F}\cdot ds = \int F_{\|} ds$, where $F_{\|}$ denotes the force component *along* the direction of motion. As just shown, no work gets done on the particle as it moves along an equipotential line. Therefore, $F_{\|}$ along an equipotential line must equal zero. In other words, no component of the electric force points along an equipotential line. Since $\mathbf{F}_{elec} = q\mathbf{E}$, it follows that no component of the electric field points along an equipotential line. In otherwords, electric field lines must point *perpendicular* to equipotential lines.

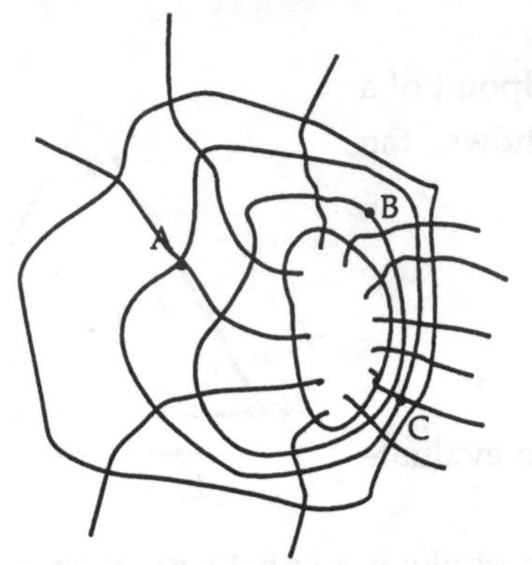

Let me phrase this argument in a different way. Suppose, hypothetically, that a component of the electric field points along an "equipotential" line. Then, you must do work to push a charged particle along that "equipotential" line, because you have to "fight" the field. By doing work on the particle, you change its potential energy. So, the "equipotential" line isn't really an equipotential line! This contradiction proves that equipotential and field lines *must* point perpendicular to each other.

If this argument seems obscure, think of contour lines on a contour map. If you walk along a contour line, you stay at the same height, and hence, at the same gravitational potential energy. So, on a hill, contour (equipotential) lines are horizontal. But the gravitational field lines point vertically, i.e., straight downward. This goes to show that field lines point perpendicular to equipotential lines.

(b) The electric field is strongest at C, where the equipotential lines are most dense. To understand why, we must think about the relationship between field, potential, and work.

Suppose the potential difference between two neighboring equipotential lines is $\Delta V = 5$ volts. Then, to push a 1-coulomb test charge from one equipotential line to the next, you must do work $W = \Delta U = q\Delta V = 5$ joules. If the equipotential lines lie close together, then you perform those 5 joules of work over a short distance. Intuitively, if you do lots of work pushing something a short distance, then you must be exerting a big force. In other words, you must be fighting against a large electric field.

By contrast, if neighboring equipotential lines are far apart, then you perform those *same* 5 joules of work over a *longer* distance. Since $W = \int \mathbf{F}\cdot ds$, and the distance is large, the force must be small. Intuitively, if you perform 5 joules of work while pushing the particle over a long distance, you must be fighting a weak electric field.

This goes to show that the electric field is strongest where the equipotential lines are closely spaced, and weakest where the equipotential lines are spread out. Mathematically, the electric field strength is proportional to the rate of change of potential over distance:

$$E_x = -\partial V/\partial_x \; E_y = -\partial V/\partial y,$$

and so on. The more quickly the potential changes (i.e., the closer the equipotential lines), the stronger the field.

Once again, the contour map analogy can help. On a contour map, closely-spaced contour lines indicate that you gain a lot of height by walking only a short distance. In other words, closely-spaced contour lines indicate steep hills. On a steep hill, hikers must "fight" a large component of gravity. By contrast, spread-out contour lines mean that you can walk a long way without going up or down too far, i.e., without fighting too much gravity.

(c) Since A and C sit on the same equipotential line, they share the same potential: $V_A = V_C$. Therefore, when the test charge moves from A to C, its potential energy ($U = qV$) neither rises nor falls. For this reason, it takes no work to push the test charge from A to C: $W = \Delta U = 0$.

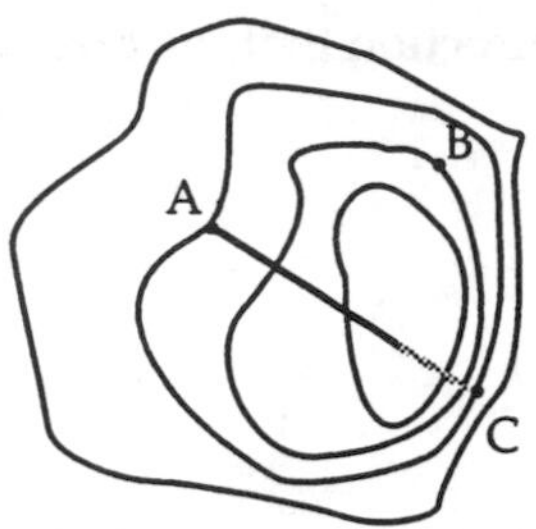

If this seems weird, look at the accompanying diagram. I divide the test charge's journey into two segments, corresponding to black and gray lines. Along the black segment, the potential rises. Therefore, you must perform positive work. By contrast, along the gray segment, the potential falls. Therefore, along this segment, you perform negative work. Intuitively, the particle naturally "falls" down the hill, dragging you with it. Since the particle does work on you, you perform negative work on it. The negative work from the gray segment cancels the positive work from the black segment.

Here's another way to reach the same conclusion. If you push the test charge from A to C along the equipotential line, you do no work. As emphasized in Chapter 35, however, the potential energy difference between two points does *not* depend on your path. Therefore, pushing the test charge from A to C along *any* path takes no work.

ANSWER 3

(a) Because the battery charged up the capacitor, the potential difference between the plates must equal the battery's voltage. For instance, a 9-volt battery creates a 9-volt potential difference between the plates.

So, we must find the plate-to-plate potential difference. Well, the charges on the plates generate an electric field, which we can figure out using Gauss' law. Integrating over that field will give us the potential difference: $V = -\int_{\text{negative plate}}^{\text{positive plate}} \mathbf{E}$

Subproblem: Use Gauss' law to find the electric field between the plates

When solving for an electric field using Gauss' law,

$$\oint \mathbf{E} \cdot d\mathbf{A} = \frac{q_{\text{encl}}}{\varepsilon_0},$$

always start by sketching in some field lines, and drawing a "good" Gaussian surface. A "good" Gaussian surface is one for which the electric field is *constant* over (part of) the surface, thereby allowing you to pull E outside the flux integral, $\int \mathbf{E} \cdot d\mathbf{A}$.

As we've seen, the field between two parallel plates points directly from one plate to the other, with approximately uniform strength (except near the fringes). Outside the space between the plates, the field drops nearly to zero. Therefore, the "pill box" drawn here is a good Gaussian sur-

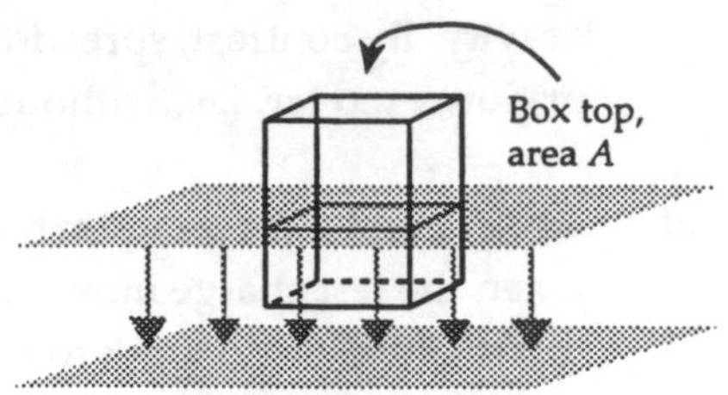

face. No flux passes through the box top, because the field vanishes there. More subtly, no flux passes through the sides of the box. The field lines skim along the sides without piercing *through* the sides. (Mathematically, each side of the box "faces" at a 90° angle to electric field, and hence $\mathbf{E}\cdot d\mathbf{A} = EdA\cos 90° = 0$.) So, *flux passes only through the bottom of the box.*

Because the field has uniform strength and points in the same direction that the box bottom "faces," we can pull E outside the flux integral:

$$\oint \mathbf{E}\cdot d\mathbf{A} \int_{\text{box bottom}} \mathbf{E}\cdot d\mathbf{A} = E\int_{\text{box bottom}} dA = EA,$$

where "A" denotes the arbitrary area of the box top and box bottom.

Now I'll simplify the other side of Gauss' law, by calculating the charge enclosed by this Gaussian box. The box "captures" area A of the top plate. So, we need to know how much charge lives in that area. Well, Q_0 spreads itself out uniformly over the whole plate, which has area L^2. So, the plate carries surface charge density (charge per area) $\sigma = Q_0/L^2$. Therefore, this Gaussian box encloses charge

$$q_{\text{encl}} = \frac{\text{charge}}{\text{area}} \times (\text{area enclosed by the Gaussian box}) = \sigma A = \frac{Q_0}{L^2}A.$$

Substitute all this information into Gauss' law to get

$$\oint \mathbf{E}\cdot d\mathbf{A} = \frac{q_{\text{encl}}}{\varepsilon_0}$$

$$EA = \frac{Q_0}{\varepsilon_0 L^2}A.$$

The A's cancel, proving that the area of the Gaussian pill box makes no difference. This leaves us with

$$E = \frac{Q_0}{\varepsilon_0 L^2}.$$

End of Gauss' law subproblem

Given this field, we can easily calculate the potential difference between the plates, which are a distance D apart. By integrating from the negative to the positive plate, you ensure that V comes out positive. Since $\mathbf{E}$ and $d\mathbf{s}$ point in opposite directions, as drawn here, the dot product $\mathbf{E}\cdot d\mathbf{s}$ reduces to a *negative* regular product:

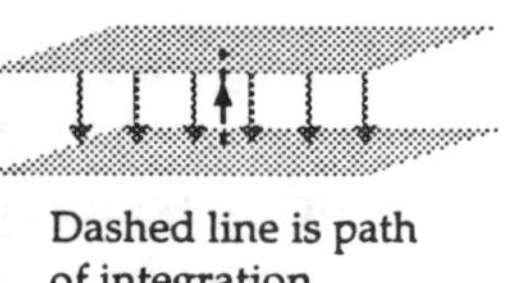

Dashed line is path of integration.

$$V = -\int_{\text{negative plate}}^{\text{positive plate}} \mathbf{E} \cdot d\mathbf{s}$$

$$= \int_0^D E ds$$

$$= E\int_0^D ds \qquad \text{[since } E \text{ is uniform between the plates]}$$

$$= ED$$

$$= \frac{Q_0}{\varepsilon_0 L^2} D.$$

That's the potential difference between the plates. Therefore, it's also the voltage of the battery used to charge up the plates.

(b) Because the point charge is negative, the downward electric field pushes it *upward*. This upward electrostatic force must cancel the downward gravitational force.

$$F_{\text{elec}} = F_{\text{grav}}$$

$$q_1 E = mg.$$

$\uparrow q_1 E$

$\downarrow mg$

Solve for m, and use our part (a) expression for the electric field, to get

$$m = \frac{q_1 E}{g} = \frac{q_1 Q_0}{\varepsilon_0 L^2 g}.$$

Students often get confused about the minus sign in $\mathbf{F} = q\mathbf{E}$. Since $\mathbf{E}$ points downward, the minus sign in front of q_1 "flips" the direction of that force, to upward. So, if we take upward as the positive direction, then $\Sigma F_y = q_1 E - mg = 0$. That's just another way of writing $q_1 E = mg$.

(c) If the battery were still connected, it would "enforce" a fixed potential difference between the plates, by adjusting the charge on the plates. But here, the battery stays disconnected. Charges have no way to flow onto or off of the plates. Therefore, the *charge* on each plate stays the same, at $\pm Q_0$. The potential difference might change, however.

In fact, inserting the dielectric *decreases* the potential difference. Here's why. The potential difference between the plates is the work needed to push a 1-coulomb test charge from one plate to the other: $V = -\int \mathbf{E} \cdot d\mathbf{s}$. Therefore, if the average electric field strength goes down, then so does V, because fighting the field takes less work. As discussed in Chapter 36, a dielectric *dampens* electric fields. More precisely, the polarized charges on the surfaces of the dielectric create their own electric field pointing *oppositely* to the external field. As a result, the total field inside the dielectric is smaller than the external field. Therefore, inserting a dielectric lowers the average electric field between the plates. This reduces the potential difference, $V = -\int \mathbf{E} \cdot d\mathbf{s}$.

(d) To solve this, think about potential energy. Let me summarize the reasoning, before diving into details. As shown in part (c), inserting the dielectric lowers the potential difference between the plates, thereby lowering the potential energy of the system. Therefore, removing the dielectric raises the system's potential energy, by some amount ΔU. For instance, if ΔU happens to equal 5 joules, then it takes 5 joules of work to pull the slab out. In general, $W = \Delta U$. So, by figuring out the system's potential energy, both with and without the slab, we can calculate the work needed to remove the slab:

$$W = \Delta U = U_{\text{withold slab}} - U_{\text{with slab inserted}}$$

You can think of the potential energy as being stored in the electric field, or as being stored in the capacitor plates. I'll work in terms of the plates.

Subproblem 1: Find $U_{\text{without slab}}$.

A capacitor "carries" potential energy $U = \frac{1}{2}QV$. (In question 36-2d, I explained where the unexpected factor of 1/2 comes from). From part (a), we know the potential difference between the plates, with no dielecteric. It's $V = \frac{Q_0}{\varepsilon_0 L^2}D$. So,

$$U_{\text{withold slab}} = \frac{1}{2}Q_0 V = \frac{1}{2}Q_0\left(\frac{Q_0}{\varepsilon_0 L^2}D\right) = \frac{Q_0^2}{2\varepsilon_0 L^2}.$$

Subproblem 2: Find $U_{\text{with slab inserted}}$

Let V' denote the potential difference between the plates with the slab inserted. Since inserting the dielectric doesn't change the charge on the plates, $U_{\text{with slab inserted}}$ equals $\frac{1}{2}Q_0 V'$. So, to finish this subproblem, we just need to find V'.

Well, in a highly symmetric situation such as this, the dielectric "dampens" the field inside of it by a factor of κ, but it doesn't affect the field outside of it. So, in the top half of the space between the plates, the electric field stays the same as it was before $E_{\text{outside dielectric}} = \frac{Q_0}{\varepsilon_0 L^2}$. By contrast, the field inside the dielectric gets damped (reduced) by a factor of κ: $E_{\text{outside dielectric}} = \frac{Q_0}{\kappa \varepsilon_0 L^2}$.

Now that we know E everywhere between the plates, let's integrate over it to find the potential difference. Break the overall path of integration into two "subpaths." The first subpath, of length $D/2$, goes from the bottom plate to the top edge of the dielectric. The second subpath, also of length $D/2$, goes through air, from the top edge of the dielectric to the top plate. In both cases, **E** points in the opposite direction from $d\mathbf{s}$, and hence the dot product reduces to a negative regular product. So,

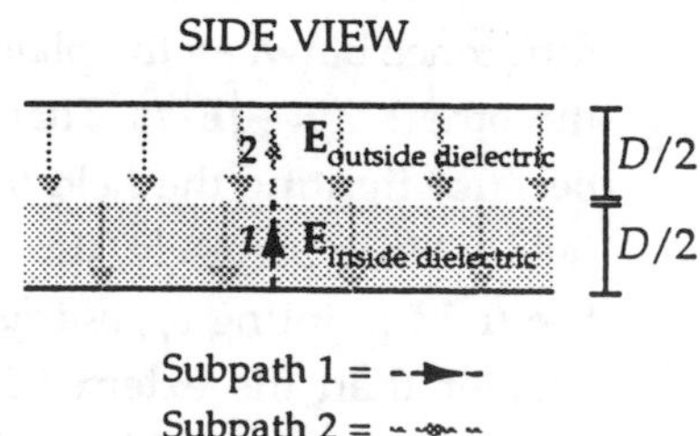

$$V' = -\int_{\text{negative plate}}^{\text{positive plate}} \mathbf{E}\cdot d\mathbf{s} = -\int_{\text{subpath 1}} \mathbf{E}\cdot d\mathbf{s} \int_{\text{subpath 2}} \mathbf{E}\cdot d\mathbf{s}$$

$$= \int_0^{D/2} E_{\text{inside dielectric}}\, ds + \int_{D/2}^{D} E_{\text{outside dielectric}}\, ds$$

$$= \int_0^{D/2} \frac{Q_0}{\kappa\varepsilon_0 L^2} ds + \int_{D/2}^{D} \frac{Q_0}{\varepsilon_0 L^2} ds$$

$$= \frac{Q_0}{\kappa\varepsilon_0 L^2} \int_0^{D/2} ds + \frac{Q_0}{\varepsilon_0 L^2} \int_{D/2}^{D} ds$$

$$= \frac{Q_0}{\kappa\varepsilon_0 L^2}\frac{D}{2} + \frac{Q_0}{\varepsilon_0 L^2}\frac{D}{2}.$$

So, with the slab inserted, the capacitor has potential energy

$$U_{\text{with slab inserted}} = \frac{1}{2}Q_0 V' = \frac{1}{2}Q_0\left(\frac{Q_0}{\varepsilon_0 L^2}\frac{D}{2} + \frac{Q_0}{\kappa\varepsilon_0 L^2}\frac{D}{2}\right) = \frac{Q_0^2}{2\varepsilon_0 L^2}\left(\frac{1}{2} + \frac{1}{2\kappa}\right).$$

Now that we know the capacitor's potential energy, both with and without the slab, we can figure out the work needed to remove the slab. It's

$$W = \Delta U = U_{\text{without slab}} - U_{\text{with slab inserted}}$$

$$= \frac{Q_0^2 D}{2\varepsilon_0 L^2} - \frac{Q_0^2 D}{2\varepsilon_0 L^2}\left(\frac{1}{2} + \frac{1}{2\kappa}\right)$$

$$= \frac{Q_0^2 D}{2\varepsilon_0 L^2}\left(\frac{1}{2} - \frac{1}{2\kappa}\right)$$

$$= \frac{Q_0^2 D}{4\varepsilon_0 L^2}\left(1 - \frac{1}{\kappa}\right).$$

So, the higher the dielectric constant, the more work it takes to yank out the slab. Intuitively, a high κ means that the dielectric dampens the electric field substantially. The bigger this damping, the *lower* the potential energy of the system becomes when you insert the slab. As a result, pulling the slab out from between the plates becomes more difficult.

This makes even more sense when you think of the system's potential energy as being stored in the electric field. By damping the field between the plates, the dielectric decreases the potential energy.

52 CHAPTER

Practice Midterm 3: Electrostatics and Circuits

Please look at my "study hints" in Chapter 47, before taking this practice exam.

Three questions, 60 minutes.

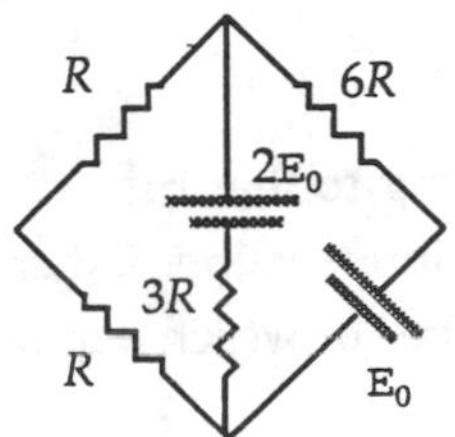

1) In this circuit, $R = 10\ \Omega$ and $\mathcal{E}_0 = 5.0$ V.

(a) What current flows through the 6*R* resistor?

(b) One of the *R* resistors is snipped in half with wire cutters. Now what current flows through the 6*R* resistor?

2) Consider this solid sphere of radius *R* and positive charge Q_0. The charge is uniformly spread out over the whole volume of the sphere.

Stuck inside this sphere, a distance $R/2$ from the center, is a tiny particle of charge Q_1. In this picture, the imaginary dashed line passes through Q_1 and through the center of the sphere.

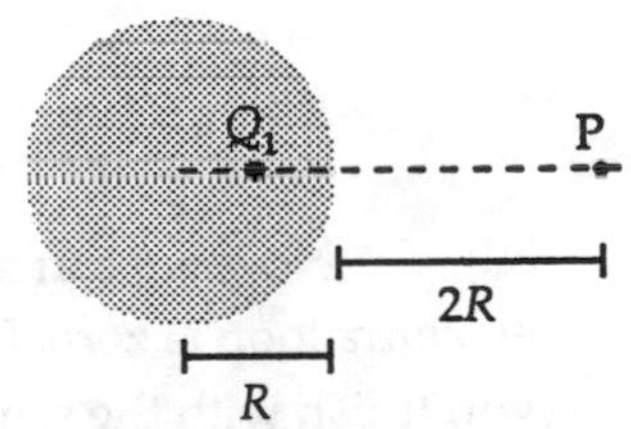

(a) What is the electric force felt by Q_1?

(b) Now a positive test charge q_2 is placed at point P. What force does it feel? Point P is on the dashed line, a distance 2*R* from the *edge* of the sphere.

(c) Now q_2 is released from rest. How fast is it moving when it gets *very* far from the charged sphere? The test charge has mass *M*. Neglect gravity.

3) In this circuit, the capacitor is initially uncharged. The battery gets "turned on" at time $t = 0$. We know that $\mathcal{E} = 9.0$ volts, $R = 1.0\ \Omega$, and $C = 2.0 \times 10^{-12}$ F.

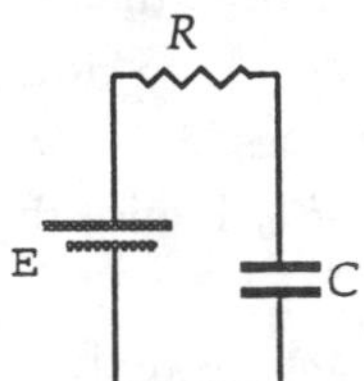

(a) *Immediately* after $t = 0$, what current flows through the resistor?

(b) At that moment, is the current through the capacitor greater than, less than, or equal to your part (a) answer? Explain.

(c) A long time later, what current flows through the resistor?

(d) At that later time, what is the charge on the capacitor?

ANSWER 1

(a) Because different currents might flow through the two batteries, we can't "reduce" the circuit using the rules about resistors in series and in parallel. Remember, circuit reduction tries to "simplify" the circuit into a single battery and a single resistor. So, when different currents flow through the two batteries, we can't "combine" them into a single battery with a single current.

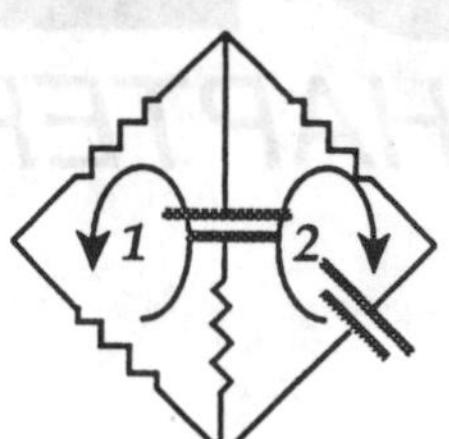

Instead, break the circuit into loops, and apply Kirchhoff's laws. For my loops, I'll choose the left and right "triangles," as indicated here. But you could equally well choose other loops.

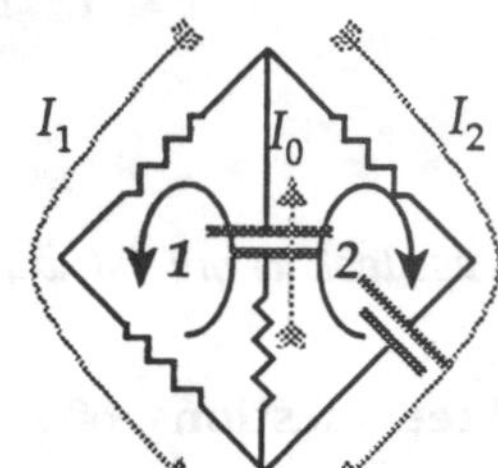

After selecting your loops, label the current in each segment of the circuit, guessing its direction. I've assumed that the $2\mathcal{E}_0$ battery "wins" in both loops, causing current to flow counterclockwise around loop 1 and clockwise around loop 2. If these guesses are wrong, I'll get minus signs in my final answers. So, an incorrect guess about a current direction can't hurt you in the long run.

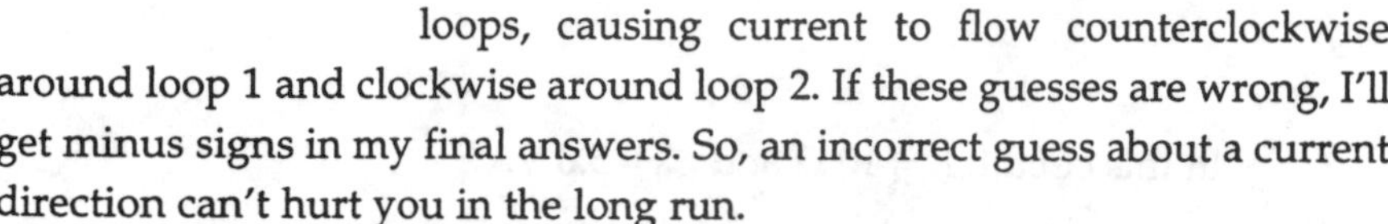

Let's start with Kirchhoff's 1st law, the "intersection rule." According to that rule, the current flowing into an intersection must equal the current flowing out of an intersection. Otherwise, charge would "built up" in the intersection, or mysteriously disappear, neither of which happens in a steady-current circuit.

Apply this rule to the intersection above the $2\mathcal{E}_0$ battery. I_0 flows into that intersection, while I_1 and I_2 flow out. So,

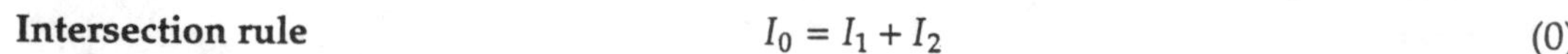

Intersection rule $$I_0 = I_1 + I_2 \qquad (0)$$

Now we can invoke Kirchhoff's 2nd law, according to which the change in potential around an *entire* loop is zero. Intuitively, if you "walk" around a loop, and return to your starting point, then you finish with the same potential energy you started with. Your overall change in potential energy—and hence, your change in potential—is zero.

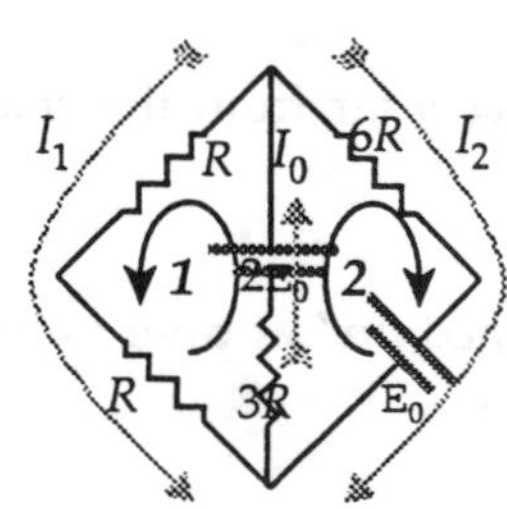

To apply this rule, I just redrew the circuit with all the resistances and voltages labeled. Though messy, this diagram contains all the information we need.

Since we're solving for I_2, let's begin with loop 2, I'll "walk" clockwise around the loop, starting from the top. So first, I pass through the $6R$ resistor, moving *with* the current. As explained in Chapter 38, this corresponds to a potential drop $\Delta V = -6RI_2$.

Next, I walk through the battery, from the positive to the negative terminal. In the car-and-roadway analogy, this corresponds to riding the car elevator down instead of up. So, my potential decreases: $\Delta V = -\mathcal{E}_0$.

Next, I walk through the $3R$ resistor, again with the current. So, $\Delta V = -3RI_0$. Notice that the current is I_0, not I_2.

Finally, I walk through the $2\mathcal{E}_0$ battery, from negative to positive terminal. Since this corresponds to riding up the car elevator, $\Delta V = +2\mathcal{E}_0$.

Since these ΔV's add up to zero,

Loop 2 $$-6I_2R - \mathcal{E}_0 - 3I_0R + 2\mathcal{E}_0 = 0. \qquad (2)$$

I apologize for writing Eq. (2) before Eq. (1), but I want to make a point: We can't yet solve for I_2. That's because Eq. (2) contains a second unknown, I_0. The intersection-rule equation doesn't resolve this problem, because it introduces a *third* unknown, I_1. Since we have three unknowns but only two equations, we need more information. To get it, focus on loop 1. I'll walk counterclockwise around that loop, starting at the top.

So first, I pass through one of the R resistors: $\Delta V = -I_1R$. Then I walk through the other R resistor, giving me another $\Delta V = -I_1R$. Next, I pass through the $3R$ resistor: $\Delta V = -3RI_0$. Finally, I "ride up" the battery, which raises my potential by $\Delta V = +2\mathcal{E}_0$. Set the sum of these potential changes equal to zero, to get

Loop 1 $$-I_1R - I_1R - 3I_0R + 2\mathcal{E}_0 = 0. \qquad (1)$$

At this point, we have three equations in three unknowns (I_0, I_1, and I_2). Only algebra remains. On a test, you should probably save messy algebra until the end, *after* you've earned at least partial credit on all the other problems. Ask your instructor about this.

Algebra starts here. We're ultimately solving for I_2. Solve Eq. (0) for I_1 to get $I_1 = I_0 - I_2$, and substitute this expression into Eq. (1) to get

$$-2(I_0 - I_2)R - 3I_0R + 2\mathcal{E}_0 = 0.$$

Solve this new equation for I_0, to get $I_0 = \dfrac{2\mathcal{E}_0 + 2I_2R}{5R}$. Finally, substitute this expression for I_0 into Eq. (2). This gives

$$-6I_2R - \mathcal{E}_0 - 3\frac{2\mathcal{E}_0 + 2I_2R}{5R}R + 2\mathcal{E}_0 = 0.$$

Expand out all the terms:

$$\begin{aligned} 0 &= -6I_2R - \mathcal{E}_0 - \frac{6}{5}\mathcal{E}_0 - \frac{6}{5}I_2R + 2\mathcal{E}_0 \\ &= -\frac{36}{5}I_2R - \frac{1}{5}\mathcal{E}_0, \end{aligned}$$

and hence

$$I_2 = -\frac{1}{36}\frac{\mathcal{E}_0}{R} = -\frac{1}{36}\frac{5.0\text{ V}}{10\ \Omega} = 0.014\text{ A}.$$

The minus sign indicates that I mis-guessed the direction of I_2. It flows "up" the circuit, not down the circuit.

(b) With one of the R resistors snipped, the left segment of the circuit becomes a "dead end" across which current can't flow. So, $I_1 = 0$. In other words, the circuit behaves as if the left segment were missing. The circuit consists entirely of loop 2.

This makes life easier. Since the current can't branch, it's the same everywhere: $I_0 = I_2$. Therefore, the loop 2 equation simplifies to

$$-6I_2R - \mathcal{E}_0 - 3I_2R + 2\mathcal{E}_0 = 0,$$

which we can immediately solve for I_2 to get

$$I_2 = \frac{1}{9}\frac{\mathcal{E}_0}{R} = \frac{1}{9}\frac{5.0\text{ V}}{10\ \Omega} = 0.056\text{ A}.$$

Notice that I_2 switched signs; it's positive instead of negative. Physically, the $2\mathcal{E}_0$ battery "overpowers" the E_0 battery, making current flow clockwise.

Notice also that, with loop 1 inactive, you can solve the remaining one-loop circuit using "circuit reduction." Specifically, the circuit has "net voltage" $\mathcal{E}_{\text{total}} = 2\mathcal{E}_0 - \mathcal{E}_0 = \mathcal{E}_0$, and equivalent resistance $R_{\text{eq}} = 3R + 6R = 9R$.

ANSWER 2

(a) A common mistake is to use Coulomb's law, $F = \dfrac{Q_1Q_0}{4\pi\varepsilon_0 r^2}$, where $r = R/2$. This shortcut would work, if Q_1 were located *outside* the charged sphere. The electric field outside a spherically symmetric charge distribution is the same *as if* the whole sphere were concentrated into a point charge at the center. But inside a charged sphere, the field gets more complicated.

Let $\mathbf{E}_1$ denote the electric field generated by the sphere at Q_1. According to the fundamental electrostatic force law, Q_1 feels a force

$$\mathbf{F} = Q_1\mathbf{E}_1.$$

So, to complete this problem, we must figure out $\mathbf{E}_1$.

As emphasized in chapters 33 and 34, we have two techniques for calculating the electric field produced by an "extended object" such as the charged sphere. Gauss' law, though always true, can help us find the field only for symmetric charge distributions. Fortunately, a sphere is plenty symmetric. (If Gauss' law didn't work, we'd break up the charge distribution into infinitesimal pieces, use Coulomb's law to calculate the field contributed by each piece, and then add up those field contributions. See Chapter 33 for examples.)

Subproblem: Use Gauss' law to find the electric field produced by the charged sphere at Q_1.

Dashed line is Gaussian surface. Q_1 is not shown, because we're finding the field produced by the sphere alone.

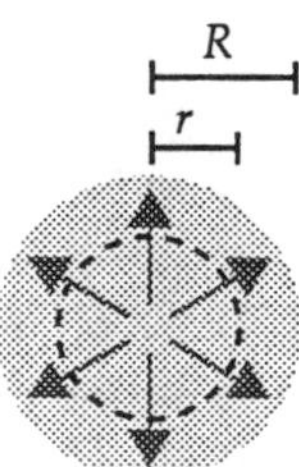

When solving for an electric field using Gauss' law,

$$\oint \mathbf{E}\cdot d\mathbf{A} = \frac{q_{\text{encl}}}{\varepsilon_0},$$

you must draw a Gaussian surface over which the electric field is constant. That way, you can pull E outside the flux integral. Since this charged sphere generates a spherically symmetric field, the field has the same strength at any two points equidistant from the center. For this reason, the field is constant over an entire "Gaussian sphere," as drawn here. Since we want the field *inside* the charged sphere, I gave my Gaussian surface radius r, where $r < R$. Later on, I'll set $r = R/2$. But for now, let me work in terms of a general r.

Because E is constant over this Gaussian surface, the flux integral simplifies as follows:

$$\oint \mathbf{E}\cdot d\mathbf{A} = E \oint_{\text{Gaussian sphere}} dA = EA_{\text{Gaussian sphere}} = E(4\pi r^2).$$

The "r" denotes the radius of the Gaussian surface, not the radius of the charged sphere.

Now I'll simplify the other side of Gauss' law. This Gaussian surface does not enclose the entire charged sphere. It only encloses a "subsphere," of radius r. To figure out how much charge lives inside the Gaussian surface, I'll first figure out the charge density, i.e., the charge per volume. Then, I'll multiply the charge density by the volume of the Gaussian surface.

Well, since the charged sphere contains total charge Q_0 uniformly distributed over volume $V_{\text{charged sphere}} = \frac{4}{3}\pi R^3$, it has charge density $\rho = \frac{Q_0}{V_{\text{charged sphere}}} = \frac{Q_0}{\frac{4}{3}\pi R^3}$. Therefore, the Gaussian surface encloses charge

$$q_{\text{encl}} = \frac{\text{charge}}{\text{volume}} \times (\text{volume enclosed}) = \rho V_{\text{Gaussian surface}} = \frac{Q_0}{\frac{4}{3}\pi R^3}\left(\frac{4}{3}\pi r^3\right) = Q_0\frac{r^3}{R^3}.$$

Throw all this information into Gauss' law, to get

$$\oint \mathbf{E}\cdot d\mathbf{A} = \frac{q_{\text{encl}}}{\varepsilon_0}$$

$$E(4\pi r^2) = \frac{Q_0 r^3}{\varepsilon_0 R^3},$$

and hence

$$E = \frac{Q_0 r}{4\pi\varepsilon_0 R^3}.$$

End of Gauss' law subproblem

We just found the field inside the charged sphere, at arbitrary distance r from the center. Here, we want the electric field acting on Q_1, which sits at $r = R/2$. So, substitute $r = R/2$ into our expression for E, to get

$$E_{\text{at } Q_1} = \frac{Q_0(R/2)}{4\pi\varepsilon_0 R^3} = \frac{Q_0}{8\pi\varepsilon_0 R^2}.$$

Therefore, Q_1 feels a radially-directed electric force of magnitude

$$F = Q_1 E_{\text{at } Q_1} = \frac{Q_1 Q_0}{8\pi\varepsilon_0 R^2}.$$

(b) A test charge at point P feels two electric fields: one from the charged sphere, and one from Q_1. According to the superposition principle, we can add these two fields to find the total field:

$$\mathbf{E}_P = \mathbf{E}_{P \text{ generated by } Q_1} + \mathbf{E}_{P \text{ generated by charged sphere}}.$$

Given $\mathbf{E}_P$, we can easily calculate the force on q_2, using $\mathbf{F} = q_2\mathbf{E}_P$. So, the meat of the problem is figuring out the two "separate" fields at point P.

Subproblem 1: Find the field at P generated by Q_1.

A point charge Q_1 creates a "Coulomb" field, $E = \dfrac{Q_1}{4\pi\varepsilon_0 r^2}$. From the diagram given in the problem, you can see that the distance between Q_1 and point P is $r = 2R + R/2 = \frac{5}{2}R$. So, at P, the field generated by Q_1 is

$$E_{P \text{ generated by } Q_1} = \frac{Q_1}{4\pi\varepsilon_0\left(\frac{5}{2}R\right)^2} = \frac{Q_1}{25\pi\varepsilon_0 R^2},$$

pointing rightward.

Subproblem 2: Find the field at P generated by the charged sphere.

Since P sits outside the charged sphere, the charged sphere "behaves" as if it were a point charge. Specifically, the charged sphere creates a Coulomb field, $E = \dfrac{Q_0}{4\pi\varepsilon_0 r^2}$. I'll show why below. Because P sits a distance $r = 3R$ from the center of the charged sphere,

$$E_{P \text{ generated by charged sphere}} = \frac{Q_0}{4\pi\varepsilon_0(3R)^2} = \frac{Q_0}{36\pi\varepsilon_0 R^2}.$$

OK, now I'll briefly review why the charged sphere "behaves" like a point charge, provided you're *outside* the charged sphere. Apply Gauss' law to the spherical Gaussian surface drawn here. This surface encloses the *entire* charged sphere. So,

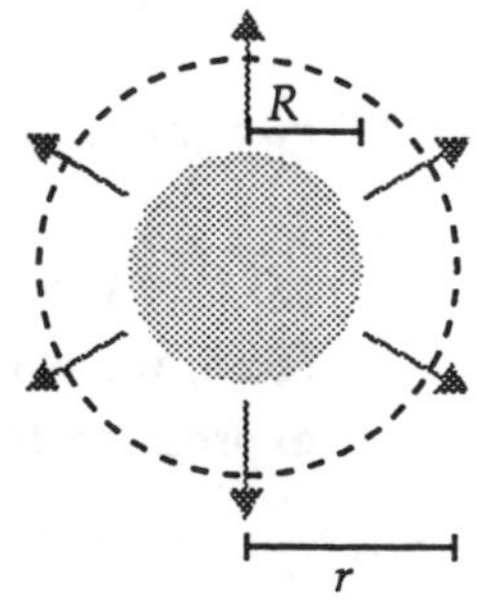

$$\oint \mathbf{E} \cdot d\mathbf{A} = \frac{q_{\text{encl}}}{\varepsilon_0}$$

$$E(4\pi r^2) = \frac{Q_0}{\varepsilon_0},$$

and hence, $E = \dfrac{Q_0}{4\pi\varepsilon_0 r^2}$, the same field produced by a point charge. As we saw in part (a), however, this result doesn't apply *inside* a charged sphere, because the relevant Gaussian surface does not enclose the entire charged sphere.

We now know the fields generated at point P by Q_1 and by the charged sphere. Since both of those fields point rightward, we can add them without worrying about vector components:

$$E_P = E_{P \text{ generated by } Q_1} + E_{P \text{ generated by charged sphere}}$$

$$= \frac{Q_1}{25\pi\varepsilon_0 R^2} + \frac{Q_0}{36\pi\varepsilon_0 R^2}.$$

Therefore, q_2 feels a rightward force of magnitude

$$F = q_2 E_P = q^2\left(\frac{Q_1}{25\pi\varepsilon_0 R^2} + \frac{Q_0}{36\pi\varepsilon_0 R^2}\right).$$

(c) As q_2 "falls" away from the charged sphere, the force on it gets smaller and smaller. Therefore, since the acceleration decreases, we can't use those old constant-acceleration kinematic formulas.

When forces get too messy, think about conservation laws. In this case, energy is conserved, with no heat dissipation or other complications. As the test charge falls away from Q_0 and Q_1, it loses potential energy and gains kinetic energy.

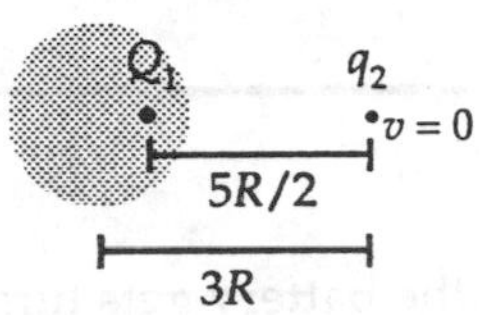

To organize my thoughts I've drawn an "initial" and "final" picture. The test charge begins motionless: $v_0 = 0$. It eventually reaches final velocity v, which is what we're solving for. So, remembering that $U = qV$, I get

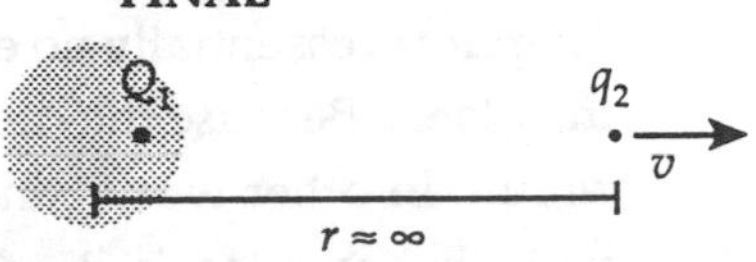

$$K_0 + U_0 = K_f + U_f$$

$$0 + q_2 V_P = \frac{1}{2}Mv^2 + q_2 V_\infty,$$

where V_P denotes the potential at point P, and V_∞ denotes the potential an effectively infinite distance away. To solve for v, we must calculate V_P and V_∞. Actually, the final velocity depends only on the *difference* between these potentials, $\Delta V = V_\infty - V_P$. We could calculate that potential difference directly, by evaluating the definite integral $\Delta V = -\int_P^\infty \mathbf{E}\cdot d\mathbf{s}$. But instead, I'll separately calculate V_P and V_∞, and subtract them.

A point charge generates a potential

$$V_{\text{point charge}} = -\int \mathbf{E}\cdot d\mathbf{s} = -\int \frac{Q}{4\pi\varepsilon_0 r^2}dr = \frac{Q}{4\pi\varepsilon_0 r},$$

where I've followed the usual convention of setting the integration constant equal to zero. We can apply this formula not only to Q_1, but also to the charged sphere. That's because a charged sphere and a point charge generate the same electric field, and hence, the same potential (if you're *outside* the charged sphere). Therefore, to find the potential at point P, just add up the potentials generated by the two "point charges," Q_1 and Q_0:

$$V_P = \frac{Q_0}{4\pi\varepsilon_0(3R)} + \frac{Q_1}{4\pi\varepsilon_0\left(\frac{5}{2}R\right)}.$$

By contrast, when q_2 gets *very* far from the charged sphere, r approaches ∞. By substituting $r = \infty$ into our formula for the potential, you can confirm that $V_\infty \approx 0$. In general, when charges are *very* far apart, they have no electrostatic potential energy.

Now that we know V_P and V_∞, let's finish solving the energy conservation equation for v:

$$K_0 + U_0 = K_f + U_f$$

$$0 + q_2 V_P = \frac{1}{2}Mv^2 + q_2 V_\infty$$

$$q_2\left[\frac{Q_0}{4\pi\varepsilon_0(3R)} + \frac{Q_1}{4\pi\varepsilon_0\left(\frac{5}{2}R\right)}\right] = \frac{1}{2}Mv^2 + 0$$

which we can immediately solve for v, to get

$$v = \sqrt{\frac{2q_2}{M}\left[\frac{Q_0}{4\pi\varepsilon_0(3R)} + \frac{Q_1}{4\pi\varepsilon_0\left(\frac{5}{2}R\right)}\right]}.$$

ANSWER 3

(a) When the battery gets turned on, charge starts to flow onto the capacitor. But *immediately* after $t = 0$, hardly any charge has had time to flow onto the capacitor. The capacitor's charge is still $Q \approx 0$. Therefore, essentially no electric field—and hence, no potential difference—gets generated between the plates. Because the capacitor generates no potential difference at this point, it plays no role in the circuit. In other words, the capacitor might as well be a resistanceless piece of wire. The circuit initially behaves as if it consisted entirely of the battery and resistor. Therefore, by Ohm's law, $i = \frac{\mathcal{E}_0}{R} = \frac{9.0\text{ V}}{1.0\ \Omega} = 9.0\text{ A}$.

If this seems too easy, apply Kirchhoff's 2nd law to the loop. Walking around this loop, you pass through a battery ($\Delta V = +\mathcal{E}_0$), a resistor ($\Delta V = -iR$), and a capacitor ($\Delta V = -Q/C$). So,

$$\mathcal{E}_0 - iR - \frac{Q}{C} = 0.$$

Immediately after $t = 0$, the capacitor carries charge $Q = 0$. Therefore, this loop equation reduces to $\mathcal{E}_0 - iR = 0$.

(b) The circuit consists of one loop, with no "branches." Therefore, at any given moment, the current through the resistor equals the current through the capacitor. In other words, at any given moment, a single current flows through the whole circuit. Here's why. If the current through the resistor were greater than the current through the capacitor, then charge would build up in the wire between the resistor and capacitor. Which can't happen.

While accepting the above argument, many thoughtful students want to say that *no* current flows through a capacitor, because charges don't "jump" across the gap from one plate to the other.

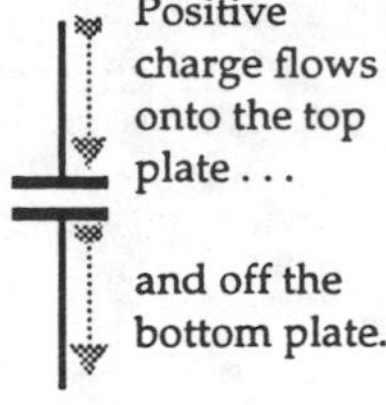
Positive
charge flows
onto the top
plate . . .
and off the
bottom plate.

Well, it's true that charges don't jump the gap. But current still flows "through" the capacitor, in the following way. As positive charges flow onto the top plate, an equal number of positive charges flow *off* the bottom plate. This leaves a deficit of positive charge, i.e., a net negative charge, on the bottom plate. So, the top plate acquires positive charge, while the bottom plate "acquires" an equal negative charge. As this diagram shows, current flows on both sides of the capacitor, even though individual charges don't jump the gap between the plates.

(c) Current flows through a capacitor *only* while the plates are charging up. Once the plates become fully charged, the top plate no longer "accepts" incoming charges, and the bottom plate no longer "spits out" charges. Therefore, the current through the capacitor stops. Because of this traffic jam, current stops flowing everywhere in the circuit. Remember, the same current flows through each circuit element. If no current flows through the capacitor, then no current flows through the resistor, either.

In summary: After a "long" time (typically a few hundredths of a second), the capacitor becomes fully charged, and hence, current stops flowing around the circuit.

(d) Intuitively, when a 9-volt battery finishes charging up the capacitor, the potential difference across the capacitor should be 9 volts. That's right. Since $C = Q/V$, we can calculate the corresponding charge on the plates:

$$CV = (2.0 \times 10^{-12}\text{ F})(9.0\text{ V}) = 1.8 \times 10^{-11}\text{ coulombs.}$$

To double check this result, look at the loop equation from part (a),

$$\mathcal{E}_0 - iR - \frac{Q}{C} = 0.$$

When the capacitor becomes fully charged, $i = 0$, as explained in part (c). Therefore, the loop equation reduces to $\mathcal{E}_0 - \frac{Q}{C} = 0$, and hence, $Q = C\mathcal{E}_0$. This agrees with the above intuitive reasoning.

Before closing, I should explain why the potential change across the capacitor is $\Delta V = -Q/C$ instead of $\Delta V = +Q/C$. Since the battery makes current flow clockwise, the top capacitor plate gains positive charge, and the bottom plate ends up with negative charge. Therefore, when "walking" clockwise around this circuit, you travel from the positive to the negative plate. A positive test charge would naturally "fall" from the positive to the negative plate, thereby losing potential energy. So, when moving from the positive to the negative plate, your potential drops.

Practice Midterm 4: More Electrostatics and Circuits

53 CHAPTER

This one is harder than most. Three questions, 75 minutes.

1) This circuit contains no battery. It has a capacitor, of capacitance $C = 5.0 \times 10^{-12}$ F. The resistors have resistance 1.0 Ω or 2.0 Ω, as labeled in the diagram.

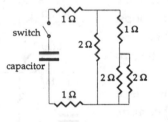

The capacitor is charged to $Q = 3.0 \times 10^{-11}$ coulombs. Then, at time $t = 0$, the switch is closed.

(a) Immediately after $t = 0$, what current flows through the switch?

(b) Sketch a rough, non-numerical graph of the current through the switch vs. time.

2) Two concentric cylindrical shells, of radius b and $2b$, both have length $L \gg b$. An unpictured battery of voltage $\mathcal{E}$ is hooked up to these shells, so that the outer shell acquires positive charge and the inner shell acquires negative charge.

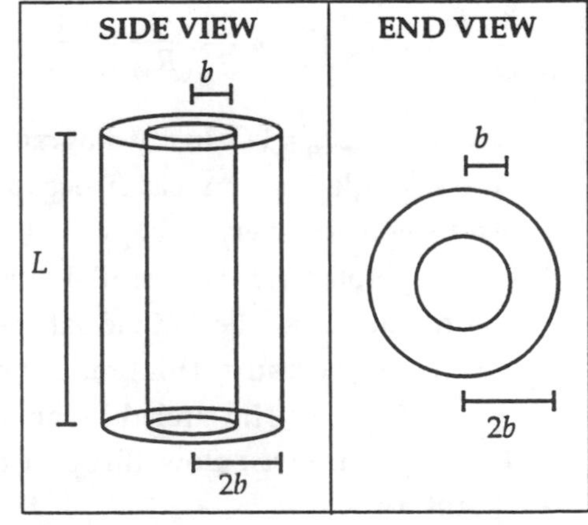

(a) How much charge ends up on the outer shell? Derive any formulas you use from basic laws.

(b) A particle of charge $-q_1$ and mass m_1 is placed between the two shells, just touching the inner shell. It is released from rest. What velocity does the particle have when it crashes into the outer shell? Neglect gravity.

(c) The particle is removed. With the battery still connected, shaving cream of dielectric constant κ is squirted between the shells, until it completely fills the space between the two shells. Is the charge on the outer shell higher than, lower than, or the same as it was in part (a)?

(d) Let Q denote the answer to part (a). What is the charge on the outer shell after the shaving cream is inserted? Express your answer in terms of Q and constants.

3) In this circuit, $V_0 = 10$ V, and $R_0 = 5.0$ Ω. The bottom resistor is a variable resistor; you can change its resistance by turning a knob.

(a) If the variable resistor is set to $R_1 = 2.0$ Ω, what power dissipates in that resistor?

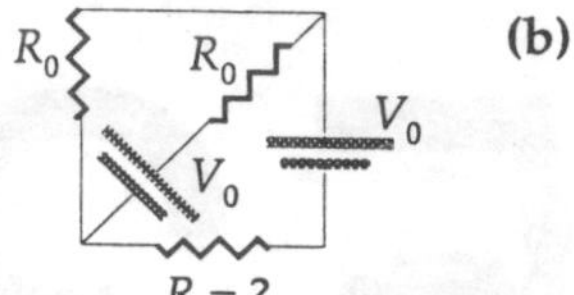

(b) If you want to maximize the power dissipated in the variable resistor, what resistance should it have? (You'll lose only 1 point for failing to complete all the math, provided you set things up properly.)

ANSWER 1

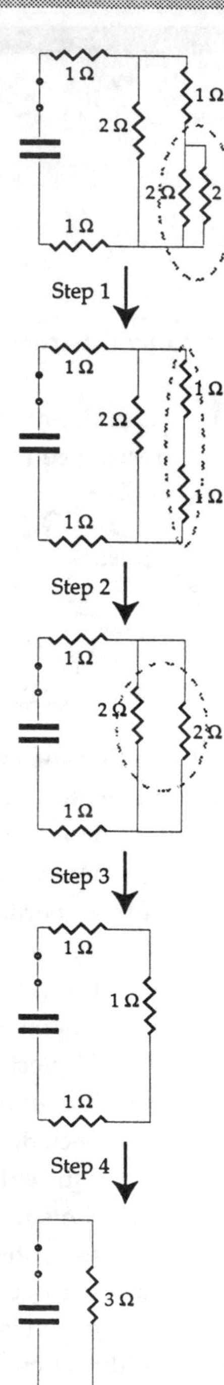

(a) By definition, capacitance is the charge held by the capacitor *per volt of potential difference*: $C = q/V$. Therefore, the potential difference across the capacitor, when it's fully charged, is

$$V = \frac{Q}{C} = \frac{3.0\times10^{-11}\text{C}}{5.0\times10^{-12}\text{F}} = 6.0 \text{ V}.$$

Immediately after $t = 0$, hardly any charge has flowed off the capacitor. Therefore, the capacitor still generates a 6-volt potential difference. In other words, the capacitor initially "behaves" like a 6-volt battery.

Given this insight, we could break the circuit into loops and apply Kirchhoff's laws. But it's faster to "reduce" the circuit, using those rules about resistors in series and in parallel. Circuit reduction works because the circuit does not contain multiple batteries or capacitors on different loops, and because the resistors don't criss-cross in complicated ways. Therefore, you can simplify the circuit, as drawn here. In each step, I "reduce" the two circled resistors into a single resistor.

Step 1: The two circled resistors are in parallel. In other words, a given bit of current flows through one *or* the other, but not both. Therefore,

$$\frac{1}{R_{\text{eq}}} = \frac{1}{2\,\Omega} + \frac{1}{2\,\Omega} = \frac{1}{1\,\Omega},$$

and hence, $R_{\text{eq}} = 1\,\Omega$. In my diagram, the two circled resistors got "combined" into a single 1-ohm resistor occupying the same location on the diagram. And so on, for every step.

Notably, for in-parallel resistors, the equivalent resistance is *less* than the resistance of the individual resistors, because the current can "spread out" over two paths. This branching allows more total current to flow.

Step 2: The circled 1-ohm resistors are in series; current that flows through one *must* flow through the other. The equivalent resistance is therefore

$$R_{\text{eq}} = 1\,\Omega + 1\,\Omega = 2\,\Omega.$$

Step 3: The circled 2-ohm resistors are in parallel. So, this step duplicates step 1! As shown above, the equivalent resistance of in-parallel 2-ohm resistors is 1 ohm.

Step 4: Since the three 1-ohm resistors are in series, they have equivalent resistance

$$R_{eq} = 1\ \Omega + 1\ \Omega + 1\ \Omega = 3\ \Omega.$$

We're done reducing the circuit. The overall circuit puts up 3.0 ohms of resistance. Therefore, since the capacitor initially supplies 6 volts, the initial current is

$$i = \frac{V}{R} = \frac{6.0\ \text{V}}{3.0\ \Omega} = 2.0\ \text{A}.$$

(b) As the capacitor discharges, the charge on the plates goes down. Since $V = Q/C$, the potential difference decreases as well. In other words, as time passes, the capacitor supplies a smaller and smaller voltage. For this reason, the current decreases.

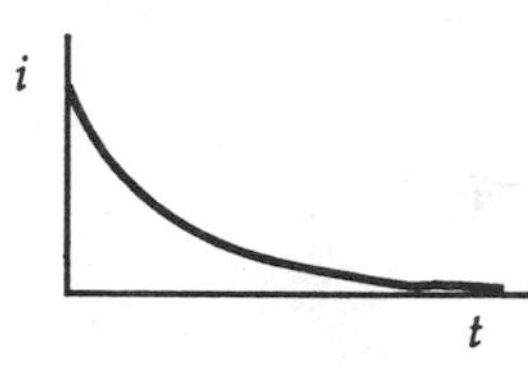

As explained in the *RC circuit* section of your textbook, the charge on the capacitor, and the current through the circuit, both decrease exponentially. Using Kirchhoff's 2nd law, you should be able to set up the relevant differential equation:

$$\frac{Q}{C} + \frac{dQ}{dt}R = 0,$$

since $i = -dQ/dt$. Here, "R" denotes the total equivalent resistance of the circuit, in this case 3.0 Ω. Using separation of variables, you can solve for $Q(t)$, and then differentiate with respect to time to get $i(t)$. Ask your professor if you're responsible for this math. In any case, you should know that the charge and current decay exponentially, as graphed here.

ANSWER 2

(a) We know the potential difference between the shells, and want the charge. How can we "get from" voltage to charge? Well, by definition, the *capacitance* specifies the charge per voltage that the system is willing to hold: $C = Q/V$. Given the capacitance, we can immediately solve for the charge, using $Q = C\mathcal{E}$.

Before continuing, let me address a common pitfall. You might try to solve "directly" for the unknown charge Q in terms of the given voltage, $\mathcal{E}$. Unfortunately, this strategy turns out to be extremely hard. It's easier to find the capacitance first, and *then* solve for Q. Furthermore, when calculating a capacitance, it's easiest to "pretend" you know the charge on the plates, and solve for the corresponding potential difference.

This should seem weird, and ironic. Even when you know the voltage but not the charge, it's easier to solve for the capacitance by *pretending* you know the charge but not the voltage. This "trick" works, because the *ratio* of charge to voltage—i.e., the capacitance—stays the same, no matter how you fiddle with the charge or voltage. In other words, the ratio q/V comes out the same, no matter whether you use the "correct" or "incorrect" charge to figure it out. Our standard capacitance-finding strategy relies on this insight:

Capacitance-finding strategy

1) Assign an arbitrary charge q to your plates.
2) Use Gauss' law or other techniques to find the electric field between the plates.
3) From that field, obtain the potential difference between the plates: $V = -\int \mathbf{E}\cdot d\mathbf{s}$.
4) Calculate $C = q/V$. The arbitrary charge q you chose in step 1 should cancel out.

I'll now implement this strategy. Remember, after finding the capacitance using an *arbitrary* charge q, we'll use that capacitance to calculate the *actual* charge Q on the outer shell.

Capacitance-finding strategy steps 1 & 2: Assign arbitrary charge to the capacitor, and find the electric field between the plates.

I've given the outer shell charge $+q$, and the inner shell charge $-q$.
To solve for the electric field using Gauss' law,

$$\oint \mathbf{E}\cdot d\mathbf{A} = \frac{q_{\text{encl}}}{\varepsilon_0},$$

we must be able to pull E outside the flux integral. So, the field must be constant over (part of) my Gaussian surface. By cylindrical symmetry, the field has the same strength at any two points equidistant from the central axis of the cylinders. For this reason, I'll draw a "Gaussian cylinder," of arbitrary radius r and length L, where r is between b and $2b$.

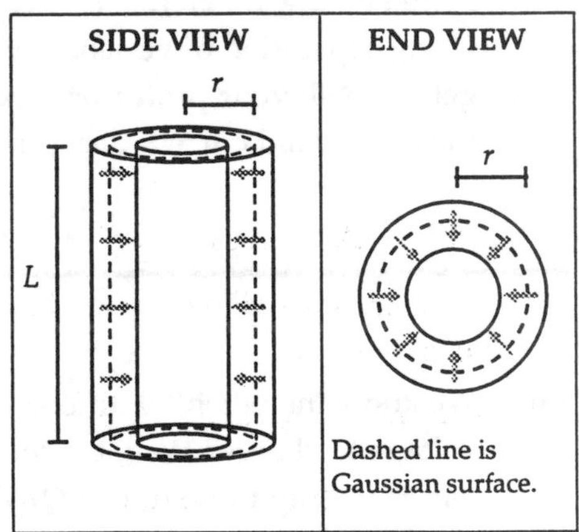

Dashed line is Gaussian surface.

Since $\mathbf{E}$ points radially, field lines flow only through the side (tube) of the Gaussian surface, not through the ends. The field strength stays constant over the whole tube, allowing us to pull E outside the integral:

$$\oint_{\text{Gaussian cylinder}} \mathbf{E}\cdot d\mathbf{A} = E \int_{\text{cylindrical tube}} dA = EA_{\text{cylindrical tube}} = E(2\pi rL),$$

where in the last step, I used the surface area of a cylindrical tube.

Since the Gaussian surface encloses the whole inner shell, but none of the outer shell, $q_{\text{encl}} = -q$. So,

$$\oint \mathbf{E}\cdot d\mathbf{A} = \frac{q_{\text{encl}}}{\varepsilon_0}$$

$$E(2\pi rL) = \frac{-q}{\varepsilon_0},$$

and hence, $E = -\dfrac{q}{2\pi r L \varepsilon_0}$. That's the field between the two shells. The minus sign indicates that it points inward.

Step 3: Find the potential difference between the plates, using the electric field just calculated.

You can ensure that $V = -\int \mathbf{E}\cdot d\mathbf{s}$ comes out positive by integrating from the *negative* to the *positive* capacitor plate. I'll integrate along a straight radial path, from the inner to the outer shell. So, the dot product reduces to a regular product:

$$
\begin{aligned}
V &= -\int_{\text{inner shell}}^{\text{outer shell}} \mathbf{E}\cdot d\mathbf{s} \\
&= -\int_b^{2b} E dr \\
&= \int_b^{2b} \frac{q}{2\pi r L \varepsilon_0} dr \\
&= \frac{q}{2\pi r L \varepsilon_0} \ln r \Big|_b^{2b} \\
&= \frac{q}{2\pi r L \varepsilon_0} (\ln 2b - \ln b) \\
&= \frac{q}{2\pi r L \varepsilon_0} \ln \frac{2b}{b} \\
&= \frac{q}{2\pi L \varepsilon_0} \ln 2.
\end{aligned}
$$

A common error is to integrate from 0 to b. But in our electric field formula, $E = \dfrac{q}{2\pi r L \varepsilon_0}$, the "$r$" means the distance *from the center of the cylinders*. Our limits of integration must "agree" with this meaning of r.

Step 4: Calculate the capacitance using $C = q/V$.

Use our result from step 3 to get

$$C = \frac{q}{V} = \frac{q}{\left(\dfrac{q}{2\pi L \varepsilon_0} \ln 2\right)} = \frac{2\pi L \varepsilon_0}{\ln 2}.$$

Notice that the arbitrary charge q canceled, as it must.

End of capacitance-finding strategy

We just found the capacitance using a "pretend" charge q on the plates. Given this capacitance, we can calculate the *actual* charge on the positive plate when the potential difference equals $\mathcal{E}$. It's

$$Q = C\mathcal{E} = \frac{2\pi L \varepsilon_0}{\ln 2} \mathcal{E}.$$

(b) Because the particle experiences a non-constant force and acceleration, we can solve most efficiently with conservation of energy. By "initial," I mean when the particle gets released from the inner shell. By "final," I mean when it reaches the outer shell. Since $U = qV$,

$$K_0 + U_0 = K_f + U_f$$

$$\frac{1}{2}mv_0^2 + (-q_1V_0) = \frac{1}{2}mv_f^2 + (-q_1V_f).$$

The particle gets released from rest: $v_0 = 0$. Isolate v_f to get

$$v_f = \sqrt{\frac{2q_1(V_f - V_0)}{m}}.$$

To finish solving for v_f, we need not worry about V_0 or V_f individually. What matters is the *difference* in potential between the two shells, $\Delta V = V_f - V_0$. But we already *know* the potential difference. It's $\mathcal{E}$. So,

$$v_f = \sqrt{\frac{2q_1\mathcal{E}}{m}}.$$

If you found yourself solving for this potential difference, then you forgot that the battery "enforces" a fixed voltage between the plates.

(c) The dielectric shaving cream *dampens* the electric field inside of it. Since the battery maintains the potential difference between the plates at $\mathcal{E}$, the charge on the shells must *increase*. Let me explain why, more carefully.

Suppose, for the sake of argument, that the charge on the plates stays the same. Then, the dielectric shaving cream *decreases* the electric field between the shells. Therefore, the potential difference between the plates ($V = -\int \mathbf{E}\cdot d\mathbf{s}$) goes down. But the battery won't let this happen. It "enforces" a fixed potential difference between the shells. Since the potential difference stays fixed, and since $V = -\int \mathbf{E}\cdot d\mathbf{s}$, the average electric field must stay fixed as well.

So far, I've shown that the battery prevents the electric field from decreasing when the dielectric gets inserted. But the dielectric dampens the field inside of it. Therefore, the charge on the shells must increase. For instance, suppose the dielectric dampens electric fields by a factor of 2. Then the charge on the shells must double. In the absence of the dielectric, these double-charged shells would create an electric field twice as big as the one before. But the dielectric damps that double-strength field, cutting it in half. As a result, the field has the same strength it did before. From this example, you can understand why the battery causes the charge on the shells to increase, when the dielectric gets inserted. The higher charge on the plates "compensates" for the field-dampening effect of the dielectric, and hence, the potential difference between the plates doesn't change.

(d) Now I'll formalize the above qualitative argument. In symmetric situations, a dielectric dampens the field inside of it by a factor of κ, the dielectric constant. Therefore, in order to keep the electric field (and the potential) between the shells fixed, the charge on the plates must increase by a factor of κ:

$$Q_{\text{with dielectric inserted}} = \kappa Q.$$

For instance, if $\kappa = 2$, then the dielectric tends to cut the field strength in half. Therefore the charge on the shells must double, to make sure the electric field between the plates stays the same.

The above reasoning gives us insight into the textbook formula $C_{\text{with dielectric}} = \kappa C_{\text{without dielectric}}$. According to this formula, if we insert shaving cream with $\kappa = 2$, then the capacitor holds twice as much charge as before. My answers to parts (c) and (d) explain why this is true.

ANSWER 3

(a) Let i_1 denote the current through R_1. That resistor dissipates power $P = i_1^2 R_1$. I'll briefly review where that formula comes from. In general, the power developed in a circuit element is $P = Vi$. Across a resistor, the voltage drops by iR. So, for resistors, $P = (iR)i = i^2R$. Therefore, finding i_1 will allow us to calculate the power.

Because different currents might flow through the two batteries, we can't "reduce" the circuit as we did in question 1. Instead, we must break it into loops and use Kirchhoff's laws. I'll choose the loops marked here.

As always, label the current in each segment of the circuit, guessing its direction. If you guess wrong, the answer will come out negative.

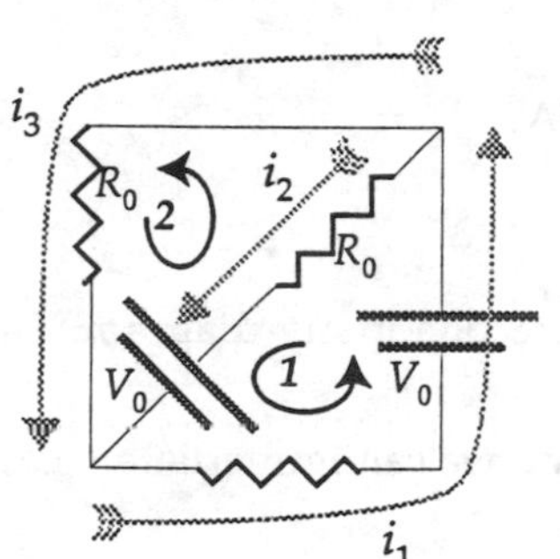

According to Kirchhoff's 1st law, the current flowing into an intersection equals the current flowing out of the intersection. Look at the upper right corner. Current i_1 flows into that intersection, while i_2 and i_3 flow out. So,

$$i_1 = i_2 + i_3 \tag{0}$$

According to Kirchhoff's 2nd law, the voltage change around a loop is zero. I'll walk around loop 1 counterclockwise, starting from the lower right corner. At the first battery, I go from the negative to the positive terminal. So, $\Delta V = +V_0$. Then I walk through the R_0 resistor with the current: $\Delta V = -i_2R_0$. Next I pass through the other battery, from positive to negative terminal: $\Delta V = -V_0$. Finally, I pass through the variable resistor, presently set to R_1. So, $\Delta V = -i_1R_1$. Setting the sum of these ΔV's equal to zero, I get

Loop 1 $$V_0 - i_2R_0 - V_0 - i_1R_1 = 0 \tag{1}$$

Even though the V_0's cancel, we still can't solve for i_1, because the equation contains a second unknown, namely i_2. To gather more information, look at loop 2. I'll walk around it counterclockwise, starting from the lower left corner.

So first, I pass through the battery, from negative to positive terminal: $\Delta V = +V_0$. Then I go through the R_0 resistor, walking *against* the current (i.e., walking in the opposite direction from my i_2 arrow). So, $\Delta V = +i_2R_0$. Finally, I walk through the other R_0, in the same direction as i_3. Therefore, $\Delta V = -i_3R_0$. Putting all this together, I get

Loop 2 $$V_0 - i_2R_0 - i_3R_0 = 0 \tag{2}$$

Given these three equations in the three unknowns (i_1, i_2, and i_3), we can solve for i_1.

Algebra starts here. From Eq. (0), $i_3 = i_1 - i_2$. Substitute that expression into Eq. (2) to get

$$V_0 - i_2R_0 - (i_1 - i_2)R_0 = 0,$$

which we can solve for i_2 to get $i_2 = \frac{i_1R_0 - V_0}{2R_0}$. Substitute this expression into Eq. (1), to get

$$V_0 - \frac{i_1R_0 - V_0}{2R_0}R_0 - V_0 - i_1R_1 = 0.$$

To simplify this equation, cancel the V_0's, and also cancel the R_0's in the numerator and denominator of the second term. This gives

$$-\frac{i_1R_0 - V_0}{2} - i_1R_1 = 0.$$

Isolate i_1 to get

$$i_1 = -\frac{V_0}{R_0 + 2R_1} = -\frac{10\,\text{V}}{5.0\,\Omega + 2(2.0\,\Omega)} = -1.11\,\text{A}.$$

End of algebra.

The minus sign indicates that I mis-guessed the current direction in my diagram. The current actually flows leftward, not rightward, through R_1.

Now that we know the current through the variable resistor, we can immediately calculate the power. It's

$$P = i_1^2R_1 = (1.11\text{ A})^2(2.0\ \Omega) = 2.5\text{ W}.$$

(b) By adjusting R_1, we want to maximize P. Let's use the standard "maximization" strategy from calculus. See question 33-4c for a full review.

First, write the variable you're maximizing, in this case P, as a function of the "free" (adjustable) variable, in this case R_1. Then, set dP/dR_1 equal to zero, and solve for R_1. On a test, most instructors award lots of credit for laying out and starting this strategy, even if you lack time to carry it through.

We nearly completed the first step in part (a). Since $i_1 = \frac{V_0}{R_0 + 2R_1}$,

$$P = i_1^2R_1 = \frac{V_0^2}{(R_0 + 2R_1)^2}R_1.$$

Crucially, you can't use "$i_1^2 R_1$" as your expression for P, because that expression contains two variables, i_1 and R_1. Before differentiating with respect to R_1, you must make sure R_1 is the *only* variable on the right-hand side. That's why I rewrote i_1 as $\frac{V_0}{R_0 + 2R_1}$. By contrast, since V_0 and R_0 are fixed constants, they are "allowed" into the equation.

Now use the chain rule to differentiate with respect to R_1, and set that mess equal to zero:

$$0 = \frac{dP}{dR_1} = \frac{V_0^2}{(R_0 + 2R_1)^2} + R_1(-2)\frac{V_0^2}{(R_0 + 2R_1)^3}(2)$$

$$= \frac{V_0^2}{(R_0 + 2R_1)^2}\left(1 - \frac{4R_1}{R_0 + 2R_1}\right).$$

From this equation, it follows that

$$1 - \frac{4R_1}{R_0 + 2R_1} = 0.$$

Multiply through by $(R_0 + 2R_1)$ to get $R_0 + 2R_1 - 4R_1 = 0$, and hence,

$$R_1 = \frac{1}{2}R_0 = 2.5\ \Omega.$$

54 CHAPTER

Practice Midterm 5. Magnetism and Induction

Three questions, 70 minutes.

1) Consider two very long cylindrical wires, both of radius R, carrying current in the same direction. The centers of the wires are separated by distance $4R$. In this picture, the dashed line is an imaginary line connecting the centers of the two wires.

Both wires have uniform current density J_0.

(a) What's the magnetic field at point W, midway between the wires?

(b) What's the magnetic field at point X, which is a distance $R/3$ from the center of the left wire, on the dashed line?

(c) What's the magnetic field at point C, the center of the left wire?

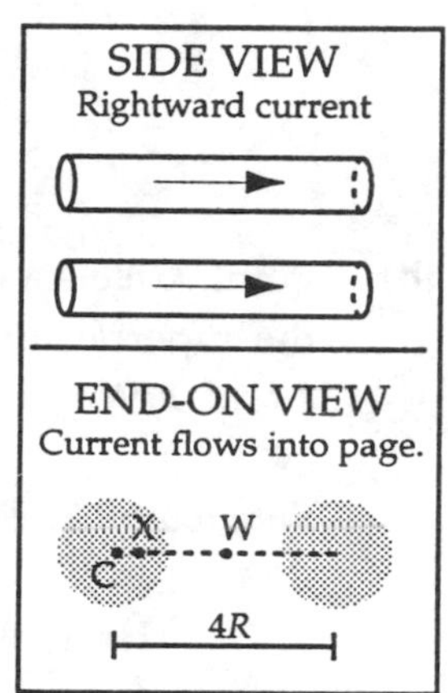

2) A particle of charge $Q = 6.0 \times 10^{-6}$ C and mass $m = 1.0 \times 10^{-10}$ kg is fired out of a "charge gun" towards a uniform magnetic field pointing into the page, as drawn here. The field has strength 0.50 T. The particle enters the field at point P, moving upward.

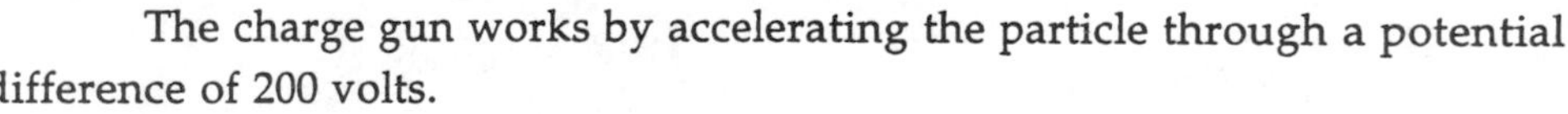

The charge gun works by accelerating the particle through a potential difference of 200 volts.

In this problem, neglect gravity.

(a) How fast is the particle moving when it leaves the charge gun?

(b) Does the particle exit the magnetic field to the left of point P or to the right of point P?

(c) How far from point P does the particle exit the magnetic field?

3) A rectangular wire loop, of length l, width w, and resistance R, "straddles" two regions of uniform magnetic field, both pointing out of the page, as drawn here. The left half of the rectangle is immersed in a uniform field of strength B_1. The right half of the rectangle is immersed in a uniform field of strength B_2, where $B_2 > B_1$. In this problem, ignore gravity.

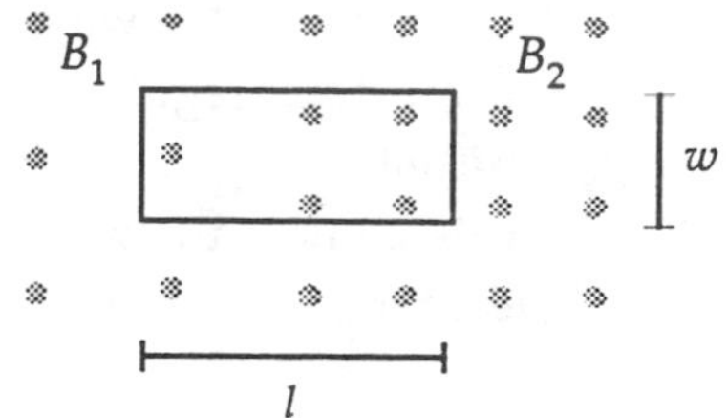

(a) Which requires more effort: Pushing the rectangle leftward, or pushing it rightward? Or do both pushes require the same effort? Explain.

(b) Which is harder: pushing the rectangle leftward, or pushing it upward? Explain.

(c) In order to push the rectangle leftward with constant speed v, what force must you apply? Answer in terms of l, w, R, B_1, B_2, v, and any universal constants you need.

ANSWER 1

(a) Midway between the wires, the field vanishes. Here's how I know.

Long wires generate circular field lines. To determine whether the field lines circulate clockwise or counterclockwise, use an abbreviated version of the right-hand rule: Point your right thumb in the direction of the current, and curl your fingertips. Your fingertips indicate the direction of the field lines. In this case, when using the end view, point your thumb into the page. Your fingertips curl clockwise. So, the field lines flow clockwise around both wires, as drawn here. Therefore, at point W, the field generated by wire 1 points down, while the field generated by wire 2 points up. Those fields cancel exactly, because point W is the same distance from both wires, and because both wires carry the same current.

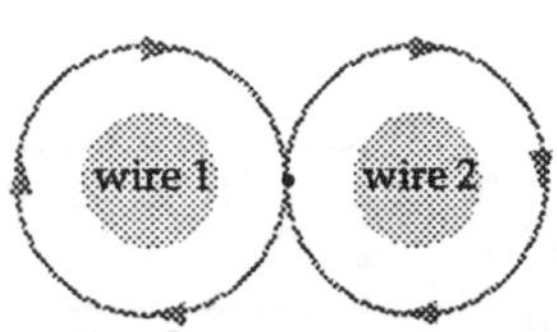

(b) Point X feels a downward field from wire 1, and an upward field from wire 2. So, according to the superposition principle, $\mathbf{B}_{\text{total}} = \mathbf{B}_1 + \mathbf{B}_2$, where $\mathbf{B}_1$ and $\mathbf{B}_2$ denote the fields generated by wires 1 and 2. We can *separately* find $\mathbf{B}_1$ and $\mathbf{B}_2$, using Ampere's law.

Subproblem 1: Find the field generated at point X by wire 1.

Here, we want the field *inside* a wire of uniform current density. So, when applying Ampere's law,

$$\oint \mathbf{B} \cdot d\mathbf{s} = \mu_0 I_{\text{encl}}.$$

we must draw the Amperian loop inside the wire. I've given it arbitrary radius r. Here, we'll set $r = R/3$. But obtaining the general result will make our lives easier in part (c).

To solve Ampere's law for the magnetic field, we need to pull B outside the line integral. This can happen only if B is constant over (part of) the Amperian loop. So, I've drawn a circular loop. Since the magnetic field lines also trace circles, $\mathbf{B}$ always points along my loop. Hence, the dot product $\mathbf{B} \cdot d\mathbf{s}$ reduces to a regular product. And since each point on my Amperian loop is equidistant from the wire's center, the magnetic field has the same strength over the entire loop (by symmetry). Therefore, we can pull B outside the Amperian line integral:

$$\oint \mathbf{B} \cdot d\mathbf{s} = \oint B ds = B \oint ds = B(2\pi r),$$

where r denotes the radius of the Amperian loop, *not* the radius of the wire.

Now I'll simplify the other side of Ampere's law. This Amperian loop encloses area $A = \pi r^2$ of the wire. Since J_0 specifies the current per area,

$$I_{\text{encl}} = \frac{\text{current}}{\text{area}} \times (\text{area enclosed by the Amperian loop}) = J_0 \pi r^2.$$

Putting all this together, we get

$$\oint \mathbf{B} \cdot d\mathbf{s} = \mu_0 I_{\text{encl}}$$

$$B(2\pi r) = \mu_0 J_0 \pi r^2,$$

and hence

$$B = \frac{\mu_0 J_0 \pi r^2}{2\pi r} = \frac{\mu_0 J_0 r}{2}.$$

To obtain the field at point X, just set $r = R/3$:

$$B_1 = \frac{\mu_0 J_0 (R/3)}{2} = \frac{\mu_0 J_0 r}{6}.$$

Since these field lines flow clockwise, as shown in part (a), $\mathbf{B}_1$ points downward.

Subproblem 2: Find the field at point X generated by wire 2.

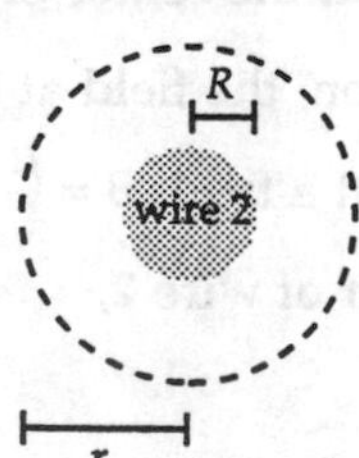

I'll duplicate the reasoning of subproblem 1, with a few modifications. Since point X is outside wire 2, I'll give my Amperian loop a radius r bigger than R. For the same reasons as before, the line integral simplifies to

$$\oint \mathbf{B} \cdot d\mathbf{s} = \oint B ds = B \oint ds = B(2\pi r).$$

But because this Amperian loop encloses *all* of wire 2, I_{encl} is the entire current in that wire. Wire 2 has cross-sectional area $A = \pi R^2$, and current per area J_0. So, the wire carries total current

$$I_{\text{encl}} = I_{\text{wire 2}} = \frac{\text{current}}{\text{area}} \times (\text{area of wire 2}) = J_0 \pi R^2.$$

Therefore, from Ampere's law,

$$\oint \mathbf{B} \cdot d\mathbf{s} = \mu_0 I_{\text{encl}}$$

$$B(2\pi r) = \mu_0 J_0 \pi r^2,$$

and hence

$$B = \frac{\mu_0 J_0 R^2}{2r}.$$

We just derived the general expression for the field outside wire 2, a distance r from its center. Let's specialize this result to the problem at hand. As this diagram shows, point X sits a distance $r = 4R - R/3 = 11R/3$ from the center of wire 2. So, at point X, wire 2 generates a field

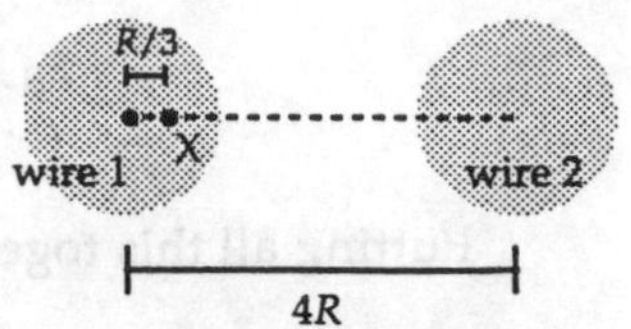

$$B_2 = \frac{\mu_0 J_0 R^2}{2(11R/3)} = \frac{3\mu_0 J_0 R}{22}.$$

Since the field lines around wire 2 flow clockwise, this field points upward.

End of subproblem 2.

At this stage, we've found the fields generated by wires 1 and 2 at point X. I'll call downward the positive direction. Since $\mathbf{B}_1$ points downward while $\mathbf{B}_2$ points upward, the total field at X has strength

$$B_{\text{total}} = B_1 - B_2 = \frac{\mu_0 J_0 R}{6} - \frac{3\mu_0 J_0 R}{22} = \frac{1}{33}\mu_0 J_0 R.$$

(c) We can solve quickly by recycling our general results from part (b). For instance, look at subproblem 1 above. We found that wire 1 generates a field $B = \frac{\mu_0 J_0 r}{2}$, at points inside wire 1. But at point C, r equals 0. So, wire 1 generates no field at its center. Intuitively, at the center of wire 1, the fields generated by all the "bits" of that wire cancel out. For this reason, the field at point C comes entirely from wire 2. From part (b), we know that wire 2 cranks out a field $B = \frac{\mu_0 J_0 R^2}{2r}$, at points outside wire 2. Since point C sits a distance $r = 4R$ from the center of wire 2,

$$B_C = \frac{\mu_0 J_0 R^2}{2(4R)} = \frac{\mu_0 J_0 R}{8}.$$

ANSWER 2

(a) This is a classic energy conservation problem. As the particle accelerates inside the gun, electrostatic potential energy converts into kinetic energy.

By definition, voltage (i.e., potential difference) specifies the change in potential energy that a 1-coulomb test charge would experience. This gun operates at 200 volts. So, a 1-coulomb charge would lose 200 joules of potential energy inside the gun. Therefore, the charge would gain 200 joules of kinetic energy.

The particle in this problem carries charge $Q = 6.0 \times 10^{-6}$ coulombs, not 1 coulomb. So, inside the gun, it loses potential energy $\Delta U = Q\Delta V = (6.0 \times 10^{-6}\text{ C})(200\text{ V}) = 1.2 \times 10^{-3}$ joules. Therefore, it gains that same amount of kinetic energy: $K = \frac{1}{2}mv^2 = 1.2 \times 10^{-3}$ joules. Solve for v to get

$$v = \sqrt{\frac{2K}{M}} = \sqrt{\frac{2(1.2 \times 10^{-3}\text{ J})}{1.0 \times 10^{-10}\text{ kg}}} = 4900\text{ m/s}.$$

In summary, I just set the total initial energy, which is entirely potential, equal to the total final energy, which is entirely kinetic: $U_0 = K_f$.

(b) Inside the magnetic field, the particle traces a semicircular path, as explained in Chapter 39. If the magnetic force initially deflects it leftward, then the particle traces a counterclockwise semicircle, and exits the field to the left of point P. By contrast, if $\mathbf{F}_{mag}$ initially points rightward, then the particle traces a clockwise semicircle, and exits the field to the right of P.

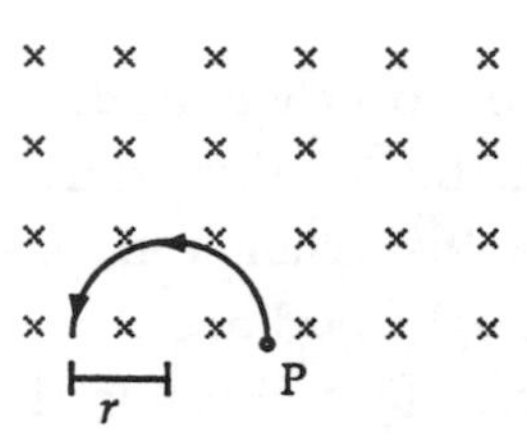

To decide whether $\mathbf{F}_{mag}$ initially points rightward or leftward, apply the right-hand rule to the cross product in $\mathbf{F}_{mag} = q\mathbf{v} \times \mathbf{B}$. Since q is positive, $\mathbf{v} \times \mathbf{B}$ tells us the direction of $\mathbf{F}_{mag}$. So, point your fingers in the direction of **v**, upward. Then curl your fingertips in the direction of **B**, into the page. Your thumb now points leftward. So, the particle deflects leftward, and eventually exits the field to the left of P.

(c) Let r denote the radius of this semicircle. As my diagram shows, the particle exits the field a distance $2r$ from point P. So, we just need to find that radius, and double it.

To derive the relevant formula, recall that the magnetic force always acts perpendicular to the direction of motion, and therefore makes the particle travel in a circle. In other words, the magnetic field provides a radial (centripetal) force. From last semester, you know that a particle moving in a circle experiences a net radial (centripetal) acceleration $a_{radial} = \frac{v^2}{r}$. So, from Newton's 2nd law,

$$F_{radial} = ma_{radial}$$

$$Q|\mathbf{v} \times \mathbf{B}| = m\frac{v^2}{r}$$

$$QvB = m\frac{v^2}{r},$$

where in the last step, I realized that **v** points perpendicular to **B**, and therefore the magnitude of the cross product reduces to a regular product.

Solve this equation for r to get

$$r = \frac{mv}{QB} = \frac{(1.0 \times 10^{-10}\text{ kg})(4900\text{ m/s})}{(6.0 \times 10^{-6}\text{ C})(0.50\text{ T})} = 0.163\text{ m}.$$

So, the particle exits the field a distance $2r = 0.33$ meters from point P.

ANSWER 3

(a) Many students think the rectangle "wants" to decrease the flux through itself, in which case pushing it leftward would require less effort than pushing it rightward. But the rectangle doesn't prefer a bigger vs. smaller flux. What the rectangle "dislikes" is a *changing* flux. The flux changes by the same amount no matter whether the rectangle moves leftward or rightward. So, pushing the rectangle in either direction requires equal effort.

Let me flesh out this rough answer. According to Faraday's law, the induced EMF—and hence, the induced current—depends not on the size of the flux, but on the *rate of change* of flux through the rectangle: $\mathcal{E}_{\text{induced}} = -d\Phi_B/dt$. By pushing the rectangle rightward, you increase the flux. By pushing it leftward at the same speed, you decrease the flux *at the same rate*. Therefore, in both cases, the same EMF gets induced. So, in both cases, the same current flows around the rectangle.

The magnetic fields act on this current, pushing the rectangle oppositely to its direction of motion, as explained in part (c) below. No matter which way you push, the left side "feels" B_1 and the right side feels B_2. Consequently, the same net magnetic force acts on the rectangle in either case. For this reason, pushing it leftward takes the same effort as pushing it rightward does.

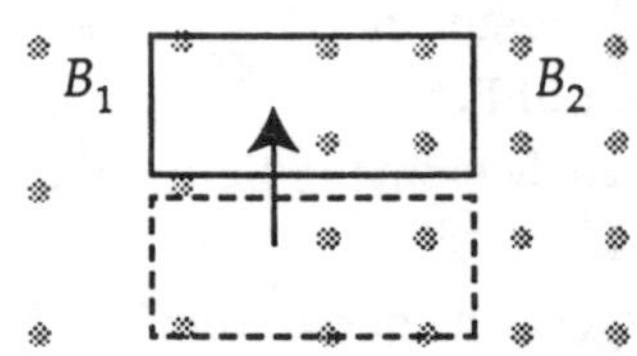

Moving the rectangle upward does *not* change the flux through it.

(b) As just explained, the rectangle "reacts" to a *changing* flux. But when the rectangle moves upward, the percentage of its area in the stronger field and the percentage of its area in the weaker field stay *constant*. So, the flux stays constant. Consequently, no EMF gets induced, and no current flows around the rectangle. Therefore, no magnetic force acts on the rectangle. With no magnetic force opposing you, it's easy to push the rectangle upward.

By contrast, when you push the rectangle leftward or rightward, the magnetic flux changes, inducing an EMF and current around the rectangle. The magnetic force acting on this current opposes your push, making it hard to maintain the rectangle's motion.

(c) When you push the rectangle leftward, a rightward magnetic force acts on it. According to Lenz' law, the rectangle "opposes" the change in flux. Here, the flux changes because you're pushing the rectangle. So, the magnetic force on the rectangle must oppose your push.

To see how your push force relates to this magnetic force, think about Newton's laws. When the rectangle moves at constant velocity, it has no acceleration. Therefore, no net force acts on it. In other words, your push force must exactly equal the magnetic force. Of course, to *get* the rectangle moving, your push force must overcome the magnetic force. But to *keep* it moving, you just need to cancel the magnetic force, ensuring a zero net force. For this reason, $F_{\text{push}} = F_{\text{mag}}$. By finding the magnetic force, we've automatically found your push force.

We can solve for F_{mag} by formalizing the qualitative reasoning in part (a). When the rectangle moves leftward, the flux changes. We can find $d\Phi_B/dt$, the rate of change of flux. Using Faraday's law and Ohm's law, we can calculate the corresponding induced current. Finally, we can figure out the magnetic force acting on this current.

Let me summarize and then implement this strategy.

1) Find the rate of change of flux, $d\Phi_B/dt$.
2) Using Faraday's law, find the induced EMF. Then use Ohm's law to find the current.
3) Figure out the magnetic force acting on that current. Since $F_{push} = F_{mag}$, you're done!

Don't get psyched out by the difficulty of some of these subproblems. You would receive lots of partial credit for using a coherent strategy, even if you couldn't complete each subproblem perfectly.

Subproblem 1: Find the rate of change of flux, $d\Phi_B/dt$.

First, I'll write down the flux through the rectangle at an arbitrary moment. Then I'll differentiate with respect to t.

Let x denote the length of the rectangle that's inside B_1, the weaker field. That portion of the rectangle has area $A_1 = xw$. From the diagram, notice that length $l - x$ of the rectangle sits inside B_2. So, the portion of the rectangle inside B_2 has area $A_2 = (l - x)w$. Therefore,

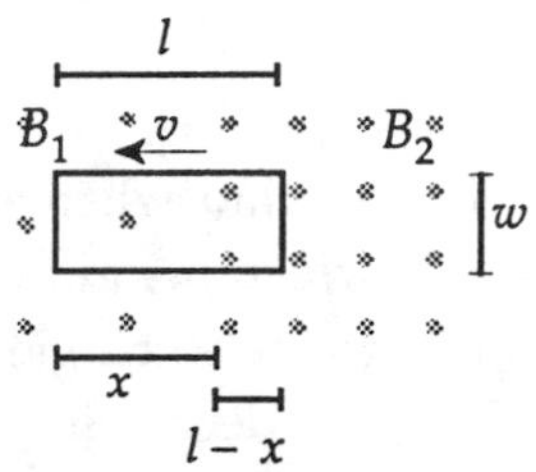

$$\begin{aligned}\Phi_B &= \int_{\text{rectangle}} \mathbf{B}\cdot d\mathbf{A} \\ &= \int\int_{\text{part of the rectangle inside } B_1} \mathbf{B}\cdot d\mathbf{A} + \int_{\text{part of the rectangle inside } B_2} \mathbf{B}\cdot d\mathbf{A} \\ &= B_1A_1 \quad + \quad B_2A_2 \\ &= B_1xw \quad + \quad B_2(l-x)w \\ &= (B_1 - B_2)xw + B_2lw,\end{aligned}$$

where in the last step, I grouped the "x" terms.

Now we must differentiate with respect to t. In this expression for Φ_B, only x changes in time. Everything else (B_1, B_2, w, and l) stays constant. So,

$$\begin{aligned}\frac{d\Phi_B}{dt} &= \frac{d}{dt}[(B_1 - B_2)xw] + \frac{d}{dt}[B_2lw] \\ &= (B_1 - B_2)w\frac{dx}{dt} \quad + \quad 0.\end{aligned}$$

But dx/dt is simply v, the rectangle's speed! Therefore, we have

$$\frac{d\Phi_B}{dt} = (B_1 - B_2)wv.$$

Since $B_1 < B_2$, this rate of change of flux is negative; the flux decreases as the rectangle moves leftward.

Subproblem 2: Use Faraday's law and Ohm's law to obtain the induced current.

According to Faraday's law,

$$\mathcal{E}_{\text{induced}} = \frac{d\Phi_B}{dt} = (B_2 - B_1)wv,$$

where I reversed the sign by switching B_1 with B_2. Ohm's law then gives us

$$i = \frac{\mathcal{E}}{R} = \frac{(B_2 - B_1)wv}{R}.$$

To find whether this current flows clockwise or counterclockwise, apply Lenz' law, according to which the induced current creates a magnetic field that tries to counteract the changing flux. Here, as the rectangle moves leftward, the out-of-the-page magnetic flux decreases. In order to counteract this decreasing flux, the rectangle "wants" to create its own magnetic field pointing out of the page. In other words, the induced current flows so as to create an out-of-the-page field, at locations inside the rectangle.

To create this field, the current must flow counterclockwise. I figured this out by applying the right-hand rule shortcut, which goes as follows: pick one of the four sides of the rectangle, and point your thumb in the direction of the current, assuming it flows counterclockwise. Your curled fingertips represent the magnetic field lines. When you curl them towards the center of the rectangle, your fingertips point out of the page—exactly the field the rectangle "wants" to generate.

So far, we've found the magnitude and direction of the induced current. Now we can calculate the magnetic force acting on that current.

Subproblem 3: Find the magnetic force acting on the induced current in the rectangle.

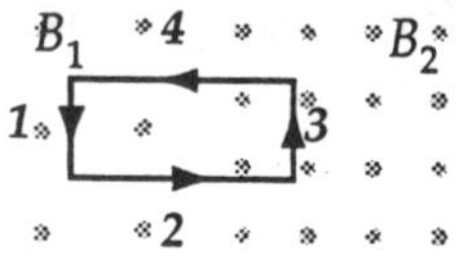

I'll calculate the force acting on each of the four sides, as labeled here. The arrowheads represent the current.

Actually, let's begin by finding the directions of those four forces. Apply the right-hand rule to the cross product in

$$d\mathbf{F}_{\text{mag}} = i d\mathbf{l} \times \mathbf{B},$$

where $d\mathbf{l}$ points in the direction of current flow. For side 1, $d\mathbf{l}$ points downward, and $\mathbf{B}$ points out of the page. So, point your fingers downward, and then curl your fingertips out of the page. Your thumb now points leftward. That's the direction of $\mathbf{F}_1$, the magnetic force on side 1.

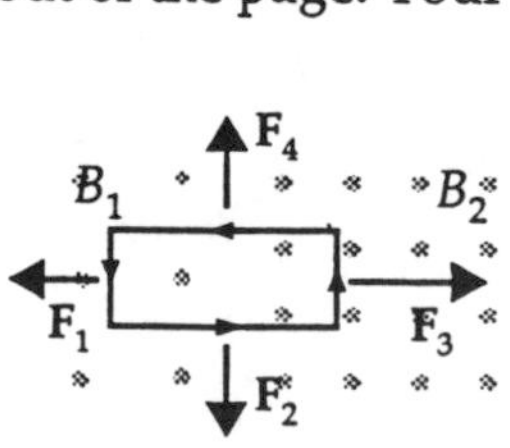

By similar reasoning, the forces on sides 2 through 4 point in the directions shown here.

At this stage, you should ask yourself whether any of these forces cancel. Since B_2 is stronger than B_1, F_3 should be larger than F_1. But F_2 and

F_4 cancel exactly. Here's why. At any given moment, sides 2 and 4 are equally immersed in B_1. For instance, if 75% of side 2 in inside B_1, then 75% of side 4 is inside B_1. Therefore, both sides experience the same average magnetic field. They also carry the same current, and have the same length. So, $F_2 = F_4$. Because those forces have the same strength but point in opposite directions, they cancel.

In summary, the rectangle experiences a net rightward magnetic force $F_{net} = F_3 - F_1$. We just need to calculate those two forces, and subtract them. Well,

$$F_1 = \int dF_1 = \int_{\text{side 1}} i|d\mathbf{l} \times \mathbf{B}|$$

$$= \int_{\text{side 1}} idlB \qquad \text{[since } \mathbf{B} \text{ and } d\mathbf{l} \text{ are perpendicular]}$$

$$= iB_1 \int_{\text{side 1}} dl \qquad \text{[since } i \text{ and } B \text{ are constant over side 1]}$$

$$= iB_1 w,$$

since side 1 has length w.

By equivalent reasoning, $F_3 = iB_2w$. So,

$$F_{mag} = F_3 - F_1 = iB_2w - iB_1w = i(B_2 - B_1)w.$$

In subproblem 2, we found the induced current. Substitute that i into this expression for the magnetic force, to get

$$F_{mag} = i(B_2 - B_1)w = \frac{(B_2 - B_1)wv}{R}(B_2 - B_1)w = \frac{(B_2 - B_1)^2 w^2 v}{R}.$$

As discussed way at the beginning of this answer, your leftward push force must cancel the rightward magnetic force, to keep the rectangle moving at constant velocity (zero acceleration). So, $F_{push} = F_{mag} = \dfrac{(B_2 - B_1)^2 w^2 v}{R}$.

55 CHAPTER

Practice Final Exam 1

This test covers chapters 33–44, but not chapters 45 and 46, because many courses don't cover displacement current or electromagnetic waves. If your exam covers those topics, please work through chapters 45 and 46.

This exam is hard, designed to help you see what topics you need to study further. 70% is an excellent score. 5 Questions, 150 minutes.

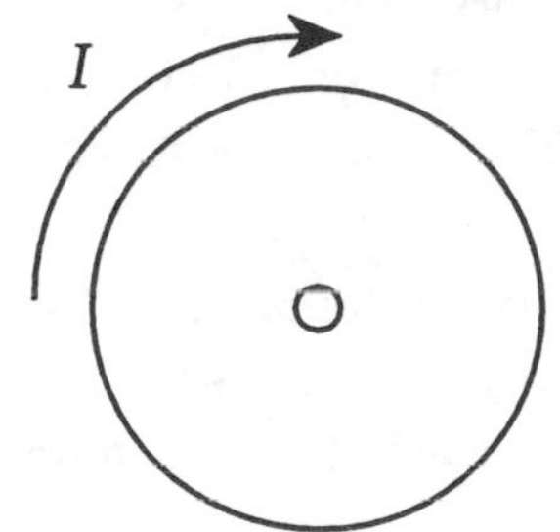

1) A small circular metal ring of radius r is concentric with a large circular metal ring, of radius $10r$. An unpictured power supply makes current flow clockwise around the large ring.

By adjusting the power supply, you can alter I, the current in the large ring. This graph shows I as a function of time. Notice that I increases from I_0 to $2I_0$ in time T.

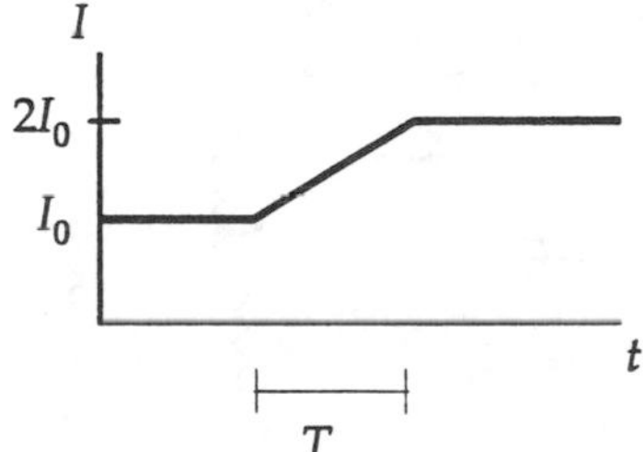

(a) Sketch a rough, qualitative graph of the current in the small ring as a function of time. Since that current flows counterclockwise (when it flows at all), let counterclockwise be the positive direction.

(b) The small ring is made of copper wire of cross-sectional area a. What is the resistance of the small ring? Your answer may include any symbolic constants you want, provided you say what they mean. Call your answer R.

(c) When the current in the large ring is $1.5I_0$, what is the current in the small ring? Neglect the small ring's self-inductance. Answer in terms of R, r, I_0, T, and universal constants. Hint: The magnetic field produced by the large ring is approximately uniform over the whole area of the small ring.

2) Consider two concentric conducting cylindrical shells, both much longer than they are wide. The inner cylinder has radius R, while the outer one has radius $4R$. Both cylinders have length L.

The inner cylinder carries charge $+2Q$, while the outer cylinder carries charge $-Q$.

A particle of charge $-q_0$ and mass M, initially a distance $5R$ from the central axis of the cylinders, is thrown directly toward the cylinders at speed v_0. This experiment happens in outer space, where you can neglect gravity.

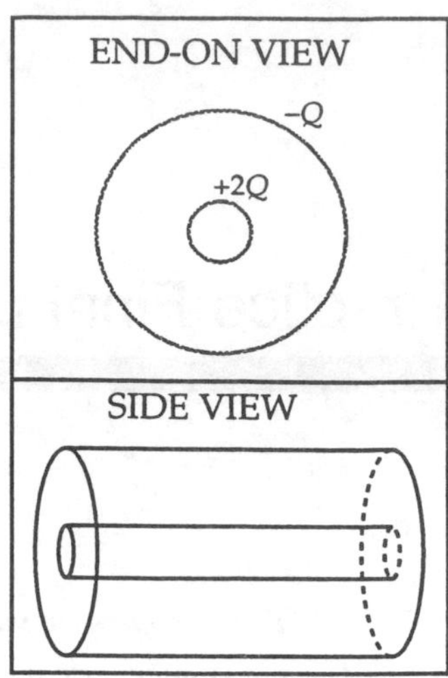

(a) How fast is the particle moving when it reaches the outer cylinder?

(b) The particle passes through a small hole in the wall of the outer cylinder. (The hole doesn't affect the charge distribution on the cylinders.) How fast is the particle moving when it reaches the inner cylinder?

(c) The whole experiment is repeated, except now the space between the two cylinders is filled with dielectric "jelly." When the particle reaches the outer cylinder, is its speed greater than, less than, or equal to the answer you obtained in part (a)? Explain your reasoning.

(d) With the dielectric jelly in place, is the particle's speed when it reaches the inner cylinder greater than, less than, or equal to the answer you obtained in part (b)? The dielectric material allows the particle to pass through frictionlessly.

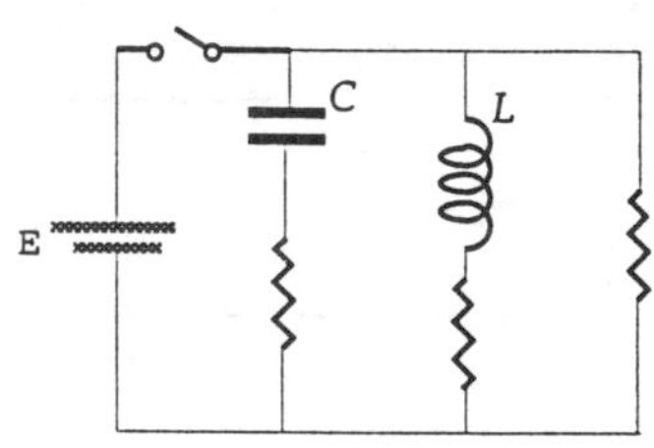

3) In this circuit, all three resistors have resistance R; the battery has EMF $\mathcal{E}$; the capacitor has capacitance C; and the inductor has inductance L. Initially, the capacitor is uncharged. The switch is closed at time $t = 0$.

(a) Let I_b denote the current through the battery. Write down the equation or equations you would need to solve for I_b as a function of time. *Do not actually solve the equation(s).*

(b) Immediately after $t = 0$, what is I_b? You can answer this *without* solving the equations generated in part (a).

(c) A very long time later, what's I_b? Again, you need incredibly little math to solve this.

(d) Without solving your part (a) equations, sketch a rough graph of I_b vs. t. Use intuitive mathematical and/or physical reasoning. Assume the capacitance is relatively small, while the inductance is relatively large (i.e., $L/R > RC$).

4) In this problem, please derive any formulas you use from basic laws.

Two parallel infinite wires, located a distance L apart, both carry current I_0. In this picture, wire 1 carries current into the page, while wire 2 carries current out of the page.

Wire 2 is directly to the right of wire 1. Point P sits midway between them.

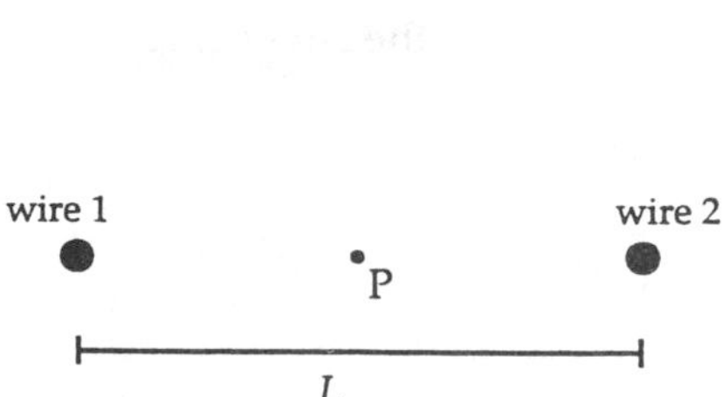

Consider a particle of charge q_1, at point P. What magnetic force acts on the particle if

(a) the particle is motionless?

(b) the particle moves into the page at speed v_1?

(c) the particle moves up the page (towards point C) at speed v_1?

(d) What's the magnetic field at point C, which is equidistant from both wires, a distance $\frac{\sqrt{3}}{2}L$ above point P?

5) Consider a square metal picture frame of side length s, mass M, and total electrical resistance R. It is dropped from rest from a height H above a region of uniform magnetic field pointing into the page. The frame accelerates downward under the influence of gravity until reaching the magnetic field. It is observed that, *while entering the magnetic field*, the frame moves with *constant velocity*.

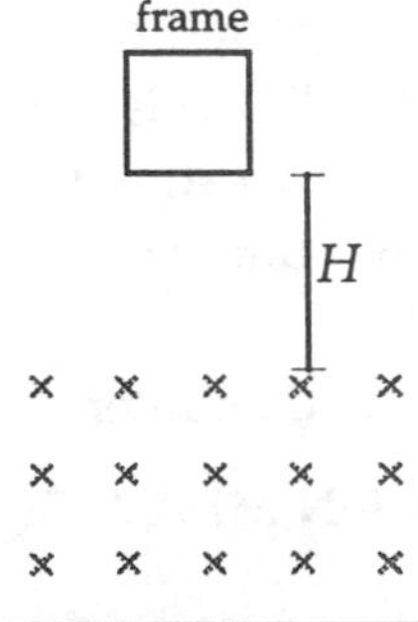

(a) What is the frame's speed when it begins to enter the magnetic field?

(b) What is the strength of the magnetic field? Solve for B in terms of the given physical quantities (s, M, R, and H), and any constants you need.

(c) After the frame has completely entered the magnetic field (i.e., after all four sides are fully "immersed" in the field), what is the frame's acceleration? Justify your answer.

ANSWER 1

(a) A common mistake is to think that current flows around the small ring whenever current flows around the large ring. But according to Faraday's law,

$$\mathcal{E}_{\text{induced}} = -\frac{d\Phi_B}{dt},$$

an EMF (voltage) gets induced in the small ring only when the magnetic flux through that ring *changes*. Here, the current in the large ring generates the magnetic field through the small ring. Therefore, the magnetic flux through the small ring *changes* only when the current in the large ring *changes*.

Let me clarify this explanation, using "I" to denote the current in the large ring. When I stays constant, the magnetic flux through the small ring stays the same. Therefore, no EMF or current gets induced in the small ring. EMF and current get induced in the small ring only when the magnetic flux through that ring *changes*, which happens only when I changes.

To avoid confusion, let i denote the current in the small ring. We just figured out that i equals zero, except when I changes. To figure out the shape of the i vs. t graph when I changes, think about how the current in the large ring relates to the flux through the small ring. According to Ampere's and Biot-Savart's law, the magnetic field created by the large ring is proportional to the current in the large ring. Therefore, the flux through the small ring is proportional to the current in the large ring: $\Phi_B \sim I$. Hence, the rate of change of flux through the small ring is proportional to the rate of change of current in the large ring: $d\Phi_B/dt \sim dI/dt$. Since the induced voltage and current in the small ring are proportional to $d\Phi_B/dt$ by Faraday's law, it follows that i is proportional to dI/dt. In words, the current induced in the small ring is proportional to the rate of change of current in the large ring. In symbols, $i \sim dI/dt$.

The previous paragraph was dense. Please reread it until you can reproduce every step of the reasoning on your own.

In summary, i is proportional to the rate of change of current in the large ring, dI/dt. According to my graph, I increases *steadily* from I_0 to $2I_0$. In other words, dI/dt stays constant during that time interval. Therefore, i remains constant during that interval.

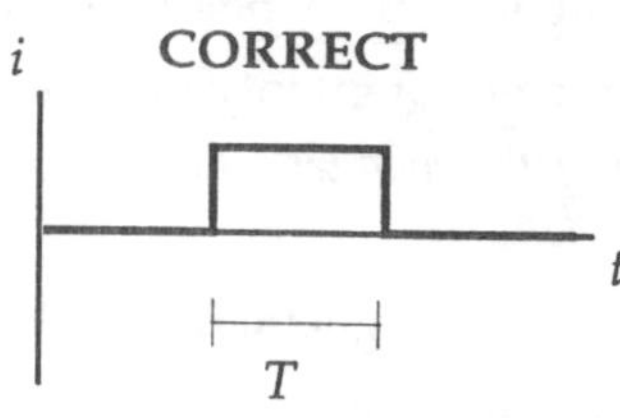

OK, time for a recap. Starting with Faraday's law, we figured out that the i vs. t graph is proportional to the *slope* of the I vs. t graph. In other words, the induced current in the small ring is proportional to the rate of change of the "source" current creating the magnetic field. For this reason, i is zero when I is constant. And i is constant when I steadily changes.

Many students graph i as increasing over the time interval T. This mistake implicitly reflects the view that induced current depends on the *size* of the magnetic flux. But according to Faraday's law, the induced EMF—and hence, the induced current—in the small ring depends not on flux itself, but on the *rate of change* of flux. When I steadily increases, the flux through the small ring changes at a constant rate. Therefore, the induced current stays constant.

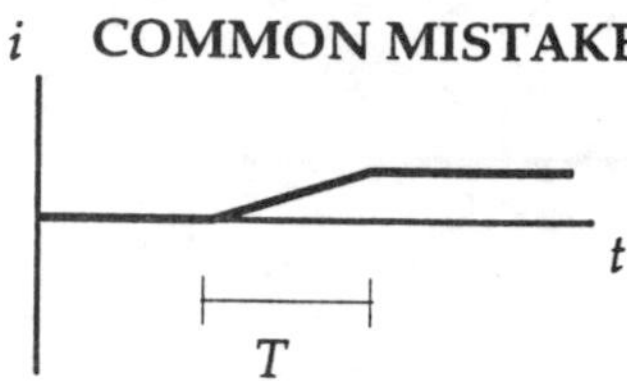

(b) As explained in Chapter 37, any material possesses an intrinsic *resistivity* to electric current. Current flowing through a wire is analogous to cars flowing over a rocky road. So, the resistivity of a wire corresponds to the rockiness (or pot-holiness) of a road. The higher the resistivity, the lower the current, other things being equal.

But the overall resistance of a wire (or roadway) depends on more than just the intrinsic resistivity (rockiness). It depends also on the length, and on the cross-sectional area. Keeping the roadway analogy in mind, you can see intuitively that current flows more easily through a wider (thicker) wire. So, the thicker the wire, the lower its resistance. Also, a longer wire puts up more resistance to current, other things being equal. So, the overall resistance of a wire is proportional to the length, L, and inversely proportional to the cross-sectional area, a:

$$R = \rho_{\text{Cu}} \frac{L}{a} = \rho_{\text{Cu}} \frac{2\pi r}{a},$$

where ρ_{Cu} denotes the resistivity of copper. I substituted in the length (circumference) of the small ring. Note that a means the cross-sectional area of the wire itself, not the area of the circle.

(c) As shown in part (a), i stays constant over the whole time interval during which I increases from I_0 to $2I_0$. There's nothing special about the moment when I reaches $1.5I_0$, because i keeps the *same* value during the entire time interval T. We're solving for i during that time interval.

Let $\mathcal{E}_{\text{induced}}$ denote the induced current in the small ring, and Φ_B denote the flux through that ring. By Faraday's law,

$$\mathcal{E}_{\text{induced}} = -\frac{d\Phi_B}{dt}.$$

Given $\mathcal{E}_{\text{induced}}$, we can invoke Ohm's law ($V = iR$) to find the corresponding induced current. So, the meat of this problem is finding $d\Phi_B/dt$, the rate of change of magnetic flux through the small ring.

Well, Φ_B gets created by the current I in the large ring. So, we can write an expression for Φ_B as a function of I. Then we can differentiate with respect to time, to obtain $d\Phi_B/dt$.

Let me flesh out this strategy.

1. Using Ampere's law or Biot-Savart, find the magnetic field through the small ring, created by the current in the large ring.
2. Using your result from step 1, write the magnetic flux through the small ring as a linear function of I, the current in the large ring: $\Phi_B = MI$, where M is a bunch of constants.
3. Differentiate both sides with respect to t to get $\frac{d\Phi_B}{dt} = M\frac{dI}{dt}$. Since $\frac{dI}{dt}$ is the slope of the I vs. t graph, and since $-\frac{d\Phi_B}{dt}$ equals the induced voltage (by Faraday's law), this step gives us $\mathcal{E}_{\text{induced}}$.
4. Invoke Ohm's law, $V = iR$, to find the induced current i in the small ring.

Here we go . . .

Step 1: Find field created at the small ring by the current in the large ring.

We want the field generated by the large ring, at its center. Unfortunately, any simple Amperian loop through the center of the ring can't help you solve for B, because B won't be constant along that loop. So, we're stuck with Biot-Savart.

Using Biot-Savart to calculate the magnetic field produced by a current is just like using Coulomb's law to calculate the electric field generated by a continuous charge distribution. In either case, you must

(i) Color in a tiny "piece" of the object,
(ii) Write the infinitesimal field due to that tiny piece, and finally
(iii) Add up the fields contributed by all those tiny pieces, to obtain the overall field.

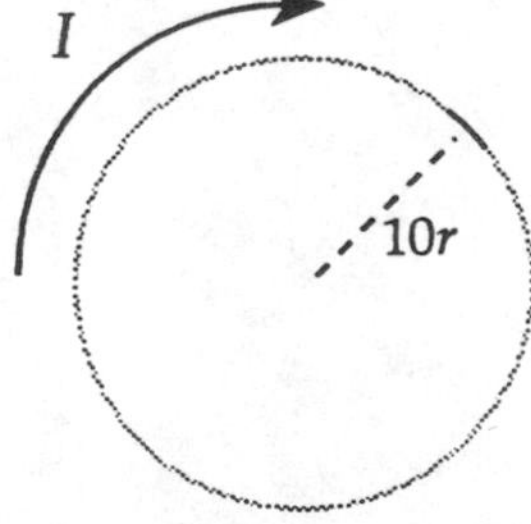

Here, I've blackened a tiny piece of the large ring, of length $d\mathbf{l}$. By Biot-Savart, that bit of wire creates a field

$$d\mathbf{B} = \frac{\mu_0}{4\pi}\frac{I d\mathbf{l} \times \hat{\mathbf{r}}}{r^2},$$

where r denotes the distance from the blackened piece to the location at which we're finding the magnetic field. Here, we want the field at the center. So, "r" is actually $10r$.

To determine the direction of this field, apply the right-hand rule to the cross product $d\mathbf{l} \times \hat{\mathbf{r}}$. First, point your fingers parallel to $d\mathbf{l}$. Then, curl your fingertips toward the point at which we're finding the field, the center of the ring. Your thumb now points into the page. That's the direction of $d\mathbf{B}$.

As you can confirm, every bit of the wire produces an into-the-page $d\mathbf{B}$. Furthermore, since $d\mathbf{l}$ points tangentially to the ring, while $\hat{\mathbf{r}}$ points radially inward, those two vectors are perpendicular. Hence, the magnitude of the cross product reduces to a regular product: $|d\mathbf{l} \times \hat{\mathbf{r}}| = dl$, because the unit vector $\hat{\mathbf{r}}$ has length 1. Putting all this together, we get

$$d\mathbf{B} = \frac{\mu_0}{4\pi} \frac{Idl}{(10r)^2}.$$

Now add up (integrate over) these tiny field contributions. Fortunately, every bit of the large ring is the same distance, $10r$, from the center. So, "$10r$" is a constant we can pull outside the integral. And since all the $d\mathbf{B}$'s point into the page, we can add them without worrying about vector components:

$$\begin{aligned} B = \int dB &= \int_{\text{large ring}} \frac{\mu_0}{4\pi} \frac{Idl}{(10r)^2} \\ &= \frac{\mu_0 I}{4\pi(10r)^2} \int_0^{2\pi(10r)} dl \\ &= \frac{\mu_0 I}{4\pi(10r)^2} [2\pi(10r)] \\ &= \frac{\mu_0 I}{2(10r)}. \end{aligned}$$

That's the magnetic field at the center of the rings, created by the current in the large ring.
Step 2: Solve for magnetic flux through small ring, as a linear function of current in large ring.

As the problem tells us, the magnetic field is approximately uniform over the whole small ring. Therefore, in the flux integral $\Phi_B = \int \mathbf{B} \cdot d\mathbf{A}$, we can pull $\mathbf{B}$ outside:

$$\begin{aligned} \Phi_B &= \int \mathbf{B} \cdot d\mathbf{A}_{\text{small ring}} \\ &= \int \mathbf{B} \cdot d\mathbf{A}_{\text{small ring}} \\ &= BA_{\text{small ring}} \qquad \text{[since the ring "faces" into the page]} \\ &= \frac{\mu_0 I}{2(10r)} \pi r^2 \\ &= \frac{\mu_0 \pi r}{20} I. \end{aligned}$$

In the second to last step, I substituted in the area of the small ring ($A = \pi r^2$), along with our step 1 expression for the magnetic field created by the large ring.

As always, Φ_B is a linear function of I: $\Phi_B = MI$. In this particular case, $M = \frac{\mu_0 \pi r}{20}$. That constant is called the "mutual inductance," for reasons we'll see below.

Step 3: Differentiate with respect to t to find the EMF in the small ring.

According to Faraday's law, the induced EMF in the small ring is $\mathcal{E}_{\text{induced}} = -\frac{d\Phi_B}{dt}$. Since $\Phi_B = \frac{\mu_0 \pi r}{20}$ I, we get

$$\mathcal{E}_{\text{induced}} = -\frac{d\Phi_B}{dt}$$
$$= -\frac{\mu_0 \pi r}{20}\frac{dI}{dt}.$$

As you know, dI/dt corresponds to the *slope* of the I vs. t graph. We're looking for the slope when I equals $1.5I_0$. That's easy, because the slope stays constant as I increases from I_0 to $2I_0$:

$$\frac{dI}{dt} = (\text{slope of } I \text{ vs. } t) = \frac{rise}{run} = \frac{2I_0 - I_0}{T} = \frac{I_0}{T}.$$

So,

$$\mathcal{E}_{\text{induced}} = -\frac{\mu_0 \pi r}{20}\frac{dI}{dt} = -\frac{\mu_0 \pi r}{20}\frac{I_0}{T}.$$

Below, I'll discuss the physical meaning of the minus sign. For now, I just want you to notice that our expression for $\mathcal{E}_{\text{induced}}$ takes the form $\mathcal{E}_{\text{induced}} = -M\frac{dI}{dt}$. Physically, the mutual inductance M specifies how strongly the small ring "reacts" to a changing current in the large ring.

Step 4: Use Ohm's law to find the current produced by this EMF.

The small ring has resistance R. So, by Ohm's law, the induced current in the small ring is

$$i = \frac{\mathcal{E}_{\text{induced}}}{R} = -\frac{\mu_0 \pi r I_0}{20RT}.$$

The minus sign indicates that the induced current opposes the changing flux. Here, as the current around the large ring increases, the into-the-page flux through the small ring increases. According to Lenz' law, the induced current creates its own magnetic field that opposes this increase. So, the induced current generates an out-of-the-page field. As you can confirm using the right-hand rule shortcut, current must flow counterclockwise around the small ring to generate an out-of-the-page field at points inside the small ring.

ANSWER 2

As always, begin by thinking physically, not by rummaging through formulas. Since the two-cylinder system carries a net positive charge, it attracts the negative particle. (The attractive force from the +2Q cylinder overcomes the repulsive force from the $-Q$ cylinder.) Therefore, the particle speeds up.

To calculate the particle's speed, you might try using forces and acceleration. Unfortunately, the electric fields and forces are not constant. The particle speeds up with non-constant acceleration. Therefore, those beloved kinematic equations, such as $v^2 = v_0^2 + 2a\Delta x$, do not apply.

You can solve more efficiently using energy conservation. As the particle "falls" toward the cylinders, it loses potential energy and gains kinetic energy.

(a) Let subscript "1" denote the moment when the particle reaches the outer cylinder. By energy conservation,

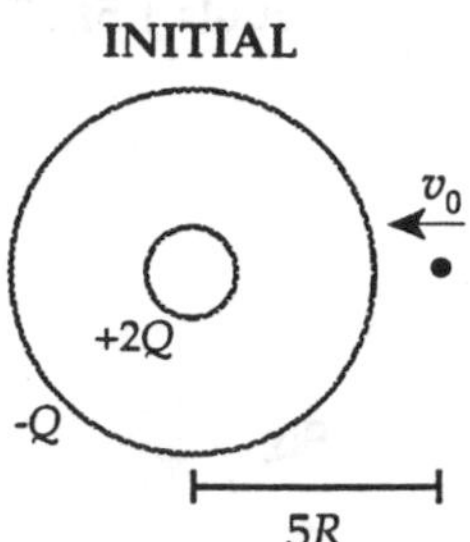

$$K_0 + U_0 = K_1 + U_1$$

$$\frac{1}{2}Mv_0^2 + (-q_0)V_0 = \frac{1}{2}Mv_1^2 + (-q_0)V_1.$$

Here, I used the relationship between potential and potential energy, $U = qV$. The particle carries charge $q = -q_0$.

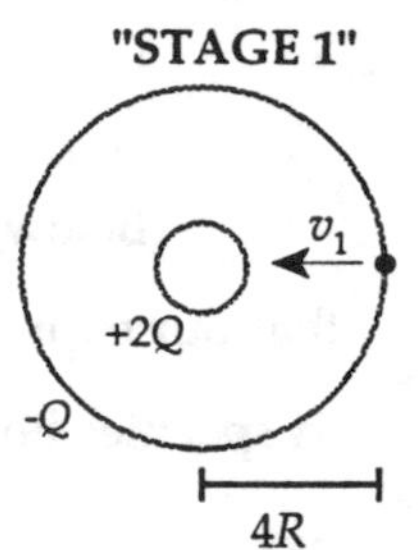

As usual, the answer—in this case, v_1—does not depend on either V_0 or V_1 individually. It depends only on the potential *difference*, $\Delta V = V_1 - V_0$. To see this explicitly, solve the energy conservation equation for v_1 to get

$$v_1 = \sqrt{v_0^2 + \frac{2q_0}{M}(V_1 - V_0)} = \sqrt{v_0^2 + \frac{2q_0}{M}\Delta V} \qquad (1)$$

So, the meat of this problem is finding the potential difference between the particle's initial position, 5R from the center, and its "stage 1" position, 4R from the center.

Well, as we've seen before, $\Delta V = -\int_{\text{initial}}^{\text{stage 1}} \mathbf{E} \cdot d\mathbf{s}$. To evaluate this integral, we need to know the electric field along a path connecting those two points. So, we need the electric field *outside* the two cylinders. Gauss' law beckons.

Let me summarize the strategy that's popped out of this reasoning.

1. Using Gauss' law, find the electric field outside the two cylinders.
2. Integrate over that field to calculate ΔV, the potential difference between the particle's initial position and its stage 1 position.
3. Substitute that potential difference into Eq. (1) above, which we found using energy conservation. This gives us v_1.

Time to dive in.

Step 1: Using Gauss' law, find the electric field outside the two cylinders.

In Chapter 34, I laid out a strategy for using Gauss' law to find electric fields. By now, you can probably combine and abbreviate the steps. Nonetheless, for the sake of review, I'll briefly present all the reasoning.

Here, you can use the superposition principle, by finding the electric field generated by each individual cylinder, and adding those fields together. Alternatively, you can treat the two cylinders as a single system. I'll use the "single system" method. But when finding the electric field created by two or more differently-shaped or non-concentric objects, you must use superposition.

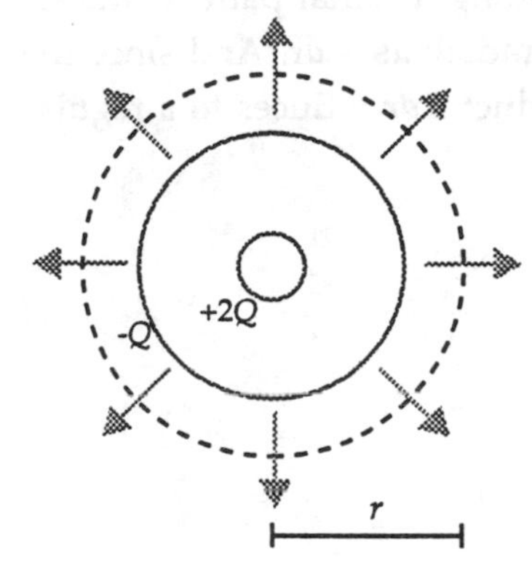

Gauss' law steps 1 and 2: Draw field lines, and draw "good" Gaussian surface.

To find the electric field using Gauss' law,

$$\oint \mathbf{E} \cdot dA = \frac{q_{\text{encl}}}{\varepsilon_0},$$

we must pull E outside that flux integral. Therefore, the field must have the same strength over (part of) the Gaussian surface. For this reason, I've drawn a dashed "Gaussian cylinder." It extends a distance L into the page, just like the charged cylinders do. By symmetry, the field has the same strength over the entire side (tube) of this Gaussian surface.

Gauss' law step 3: Solve for the electric field using Gauss' law.

Therefore, we can pull the field outside the Gaussian flux integral:

$$\oint \mathbf{E} \cdot d\mathbf{A} = E \int_{\text{tube}} dA = EA_{\text{tube}} = E(2\pi rL).$$

I integrated over the tube only, because essentially no field lines flow through the ends of this Gaussian surface. See question 34-3b for a fuller explanation.

Since my Gaussian surface completely surrounds both cylinders, it encloses charge

$$q_{\text{encl}} = 2Q + (-Q) = +Q.$$

So, by Gauss' law,

$$\oint \mathbf{E} \cdot d\mathbf{A} = \frac{q_{\text{encl}}}{\varepsilon_0}$$

$$E(2\pi rL) = \frac{Q}{\varepsilon_0},$$

and hence

$$E = \frac{Q}{L(2\pi\varepsilon_0 r)}.$$

That's the electric field a distance r from the center of the cylinders, where $r > 4R$. Now let's return to our overall strategy. Remember, we're finding the potential difference between the particle's initial position and its stage 1 position, so that we can substitute ΔV into Eq. (1) above.

Step 2: Using this field, find the potential difference, ΔV, between "initial" and "stage 1."

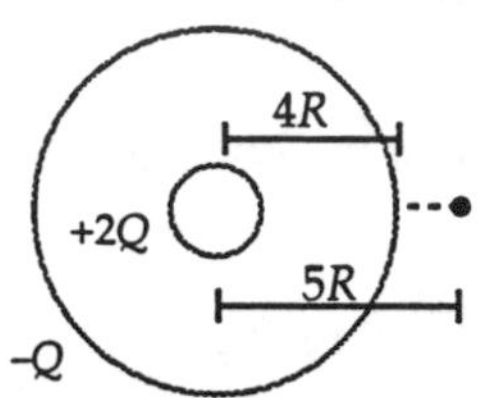

Dashed line is path of integration.

We just found the field outside the cylinders. It's $E(r) = \frac{Q}{L(2\pi\varepsilon_0 r)}$. Now we must integrate this field along a path from the particle's initial position to its stage 1 position. I'll choose a straight radial path, as drawn here. So, our infinitesimal length element is radial: $d\mathbf{s} = d\mathbf{r}$. And since the electric field also points radially, the dot product $\mathbf{E}\cdot d\mathbf{r}$ reduces to a regular product. Putting all this together, I get

$$\begin{aligned}\Delta V &= -\int_{\text{initial}}^{\text{stage}} \mathbf{E}\cdot d\mathbf{s} \\ &= -\int_{5R}^{4R} E\,dr \\ &= -\int_{5R}^{4R} \frac{Q}{L(2\pi\varepsilon_0 r)}dr \\ &= -\frac{Q}{2\pi\varepsilon_0 L}\ln r\Big|_{5R}^{4R} \\ &= +\frac{Q}{2\pi\varepsilon_0 L}\ln\frac{5}{4}.\end{aligned}$$

Let me discuss two common errors. Many students integrate from 0 to R, or something like that. But remember, the "r" in $E(r) = \frac{Q}{L(2\pi\varepsilon_0 r)}$ refers to the distance from the *center* of the cylinders. Therefore, when integrating over r, you're committed to setting $r = 0$ at the center, not at some other point. A second mistake is to plug a specific value of r, such as $4R$ or $5R$, into your expression for the electric field. In that case, you end up with $\Delta V = -\int \frac{Q}{L2\pi\varepsilon_0(5R)}dr$, or something like that. But as the particle travels from $r = 5R$ to $r = 4R$, it "feels" the electric field at all points between those limits. That's why $5R$ and $4R$ show up as limits of integration, not as part of your expression for electric field. You must integrate over the *general* expression for $E(r)$, without a specific r substituted in. The integral $-\int \frac{Q}{L2\pi\varepsilon_0(5R)}dr$ tells you the potential change the particle would experience if it "felt" the *same* electric field ($E = \frac{Q}{L2\pi\varepsilon_0(5R)}$) during its entire journey. But the particle feels a *different* electric field at each different r.

Step 3: Use this potential difference in your energy conservation equation.

Way back at the beginning of this problem, we invoked energy conservation to find the particle's speed at the outer cylinder. Our answer depended on ΔV, the potential difference between "initial" and "stage 1." We just found that potential difference. Substitute it into Eq. (1), to get

$$v_1 = \sqrt{v_0^2 + \frac{2q_0}{M}\Delta V} = \sqrt{v_0^2 + \frac{2q_0}{M}\left(\frac{Q}{2\pi\varepsilon_0 L}\ln\frac{5}{4}\right)}.$$

(b) Here, we can replicate the reasoning of part (a). Let subscript "f" denote the moment at which the particle reaches the inner cylinder. According to energy conservation,

$$K_1 + U_1 = K_f + U_f$$

$$\frac{1}{2}Mv_1^2 + (-q_0)V_1 = \frac{1}{2}Mv_f^2 + (-q_0)V_f.$$

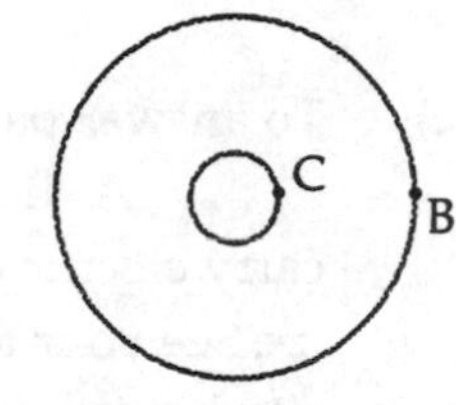

(You could also set the final energy equal to the initial energy, $K_0 + U_0$.) Solve for v_f to get

$$v_f = \sqrt{v_1^2 + \frac{2q_0}{M}\Delta V_{BC}},$$

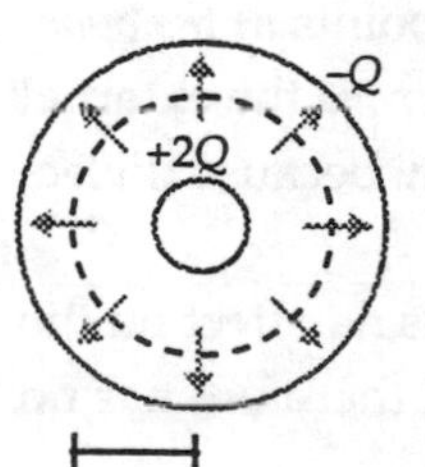

where $\Delta V_{BC} = V_f - V_1$ denotes the change in potential between points B and C on this diagram.

To evaluate $\Delta V_{BC} = -\int_B^C \mathbf{E}\cdot d\mathbf{s}$, we need the electric field *between the two cylinders*. So, draw a cylindrical Gaussian cylinder between the two charged cylinders, of arbitrary radius r. This Gaussian surface, unlike the one in part (a), encloses only the inner cylinder. Therefore, $q_{encl} = +2Q$. So,

$$\oint \mathbf{E}\cdot d\mathbf{A} = \frac{q_{encl}}{\varepsilon_0}$$

$$E(2\pi rL) = \frac{2Q}{\varepsilon_0},$$

and hence

$$E = \frac{2Q}{2\pi\varepsilon_0 rL} = \frac{Q}{\pi\varepsilon_0 rL}.$$

That's the field between the two cylinders.

To find the potential difference between points B and C, integrate that field from point B, a distance $4R$ from the center, to point C, a distance R from the center:

$$\Delta V_{BC} = -\int_B^C \mathbf{E}\cdot d\mathbf{s}$$
$$= \int_{4R}^{R} \frac{Q}{\pi r \varepsilon_0 L} dr$$
$$= \ldots \text{ math steps equivalent to those in part (a)} \ldots$$
$$= +\frac{Q}{\pi \varepsilon_0 L} \ln 4.$$

Substitute this potential change into the above formula for v_f, to get

$$v_f = \sqrt{v_1^2 + \frac{2q_0}{M}\Delta V_{BC}} = \sqrt{v_1^2 + \frac{2q_0}{M}\left(\frac{Q}{\pi \varepsilon_0 L}\ln 4\right)},$$

where v_1 is our part (a) answer. On a test, most instructors would award full credit for expressing v_f as a function of v_1, even if your answer for v_1 was wrong.

(c) To answer parts (c) and (d), we must review the behavior of dielectrics.

A dielectric is electrically neutral. It carries no net positive or negative charge. Nor does it carry electric current. In the presence of an electric field, however, the dipoles in the dielectric *polarize.* See your textbook for details. The polarized dipoles create an electric field that points *in the opposite direction* from the original (external) electric field. In this way, the dielectric *dampens* the external electric field. The field inside a dielectric is *weaker* than it otherwise would have been, because dielectric's field partially cancels the external field.

Nonetheless, in highly symmetrical situations such as this, the dielectric has no effect on the electric field *outside* of itself. That's because the dielectric carries no net charge, and therefore has no effect on q_{encl} when the Gaussian surface completely surrounds the dielectric.

Using these insights, we can immediately solve parts (c) and (d). The dielectric fills the space *between* the cylinders. Since the dielectric does not affect the electric field outside of itself, the field *outside* both cylinders is the same as it was before. Therefore, the potential difference between "initial" and "stage 1" is the same as it was in part (a). So, the particle reaches the outer cylinder with speed v_1, the same speed we found in part (a).

(d) By contrast, the dielectric dampens the electric field between the cylinders. Therefore, while between the cylinders, the particle gets pushed less hard than it did before. Consequently, it reaches the inner cylinder with less speed than it had in part (b).

Let me re-explain this in terms of energy. Since the dielectric decreases **E** between the cylinders, it decreases $\Delta V_{BC} = -\int_B^C \mathbf{E}\cdot d\mathbf{s}$. As shown above, the particle's speed at the inner cylinder depends on ΔV_{BC}. By reducing the potential difference between B and C, the dielectric reduces how much potential energy the particle loses while "falling" from B to C. For this reason, the particle gains less kinetic energy during the fall, and hits the inner cylinder with less speed than it had in part (b).

ANSWER 3

(a) To solve a multi-loop circuit, use Kirchhoff's laws, as explained in Chapter 38. Here's the condensed strategy.

> **Multi-loop circuit strategy**
>
> 1) Choose and label your loops. Also, "guess" the current's direction in each segment of the circuit, and give those currents names, whether they're known or unknown.
>
> 2) Apply Kirchhoff's 1st law, the "intersection rule," to relate the currents in the different parts of the circuit: *The current flowing into an intersection equals the current flowing out of that intersection.*
>
> 3) Use Kirchhoff's 2nd law on each loop individually: *The total change in potential around a loop is zero.* Apply this rule by "walking around the loop" and writing down the potential change across each circuit element. Using the car elevator and pot-holed hill analogies for batteries and resistors, we saw that
>
> (a) If you walk "up" the battery (from negative to positive terminal), the potential changes by $+\mathcal{E}$. If you walk from positive to negative terminal, it's $-\mathcal{E}$.
>
> (b) If you walk across a resistor with the current, then you're walking "downhill," and the potential change is $-IR$. If you walk against the current, it's $+IR$.
>
> (c) If you walk across a capacitor, your change in potential is plus or minus Q/C.
>
> (d) If you walk across an inductor with the current, the induced voltage is *negative* $L\,di/dt$.

I'll now implement this strategy.

Step 1: Label your currents and loops.

Let i_1, i_2, and i_3 denote the current through the capacitor, inductor, and "end" resistor, respectively. I_b denotes the current through the battery.

To set up the relevant equations, we'll need to consider three loops. The equations come out simplest if you include the battery in each loop. So, my loop 1 includes the battery and capacitor. My loop 2 goes up through the battery and down through the inductor. Loop 3 goes around the perimeter of the circuit, up through the battery and down through the "end" resistor on the far right side. Of course, you could equally well choose other loops, such as the three small ones. A different choice of loops leads to a different set of equations. But those different sets of equations are equivalent; they yield the same answers for I_b, i_1, i_2, and i_3.

MY CHOICE OF LOOPS

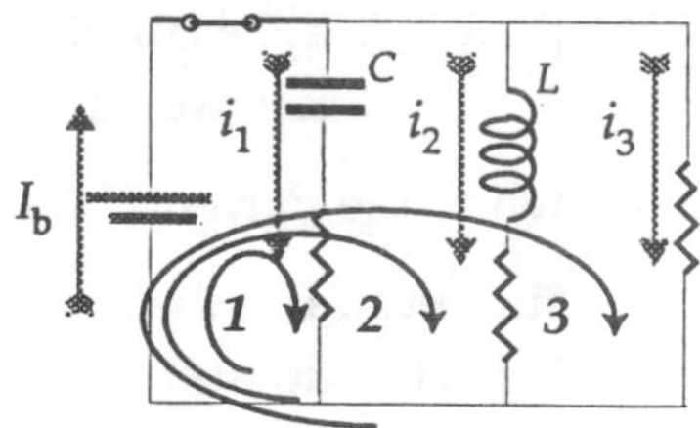

ALTERNATE CHOICE OF LOOPS

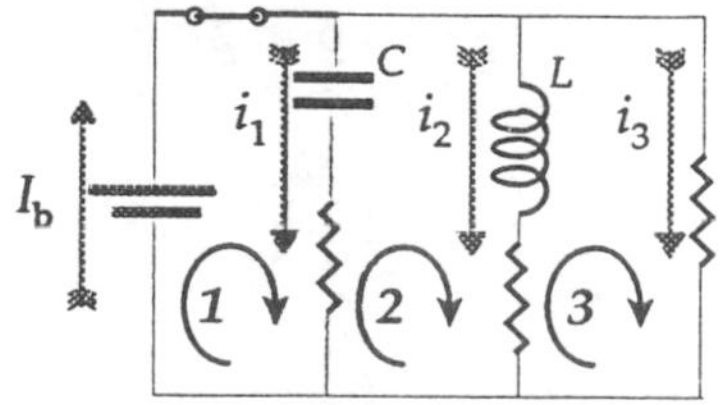

Step 2: Apply Kirchhoff's 1st law, the "intersection rule," to relate the currents.

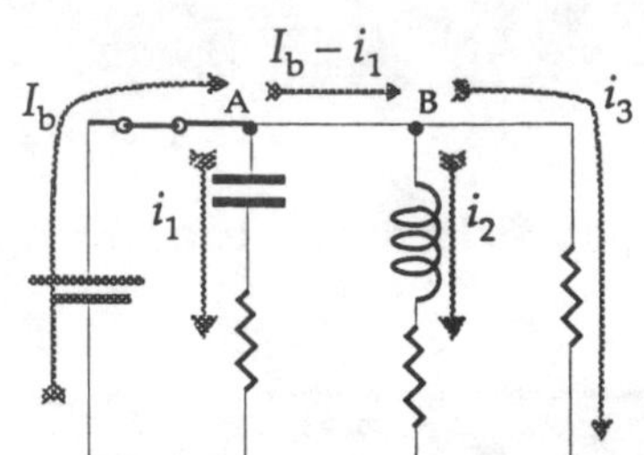

Focus first on point A, the junction above the capacitor. Since the current flowing into the intersection (I_b) must equal the current flowing out of that intersection ($i_1 + i_{\text{rightward from A}}$), the current flowing from point A to point B must be $I_b - i_1$.

Apply similar reasoning to point B. Current $I_b - i_1$ flows into that intersection. Currents i_2 and i_3 flow out. So, again by Kirchhoff's 1st law,

$$I_b - i_1 = i_2 + i_3 \qquad \text{[Kirchhoff' s 1st law at point B]}$$

To make this equation more intuitive, move i_1 to the right-hand side. This gives

Intersection rule $$I_b = i_1 + i_2 + i_3. \tag{0}$$

Eq. (0) says that the overall current I_b splits into three smaller currents, which eventually re-merge.

Step 3: Write Loop equations using Kirchhoff's 2nd law.

Consider my original choice of loops as drawn above, not the alternate choice of loops. I'll walk clockwise around loop 1, starting from below the battery. So first, I "climb" through the battery (from negative to positive terminal), which changes my potential by $+\mathcal{E}$. Next, I walk through the capacitor. Since current flows onto the top plate, the top plate acquires positive charge. So, I'm walking from the positive to the negative plate—a negative change in potential. Specifically, my potential changes by $-Q/C$, where Q denotes the charge on the plates.

Next, I walk through the resistor *with* the current (i.e., "downhill"). Hence, my potential changes by $-i_1R$. And I encounter no other circuit elements before completing the loop. Since my overall change in potential is zero,

Loop 1 $$\mathcal{E} - Q/C - i_1R = 0 \tag{1}$$

This equation contains two physical quantities that change in time: The charge Q on the capacitor plates, and the current i_1 flowing "through" the capacitor.

Since we still lack sufficient information to solve for I_b, consider loop 2. By Faraday's law, when the current through an inductor *changes*, an EMF $\mathcal{E}_{\text{induced}} = -L\frac{di}{dt}$ gets induced. The minus sign indicates that the induced voltage acts contrary to the change in the current. Roughly put, the inductor doesn't want the current through it to change. Combining this term with the voltage changes across the other elements of loop 2, I get

Loop 2 $$\mathcal{E} - L\frac{di_2}{dt} - i_2R = 0 \tag{2}$$

Loop 3 is simpler, because it contains only a battery and a resistor:

Loop 3 $$\mathcal{E} - i_3R = 0 \tag{3}$$

In summary, we've used Kirchhoff's 1st and 2nd laws to generate four equations. Is this enough information to solve for I_b? Well, we don't know I_b, i_1, i_2, or i_3. But that's not all. We also don't know Q, the charge on the capacitor as a function of time. Sure, we know Q at $t = 0$. But we don't know Q at arbitrary later time. So, it's a fifth unknown. And we have only four equations. Uh oh!

Fortunately, we can generate another equation by thinking physically about the capacitor. It charges up *because* current flows onto the top plate (and off the bottom plate). Indeed, since current is charge per time, i_1 tells us the rate at which the capacitor gets charged up:

Loop 1 additional information $$\frac{dQ}{dt} = i_1. \qquad (4)$$

For instance, if $i_1 = 3$ coulombs per second, then three coulombs of charge get dumped onto the capacitor each second. Let me explain why this equation contains no minus sign. When i_1 is positive, the charge on the top plate *increases*, and hence, dQ/dt is positive. By contrast, when a capacitor discharges, i is positive when Q *decreases*, and hence, $i = -dQ/dt$.

At this point, we have five equations in the five unknowns. So, assuming differential equations (2) and (4) can be dealt with—which they can!—we have enough information.

(b) Parts (b) through (d) rely on the following insight: i_1, i_2, and i_3 are independent. Mathematically, this follows from the above equations. Eqs. (1) and (4) allow us to solve for i_1, without having to worry about i_2 or i_3. From Eq. (2) alone, we can solve for i_2. And from Eq. (3) alone, we can solve for i_3. Physically, each of my three loops behaves independently. For instance, $i_1(t)$ does not depend on the inductor, and $i_2(t)$ does not depend on the capacitor. Each loop "does its own thing," unaffected by the other loops. That's unusual, and would not be true if any of the resistors were "shared" by two loops.

Here's the point: We can *separately* figure out i_1, i_2, and i_3 at $t = 0$. Then, we can add up those three current to obtain I_b at $t = 0$:

$$I_b(0) = i_1(0) + i_2(0) + i_3(0).$$

Subproblem 1: Find $i_1(0)$, the current through the capacitor at $t = 0$.

At $t = 0$, essentially no charge has had time to flow onto the capacitor. So, $Q \approx 0$. Therefore, the capacitor creates essentially no voltage: $V = Q/C \approx 0$. So, at $t = 0$, the capacitor plays no role in loop 1. In other words, loop 1 initially behaves *as if* the capacitor were a resistanceless wire. At $t = 0$, loop 1 might as well contain nothing except the battery and resistor. Therefore, $\mathcal{E} - i_1R = 0$, and hence $i_1 = \mathcal{E}/R$. That's the current through the capacitor at $t = 0$.

Subproblem 2: Find $i_2(0)$, the current through the inductor at $t = 0$.

Physically, an inductor "fights" a change in current. It induces a backward EMF that tries to prevent the current from changing. Here, when the switch closes, the battery wants i_2 to change from

0 to some positive value. The inductor makes this transition happen gradually, as opposed to abruptly. So, immediately after $t = 0$, the current through the inductor is still $i_2(0) \approx 0$.

Subproblem 3: Find $i_3(0)$, the through the "end" resistor at $t = 0$.

Loop 3 is easy. At *any* time, $i_3 = \mathcal{E}/R$, a result you can obtain from Eq. (3) above.

Add these three little i's to get

$$\begin{aligned} I_b(0) &= i_1(0) + i_2(0) + i_3(0) \\ &= \mathcal{E}/R \;+\; 0 \;+\; \mathcal{E}/R \\ &= 2\mathcal{E}/R. \end{aligned}$$

(c) Here, we want the current through the battery a long time later, at $t \approx \infty$. As in part (b), we can *separately* calculate the currents through the capacitor, inductor, and "end" resistor at $t = \infty$. Then we can add those three currents to obtain $I_b(\infty)$.

Subproblem 1: Find $i_1(\infty)$, the current through the capacitor a long time later.

As the capacitor becomes more and more charged up, it creates a voltage that partially counteracts the battery's voltage. Consequently, i_1 decreases. When the capacitor finishes charging up, its voltage "cancels" the battery's voltage, and therefore no more current flows through the capacitor. Intuitively, once the capacitor "fills up" with charge, no more charge can flow onto it; the capacitor becomes a "dead end" through which no current flows. So, $i_1(\infty) = 0$. Remember, as explained in Chapter 36, current flows "through" a capacitor because charges flow onto the positive plate and off the negative plate. When the positive plate no longer "accepts" charges, and the negative plate no longer "spits out" charges, the current stops.

Subproblem 2: Find $i_2(\infty)$, the current through the inductor a long time later.

As we saw above, the inductor delays the process by which the current around loop 2 increases from 0 to some final value. But in the long run, the inductor can't prevent the current from reaching whatever value the battery wants it to reach. Here's why. When the current finally reaches its maximum value and becomes steady, $di_2/dt = 0$. But the induced EMF depends on the *rate of change* of the current, di_2/dt. So, once the current stops changing, the inductor no longer induces an EMF; it becomes "inactive," and might as well be a straight piece of wire. When i_2 "settles" to its final value, loop 2 behaves as if it contains only the battery and resistor. So, $i_2(\infty) = \mathcal{E}/R$.

Subproblem 3: Find $i_3(\infty)$, the current through the end resistor a long time later.

As noted above, $i_3 = \mathcal{E}/R$ at all times.

Putting all this together yields

$$I_b(\infty) = i_1(\infty) + i_2(\infty) + i_3(\infty)$$
$$= 0 + \mathcal{E}/R + \mathcal{E}/R$$
$$= 2\mathcal{E}/R.$$

Compare this to our part (b) answer. The initial current through the circuit equals the current a long time later! But the current doesn't stay constant between $t = 0$ and $t = \infty$. As we'll see below, the current takes a "dip" before eventually returning to its initial value.

(d) Since $I_b = i_1 + i_2 + i_3$, we can graph I_b vs. t by modifying our strategy from parts (b) and (c). First, consider each loop *separately*, and graph i_1 vs. t, i_2 vs. t, and i_3 vs. t. Then, add those three graphs to get I_b vs. t.

Subproblem 1: By considering loop 1, graph i_1 vs. t.

Loop 1 is a standard RC circuit with a battery. At $t = 0$, when the capacitor is uncharged, it creates no voltage. Therefore, current passes through the capacitor *as if* it were a resistanceless wire. By contrast, as the capacitor acquires more and more charge, it generates a voltage that partially counteracts the battery's voltage. As a result, the current decreases. When the capacitor finishes charging up, its voltage "cancels" the battery's voltage, and therefore no more current flows through.

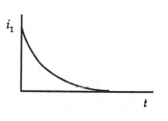

By translating this qualitative description into a graph, you get the "decaying current" drawn here. The decay turns out to be exponential, as you could confirm by solving Eqs. (1) and (4). (You should know that RC circuits charge up and decay exponentially.) Because of the small capacitance, the capacitor "fills up" quickly. Therefore, this exponential decay happens quickly.

Subproblem 2: By considering loop 2, graph i_2 vs. t.

Loop 2 is a standard LR circuit with a battery. As explained above, when the switch closes, the battery "wants" the current to shoot up from 0 to $\mathcal{E}/R$. The inductor can't prevent this from happening. But it slows down the process. The current shoots up gradually instead of abruptly. But eventually, i_2 settles to its final value and stops changing. Because the inductance is large, this process takes a "long" time, maybe a few hundredths or tenths of a second instead of a few milliseconds.

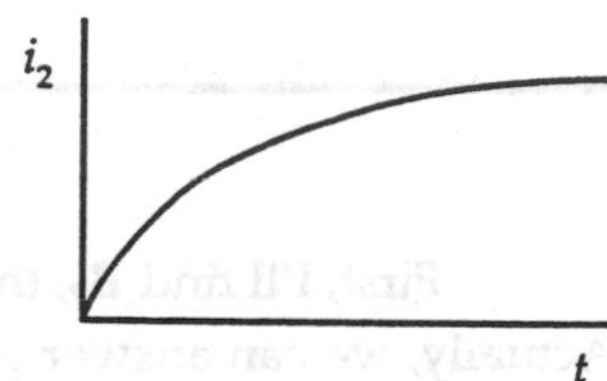

Subproblem 3: By considering loop 3, graph i_3 vs. t.

Loop 3 contains no inductor to delay the current from reaching its final value. Therefore, in about the time needed for light to travel from one end of the circuit to the other, i_3 attains its final value. For all practical purposes, i_3 reaches its final value instantaneously.

We've graphed i vs. t for each "leg" of the circuit. Since $I_b = i_1 + i_2 + i_3$, we can find I_b vs. t by adding the three graphs:

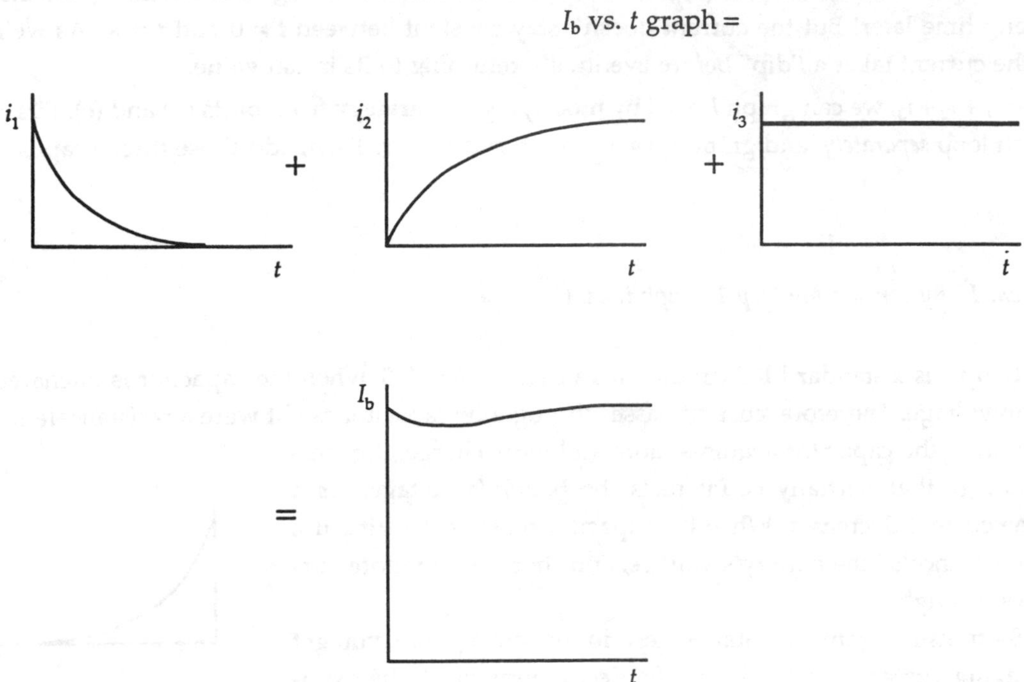

Because the capacitance is comparatively small while the inductance is comparatively large, i_1 falls more rapidly than i_2 rises. That's why I_b initially "dips." But as we found in part (c), the current eventually converges towards its initial value.

ANSWER 4

First, I'll find $\mathbf{B}_P$, the magnetic field at point P. Using that field, I'll then answer parts (a) through (c). (Actually, we can answer part (a) without knowing the field. More on this later.)

A common mistake is to try to calculate $\mathbf{B}_P$ in one step. To figure out $\mathbf{B}_P$ more easily, first find the field generated by each *separate* wire. Then use superposition: $\mathbf{B}_P = \mathbf{B}_1 + \mathbf{B}_2$, where $\mathbf{B}_1$ and $\mathbf{B}_2$ denote the field generated at point P by wires 1 and 2, respectively. To carry out this strategy, we need to calculate the field generated by a *single* infinite wire. But we've done that before . . .

Subproblem: Use Ampere's law to find the field produced by a single infinite wire.

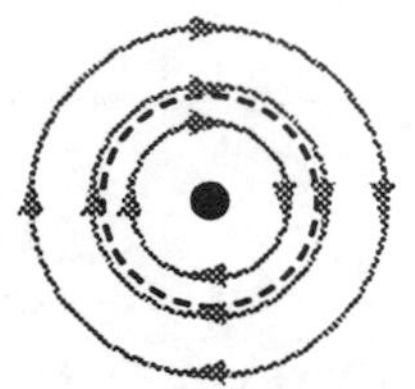

The dashed Amperian loop runs along a field line.

In order to solve for B using Ampere's law,

$$\oint \mathbf{B} \cdot d\mathbf{s} = \mu_0 I_{\text{encl}},$$

you must be able to pull B outside the line integral. Therefore, the field strength must be constant along (part of) your Amperian loop. For this reason, I've drawn a circular loop. By symmetry, the field has the same strength at any two points equidistant from the wire. Therefore, B is constant over this entire loop.

For now, I'll give the Amperian loop arbitrary radius r. In parts (a) through (c), we'll set $r = L/2$. But in part (d), r will take on a new value.

As discussed in Chapter 40, an infinite wire generates circular field lines. So, the field lines point along my Amperian loop, reducing the dot product $\mathbf{B} \cdot d\mathbf{s}$ to a regular product:

$$\oint \mathbf{B} \cdot d\mathbf{s} = \int_0^{2\pi r} B ds = B \int_0^{2\pi r} ds = B(2\pi r).$$

Since the Amperian loop encloses the whole wire, $I_{\text{encl}} = I_0$. So, from Ampere's law,

$$\oint \mathbf{B} \cdot d\mathbf{s} = \mu_0 I_{\text{encl}}$$

$$B(2\pi r) = \mu_0 I_0,$$

and hence

$$B = \frac{\mu_0 I_0}{2\pi r}.$$

End of Ampere's law subproblem

Both wires are a distance $r = L/2$ from point P. Therefore, at point P, they both produce a field of strength

$$B_1 = B_2 = \frac{\mu_0 I_0}{2\pi(L/2)}.$$

Do those fields reinforce each other, or cancel? To find out, we must determine whether the field lines flow clockwise or counterclockwise around each wire.

According to the right-hand rule shortcut, you can reason as follows. Point your thumb in the direction of the current. Your fingertips now curl in the direction of the field lines. For instance, the current in wire 1 flows into the page. When you point your thumb into the page, your fingertips curl clockwise. So, the field lines around wire 1 flow clockwise. By similar reasoning, the field lines around wire 2 flow counterclockwise. Therefore, as this diagram shows, *both* wires produce a downward field at point P. The combined field has magnitude

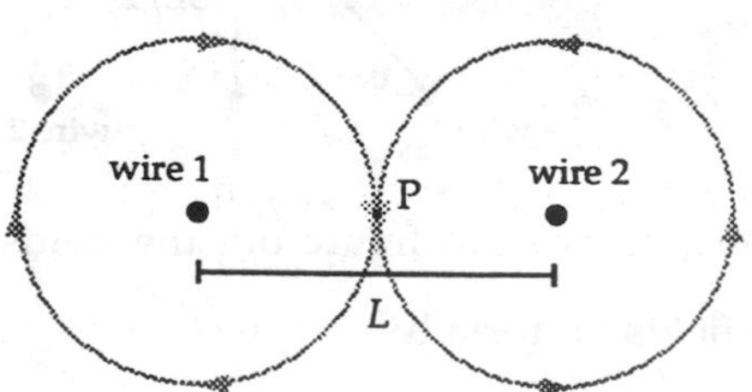

$$B_P = B_1 + B_2$$
$$= \frac{\mu_0 I_0}{2\pi(L/2)} + \frac{\mu_0 I_0}{2\pi(L/2)}$$
$$= \frac{2\mu_0 I_0}{\pi L}.$$

Given this field, we can now address parts (a) through (c).

(a) In a magnetic field, a particle of charge q feels a force

$$\mathbf{F}_{mag} = q\mathbf{v} \times \mathbf{B},$$

where $\mathbf{v}$ denotes the particle's velocity. Here, $\mathbf{v} = 0$. So, the particle feels no magnetic force. Magnetic fields act only on *moving* charges. In this way, they differ from other fields yŏu've encountered.

(b) As just mentioned, when q_1 has velocity $\mathbf{v}_1$ at point P, it experiences a magnetic force

$$\mathbf{F}_{mag} = q_1\mathbf{v}_1 \times \mathbf{B}_P.$$

In this case, $\mathbf{v}_1$ points perpendicular to $\mathbf{B}_P$. Therefore, the magnitude of the cross product reduces to a regular product:

$$F_{mag} = q_1 v_1 B_P = q_1 v_1 \frac{2\mu_0 I_0}{\pi L}.$$

To find the direction of this force, apply the right-hand rule to the cross product in $\mathbf{F}_{mag} = q_1\mathbf{v}_1 \times \mathbf{B}_P$. Point your fingers with $\mathbf{v}_1$, into the page. Then curl your fingertips with $\mathbf{B}_P$, down the page. Your thumb now points leftward. That's the direction of $\mathbf{F}_{mag}$. The particle deflects toward wire 1.

(c) Since $\mathbf{v}_1$ and $\mathbf{B}_P$ are parallel, $\mathbf{v}_1 \times \mathbf{B}_P = 0$, and hence, $\mathbf{F}_{mag} = 0$. Remember, a cross product has magnitude $|\mathbf{a} \times \mathbf{b}| = ab\sin\phi$, where ϕ denotes the angle between the two vectors. For two parallel or antiparallel vectors, ϕ equals 0° or 180°, and hence, $\sin\phi = 0$.

In physical terms, a particle feels no magnetic force when it travels *along* a field line. The magnetic force comes from the component of the particle's velocity *perpendicular* to the field.

(d) Applying the Pythagorean theorem to this triangle diagram, we see that point C sits a distance

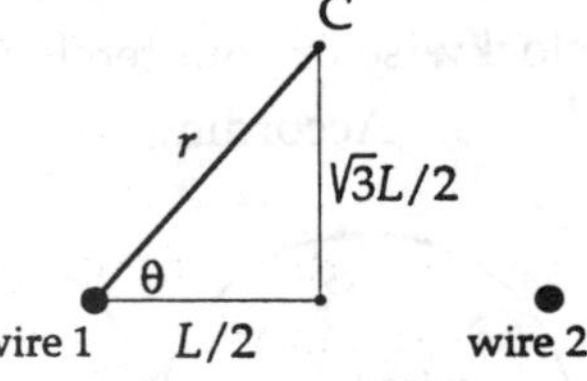

$$r = \sqrt{\left(\frac{L}{2}\right)^2 + \left(\frac{\sqrt{3}L}{2}\right)^2} = L$$

from both wires. Also, $\theta = \tan^{-1}\frac{\sqrt{3}L/2}{L/2} = 60°$. Using these insights, we can figure out the fields generated by wires 1 and 2 at point C. Then, we can add those fields vectorially.

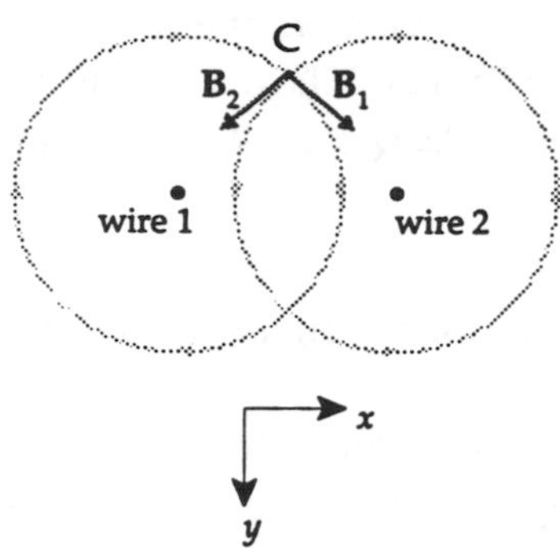

As shown above, each individual wire generates a field of strength $B = \frac{\mu_0 I_0}{2\pi r}$. From now on, B_1 and B_2 denote the fields generated by wires 1 and 2 *at point* C. Since both wires are distance L from point C,

$$B_1 = B_2 = \frac{\mu_0 I_0}{2\pi L}.$$

But those fields point in different directions. So, B_C does not equal $B_1 + B_2$. We must add $\mathbf{B}_1$ and $\mathbf{B}_2$ vectorially.

Well, $\mathbf{B}_1$ and $\mathbf{B}_2$ point tangentially to the "field line circles," as drawn here. As a result, their vector sum simplifies in two ways. First, the rightward component of $\mathbf{B}_1$ cancels the leftward component of $\mathbf{B}_2$. Therefore, $\mathbf{B}_C$ points entirely in the y-direction. Second, the y-component of $\mathbf{B}_1$ equals the y-component of $\mathbf{B}_2$. Consequently,

$$B_C = B_{1y} + B_{2y} = 2B_{1y}.$$

So, to complete the problem, we just need to find the y-component of $\mathbf{B}_1$.

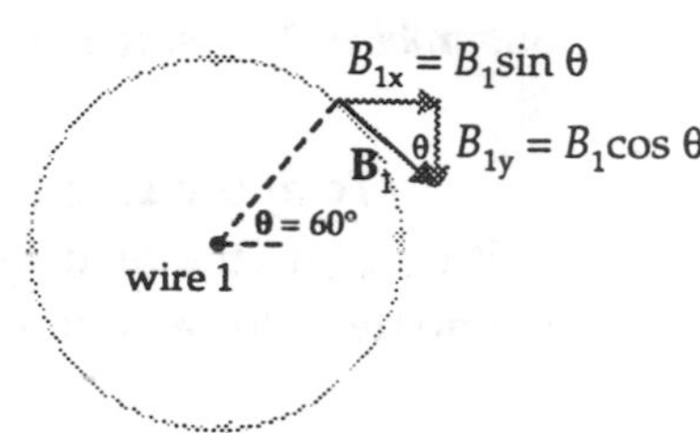

Well, as this new diagram shows, $B_{1y} = B_1 \cos\theta = \frac{\mu_0 I_0}{2\pi L}\cos 60°$. So,

$$B_C = 2B_{1y} = \frac{\mu_0 I_0}{\pi L}\cos 60° = \frac{\mu_0 I_0}{2\pi L}.$$

ANSWER 5

(a) You can solve using old-fashioned kinematics, or energy conservation. I'll use kinematics. As we saw last semester, an object accelerating at rate a through distance Δy acquires final speed $v^2 = v_0^2 + 2a\Delta y$. I'll take downward as the positive direction. So, $v_0 = 0$, and $a = g$. Solve for v to get

$$v = \sqrt{2g\Delta y} = \sqrt{2gH}.$$

The same answer pops out of energy conservation, according to which gravitational potential energy mgH converts into kinetic energy, $\frac{1}{2}mv^2$.

(b) The hardest part of this problem is figuring out where to begin. We can gain insight by thinking physically. As it enters the magnetic field, the frame moves at constant velocity. In other words, its acceleration goes to zero. Therefore, according to Newton's 2nd law, the net force goes to zero. So, an upward magnetic force, F_{mag}, must cancel the downward gravitational force. The magnetic field must be strong enough to make sure this happens. Since the magnetic force depends on the magnetic field, we can solve for that field by setting $F_{mag} = Mg$.

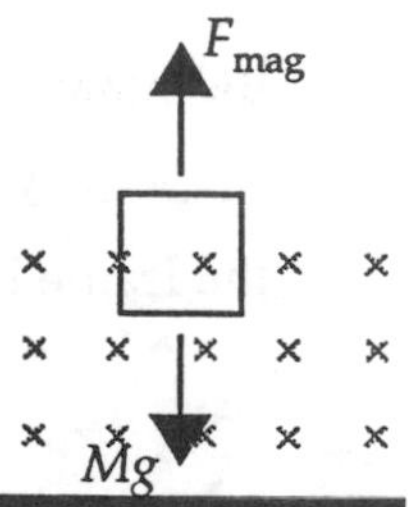

Since the frame has no acceleration, these forces must cancel.

So first, we need an expression for the magnetic force acting on the frame. Well, the magnetic force on each side depends on the current. And according to Faraday's law, a voltage and current get induced because the magnetic flux through the frame *changes* as it enters the field.

Let me tie all this together, in the form of a strategy.

1) Using Faraday's law, calculate the induced EMF—and the corresponding induced current—in the frame, as it enters the magnetic field.
2) Given that current, use magnetic force reasoning to find the force on each side of the frame. Add them up to obtain the resulting total magnetic force, F_{mag}. That total force will depend on B, the magnetic field strength.
3) Finally, set $F_{mag} = Mg$, and solve for B. This gives you the field strength needed to ensure that the frame has no acceleration (constant velocity) while entering the field.

We've completed the trickiest part of the problem, formulating an overall strategy. Before reading the rest of this answer, see if you can solve everything on your own.

Subproblem 1: Using Faraday's law, find the induced EMF and current as the frame enters the field.

To apply Faraday's law, we must find the rate at which the flux through the frame *changes* as it falls into the field. So first, I'll write an expression for the flux through the frame. Then, I'll differentiate with respect to time, to obtain $d\Phi_B/dt$.

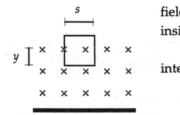

Let y denote the "height" of the frame that's inside the magnetic field at an arbitrary moment. So, y changes with time. The area of the frame inside the field is $A_{\text{in field}} = ys$.

Since the magnetic field is constant, we can pull it outside the flux integral:

$$\Phi_B = \int \mathbf{B} \cdot d\mathbf{A} = BA_{\text{in field}} = Bys.$$

As the frame falls into the field, y increases, and therefore Φ_B increases. To find the rate of change of Φ_B, differentiate it with respect to time. Since B and s are constant, the differentiation only "hits" y:

$$\frac{d\Phi_B}{dt} = \frac{d(Bys)}{dt} = Bs\frac{dy}{dt} = Bsv.$$

In the last step, I remembered that velocity is the rate at which position changes: $v = dy/dt$.

By Faraday's law, $\mathcal{E}_{\text{induced}} = -\frac{d\Phi_B}{dt} = -Bsv$. Therefore, by Ohm's law, the induced current in the frame is

$$i = \frac{\mathcal{E}_{\text{induced}}}{R} = -\frac{Bsv}{R},$$

where $v = \sqrt{2gH}$, as we found in part (a). The minus sign indicates that the induced current "fights" the change in flux. Here, as the frame falls into the field, the into-the-page flux through it increases.

Therefore, the induced current "wants" to generate an out-of-the-page field, at points inside the frame. To generate such a field, the induced must flow counterclockwise. I figured this out using the right-hand rule shortcut, as follows. Pick any side of the square, and point your right thumb with the current, assuming a counterclockwise flow. Your curled fingertips represent the field lines produced by that current. When you curl them towards the center of the square, they point out of the page. This goes to show that a counterclockwise current creates an out-of-the-page field, at points inside the square.

Subproblem 2: Write an expression for the net magnetic force on the frame

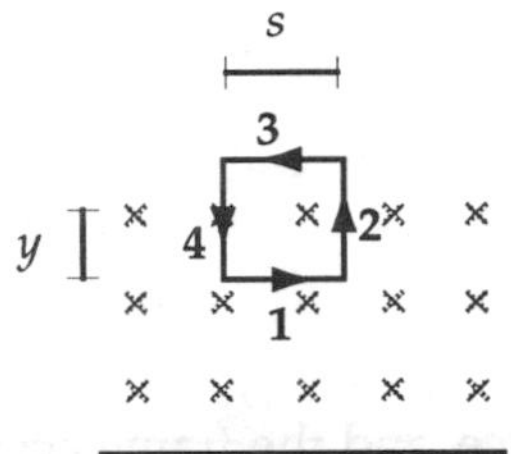

The magnetic field can exert a force only on sides 1, 2 and 4, as labeled here. Starting from the basic magnetic force law $\mathbf{F}_{\text{mag}} = q\mathbf{v} \times \mathbf{B}$, you can derive the infinitesimal force acting on an infinitesimal length of wire:

$$d\mathbf{F}_{\text{mag}} = i d\mathbf{l} \times \mathbf{B}.$$

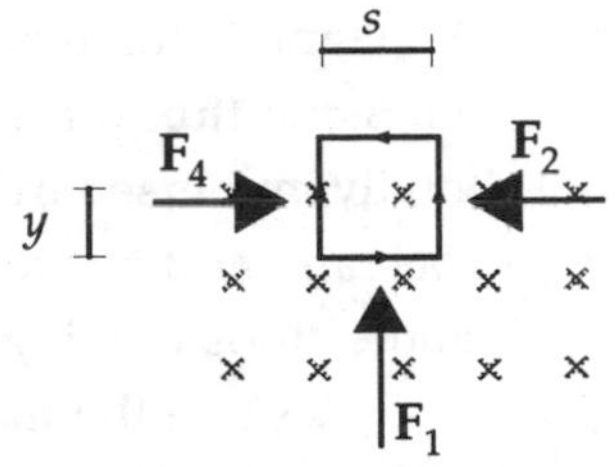

So, we'll need to integrate over $d\mathbf{l}$, for sides 1, 2, and 4. But wait. $\mathbf{F}_2$ and $\mathbf{F}_4$ must cancel. Since those two sides carry the same current, and since the same length of those sides (namely, length y) is immersed in the field, $\mathbf{F}_2$ and $\mathbf{F}_4$ are equally strong. But along side 2, $d\mathbf{l}$ points downward; while along side 4, $d\mathbf{l}$ points upward. Therefore, $\mathbf{F}_2$ points in the opposite direction from $\mathbf{F}_4$. Since those two forces have the same strength, but point in opposite directions, they cancel.

For this reason, the total magnetic force felt by the frame is simply $\mathbf{F}_1$. Let's figure out that force. Along side 1, $d\mathbf{l}$ and $\mathbf{B}$ are perpendicular. So, the magnitude of the cross product reduces to a regular product: $dF_1 = iBdl$. Since that whole side is immersed in the field,

$$F_1 = \int dF_1 = \int iBdl = iB\int dl = iBs,$$

because side 1 has length s.

To find the direction of that force, apply the right-hand rule to the cross product in $\mathbf{F}_1 = i d\mathbf{l} \times \mathbf{B}$. Point your fingers rightward with $d\mathbf{l}$. Then curl your fingertips into the page with $\mathbf{B}$. Your thumb now points upward. That's the direction of $\mathbf{F}_1$. Indeed, we already knew that $\mathbf{F}_1$ points upward, since the magnetic force must cancel the downward gravitational force. Without this cancellation, the frame would accelerate downward, instead of drifting downward at steady speed.

Before continuing, let me express F_1 in terms of the given physical quantities. From subproblem 1, $i = \frac{Bvs}{R} = \frac{B\sqrt{2gH}s}{R}$, since the frame enters the field with speed $v = \sqrt{2gH}$. Therefore, the upward magnetic force has strength

$$F_1 = iBs = \left(\frac{B\sqrt{2gH}s}{R}\right)Bs = \frac{B^2 s^2 \sqrt{2gH}}{R}.$$

Subproblem 3: By balancing the forces, solve for the magnetic field, B.

Since the frame enters the field at constant velocity (zero acceleration), the net force on it must vanish. In other words, the upward magnetic force cancels the downward gravitational force:

$$F_1 = Mg$$
$$\frac{B^2 s^2 \sqrt{2gH}}{R} = Mg.$$

Solve for B to get

$$B = \sqrt{\frac{MgR}{s^2 \sqrt{2gH}}}.$$

If B is weaker than this, then gravity overcomes the upward magnetic force, and the frame accelerates downward.

(c) A magnetic force acts on the frame *because* current flows around it. And current flows *because* the magnetic flux through the frame *changes*, thereby inducing an EMF. But once the frame becomes totally immersed in the magnetic field, the flux through it no longer increases. At that stage, Φ_B stays constant, and hence, $d\Phi_B/dt = 0$. Therefore, no EMF gets induced. Therefore, the current around the frame stops flowing. As a result, no magnetic force acts on the frame.

Once the magnetic force vanishes, only gravity acts on the frame. It accelerates downward, with acceleration g.

56 CHAPTER

Practice Final Exam 2

6 Questions, 3 hours. These questions range in difficulty from medium to very hard. You're not expected to get everything right. 70% is a very good score.

1) Two conducting spheres, one with radius $s = 0.10$ m, the other with radius $2s = 0.20$ m, are sitting far apart. Both spheres carry charge $Q = 1.0 \times 10^{-11}$ coulombs.

(a) What is the potential right at the surface of the smaller sphere? The bigger sphere contributes negligibly to this potential, since it's so far away. Please set the "zero" of potential at infinity: $V = 0$ where $r = \infty$.

(b) A particle of charge $q = 1.0 \times 10^{-6}$ coulombs and mass $M = 3.0 \times 10^{-7}$ kg is released from rest from the surface of the smaller sphere. How fast is it moving when it reaches a distance $3s = 0.30$ m from the center of the smaller sphere? Neglect gravity, and neglect the larger sphere.

(c) Now the two spheres are connected by a long wire, of resistance $R = 2.0\ \Omega$. Immediately after the wire gets connected, what current flows through it?

(d) As time passes, does this current increase, decrease, or stay constant? Explain.

(e) When current stops flowing, what is the charge on the smaller sphere?

2) An AC power supply is connected to a resistor, capacitor, and inductor in series, as drawn here. The power supply has adjustable frequency. The resistance, capacitance, and inductance are $R = 0.030\ \Omega$, $C = 0.10$ F, and $L = 0.20$ H.

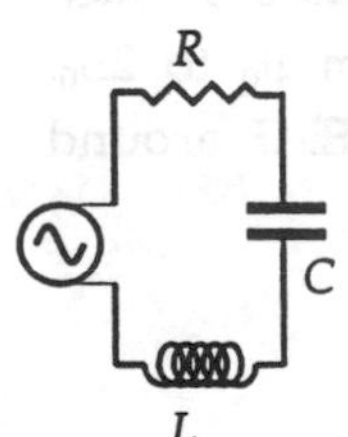

(a) On what frequency should we set the AC power supply, if we want as much heat as possible to dissipate in the resistor? Hint: You don't need too much math. Think physically.

(b) If the frequency is actually set to 60 Hz, and the rms voltage is 120 V, what will be the average power dissipated in the resistor?

(c) At time t_1, the power supply is turned down to 0 volts; but it is left connected, so that current can flow through it. Sketch a rough graph of the current through the power supply as a function of time. Your graph should show the current both before and after the power supply was turned to 0 volts. The graph need not be numerically accurate.

3) Consider a long thin pole, with uniform linear charge density λ_0. The pole is stuck into the ground, and points straight up. Attached to the pole, by an essentially massless string, is a very small ball of mass M and charge Q. The string has length L. In this problem, gravity matters.

When the ball settles to its final position, what's the angle, θ, between the pole and the string? Set up, *but do not solve,* the equation or equations needed to find θ. Please figure out any formulas you use from basic laws.

Hint: Figure out the electric field generated by the pole. The pole is very long as compared to the ball's distance from the pole.

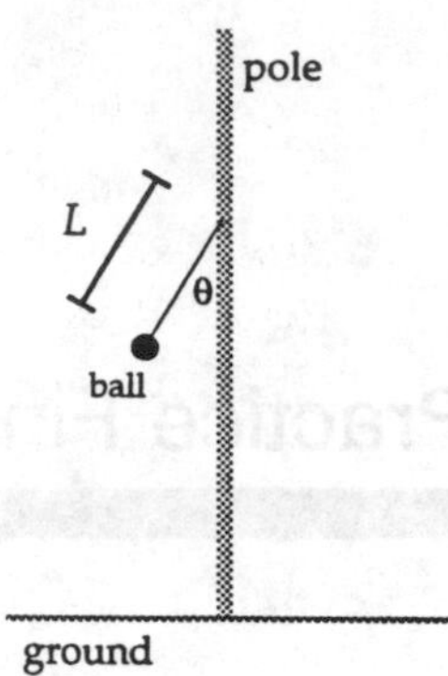

4) In the x-y plane, a large plate of uniform surface charge density σ_0 moves in the x-direction with constant speed v_0. At the moment the middle of the plate passes under point P, a particle of charge q_1 is at point P. That point is a distance s above the plate, in the z-direction.

Neglecting gravity, find the net force on the particle if its velocity is

(a) zero;

(b) v_1 in the y-direction;

(c) v_1 in the x-direction.

Derive any formulas you use from fundamental laws; don't just plug in pre-derived textbook formulas about infinite plates of charge. As I just hinted, you may treat the plate as infinitely big.

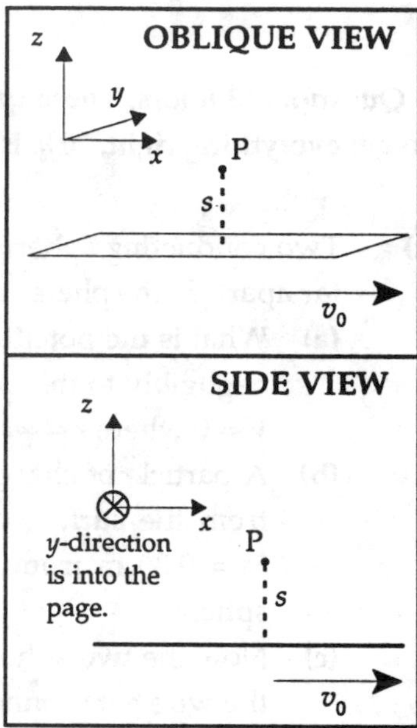

5) Consider this circuit, which contains no battery. It consists of two squares, each of side length s. Two of the resistors have resistance R, while the third has resistance $2R$. As drawn here, the circuit is partially "immersed" in a uniform magnetic field of strength B_0, pointing out of the page. The entire left square, *but only half of the right square,* is inside the field.

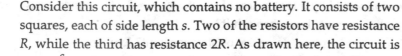

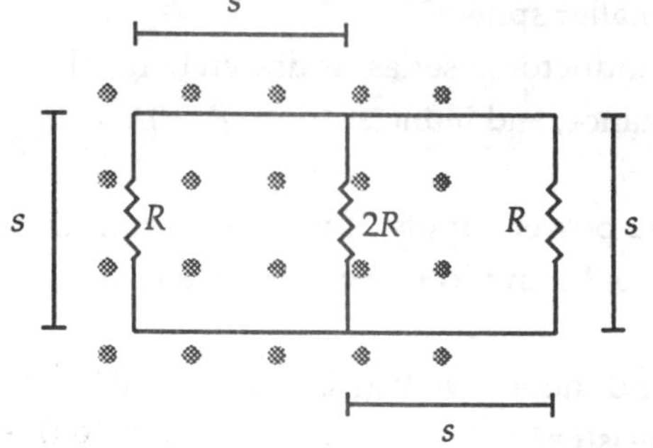

(a) While the magnetic field stays constant, what current flows through the $2R$ resistor?

(b) Now the magnetic field steadily increases. In fact, over a time t_1, the field strength doubles from B_0 to $2B_0$. During that time interval, what is the induced EMF around the right square?

(c) During that time interval, what current flows through the $2R$ resistor?

6) Consider this RLC circuit. The capacitor initially carries charge Q_0.

Jan has at her disposal some dielectric jelly and some paramagnetic putty. If she wants, she can stuff the jelly or the putty inside the capacitor or inside the inductor. Or both.

After inserting dielectric and/or paramagnetic material, Jan closes the switch at $t = 0$. Let t_1 denote the earliest time at which the capacitor carries no charge.

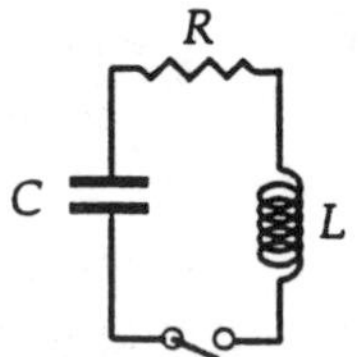

You can answer all the following questions qualitatively, with minimal math.

(a) Suppose Jan wants to make t_1 as small as she can. Should she use the dielectric or the paramagnetic substance? Or both? If so, where should she put them? Explain.

(b) Now Jan wants to make t_1 as large as she can. What should she do? Why?

(c) When Jan closes the switch, current flows back and forth for a while, but gradually dies out, due to the resistor. Suppose Jan wants to generate as much heat in the resistor as she can. Should she insert paramagnetic putty into the inductor, or not? Why?

ANSWER 1

(a) At points outside a spherically symmetric charge distribution, the sphere generates the same potential created by a point charge: $V = \frac{Q}{4\pi\varepsilon_0 r}$. Here, r denotes the distance from the center of the sphere. At the surface of the sphere, $r = s$. So,

$$V = \frac{Q}{4\pi\varepsilon_0 s} = \frac{1.0\times10^{-11}\ \text{C}}{4\pi(8.85\times10^{-12}\ \text{F/m})(0.10\ \text{m})} = 0.90\ \text{V}.$$

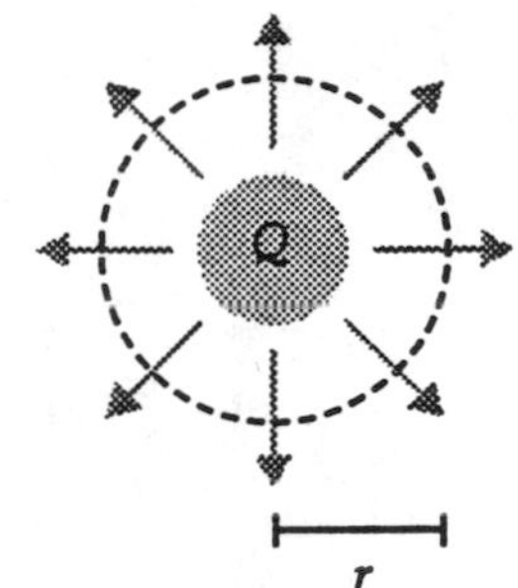

Let me explain why this point-charge formula also applies to charged spheres. In general, $V = -\int \mathbf{E}\cdot d\mathbf{s}$. So, if two objects produce the same electric field, then they create the same potential. As we've seen before, a point charge and a charged sphere generate the same electric field, at points outside the sphere. To see why, surround the charged sphere with a spherical Gaussian surface of radius r, and apply Gauss' law, to get

$$\oint \mathbf{E}\cdot d\mathbf{A} = \frac{q_{\text{encl}}}{\varepsilon_0}$$

$$E\oint_{\text{Gaussian sphere}} dA = \frac{Q}{\varepsilon_0}$$

$$E(4\pi r^2) = \frac{Q}{\varepsilon_0},$$

and hence, $E = \frac{Q}{4\pi\varepsilon_0 r^2}$, the same field produced by a point charge. In summary, since a point charge and a spherically symmetric charge distribution create the same electric field, they also create the same potential (at points outside the sphere).

(b) Many students figure out the force on the particle, calculate its acceleration, and then use an old kinematic equation. But those old kinematic formulas apply only to *constant* accelerations. Here, as the particle gets farther from the charged sphere, it feels a decreasing force, and hence, a decreasing acceleration.

Fortunately, conservation laws allow us to sidestep messy forces. Here, as the particle rushes away from the charged sphere, no heat or other "dissipative" energy gets produced. So, energy conservation gets us to the answer.

The particle starts at rest, with no kinetic energy. Using $U = qV$, we get

$$K_0 + U_0 = K_f + U_f$$

$$0 + qV_0 = \frac{1}{2}Mv^2 + qV_f.$$

To solve for v, the "final" velocity when the particle reaches a distance 3s from the center, we just need to know V_0 and V_f, the initial and final potential. (Actually, v depends only on the *change* in potential. But here, we can find that change most easily by calculating V_f and V_0 separately, and then subtracting.) From part (a), we already know the potential created by the small sphere. It's $V = \frac{Q}{4\pi\varepsilon_0 r}$. Since the particle starts at $r = s$, and reaches its "final" position at $r = 3s$,

$$V = \frac{Q}{4\pi\varepsilon_0 s} \text{ and } V_f = \frac{Q}{4\pi\varepsilon_0 (3s)}.$$

Solve the above energy conservation equation for v, and substitute in these potentials, to get

$$v = \sqrt{\frac{2q(V_0 - V_f)}{M}} = \sqrt{\frac{2q\left[\frac{Q}{4\pi\varepsilon_0 s} - \frac{Q}{4\pi\varepsilon_0(3s)}\right]}{M}}$$

$$= \sqrt{\frac{qQ}{3\pi\varepsilon_0 sM}}$$

$$= \sqrt{\frac{(1.0\times10^{-6}\ \text{C})(1.0\times10^{-11}\ \text{C})}{3\pi(8.85\times10^{-12}\ \text{F/m})(0.10\ \text{m})(3.0\times10^{-7}\ \text{kg}}}$$

$$= 2.0\ \text{m/s}.$$

(c) A common mistake is to think that no current flows, because both spheres carry the same charge. But the spheres have different *potentials*. Remember, current flows through a wire *because* of the potential difference (voltage) across the wire. Usually, a battery generates the voltage. But here, the relevant voltage is simply the potential difference between the two spheres. Current flows from the higher-potential sphere to the lower-potential sphere. Indeed, according to Ohm's law ($V = IR$), the current is proportional to this potential difference.

To figure out the potential difference, I'll just calculate the potential of each separate sphere, and subtract. In part (a), we already found the potential on the surface of the small sphere. It's $V_{\text{small sphere}} = \frac{Q}{4\pi\varepsilon_0 s}$. Since the big sphere also carries charge Q, but has radius $r = 2s$, the potential on its surface is $V_{\text{big sphere}} = \frac{Q}{4\pi\varepsilon_0(2s)}$. Notice that the smaller sphere has higher potential. That's because the charges on the small sphere are more concentrated, as drawn here. Consequently, charges "want" to move from the small sphere to the big one. Current flows *as if* driven by a battery of voltage

wire

$$V = V_{\text{small sphere}} - V_{\text{big sphere}} = \frac{Q}{4\pi\varepsilon_0 s} - \frac{Q}{4\pi\varepsilon_0(2s)} = \frac{Q}{8\pi\varepsilon_0 s}.$$

Therefore, according to Ohm's law, the wire initially carries current

$$I_0 = \frac{V}{R} = \frac{Q}{8\pi\varepsilon_0 sR} = \frac{1.0\times10^{-11}\text{ C}}{8\pi(8.85\times10^{-12}\text{ F/m})(0.10\text{ m})(2.0\ \Omega)} = 0.22\text{ A}.$$

(d) As just explained, the current gets "driven" by the potential difference between the two spheres. But as charge flows from the small sphere to the big sphere, this potential difference decreases. Here's why. As the small sphere loses charge, its potential goes down. And as the big sphere gains charge, its potential goes up. Therefore, the potential *difference*, $V = V_{\text{small sphere}} - V_{\text{big sphere}}$, gets smaller. As a result, the current gets smaller.

A numerical example clarifies this reasoning. Initially, both spheres carry $Q = 1.0 \times 10^{-11}$ coulombs. So, the spheres initially generate potentials

$$V_{\text{small sphere, initial}} = \frac{Q}{4\pi\varepsilon_0 s} = \frac{1.0\times10^{-11}\text{ C}}{4\pi(8.85\times10^{-12}\text{ F/m})(0.10\text{ m})} = 0.90\text{ volts}$$

$$V_{\text{large sphere, initial}} = \frac{Q}{4\pi\varepsilon_0 (2s)} = \frac{1.0\times10^{-11}\text{ C}}{4\pi(8.85\times10^{-12}\text{ F/m})(0.20\text{ m})} = 0.45\text{ volts}.$$

Therefore, the initial potential difference across the wire is

$$V = V_{\text{small sphere}} - V_{\text{big sphere}} = 0.90\text{ V} - 0.45\text{ V} = 0.45\text{ V}.$$

But at a later time, after 0.1×10^{-11} coulombs have flowed from the small sphere to the big one, the small sphere carries only 0.9×10^{-11} coulombs. So,

$$V_{\text{small sphere, later}} = \frac{0.9\times10^{-11}\text{ C}}{4\pi(8.85\times10^{-12}\text{ F/m})(0.10\text{ m})} = 0.81\text{ volts}.$$

And now the large sphere carries 1.1×10^{-11} coulombs, which generates a potential

$$V_{\text{large sphere, later}} = \frac{1.1\times10^{-11}\text{C}}{4\pi(8.85\times10^{-12}\text{ F/m})(0.20\text{ m})} = 0.50\text{ volts}.$$

Therefore, the potential difference between the two spheres is only 0.81 V – 0.50 V = 0.31 volts, less than the original potential difference of 0.45 volts. Since the potential difference goes down, so does the current.

(e) Eventually, the current stops entirely. This happens when the potential difference drops to zero—i.e., when both spheres have the same potential. A common mistake is to say the current vanishes when both spheres carry the same surface charge density. But for reasons too subtle to explain here, even when the spheres carry the same surface charge density, they have slightly different potentials. And potentials are what matters. Ask your instructor about this, if you're interested.

So, this problem wants to know the charge on the small sphere when $V_{\text{small sphere}} = V_{\text{big sphere}}$. To figure it out, let Q_{small} and Q_{big} denote the final charge on the small and big sphere. Do we know the total charge, $Q_{small} + Q_{big}$? Yes. Intuitively, every bit of charge that flows off the small sphere flows onto the large sphere. Therefore, the *total* charge on the two spheres stays constant. Initially, the two spheres carry total charge $2Q$. So, that's the total final charge, too: $Q_{small} + Q_{big} = 2Q$.

Given this insight, we can set the final potential of the small sphere equal to the final potential of the large sphere:

$$V_{\text{small sphere}} = V_{\text{big sphere}}$$

$$\frac{Q_{small}}{4\pi\varepsilon_0 s} = \frac{Q_{big}}{4\pi\varepsilon_0 (2s)},$$

and hence, $Q_{big} = 2Q_{small}$. In words, the two spheres generate the same potential when the larger one carries twice as much charge. Since the total charge is $2Q$,

$$\begin{aligned} 2Q &= Q_{small} + Q_{big} \\ &= Q_{small} + 2Q_{small} \\ &= 3Q_{small}, \end{aligned}$$

and hence

$$Q_{small} = \frac{2}{3}Q = \frac{2}{3}(1.0 \times 10^{-11}\ \text{C}) = 6.7 \times 10^{-12}\ \text{C}.$$

ANSWER 2

(a) In general, power is the rate at which energy gets transferred from one form to another. The power developed in a resistor is the heat dissipated per time. So, we're trying to maximize the power. As shown in Chapter 38, a resistor dissipates power $P = i^2R$. Hence, for a given resistance, we maximize the power by maximizing the current. In other words, we want to make i as high as possible.

I see two ways of approaching this. On the one hand, we can write a general expression for i as a function of the frequency (ν), and then maximize i by setting $di/d\nu = 0$. But we can also reason more *physically*, in terms of resonance. Let me explain.

An oscillating system, such as an LC circuit or a pendulum, has a natural, "resonant" frequency of oscillation. For instance, in an unforced (battery-less) RLC circuit, charges naturally slosh back and forth between the two capacitor plates. But let's think about a pendulum, which is easier to visualize. Suppose that, instead of letting it swing freely, you push the pendulum periodically, to make it swing as high as possible. With what frequency should you push it? If you've pushed a kid on a swing, then you intuitively know that it's "best" to push once per oscillation. For instance, if the pendulum naturally swings back and forth 30 times per minute, then you should push it 30 times per minute, to make it swing as high as possible.

This conclusion applies not just to pendulums, but to all oscillating systems, including RLC circuits. So, we can generate the biggest current (i.e., the biggest oscillations) by setting the driving frequency of the AC power supply equal to the natural (resonant) frequency of the RLC circuit.

To complete this problem, we just need to find the natural frequency of the circuit. In Chapter 44, using a deep analogy between an RLC circuit and a block on a spring, we found that an RLC circuit naturally oscillates with angular frequency $\omega = \frac{1}{\sqrt{LC}}$. This corresponds to frequency

$$\nu = \frac{\omega}{2\pi} = \frac{1}{2\pi\sqrt{LC}} = \frac{1}{2\pi\sqrt{(0.20\text{ H})(0.10\text{ F})}} = 1.1\text{ s}^{-1}.$$

By setting the AC power supply to 1.1 s^{-1}, we achieve "resonance"; the power supply "pushes" the current in sync with its natural oscillations, thereby causing the peak current to become as large as possible. This maximizes the power dissipated in the resistor.

(b) In part (a), by thinking about resonance, we didn't need to write the current as a function of frequency. But now, we have no choice. After finding the rms current, we can calculate the average power dissipated in the resistor, using $\overline{P} = i_{rms}^2 R$. Remember, the rms current or voltage tells you the average of the *square* of the quantity. So, i_{rms}^2 is the average value of i^2, and $\mathcal{E}_{rms}^2$ is the average value of $\mathcal{E}^2$.

By the way, a common mistake is to write $\overline{P} = \mathcal{E}_{rms}^2 / R$. This would work if $\mathcal{E}_{rms}$ referred to the voltage *across the resistor*. But $\mathcal{E}_{rms}$ tells us the voltage *across the power supply*. These two voltages differ, for the following reason. By Kirchhoff's 2nd law, the power supply's voltage equals the sum of the voltage drops across all the other circuit elements (in series). So here, $\mathcal{E}_{\text{power supply}}$ equals the voltage drop across the resistor *plus* the voltage drop across the capacitor *plus* the voltage drop across the inductor. That's why $\mathcal{E}_{\text{power supply}}$ does not equal V across the resistor. Therefore, to solve this problem, we must find the rms current through the resistor.

From Chapter 44, recall the five central AC circuit equations:

$$\mathcal{E} = \mathcal{E}_m \sin \omega t \qquad (1)$$

$$Z = \sqrt{R^2 + (\omega L - 1/\omega C)^2} \qquad \textbf{Impedance} \qquad (2)$$

$$i_m = \frac{\mathcal{E}_m}{Z} \qquad \textbf{Maximum current} \qquad (3)$$

$$i = i_m \sin(wt - \phi) \qquad \textbf{Current} \qquad (4)$$

$$\phi = \tan^{-1}\frac{\omega L - 1/\omega C}{R}. \qquad \textbf{"Lag constant"} \qquad (5)$$

Here, we only need (2) and (3). Eq. (3) plays the role of Ohm's law. The impedance, Z, is kind of like an "AC resistance," which we can calculate from Eq. (2). When $\nu = 60\text{ s}^{-1}$, the circuit oscillates with angular frequency $\omega = 2\pi\nu = 2\pi(60\text{ s}^{-1}) = 377\text{ s}^{-1}$. So,

$$Z = \sqrt{R^2 + \left(\omega L - \frac{1}{\omega C}\right)^2} = \sqrt{(0.030\ \Omega)^2 + \left[(377\text{ s}^{-1})(0.20\text{ H}) - \frac{1}{(377\text{ s}^{-1})(0.10\text{ F})}\right]^2}$$

$$= \sqrt{(0.030\ \Omega)^2 + [(75.4\text{ H/s}) - (0.03\text{ F}^{-1}\text{s}^{-1})]^2}$$

$$= 75.4\ \Omega.$$

Notice that the inductor "causes" almost all the impedance. That's not a general result. It just happened to work out that way, in this case.

Given the impedance, we can use a modified version of Eq. (3), to relate the rms voltage to the rms current:

$$i_{\text{rms}} = \frac{\mathcal{E}_{\text{rms}}}{Z} = \frac{120 \text{ V}}{75.4 \ \Omega} = 1.59 \text{ A}.$$

(As proven in your textbook, the rms value of $\mathcal{E}$ or i is $1/\sqrt{2}$ times the maximum value. Therefore, since i_m and $\mathcal{E}_m$ are proportional, so are i_{rms} and $\mathcal{E}_{\text{rms}}$, with the same proportionality constant.)

Using this current, we can immediately calculate the average power dissipated in the resistor:

$$\overline{P} = i_{\text{rms}}^2 R = (1.59 \text{ A})^2 (0.030 \ \Omega) = 0.076 \text{ W}.$$

(c) Before the power supply gets turned to 0 volts, it makes the current oscillate 60 times per second. These oscillations neither increase nor decrease in size (amplitude), because the power supply replenishes the energy dissipated in the resistor. So, before t_1, the current vs. time graph is a high-frequency sine wave. You may have seen this in lab, using an oscilloscope.

After the power supply voltage goes to 0, current continues to oscillate. But two things change. First, because the energy dissipated in the resistor no longer gets replenished, these natural oscillations gradually decrease in amplitude, until the current dies completely. (Similarly, a pendulum's oscillations eventually die away, due to air resistance.) Second, with the power supply no longer driving the system, the current oscillates at the natural (resonance) frequency of the circuit. In part (a), we found this frequency to be 1.1 Hz, much less than the driving frequency of 60 Hz. So, after t_1, the frequency gets *much* smaller, which means the period gets much larger.

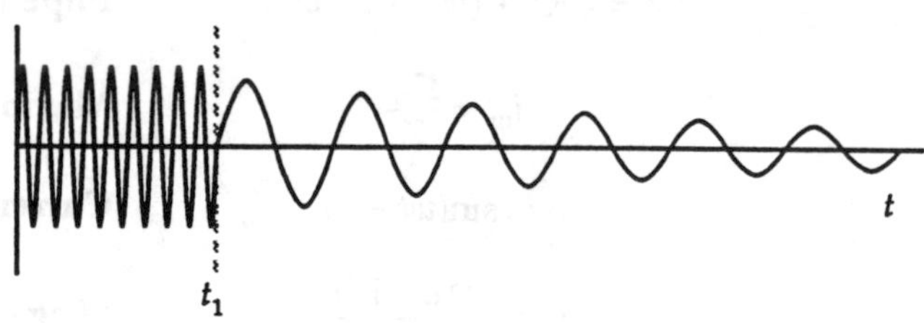

ANSWER 3

If the pole carried no charge, the ball would hang straight down. But the pole generates an electric field that pushes the ball away. Consequently, the ball swings out to some angle, θ.

So, we need to find E, the electric field produced by the pole. Given E, we can calculate the electric force on the ball, $F_{\text{elec}} = QE$. By incorporating this force into a free-body diagram, we can then solve for θ.

Notice how I outlined a general strategy before diving into details. Doing so keeps you organized, and earns you partial credit even if you can't complete every step.

Subproblem 1: Find the electric field generated by the pole.

We can treat the pole as infinitely long, since it's very long as compared to the ball's distance from the pole. This approximation introduces enough symmetry that we can solve for E using Gauss' law.

To find E, we'll need to pull it outside the flux integral in Gauss' law,

$$\oint \mathbf{E} \cdot d\mathbf{A} = \frac{q_{\text{encl}}}{\varepsilon_0}.$$

Therefore, E must be constant over (part of) the Gaussian surface. Here, by radial symmetry, the field has the same strength at any two points equidistant from the pole. So, if we enclose part of the pole with a Gaussian cylinder, as drawn here, the field is constant over the entire side (tube) of the cylinder. Furthermore, no field lines pierce through the ends (top or bottom) of this Gaussian surface. The field lines merely "skim" those surfaces. Therefore, the Gaussian flux integral simplifies to

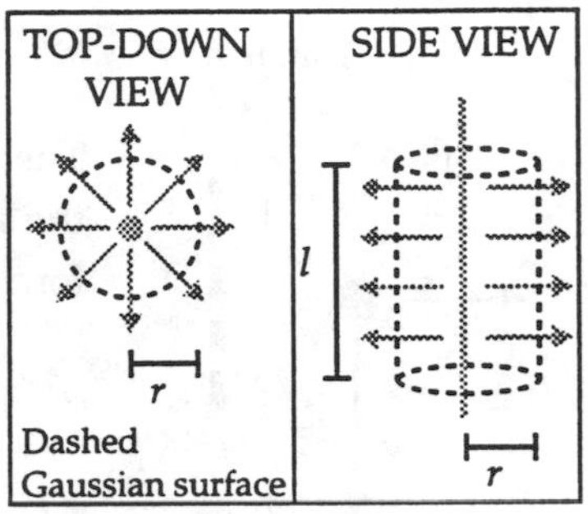

$$\oint \mathbf{E} \cdot d\mathbf{A} = E \int_{\text{side (tube) of Gaussian surface}} dA = E(2\pi r l),$$

where I used the surface area of a cylindrical tube of radius r and length l.

Now I'll simplify the other side of Gauss' law. Linear charge density is charge per length. Since my Gaussian surface encloses length l of the pole,

$$q_{\text{encl}} = \frac{\text{charge}}{\text{length}} \times (\text{length of pole enclosed}) = \lambda_0 l.$$

Substitute all this information into Gauss' law, to get

$$\oint \mathbf{E} \cdot d\mathbf{A} = \frac{q_{\text{encl}}}{\varepsilon_0}$$

$$E(2\pi r l) = \frac{\lambda_0 l}{\varepsilon_0}.$$

The l's cancel, proving that the length of your Gaussian surface makes no difference. Solve for E to get

$$E = \frac{\lambda_0}{\pi \varepsilon_0 r}.$$

That's the field generated by the pole, a radial distance r from the center. We just derived the general formula for the field generated by a line charge or by a charged cylinder.

Subproblem 2: Find the electric force on the ball.

Since $\mathbf{F}_{elec} = Q\mathbf{E}$, the electric field pushes the ball away from the pole, with force

$$F_{elec} = QE = Q\frac{\lambda_0}{2\pi\varepsilon_0 r}.$$

Subproblem 3: By drawing a force diagram, generate equations that can be solved for θ.

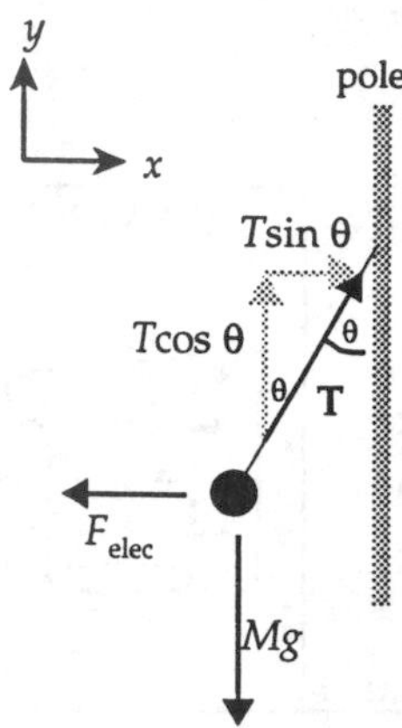

As just mentioned, the electric force pushes the ball away from the pole. Gravity pulls it straight down, with force Mg. And tension pulls the ball along the direction in which the string is stretched. I've broken tension into its horizontal and vertical components.

The ball, after settling into its final position, has no acceleration. Therefore, the net horizontal and vertical forces vanish. Actually, you can reach the answer faster by setting the net *torque* equal to zero. But I'll work in terms of forces, because it's more intuitive for most students. I get

$$\Sigma F_x = T\sin\theta - Q\frac{\lambda_0}{2\pi\varepsilon_0 r} = 0. \tag{1}$$

$$\Sigma F_y = T\cos\theta - Mg = 0. \tag{2}$$

We can't yet solve for θ, because these two equations contain three unknowns: θ, T, and r. Fortunately, we can eliminate r by looking at the original diagram of the problem. From that diagram, $r = L\sin\theta$. Substitution this expression into Eq. (1), to get

$$T\sin\theta - Q\frac{\lambda_0}{2\pi\varepsilon_0 L\sin\theta} = 0. \tag{1*}$$

Now we have two equations in two unknowns, θ and T. So, we can solve for θ.

ANSWER 4

Since the plate carries charges, it creates an electric field. And since those charges are moving, the plate also creates a magnetic field. So, a particle at point P can feel electric *and* magnetic forces. For this reason, we need to figure out the electric and magnetic fields created by the plate at point P. Given those fields, we can find the overall force on the particle using the Lorentz force law, $\mathbf{F} = q(\mathbf{E} + \mathbf{v} \times \mathbf{B})$.

Subproblem 1: Find the electric field generated by the plate at point P.

Welcome to a classic Gauss' law problem, the infinite plane of charge. Since your textbook presents this example, I'll whip through the reasoning.

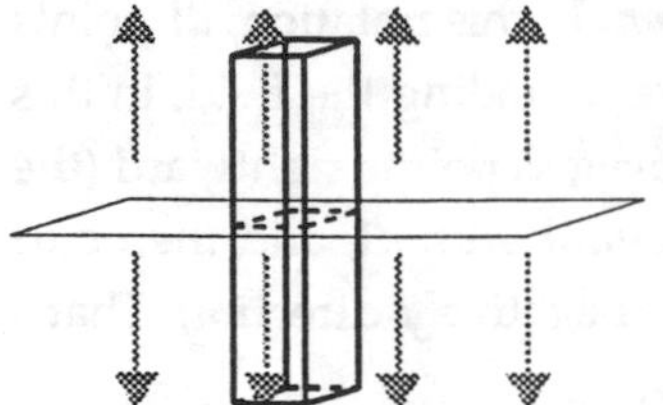

By symmetry, the electric field points perpendicular to the plate, and has the same strength at any two points equidistant from the plate. Therefore, if we can draw a Gaussian surface consisting of points equidistant from the charged plate, the field will be constant over that surface, allowing us to pull **E** outside the flux integral, $\int \mathbf{E}\cdot d\mathbf{A}$.

Well, the "top" and "bottom" of this Gaussian pillbox consist of points equidistant from the plate. And no flux passes through the sides of the pillbox, because the field lines skim along the sides without piercing *through*. (Mathematically, $\mathbf{E}\cdot d\mathbf{A} = 0$ on the sides, because **E** points in the *z*-direction while the sides of the box face the *x*- or *y*-direction.)

Since the field below the plate points in the negative *z*-direction, you might think the flux through the bottom of the pillbox is negative. But remember, positive flux means that the field lines flow *out* of the Gaussian surface; while negative flux means that the field lines flow *into* the Gaussian surface. So here, the top and bottom of the pillbox both contribute *positive* flux. Indeed, letting "*A*" denote the area of the top and bottom, we get

$$\begin{aligned}\oint \mathbf{E}\cdot d\mathbf{A} &= \int_{\text{top}} \mathbf{E}\cdot d\mathbf{A} + \int_{\text{bottom}} \mathbf{E}\cdot d\mathbf{A} \\ &= EA_{\text{top}} + EA_{\text{bottom}} \\ &= 2EA.\end{aligned}$$

To use Gauss' law, we also need to know the enclosed charge. As the dashed line on the diagram shows, the Gaussian surface encloses area *A* of the plate. Therefore,

$$q_{\text{encl}} = \frac{\text{charge}}{\text{length}} \times (\text{length of pole enclosed}) = \sigma_0 A.$$

Putting all this information into Gauss' law, we get

$$\begin{aligned}\oint \mathbf{E}\cdot d\mathbf{A} &= \frac{q_{\text{encl}}}{\varepsilon_0} \\ 2EA &= \frac{\sigma_0 A}{\varepsilon_0},\end{aligned}$$

and hence

$$E = \frac{\sigma_0}{2\varepsilon_0}.$$

Surprisingly, the field doesn't depend on your distance from the plate, assuming you're close enough that the "infinite plate" approximation holds.

So far, we've found the electric field at point P. But remember, the moving charged plate also generates a magnetic field.

Subproblem 2: Find the magnetic field at point P produced by the moving plate.

First, let's figure out the direction of the magnetic field above the plate. To do so, apply the right-hand rule to the cross product in the Biot-Savart law, $d\mathbf{B} = \frac{\mu_0}{4\pi}\frac{I d\mathbf{l} \times \hat{\mathbf{r}}}{r^2}$. When we're finding the *strength* of this field,

Ampere's law works more efficiently. But for now, we're just finding the *direction*. In this notation, $d\mathbf{l}$ points in the direction of current flow, and $\hat{\mathbf{r}}$ points from $d\mathbf{l}$ to the location at which we're finding the field, in this case P. As our "$d\mathbf{l}$," consider a tiny sliver of the plate directly under P. For that sliver, $d\mathbf{l}$ points rightward (the x-direction), and $\hat{\mathbf{r}}$ points upward (the z-direction). So, point your fingers rightward with $d\mathbf{l}$, and then curl your fingertips upward toward P. Your thumb now points out of the page, in the negative y-direction. That's the direction of **B**.

Subtle digression. Actually, I just glossed over some complications, by considering the sliver of plate directly below point P. Other slivers of the plate generate magnetic fields pointing in different directions. Fortunately, none of these field contributions has a z-component. And the x-components cancel out. See your textbook for details. The overall field above the plate does indeed point in the $-y$ direction. *End of subtle digression.*

By equivalent reasoning, the field below the plate points in the positive y-direction.

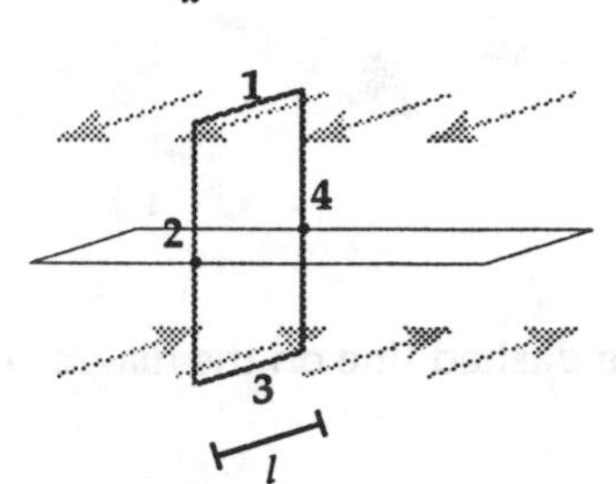

MAGNETIC FIELD LINES, AND AMPERIAN LOOP

Now we can invoke the magnetic analog of Gauss' law, namely Ampere's law, to calculate the magnetic field at P. According to Ampere's law, the line integral of the magnetic field around an Amperian loop is proportional to the current enclosed by that loop:

Ampere's law $$\oint_{\text{loop}} \mathbf{B}\cdot d\mathbf{s} = \mu_0 I_{\text{encl}}.$$

To solve this equation for the field, we need to pull B outside the line integral, which can happen only if B is constant along (part of) the Amperian loop. The rectangular loop drawn here accomplishes this aim. By symmetry, the magnetic field has the same strength at any two points equidistant from the plate. So, B is constant along sides 1 and 3. Indeed, B has the same strength along both those sides, because I drew them equidistant from the charged plane. Furthermore, sides 1 and 3 are parallel to the magnetic field lines. If you travel counterclockwise around the loop, then along side 1 you're walking *with* the field (parallel to it). And along side 3, you're *also* walking *with* the field. Mathematically speaking, along sides 1 *and* 3, **B** points in the same direction as $d\mathbf{s}$. Therefore, $\int \mathbf{B}\cdot d\mathbf{s}$ is positive for both sides.

By contrast, sides 2 and 4 are perpendicular to the field. Therefore, $\mathbf{B}\cdot d\mathbf{s} = Bds\cos 90° = 0$.

Putting all this together, and noticing that I've given sides 1 and 3 length l, we can simplify the left-hand side of Ampere's law:

$$\begin{aligned}\int_{\text{loop}} \mathbf{B}\cdot d\mathbf{s} &= \int_1 \mathbf{B}\cdot d\mathbf{s} + \int_2 \mathbf{B}\cdot d\mathbf{s} + \int_3 \mathbf{B}\cdot d\mathbf{s} + \int_4 \mathbf{B}\cdot d\mathbf{s} \\ &= B\int_1 ds \quad + \quad 0 \quad + \quad B\int_3 ds \quad + \quad 0 \\ &= Bl + 0 + Bl + 0 \\ &= 2Bl.\end{aligned}$$

Remember, we're using Ampere's law to find the magnetic field produced by the moving plate o' charge. Now I'll simplify the other side of Ampere's law, by finding the current enclosed by my Amperian loop. This is hard, because the "current" takes an unusual form. We'll need to picture things carefully.

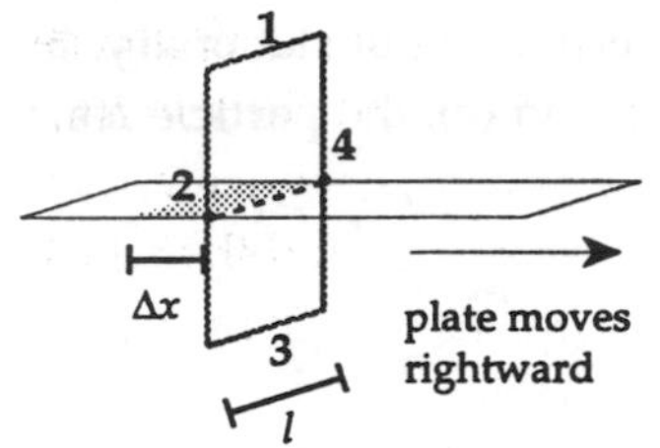

The gray area of the plate passes through the Amperian loop in time Δt.

By definition, current is charge per time. So, we need the charge per time passing through my Amperian loop—i.e., the charge per time passing the dashed line on this diagram.

Well, the plate moves at constant velocity. So, $v = \Delta x/\Delta t$. Therefore, the length of plate passing the dashed line in time Δt is

$$\Delta x = v_0 \Delta t.$$

I've marked Δx on the diagram, and colored in the area of the plate that passes the dashed line in time Δt. It's

$$\Delta A = l\Delta x = lv_0\Delta t.$$

Therefore, the charge passing the dashed line in time Δt is

$$\begin{aligned}\Delta q &= \frac{\text{charge}}{\text{area}} \times (\text{area that passes dashed line in time } \Delta t) \\ &= \sigma_0 \Delta A \\ &= \sigma_0 l v_0 \Delta t.\end{aligned}$$

Since current is charge per time, the current passing through the Amperian loop is

$$I_{\text{encl}} = \frac{\Delta q}{\Delta t} = \frac{\sigma_0 v_0 l \Delta t}{\Delta t} = \sigma_0 v_0 l.$$

OK, we've simplified both sides of Ampere's law. Let's solve for the magnetic field:

$$\oint_{\text{loop}} \mathbf{B}\cdot d\mathbf{s} = \mu_0 I_{\text{encl}}$$

$$2Bl = \mu_0 \sigma_0 v_0 l.$$

Cancel the l's and isolate B to get

$$B = \frac{\mu_0 \sigma_0 v_0}{2}.$$

Compare this to the electric field produced by the plate, $E = \frac{\sigma_0}{2\varepsilon_0}$. In both cases, the field does not depend on your distance from the plate, provided you're close enough that the infinite plate approximation holds. And both fields are proportional to the surface charge density.

At this stage, we've used Gauss' law to find the electric field, and Ampere's law to find the magnetic field, produced by the plate at point P. Given these fields, we can calculate the force on the particle.

(a) Since the particle sits still, *it can't experience a magnetic force.* Only *moving* charges react to magnetic fields. Mathematically, this follows from the formula $\mathbf{F}_{\text{mag}} = q\mathbf{v} \times \mathbf{B}$. When $\mathbf{v} = 0$, the particle feels no magnetic force.

By contrast, the electric field "doesn't care" whether the particle is moving or stationary. In either case, the particle feels a force $\mathbf{F}_{elec} = q\mathbf{E}$. Therefore, in parts (a), (b), and (c), the particle feels the *same* electric force. And here, the electric force is the only force:

$$\mathbf{F}_{total} = \mathbf{F}_{elec} = q\mathbf{E} = q_1 \frac{\sigma_0}{2\varepsilon_0}\hat{\mathbf{z}},$$

where the unit vector $\hat{\mathbf{z}}$ indicates the direction.

(b) The magnetic force has magnitude $F_{mag} = q|\mathbf{v} \times \mathbf{B}| = qvB\sin\theta$, where θ denotes the angle between the particle's direction of motion ($\mathbf{v}$) and the magnetic field direction ($\mathbf{B}$). Here, the particle moves antiparallel to the magnetic field: $\theta = 180°$. Since $\sin 180° = 0$, the particle feels no magnetic force.

In general, a charged particle moving *along* a magnetic field line experiences no magnetic force. $\mathbf{F}_{mag}$ kicks in only when part of the particle's motion is perpendicular to the magnetic field.

By contrast, the electric force depends only on the particle's charge and the electric field, not on the particle's motion. So once again,

$$\mathbf{F}_{total} = \mathbf{F}_{elec} = q\mathbf{E} = q_1 \frac{\sigma_0}{2\varepsilon_0}\hat{\mathbf{z}},$$

(c) Finally, it feels a magnetic force. Since the particle's velocity is perpendicular to the magnetic field, the cross product $\mathbf{v} \times \mathbf{B}$ reduces to a regular product: $|\mathbf{v} \times \mathbf{B}| = v_1 B\sin 90° = v_1 B$. Notice that v denotes the speed of the particle that "feels" the magnetic field, not the speed of the charges *creating* the magnetic field.

To find the direction of $\mathbf{F}_{mag} = q\mathbf{v} \times \mathbf{B}$, apply the right-hand rule to the cross product. First, point your fingers in the direction of $\mathbf{v}_1$. Then, curl your fingertips in the direction of $\mathbf{B}$. Your thumb now points in the $-z$ direction. So, $\mathbf{F}_{mag}$ pushes the particle toward the plate, assuming q_1 is positive.

As we saw in parts (a) and (b), the electric force pushes the particle away from the plate, assuming q_1 is positive. So, the magnetic force opposes the electric force. The total force is simply

$$\begin{aligned}\mathbf{F}_{total} &= \mathbf{F}_{elec} + \mathbf{F}_{mag} \\ &= q_1\mathbf{E} + q_1\mathbf{v} \times \mathbf{B} \\ &= q_1 \frac{\sigma_0}{2\varepsilon_0}\hat{\mathbf{z}} + q_1 v_1 \frac{\mu_0 \sigma_0 v_0}{2}(-\hat{\mathbf{z}}) \\ &= q_1\left(\frac{\sigma_0}{2\varepsilon_0} - v_1 \frac{\mu_0 \sigma_0 v_0}{2}\right)\hat{\mathbf{z}},\end{aligned}$$

where I used our expressions for the fields, obtained in subproblems 1 and 2 above. By the way, even if q_1 is negative, the electric and magnetic forces still oppose one another. In that case, $\mathbf{F}_{elec}$ points down and $\mathbf{F}_{mag}$ points up.

From this calculation, notice that the sum of the electric and magnetic forces is $\mathbf{F}_{total} = q[\mathbf{E} + \mathbf{v} \times \mathbf{B}]$. Your textbook calls this equation the Lorentz force law.

Advanced theoretical digression for interested students. Given that the magnetic force opposes the electric force, you might wonder which force is bigger. To see why the electric force must "win," imagine yourself "riding along" on the plate. From your reference frame, the plate doesn't move, and therefore creates no magnetic field. So, in your reference frame, the particle feels no magnetic force; the electric force is the *only* force. For this reason, the particle gets pushed away from the plate, not towards it.

Of course, in the original (lab) reference frame, the particle feels a magnetic force. So, different reference frames "disagree" about the existence of the magnetic field. But different reference frames must *agree* about whether the particle gets pushed towards the plate or away from the plate. And indeed, for any values of v_0 and v_1 (less than the speed of light), $v_1 \frac{\mu_0 \sigma_0 v_0}{2}$ is less than $\frac{\sigma_0}{2\varepsilon_0}$. The electric force wins, and hence, the particle gets pushed away from the plate. If all this sounds interesting, you should become a physics major.

ANSWER 5

(a) According to Faraday's law,

$$\mathcal{E}_{\text{induced}} = -\frac{d\Phi_B}{dt},$$

an EMF gets induced by a *changing* magnetic flux. As long as the magnetic field stays constant, so does the flux through both squares. Therefore, no voltage gets induced in either square, and no current flows.

(b) To apply Faraday's law, we must figure out the rate of change of flux through the right square. Let's begin by finding the initial flux and the "final" flux (after time t_1). Because the flux changes steadily over time interval $\Delta t = t_1$, we can then use

$$\mathcal{E}_{\text{induced}} = -\frac{\Delta\Phi_B}{\Delta t} = -\frac{\Phi_{B\text{ final}} - \Phi_{B\text{ init}}}{t_1}.$$

Find the initial flux. Since the magnetic field and the right square both "face" the same direction, the dot product in $\Phi_B = \int \mathbf{B} \cdot d\mathbf{A}$ reduces to a regular product. Furthermore, since $\mathbf{B}$ is spatially uniform, we can pull it outside the integral. (Although B changes in time, at any given moment it's uniform in space. So, you can pull it outside an area integral, but not a time integral.)

These simplifications lead some students to write $\Phi_B = \int \mathbf{B} \cdot d\mathbf{A} = B_0 \int d\mathbf{A} = B_0 s^2$. But remember, the magnetic field "covers" only *half* of the right square. So,

$$\Phi_{B\text{ init}} = \int \mathbf{B} \cdot d\mathbf{A} = B_0 \int_{\text{half the square}} dA = 2B_0 A_{\text{half the square}} = B_0\left(\frac{1}{2}s^2\right) = \frac{1}{2}B_0 s^2.$$

Find the final flux. After the field has increased to $2B_0$, the flux through the right square becomes

$$\Phi_{B\text{ final}} = 2B_0 \int_{\text{half the square}} dA = 2B_0\left(\frac{1}{2}s^2\right) = B_0 s^2.$$

Substitute this initial and final flux into Faraday's law, to get

$$\mathcal{E}_{\text{induced}} = -\frac{\Phi_{B\text{ final}} - \Phi_{B\text{ init}}}{\Delta t} = -\frac{B_0 s^2 - \frac{1}{2}B_0 s^2}{t_1} = -\frac{B_0 s^2}{2t_1}.$$

The minus sign reminds us that the induced voltage and current "counteract" the changing flux. Here, the external magnetic field through the right square increases, and points out of the page. Therefore, the right square "wants" to generate its own magnetic field pointing *into* the page. To produce this field, the induced current around the right square must flow *clockwise*. I figured that out using the right-hand rule shortcut introduced in Chapter 39: when you point your thumb along the direction of current flow through a wire, your curled fingertips represent the field lines produced by that wire. In this case, when I pick any side of the right square and point my thumb with the current (assuming a clockwise flow), and then curl my fingertips towards the center of the square, my fingertips point into the page. So, current must flow clockwise around the right square, in order to produce an into-the-page field at points inside the square.

(c) When the flux through a square changes, a voltage (EMF) gets induced in the square. This voltage makes current flow. For instance, since a clockwise voltage $\mathcal{E}_{\text{induced}} = B_0 s^2/2t_1$ gets induced in the right square, current flows *as if* the square contained a battery of voltage $B_0 s^2/2t_1$. The induced EMF acts just like a battery

This insight hands us a problem-solving technique. First, we can calculate $\mathcal{E}_{\text{right}}$ and $\mathcal{E}_{\text{left}}$, the induced EMF around each square. (We already found $\mathcal{E}_{\text{right}}$ in part b.) Then, we can redraw the circuit, including batteries of strength $\mathcal{E}_{\text{right}}$ and $\mathcal{E}_{\text{left}}$. These "pretend" batteries remind us that the induced voltages behave just like real batteries. Finally, using Kirchhoff's laws, we can solve for the current through the $2R$ resistor.

Subproblem 1: Find the induced EMF around both squares.

As found in part (b), an EMF

$$\mathcal{E}_{\text{right}} = -\frac{B_0 s^2}{2t_1}$$

gets induced in the right square. The minus sign indicates clockwise.

We still need to deal with the left square. Since the magnetic field "covers" the entire square (of area $A = s^2$), the initial flux has magnitude $\Phi_{B\text{ init}} = \int \mathbf{B}\cdot d\mathbf{A} = B_0 s^2$. When the field doubles, the flux through the left square increases to $\Phi_{B\text{ final}} = 2B_0 s^2$. So,

$$\mathcal{E}_{\text{left}} = -\frac{\Phi_{B\text{ final}} - \Phi_{B\text{ init}}}{\Delta t} = -\frac{2B_0 s^2 - B_0 s^2}{t_1} = -\frac{B_0 s^2}{t_1}.$$

Using the right-hand rule shortcut explained in part (b), you can confirm that the minus sign once again indicates clockwise. So, $\mathcal{E}_{\text{left}}$ and $\mathcal{E}_{\text{right}}$ both try to make current flow clockwise around their respective loops.

Subproblem 2: Solve the circuit using Kirchhoff's law.

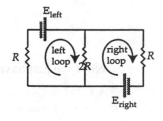

Since these induced EMFs "behave" just like batteries, the circuit acts as if it contains batteries of strength $\mathcal{E}_{\text{left}}$ and $\mathcal{E}_{\text{right}}$. Both batteries "try" to make current flow clockwise, as drawn here.

Always label the currents in the different segments. We're solving for I_2, the current through the $2R$ resistor. As always, I've guessed the direction of each current. A wrong guess can't hurt me. For instance, if I_2 actually flows "up" the $2R$ resistor, I'll obtain a negative answer for I_2. No big deal.

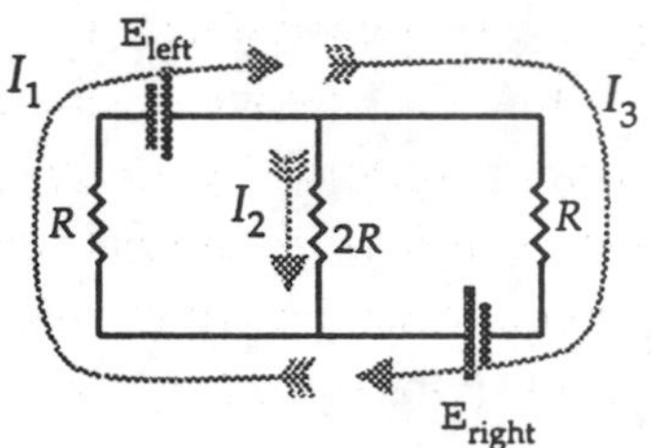

According to Kirchhoff's 1st law, the current flowing into an intersection equals the current flowing out of the intersection. Look at the intersection above the $2R$ resistor. Current I_1 flows in, while I_2 and I_3 flow out. So,

Intersection rule $$I_1 = I_2 + I_3 \qquad (0)$$

Now use Kirchhoff's 2nd law: The total change in potential around a loop is zero. Let's start with the left loop. I'll walk around it clockwise, starting from the upper left corner. So first, I pass through the battery, from negative to positive terminal. This raises my potential by $\mathcal{E}_{\text{left}}$. Then I walk with the current through the $2R$ resistor. Therefore, my potential drops: $\Delta V = -I_2(2R)$. Finally, I walk with the current through the R resistor on the left leg of the circuit: $\Delta V = -I_1R$. Since these potential changes sum to zero,

Left loop $$\mathcal{E}_{\text{left}} - I_2(2R) - I_1R = 0 \qquad (1)$$

Be careful not to mix up I_1 with I_2.

We now have two equations in three unknowns (I_1, I_2, and I_3). So, we need one more equation. To obtain it, apply Kirchhoff's 2nd law to the right loop. I'll walk clockwise, starting from the upper right corner. So first, I walk with the current through the R resistor: $\Delta V = -I_3R$. Next, I pass through the battery: $\Delta V = +\mathcal{E}_{\text{right}}$. Finally, while walking up the center leg of the circuit, I pass through the $2R$ resistor. But this time, I'm walking ***against*** the current—swimming upstream, so to speak. Therefore, my potential changes by $\Delta V = +I_2(2R)$.

Set the sum of these potential changes equal to zero:

Right loop $$-I_3R + \mathcal{E}_{\text{right}} + I_2(2R) = 0 \qquad (2)$$

Since we found $\mathcal{E}_{\text{left}}$ and $\mathcal{E}_{\text{right}}$ in subproblem 1 above, these three equations contain only three unknowns, namely the three currents. We're done, except for algebra.

Algebra starts here. From Eq. (0), $I_1 = I_2 + I_3$. Substitute this expression for I_1 into Eq. (1), to get

$$\mathcal{E}_{\text{left}} - 2I_2R - (I_2 + I_3)R = 0.$$

Solve this equation for I_3 to get $I_3 = \frac{\mathcal{E}_{\text{left}} - 3I_2R}{R}$. Finally, substitute this expression into Eq. (2):

$$-\frac{\mathcal{E}_{\text{left}} - 3I_2R}{R}R + \mathcal{E}_{\text{right}} + I_2(2R) = 0.$$

Now we just have to isolate I_2. Expanding out all the terms gives me

$$-\mathcal{E}_{\text{left}} + 3I_2R + \mathcal{E}_{\text{right}} + 2I_2R = 0,$$

and hence

$$I_2 = \frac{\mathcal{E}_{\text{left}} - \mathcal{E}_{\text{right}}}{5R} = \frac{\frac{B_0s^2}{t_1} - \frac{B_0s^2}{2t_1}}{5R} = \frac{B_0s^2}{10t_1R},$$

where I used our expressions for $\mathcal{E}_{\text{left}}$ and $\mathcal{E}_{\text{right}}$ from subproblem 1. Actually, I used the absolute values of those EMFs, because I already built their directions (clockwise) into loop equations (1) and (2).

ANSWER 6

I'll first discuss how a dielectric or paramagnetic substance affects a capacitor or inductor. Using these results, I'll then solve parts (a) through (c).

What does the dielectric jelly do?

Roughly speaking, a dielectric *dampens* electric fields. So, if we fill the capacitor with dielectric jelly, the field between the plates decreases. Therefore, the potential difference between the plates decreases, other things being equal. In other words, for a given charge Q on the capacitor, the potential difference V is *less than* it otherwise would have been. In this way, the dielectric increases the ratio of charge to voltage, Q/V. But the ratio of charge to voltage *is* the capacitance: $C = Q/V$. In summary, inserting dielectric jelly raises the capacitance. It makes the capacitor "willing" to hold more charge per volt.

By contrast, dielectric jelly has negligible effect on the inductor. An inductor generates a magnetic field. But a dielectric dampens only electric fields. (From Chapter 45, you know that the oscillating magnetic field inside the inductor induces an electric field. But this electric field is tiny. You aren't expected to deal with it, in this sort of problem.)

What does the paramagnetic putty do?

As explained in Chapter 42, a paramagnetic material *amplifies* the magnetic field inside of it. So, the putty would hardly affect the capacitor, which is filled with electric field. (Actually, the "displacement current" induces a small magnetic field. But here, we can neglect it.) By contrast, filling the inductor with the paramagnetic putty *increases* the inductance. Here's why.

By definition, inductance relates the rate of change of current through the inductor to the induced EMF: $\mathcal{E}_{\text{induced}} = -L\,di/dt$. Let me review what this formula means. By changing the current through the inductor, you change the magnetic field and flux inside the inductor. According to Faraday's law, this changing flux generates an EMF. The inductance specifies how big an EMF gets induced—i.e., how quickly the magnetic flux changes—for a given rate of change of the current.

By filling the inductor with paramagnetic putty, we increase the magnetic flux, other things being equal. Therefore, we increase the rate at which the flux changes, other things being equal. Consequently, for a given di/dt, a putty-filled inductor generates a bigger EMF than an unfilled inductor does. In other words, filling the inductor with paramagnetic putty increases the inductance.

Summary of the effects of dielectric jelly and paramagnetic putty

Inserting dielectric jelly into the capacitor increases the capacitance. Inserting paramagnetic putty into the inductor increases the inductance. Using these results, we can address parts (a) through (c).

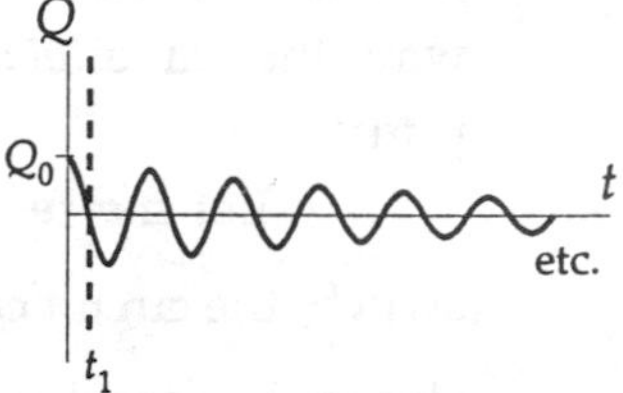

(a) Recall that an RLC circuit behaves just like a damped harmonic oscillator, such as a block on a spring with air resistance. Charge sloshes back and forth between the two capacitor plates. This graph shows the charge on the top plate, as a function of time, assuming the top plate initially carries charge Q_0. Notice that t_1 is one quarter of a complete period. So, *making t_1 small is the same thing as making the period small.* Jan wants to make the period as small as possible. Should she use the dielectric jelly, or the paramagnetic putty, or both?

Because I have trouble thinking intuitively about RLC circuits, let's reason in terms of the analogy with a block on a spring. To decrease the period (and hence, to decrease t_1), we can decrease the mass of the block. A lighter block whips back and forth more quickly, other things being equal. In addition, we should raise the spring constant (k), because a stiffer spring causes quicker vibrations. The formula for period confirms these intuitions. Decreasing m and increasing k makes the period smaller. Of course, to apply these conclusions to the RLC circuit, we must review the analogy between blocks and inductors, and between springs and capacitors.

As explained in Chapter 44, the inductor "plays the role" of the block. Just as a block resists having its velocity changed, an inductor "resists" having its current changed. This goes to show that L is analogous to m. Therefore, since decreasing the block's mass leads to quicker (lower period) oscillations, decreasing the inductance has the same effect on an RLC circuit. To make the period (and hence, t_1) as small as she can, Jan should lower the inductance. But inserting the paramagnetic putty would *raise* the inductance, as explained above. So, **Jan should not use the paramagnetic putty.**

ANALOGY TABLE

block on spring	RLC circuit
m	L
k	$1/C$
air resistance constant	R

What about the dielectric jelly? Inserting it into the capacitor would raise the capacitance. In other words, the dielectric would make the capacitor more "willing" to hold charge. This corresponds to making a spring more "willing" to be stretched, i.e., looser. A looser spring has lower k. So, increasing the capacitance corresponds to decreasing the spring constant. And vice versa. In mathematical terms, C is analogous to $1/k$, not to k itself.

To decrease the period of a block on a spring, you must raise k, making the spring stiffer. As just shown, raising k corresponds to lowering the capacitance, C. Since inserting the dielectric jelly would raise the capacitance, **Jan should not use the dielectric jelly,** either.

(b) Now Jan wants to make t_1, and hence the period, as big as she can. Again, let's think in terms of the block-on-spring analogy. To increase the period, Jan would make the block heavier (by raising m), and the spring looser (by lowering k). Since L is analogous to m, she should raise the inductance. And since C is analogous to $1/k$, she should increase the capacitance, too. For these reasons, she **should insert paramagnetic putty into the inductor, and dielectric jelly into the capacitor.**

(c) If Jan inserts paramagnetic putty, the inductance increases. Therefore, current oscillates back and forth more slowly, and takes longer to die out. As a result, you might think that more heat gets generated. But the increased inductance also leads to a smaller average current, just like a heavier block leads to a smaller average velocity. Because of the smaller current, heat gets generated at a lower rate. In summary, if Jan leaves out the paramagnetic putty, then heat dissipates at a high rate, but for a short time. If she inserts the putty, then heat dissipates at a lower rate, but for more time. Either way, the same *total* heat gets produced. **It doesn't matter whether Jan inserts the paramagnetic putty.**

Let me re-express this argument in terms of energy conservation. Before Jan closes the switch, the circuit carries a certain potential energy, all stored in the capacitor: $U_0 = \frac{Q_0^2}{2C}$. As charge sloshes back and forth, this energy "trades" back and forth between the capacitor and the inductor. But some energy dissipates away as heat in the resistor. In fact, by the time the circuit dies out, *all* the initial energy has dissipated away as heat. For instance, if the capacitor initially stores 2 joules of potential energy, then eventually those 2 joules convert into heat dissipated in the resistor.

By inserting the paramagnetic putty, you slow down the conversion of potential energy into heat. In other words, you reduce the *power* developed in the resistor. But eventually, all that stored energy turns into heat. The same total heat gets produced, with or without the putty.